Safety Engineering

Stefan Bracke · Ralf Pieper
Editors

Safety Engineering

Fundamentals, Methods, Research Topics

Springer

Editors
Stefan Bracke
FB D Sicherheitstechnik c/o Bergische
Universität Wuppertal
Wuppertal, Germany

Ralf Pieper
FB D Sicherheitstechnik c/o Universität
Wuppertal
Wuppertal, Germany

ISBN 978-3-658-47472-0 ISBN 978-3-658-47473-7 (eBook)
https://doi.org/10.1007/978-3-658-47473-7

This Springer imprint is published by the registered company Springer Fachmedien Wiesbaden GmbH, part of Springer Nature.
The registered company address is: Abraham-Lincoln-Str. 46, 65189 Wiesbaden, Germany

Preface

The discipline of safety engineering is both an overarching and specialised field of research and teaching in the engineering sciences and deals with hazards and dangers associated with the development and application of technologies and engineering or organisation for people and the environment, as well as their evaluation.

Safety engineering is an engineering discipline with an interdisciplinary and application-orientated concept. The main focus is on the knowledge-based development of methods to avoid or minimise unacceptable or intolerable risks to the safety and health of people and the environment. In addition to engineering know-how from the fields of mechanical engineering, process engineering, electrical engineering, computer science, civil engineering and the natural sciences, findings from the labour, legal and social sciences are also incorporated and their development promoted.

The safety science terminology and methodology form the basis of the concepts, principles and instruments of safety engineering. With integrative, interdisciplinary and transdisciplinary intentions, these refer to corresponding axioms, in particular from the natural, human, labour, humanities and social sciences. The correspondingly comprehensive and overarching objective of safety engineering is reflected in its diverse application in the field of socio-ethical, political-economic and ecological interests.

Based on the principles of efficient and effective prevention and reduction of remaining hazards, including hazard defence and the potential and resources for safety and health, safety engineering in research and teaching at the University of Wuppertal is particularly effective in the prevention of technically complex and risky products and systems in the world of work and in the environment. This includes the teaching and research fields of plant and product safety, substance and material safety, occupational safety, environmental safety, traffic safety, reliability engineering, energy efficiency, consumer protection, civil protection and disaster control as well as fire and explosion protection.

In 1975, the University of Wuppertal established the first international university department with an interdisciplinary safety science concept. Today it is an integral part of the Faculty of Mechanical Engineering and Safety Engineering. The institutional link with the engineering discipline of mechanical engineering has been in place at the University of Wuppertal since its foundation in 1972.

From 1975, the first established subject areas of the initially organisationally independent Department of Safety Engineering were 'General Safety Engineering' and, subsequently, 'Occupational Safety Engineering/Ergonomics', 'Fire Protection' and 'Traffic Safety Engineering'. The changes that took place in the following decades are documented institutionally in particular by the addition of the chairs of 'Occupational Science', 'Civil Protection', 'Chemical Safety', 'Product Safety and Quality', 'Process and Plant Safety', 'Safety and Quality Law', 'Environmental Safety' and 'Reliability Engineering and Risk Analysis'. In addition, further institutes and spin-offs have been established at the University of Wuppertal.

The dynamic changes in the world of work and in living environments with the associated challenges and opportunities emphasise the need to further develop the research and teaching field of safety engineering at the University of Wuppertal within the network of safety science. The task and goal is to provide scientific impetus for the humane and environmentally friendly design of working and living conditions. The present conference proceedings with over 20 contributions from the specialist areas of the Faculty of Mechanical Engineering and Safety Engineering emphatically demonstrate this and should also be a starting point for a transfer to other specialist disciplines.

These conference proceedings were produced on the occasion of the forum '50 Years of Safety Engineering at the University of Wuppertal (Bergische Universität Wuppertal)' organised by the Faculty of Mechanical Engineering and Safety Engineering (University of Wuppertal) in cooperation with the Safety Science Forum of the University of Wuppertal and the VDSI Forum NRW. The editors would like to thank all the authors for their comprehensive contributions and their commitment to these conference proceedings. Special thanks go to Philipp Heß for his support in organising the conference proceedings; his patience and accuracy have made a significant contribution to the success of this publication. The editors would also like to thank the Dean of the Faculty of Mechanical Engineering and Safety Engineering, Eberhard Schmidt, whose support in the organisation of this publication enabled the diversity of safety engineering to be comprehensively represented.

We would like to thank the ladies and gentlemen of Springer Verlag for their dedicated expert advice in the design of this book and their patience; in particular Eric Blaschke for his organisation, suggestions and advice.

Wuppertal, Germany
December 2024

The editors
Prof. Dr.-Ing. Stefan Bracke,
Chair of Reliability Engineering and Risk Analytics,
University of Wuppertal
Prof. Dr. rer. pol. Ralf Pieper,
Chair of Safety and Quality Law,
University of Wuppertal

Contents

Safety Engineering and Reliability Engineering

The German and Japanese Product Development Process and the Usage Phase of Complex Technical Goods: Innovation, Optimisation and Reliability Engineering

Stefan Bracke and Masato Inoue

Contents

Abstract

The industrialisation of Europe and Asia leads to the development of innovative and reliable technical complex products. Industrial nations like Germany and Japan were going separate ways to highly developed industrial nations. This leads to different goals and philosophy in terms of reliability and safety design

S. Bracke (✉)
Chair of Reliability Engineering and Risk Analytics,
University of Wuppertal, Wuppertal, Germany
e-mail: bracke@uni-wuppertal.de

M. Inoue
Department of Mechanical Engineering Informatics,
Meiji University, Kanagawa, Japan

S. Bracke and R. Pieper (eds.), *Safety Engineering*,
https://doi.org/10.1007/978-3-658-47473-7_1

regarding product, process development and the efficiency of the product in the use phase. This paper shows essential aspects of the differences in German and Japanese reliability and safety engineering. Base of operations is the fundamental difference of the realisation of the principles "innovation" versus "optimisation": German products functionality contains often innovative solutions, Japanese products functionality shows often optimised solutions. These different designs are caused by different impacts: Historical impact, society and social interaction impact, education System and work environment impact and geographical and location impact. This leads to the different technical reliability and safety engineering solutions in engineering and product use in Germany and Japan. Finally, the findings are substantiated using the example of the long-distance train transport system in Japan and Germany.

Keywords

Reliability engineering · Japanese engineering · German engineering · Innovation versus optimization · Impacts on engineering · Railway system

1 Introduction

The industrialisation of Europe and Asia leads to the development of innovative and reliable technical complex products and the efficient use of technical products in the use phase. Industrial nations like Germany and Japan were going separate ways to highly developed industrial nations. Both countries have the ability to develop and manufacture technical complex products with a high reliability and safety standard and a high efficiency in the application within the usage phase. But a detailed look at Japanese and German products shows differences regarding the product characteristics "Innovation" and "Optimisation", especially in case of reliability aspects and efficient use. New in Germany developed product generations including often innovations regarding func-tionality. Japanese product engineering considers very often the optimisation in relation to the previous product generation. An example for illustration is the comparison of the Japanese and German long-distance transport system (Deutsche Bahn, System ICE, IC versus Japan Railway, Shinkansen) with the focus on criterions like innovation, optimisation, efficiency in use and reliability in use. The railway system was invented in Europe (Great Britain, afterwards also Germany; cf. (Griffin 2010) and (Pollard 1981)). Subsequently it was imported by the U.S. to Japan after opening the country for trade. The Japanese engineers adapted and optimised the railway system: The concept (e.g.: Separation of high-speed, regional and goods train system), the logistics (e.g.: boarding/getting out); cf. (Nussbaum and Roth 2005).

The reasons for the differences in Japanese and German engineering and product use are manifold: This paper elaborates on the differences that lead to product differences (innovation versus optimisation) and efficient product use. This includes in detail:

- Product development processes in Germany (Sect. 3) and Japan (Sect. 4)
- Problems and solutions within product development
- Impacts on Japanese and German engineering (Sect. 5):
  - Historical impact (Sect. 5.1),
  - Education System and work environment impact (Sect. 5.2),
  - Society and social interaction impact (Sect. 5.3)
  - Geographical and location impacts (Sect. 5.4).
- Example railway system: Comparison of German and Japanese railway system

This paper builds on the following publications: Bracke (2024), esp. Sect. 5.3 focuses on product development process in Germany; Inoue et al. (2011), esp. Sect. 5.4 focuses esp. on product development in Japan; Bracke and Inoue (2018), esp. Sects. 5.1, 5.2 and 5.4 focusing on innovation versus optimisation. The aforementioned explanations were analysed and expanded to include aspects of the influence of society and social interaction in Germany and Japan. Furthermore, the findings are substantiated using the example of the long-distance train transport system in Japan and Germany.

2 Goal of Research Collaboration: Impacts on Reliability Engineering in Product Development Process

The underlying question of the research cooperation (University of Wuppertal, Chair of Reliability Engineering and Risk Analytics and Meiji University, Department of Mechanical Engineering Informatics) is to work out the differences between the highly developed industrialised nations of Germany and Japan with regard to product development and product operation. The following perspectives and questions take centre stage:

1. Product development: How do the procedures for developing technically complex products differ in Germany and Japan (cf. Sects. 3 and 4)?
2. Innovation versus optimisation: Why are German products often characterised by innovative functions, whereas Japanese products are characterised by optimised functions?
3. Influences on product development and operation in the context of reliability engineering in Germany and Japan: What influence do history, education system, society and geography have on technical reliability, innovation and optimisation in the context of product development and product use?

In this article, these three perspectives respectively questions are examined and the differences between the industrialised nations of Germany and Japan are highlighted using examples from automotive industry and railway systems.

3 Product Development Process: Germany

The product emergence process or product life cycle in Germany is divided into different phases (e.g.: series development), subphases (e.g.: preparation for series production) and activities (e.g.: component testing), cf. (Bracke 2024) and (DIN 2017). The product emergence process comprises the phases between concept development and sales (cf. Fig. 1). In common usage, a part of the use phase is often included here, since field support is also a part of the field phase (after sales). If the field phase with the subphase use is mapped comprehensively up to the end of the product use, then the term product life cycle should be chosen instead of product emergence process, since the use phase can cover a longer period than the field support.

Furthermore, the product life cycle also includes the recycling phase, which follows the use phase; cf. also (Bender and Gerricke 2021). In summary, it is stated that the product life cycle, in contrast to the product emergence process, additionally includes the complete use phase as well as the recycling phase. In the development of a complex, technical product, the phases often take place in parallel in superordinate specialist disciplines. In the example of automotive engineering, this can be the six specialist disciplines of vehicle, body, chassis, powertrain, electrics/electronics and software (Fig.; cf. also Braess and Seiffert 2013: 1133). In case of automotive engineering, the following systems or component scopes (excerpt) belong to the above-mentioned disciplines:

1. Vehicle: Subject areas that relate to the entire vehicle (e.g. fuel consumption, noise emissions; cf. further explanations) and cannot be assigned exclusively to one of the other disciplines mentioned.
2. Body: body frame, doors, covers, interior (assembly peripheries such as seats, instrument panel), exterior (add-on parts such as mirrors, trim panels).

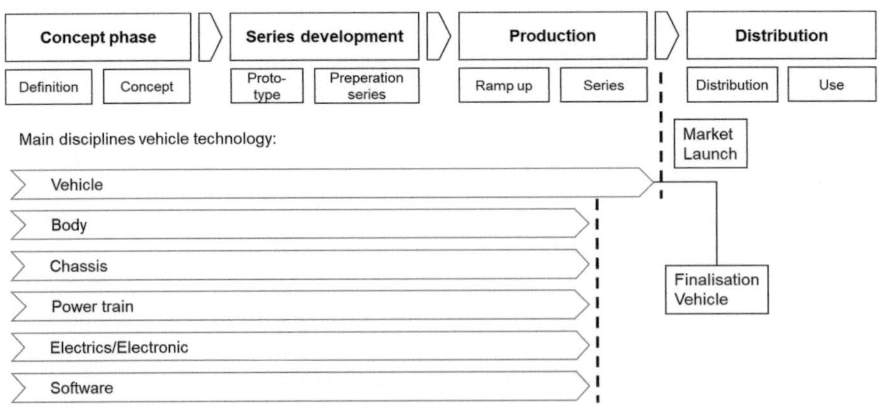

Fig. 1 Product emergence process using the example of automotive engineering (Bracke 2024)

3. Chassis: steering, axles, trailing arms, transverse control arm, braking system.
4. Power train: engine, gearbox, clutch, attachments (e.g. coolant pump).
5. Electrics/electronics: actuators, control units, bus systems, networking.
6. Software: programs, algorithms, control codes, characteristic diagrams.

If a target value is defined at vehicle level (e.g. fuel/electricity consumption), several disciplines may be involved in its implementation. As an example, the characteristic vehicle weight (impact on fuel/electricity consumption) can be assigned to the vehicle discipline: The definition of the vehicle function spectrum (equipment), the selection of materials as well as the geometric design of the components have a significant influence on the total weight of the vehicle.

3.1 Concept and Series Development Phases

The concept phase essentially consists of the sub-phases pre-development (development of new technologies and functional principles, respectively), definition and confirmation of the product design (concept confirmation); cf. also (Braess and Seiffert 2013). The aim of the concept phase is essentially to define the product structure, the styling and the development of the performance specifications for systems, assemblies and components on system, assembly and component level. Furthermore, a decision is made as to whether systems, assemblies, components are to be developed respectively produced in-house (OEM; Original Equipment Manufacturer) or developed respectively produced in outsourcing (supplier).

The series development phase comprises the two subphases of design and prototypes and series preparation. The objective of series development is the development of all systems, assemblies, components as well as the preparation of series production (procurement and construction of systems, machines, tools).

In addition to component development, the design and prototype subphase also includes simulation as well as the design and implementation of system, assembly and component testing. Simulation and testing have the goal of testing functions and reliability. The procedures include, among others, laboratory testing, Accelerated Testing, Highly Accelerated Life Test (HALT) or the vehicle test. Prototypes are developed with progressive design status 1 … n, which reflect the respective degree of maturity of the product. Within the subphase series preparation the production facilities for vehicle manufacture are implemented; approximately 95% of components are produced using series tools. If the manufacture of small or large series is planned, the activity Confirmation of the capabilities of product and process provides an essential contribution with regard to the preparation of a controlled and quality-capable manufacturing process, in particular for the realization of safety and function-critical product features. This includes e.g. the activities of measurement system and process capability, machine capability as well as process capability investigation. Once series preparation has been completed, it is possible to manufacture a vehicle under near-series conditions and to use series tools.

3.2 Production and Distribution Phases

The production phase after point of time "start of production (SOP)" comprises the two subphases series ramp-up and series production. The series ramp-up (in short "ramp-up") refers to the number of vehicles to be produced in relation to the first weeks of production. The number of units is steadily increased until the so-called operation line is reached. The operation line is the upper limit of the potential capacity utilization of a production. When the operation line is reached, series production starts (short: "series"). With regard to ensure product reliability, the activities of inline testing during series production (system, subassembly, component level), final product testing (entire product) and external testing processes (e.g. laboratory or supplier) are of particular interest here in addition to proven process capability; cf. in particular also series preparation. Safety–critical features are often subjected to a full inspection (also referred to as 100% inspection), provided that non-destructive testing is possible. Function-critical features can be checked by means of full inspection or sampling inspection, also in connection with Statistical Process Control (SPC). Safety and function-critical product characteristics can be inspected directly or indirectly. The production phase ends with the EOP (End of Production): The production of the vehicle ends, spare parts continue to be manufactured for the time being. For this reason, some of the production engineering equipment continues to be operated for the supply of spare parts to the field.

The sales and field support phase comprises the sub-phases of the actual product distribution and the field support related to product usage. The sales sub-phase starts with the sale of the products usually directly after the ramp-up and ends with the discontinuation of the production of the current series (End of Production (EOP)). The use phase is understood as the complete period of product use by the customer until the end of the life cycle. The spare parts supply can be ensured for the entire scope of components for X years (automotive engineering: e.g. 15 years), then to a reduced extent – e.g. function- and safety–critical components – for X+Y years (e.g. automotive engineering: X+Y can cover 30 to 40 years). Field support begins with the sale and use of the first products and also includes the period after the EOP.

3.3 Typical Problems and Weak Points Within Product Development

Due to the contradiction outlined between the processes of decision certainty and influence in relation to the product emergence process (Fig. 1) and the connection between influence and value creation, the efforts for risk elimination or risk prevention increase significantly in many projects of OEMs and/or suppliers in the realization phase (series preparation and production phase). The illustration in Fig. 2 shows the expenditure for risk elimination or risk prevention along the product emergence process as well as the curve for decision certainty. The concept phase plays a decisive role here: The basic project knowledge is available in this phase, which in many cases leads to decisions based on little knowledge. If these decisions are wrong, the effort for failure elimination (risk elimination) increases significantly within the phase series preparation, explicitly in the prototype phase as well as in the production of the pre-series: Through realization and testing, the product weaknesses become apparent and the effort for a re-design also increases significantly. The load peak occurs when the assembly line is reached and leads to high resource input and improvised product and process solutions. The consequences are start of production (SOP) problems, series maturity problems and field problems in the early use phase.

3.4 General Principles for Developing Safe and Reliable Products from a German Perspective

In the so-called frontloading approach, resources are increasingly used for reliability planning and for the application of technical reliability methods in the concept and series development, since a large part of the product costs is determined here with regard to production. In contrast to the outlined efforts for risk elimination with low resource input in the concept phase (Fig. 2), Fig. 3 shows the curve with regard to the efforts for risk avoidance when focusing on concept development and careful reliability planning in the concept phase. The expenses for risk avoidance and elimination are on a high level in the concept development. Furthermore, the development work is systematically accompanied in the series development by methods of the technical reliability, such as Fault Tree Analysis (FTA), Failure

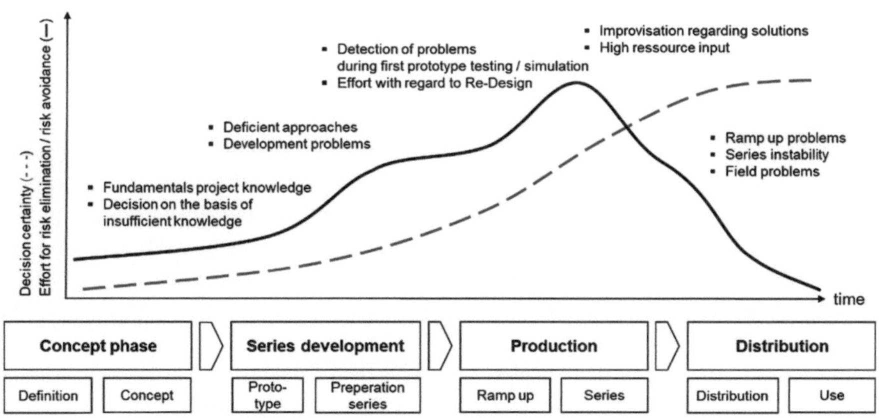

Fig. 2 Efforts for risk elimination and avoidance as well as decision certainty in the product emergence process with insufficient project knowledge in the concept phase (Bracke 2024)

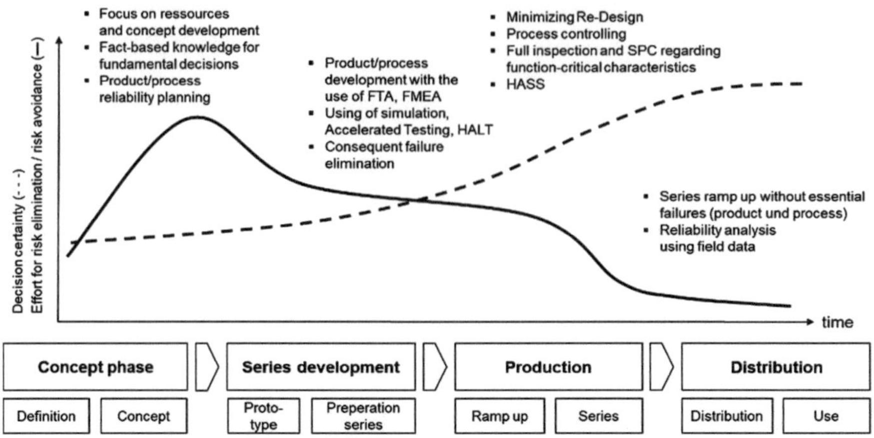

Fig. 3 Efforts for risk elimination and avoidance as well as decision certainty in the product emergence process with careful reliability planning in the concept phase (Bracke 2024)

Mode and Effects Analysis (FMEA), Simulation, Accelerated Testing, Highly Accelerated Life Test (HALT).

This reduces the efforts for a potential re-design. Early process control and quality capability in the start of series production are made possible. The increased use of resources with well-founded reliability planning in the concept phase and the use of technical reliability methods during development can lead to a low-fault series start of production. In addition, the level of decision reliability is explicitly raised in the early phases of product development.

4 Product Development Process: Japan

4.1 Fundamentals: Phases, Activities

In general, it is remarkable that a product development process of Japan is different from all the others. In Euro-American product development, the design phases are broadly divided definitely: the initial design phase and the detail design phase. Engineers' responsibilities are also divided definitely. In the initial design phase, the concept of a product can be allocated to the requirements of each part. In addition, design environment does not too often change at the detail design phase. Therefore, designers can handle own parts under their own responsibilities at the detail design phase.

In contrast, the initial design phase and the detail design phase are not divided definitely in Japanese product development. One of the features of Japanese product development is a product integrity development which can obtain a multi-objective satisfactory design solution by "*Suriawase*" in Japanese: Japanese way for negotiation among several teams of engineers including design teams, production divisions, and parts manufacturers (suppliers) from the initial design phase to the detail design phase concurrently. In the initial design phase, uncertainties arise from many sources of variations. The initial design phase goes through the transition to the detail design phase with uncertain design concept for each part of a product. That is to say, the design requirements of each part cannot be defined definitely even at the detail design phase when the concept of the product is allocated to the requirement of each part. Therefore, the tentative design requirements of each part are defined by *Suriawase* among related teams, and then each team aims for the tentative design requirements at the detail design phase. This enables designers to go forward under the uncertainties at the detail design phase without a letup and shortens the product development time accordingly.

In addition, a design environment of Japan in general often changes even in the detail design phase for reflecting requirements of the customers and the marketplace in a product as much as possible before production process especially in an industry which has a heavy competition for development race such as automotive industry. For example, an unplanned design requirement sometimes must be considered as an additional requirement even though the design process proceeds to the detail design phase. The *Suriawase* development can delay the final decision by definition of the design requirements at the detail design phase. It is possible to adapt the change of the design environment flexibly at the detail design phase.

However, the *Suriawase* development needs to define the single-point value of the tentative design requirements of each part among related teams, consider the feasibility of the value as a whole, and repeat this procedure many times until the tentative design requirements involve feasible design domain. This repetitive operation depends on the contributions in personnel by communication and negotiation (*Suriawase*) among several teams because of the confliction among them, and they bear a heavy burden. They also have to redefine the tentative design requirements in case of the changes to the product specifications or the addition of the design requirements along the way.

Therefore, a design support method for *Suriawase* to obtain the flexible and robust design solution which can satisfy all requirements of several teams under the uncertain design environment and to provide a suggestion of the direction of design modification for *Suriawase* among several teams is required. This paper focuses on the uncertainties accompanied with the design process, cf. Sects. 4.2 and 4.3.

4.2 Typical Problems and Weak Points Within Product Development

Japanese companies have had their engineers trained for multiple tasks and skills (called "*Tanoko*" in Japanese) based on long-term employment and long-term trading with suppliers because of the labor shortage after World War II. That is why Japanese engineers are good at works of many different teams and can find an optimum design solution faster than other countries through a trial-and-error processes among several teams. Therefore, the product development based on *Suriawase* takes root in Japanese culture.

The *Suriawase* development has an advantage in a product with integral architecture which individual parts of a structure are complexly intertwined with related functions as shown in the Fig. 4 such as automobiles, motorcycles, or light and nimble home electrical appliances (Takeishi and Fujimoto 2001). It does not allow one-to-one correspondence between the structure (parts) and their functions, and does not enable, for example, the designer for part [s1] to focus solely on function [f1].

The present procedure of the product development based on *Suriawase* is as follows:

1. The designers who take charge of the parts (e.g. [s1]–[s3] in Fig. 4) associated with a function (e.g. [f2]) get together in a real large room.
2. When they do not have the value of design requirement for the function which satisfy all requirements of the product functions (e.g. [f1] and [f2] for the part [s1], [f1]–[f3] for the part [s2], [f2]–[f4] for the part [s3]), each designer negotiates with the other designers to understand the situation and background of the others each other to share the problem as shown in Fig. 5(a).

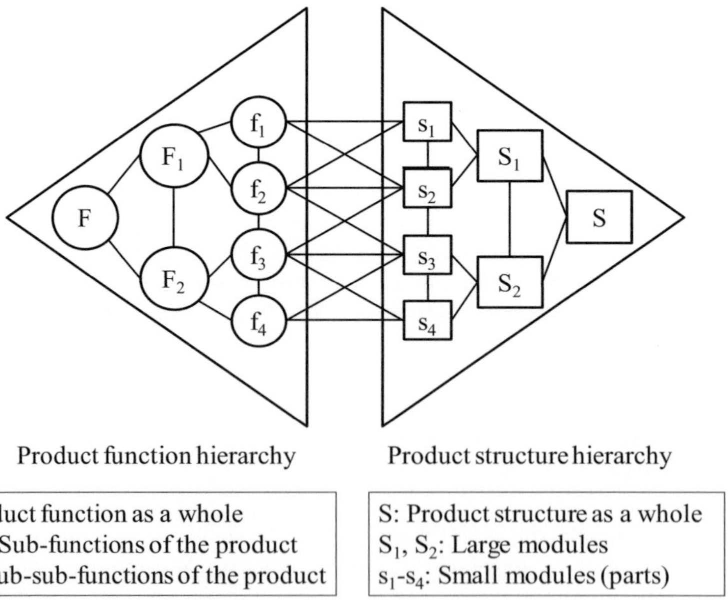

Fig. 4 Product with integral architecture

3. They define the tentative design requirements of each part (e.g. [f2]) which "are supposed to" satisfy all requirements of the product functions and aim for the tentative design requirements as shown in Fig. 5(b).
4. They consider the feasibility of the tentative design requirements and repeat the consideration and negotiation many times until the tentative design requirements involve feasible design domain as shown in Fig. 5(c).
5. After they found a tentative design solution which is located in the feasible design domain, they can find an optimum design solution (final design solution) as shown in Fig. 5(c).

The first difficulty of the *Suriawase* development arises in the definition of the well-chosen tentative design requirements of each part which satisfy all requirements of the product functions under the uncertain design environment. If they do not define the well-chosen tentative requirement of the function, the numerous repetitive negotiation operations are needed. Therefore, the designers are required to capture own and the other design domains each other when they negotiate with the other designers to understand the situation and background of the others each other to share the problem.

The second difficulty of the *Suriawase* development arises in the variability in the design environment by the changes to the product specifications or the addition of the design requirements. Therefore, the designers are required to comprehend the impact on the other team (designer) coming from the own design modification and need to adapt the change of the design environment flexibly.

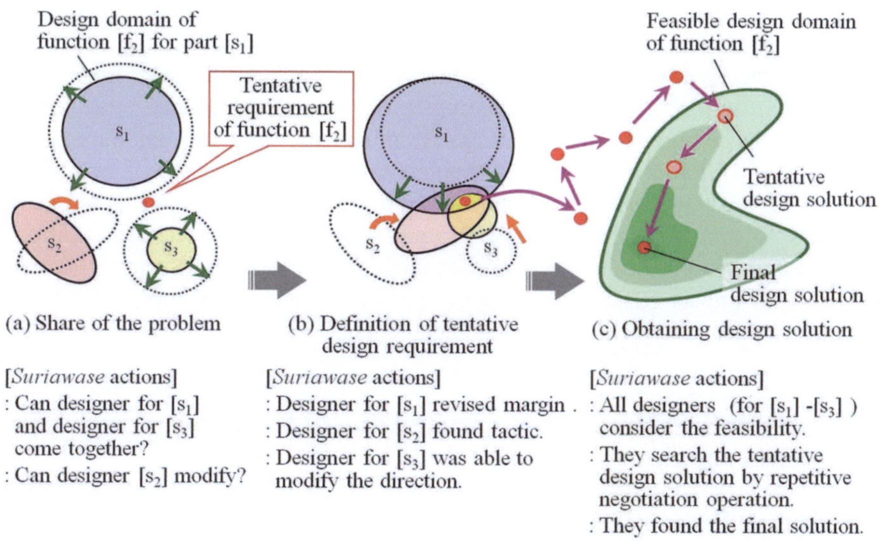

Fig. 5 Present situation of product development process based on *Suriawase*

4.3 General Principles for Developing Safe and Reliable Products from a Japanese Perspective

It is thought that we need to support before *Suriawase* among several teams. If the designers can capture the feasible design domain which satisfy all required functions under the uncertain design environment and can comprehend the impact on the other teams coming from the own design modification in case of the changes of the design environment, it enables the designers to lead to more efficiency for *Suriawase*.

We propose a design support for *Suriawase* by following two phases:

1. Phase I: Display of the feasible design domain
   We propose a design support method to obtain the multi-objective satisfactory design domain. We display the feasible design domain to the designers by applying the set-based design method which can obtain a set of design solution under the uncertain design environment; cf. (Sobek et al 1999) and (Inoue et al. 2011).
2. Phase II: Display of the impact of the design modification on the other teams
   We propose a design support method to display the impact of the design modification on the other teams (the other parts and functions) to the designers by applying the quantitative dependency. The designers can modify the own part of the product through a trial-and-error processes before *Suriawase* among several teams.

5 Impacts on German and Japanese Engineering

From the historical point of view, industrial nations like Germany and Japan were going separate ways to highly developed industrial nations (Bracke and Inoue 2018). This leads to different focusing with regard to the approaches "Innovation" and "Optimisation" within engineering processes. Furthermore, the technical reliability of high-tech products based on Japanese and German engineering is pushed by different factors. This chapter works out differences of both countries, which leads to different philosophy and design in terms of the characteristics "Innovation" and "Optimisation" and in terms of reliability of technical complex products and processes: historical impact (Sect. 5.1), Education System and work environment impact (Sect. 5.2), Society and social interaction impact (Sect. 5.3) and geographical and location impacts (Sect. 5.4).

5.1 Historical Impact: Industrialisation in Germany and Japan

The industrialisation of Europe and Asia leads to the development of innovative and reliable technical complex products. Industrial nations like Germany and Japan were going separate ways to the level of highly developed industrial nations. The industrialisation starts in Europe within the second half of eighteenth century in England, and subsequently in Germany; cf. (Griffin 2010) and (Pollard 1981). Within the German empire time, the industrialisation was growing rapidly. Exemplary, the industrialisation process is explained using the example railway system. Begin of the railway history was in 1804 with the start up of the first steam locomotive by Richard Trevithick. Short time later in 1825, the first public railway system was opened by Stockton and Darlington Railway in England (Tomlinson 1915), the transport of people and goods was possible. Traveling times became short in Europe and North America, therefore the industrial revolution itself was supported by the railway system. Furthermore, it was pushed by the European wars, especially the German–French war from 1870 to 1871, where the transport of war goods was important. Today, the high-speed train ICE ("Inter City Express") is an essential part of the German traffic system: The four ICE generations were developed between 1990 and 2017 (Jehle et al. 2006).

However, in Japan the industrialisation starts in the end of nineteenth century caused by external pressure from the United States. The external pressure leads to the end of the Edo era (1603 until 1867) and the new Meiji area (1868 until 1912) started, where the industrialisation grows rapidly (Nussbaum and Roth 2005). Within the Edo era, Japan was encapsulated to the rest of the world because of the isolation strategy of the country. Entry and departure for foreigners were not allowed (exception: small exchange with China and Netherlands; colony island Dejima) and contact to other states were on a minimum level. In 1853, four war ships, leaded by Matthew Perry, were running in the harbour of Uraga. Based on a message from US president Millard Fillmore, the Tokugawa government was pressurised to start trade with the United States (otherwise U.S. go for war); cf. (Morinosuke 1976). The trade includes the opening of the harbours Shimoda and Hakodate and the possibility for trading. Parallel, the Edo era ended and the reform of the empire leads to the Meiji era, starting in 1868. The war gentry ended, the constitution was modernised and the development of the country – especially from the engineering point of view started. This point is exemplary illustrated with the Japanese railway system; cf. (Givoni 2006): Within the Meiji era, the railway system was developed – but limited to a track range of 1067 mm. This leads to reduced reachable speed level of 100 km/h until 1950. Huge steps of the railway system development followed in the sixties and subsequently, the high-speed trains ("Shinkansen" – high speed) were developed in the nineties parallel to Europe (1992: Series type 400, 1997: E3; 1999 E3-1000). Today the railway system in Japan is the most reliable and accurate system worldwide (Givoni 2006).

The other historical aspect of Japanese product development style is World War II. Japanese companies have had their engineers trained for multiple tasks and skills (called "Tanoko" in Japanese) based on long-term employment and long-term trading with suppliers because of the labour shortage after World War II. That is why Japanese engineers are good at works of many different teams and can find an optimum design solution faster than other countries through a trial and error processes by "*Suriawase*" in Japanese. Japanese way is the negotiation among several teams of engineers including design teams, production divisions, and parts manufacturers (suppliers) from the initial design phase to the detail design phase concurrently (Inoue et al. 2011).

As a comparison between Germany and Japan: In Germany, the innovative engineering was pushed by the industrialisation under environmental conditions, which allows free thinking. Prussia and the subsequently following German empire's education system ("Humboldtian Model"; (Baumgart 1990)) allows free thinking and development of creative solutions (cf. Sect. 4.2). However Japanese engineering was pushed to open the country, caused by external pressure from the U.S. Many technologies were imported during Meiji era from Europe and U.S. and adapted to Japanese environment (Nussbaum and Roth 2005). This was a part of the base of operations for Japanese focus on optimisation of imported technical products and processes.

5.2 Education System and Work Environment Impact: Germany versus Japan

The culture and society of Germany and Japan is very different. Germany society is embossed by a hierarchical system parallel to an established social market economy. As mentioned in Sect. 4.1, the Prussia educations system, founded by Wilhelm Humbold, focusses on a comprehensive education based on a combination of research and studies connected to the study course. The "Humboldtian Model" (Baumgart 1990) model includes arts and sciences with research. The goal is comprehensive general learning and cultural knowledge. The model is still today base of the German education system and is the fundament for creative, innovative thinking. Furthermore, in Germany, the social market economy is established since 1949. Social market economy allows free entrepreneurial acting and parallel, the social compensation is considered. The social market economy was pushed by Chancellor Ludwig Erhard and is an important factor to ensure free thinking in a secured environment.

The Japanese society is embossed by a strong hierarchical structure. Acting within the society is characterised by discipline. Furthermore, education has a high value within Japanese society (like in all Confucian orientated Asian countries).

Important character of the Japanese working environment is the principal of life long employment (until twentieth century); cf. (Japan 2012). Furthermore, the loyalty regarding the employer is very important. Loyalty, discipline and healthy food leads to low number of employees away sick (Japan: Average 1%; Germany: Average 3.2%). The strong connection between working environment and personal life style leads to a strong identification of Japanese employees with their work.

Both countries have an education system, which ensures a high education qualification level. The differences are the social market economy, hierarchical system, employee identification with work, discipline and have highly influence on different engineering philosophies regarding product innovation and product optimisation. One the one side strong hierarchical structure, discipline and loyalty supports product optimisation. On the other side, free thinking within a social market economy supports creative, product innovations. Precondition for the ability to create product innovations or optimisation is the education fundament, which is given in both countries.

5.3 Society and Social Interaction Impact: Comparison of Japan and Germany

This chapter deals with the basic characteristics and features of Japanese and German society in terms of the behavior and interaction of people within each society. It forms the basis for efficiency and effectiveness in the use of complex technical systems by the end user; see also Sect. 6.

5.3.1 Japan

Social interaction in Japan is strongly characterised by respect, consideration and harmony. These values are deeply rooted in Japanese culture and influence many aspects of social and public life. The following is a brief outline of the main characteristics of social interaction in Japan:

5.3.1.1 Hospitality and Service Culture (Principle "Omotenashi")

Warmth and attentiveness: *Omotenashi* is the Japanese concept of hospitality. It describes a deep culture of service in which hosts and service providers are proactive and attentive to the needs of others without expecting anything in return.

Consideration for comfort: Japanese society attaches great importance to offering others a pleasant experience without ostentatious self-promotion, for example in restaurants, hotels or on public transport.

5.3.1.2 Harmony and Group Orientation (Principle "Wa")

Avoiding conflict: Harmony, known as 'Wa', is a central social principle. It is considered important to avoid conflict and to prioritise the good of the group over one's own interests.

Group-orientation: In Japanese culture, the good of the group is often considered more important than individual needs. Decisions and behaviour are based on what best supports the group, which often leads to smoother social dynamics.

5.3.1.3 Politeness and Respect (Principle "Reigi")

Formalities and etiquette: Politeness and a strong understanding of etiquette play a major role with set rules of behaviour and formalities as a sign of respect and humility.

Respect for elders and authority figures: Elderly people and authority figures are traditionally highly respected in Japan, which is also reflected in the language, as the Japanese language recognises special forms of politeness here.

5.3.1.4 Consideration in Public Spaces

Quiet communication and restraint: In Japanese culture, it is customary to behave quietly in public spaces. Phone calls are often avoided on trains and buses, and communication tends to take place at a lower volume.

Order and cleanliness: Japanese society attaches great importance to cleanliness and order. Example: Japanese people do not leave rubbish in public spaces (public transport, parks, streets); delays in public transport due to extraordinary cleaning activities are extremely rare (cf. Sect. 6.1).

5.3.1.5 Collective Responsibility and Discipline

Self-discipline and responsibility: There is a high degree of self-discipline in Japan. Most people adhere to rules and social expectations, even if they are not enforced hierarchically.

Apology culture: The willingness to take responsibility is very pronounced in Japan. Apologising, even for minor mishaps, is normal in Japanese culture and shows respect and remorse.

5.3.1.6 Loyalty and Sense of Duty

Commitment to company and family: Loyalty to employers, family and friends is very important in Japan. This can be seen, for example, in the strong loyalty of many Japanese to their employers and the high level of willingness to take responsibility for their own family.

Team spirit and cooperation: Japanese schools and companies place great value on teamwork and the ability to cooperate. Learning and working in a team is seen as an important building block for success.

5.3.1.7 Self-Restraint (Principle "Ken")

Modesty and restraint: It is customary to be reserved and modest. Self-praise or emphasising one's own achievements is considered impolite and is avoided; the focus is on the performance and results of the collective.

5.3.1.8 Interim Summary

These values and behaviours have contributed to Japan being a very orderly, disciplined and respectful society in which harmony and consideration are paramount. The behaviours outlined lead to extreme efficiency and effectiveness in an engineering context, e.g. in the operation of production processes, maintenance measures and public transport (cf. Sect. 6). Furthermore, the outlined behaviours promote the continuous optimisation of technical solutions and/or processes.

5.3.2 Germany

Social interaction in German society is characterised by values such as individuality, openness, reliability and a certain directness. These values are reflected in many aspects of social life and characterise German culture and the way people interact with each other. Important characteristics of social interaction in Germany are outlined below:

5.3.2.1 Individualism and Independence

Personal freedom: Germans attach great importance to individuality and personal freedom. Self-realisation and the pursuit of one's own interests are valued and encouraged, which is also evident in education and the world of work.

Personal responsibility: People in Germany are often independent and responsible. The state and society attach importance to people being able to look after themselves and make their own decisions.

5.3.2.2 Directness and Honesty

Open communication: In Germany, directness is seen as a sign of honesty and respect. It is appreciated to address things openly and clarify problems directly, without much ado.

Ability to take criticism: In Germany, criticism and feedback are seen as important tools for improvement and personal development. People are often open to constructive criticism and see it as part of social and professional interaction.

5.3.2.3 A Sense of Duty and Reliability

Discipline and structure: Punctuality, discipline and structured work are highly regarded in German society. Reliability in adhering to rules and agreements is seen as a sign of respect and commitment.

Honouring contracts: Reliability is also reflected in the appreciation of contracts and agreements. Agreements are often taken very seriously in Germany and are seen as a moral obligation.

5.3.2.4 Consideration and Tolerance

Acceptance of diversity: Germany is a multicultural society and there is an increasing emphasis on tolerance and acceptance of different lifestyles, cultures and opinions.

Consideration for public spaces: In public spaces, Germans often emphasise consideration, cleanliness and order. Rules such as observing quiet hours or separating rubbish are observed and seen as an expression of mutual consideration.

5.3.2.5 Privacy and Distance

Respect for privacy: In Germany, privacy is a valuable asset. Personal spaces are respected and it is customary to maintain distance until trust has been established.

Value on the personal sphere in social interaction: Professional and private matters are kept separate. It is expected that private matters are treated with restraint.

5.3.2.6 Participation and Co-Determination

Civic participation and understanding of democracy: German society attaches great importance to co-determination and democracy. Many people are involved in associations, citizens' initiatives and political bodies in order to actively contribute to society.

Culture of discussion: Discussion and freedom of opinion are valued, even when different opinions clash. The exchange of arguments and the search for compromises are integral parts of the German culture of debate.

5.3.2.7 Structure and Rules

Order and awareness of rules: German society has a strong awareness of order and rules. It is perceived as positive to abide by rules, whether in traffic, laws or social agreements.

Regulated procedures: German everyday life is often highly structured and it is appreciated when procedures are predictable and reliable. This is noticeable in administration, at work and in other areas of life.

5.3.2.8 Orientation Towards the Common Good and Solidarity

Solidarity and the welfare state: Solidarity is deeply rooted in Germany, which is reflected in the welfare state, among other things. Systems such as health insurance, social welfare and other social security schemes are an expression of this orientation.

Commitment to the weak: Especially in times of crisis or social challenges, there is a great willingness to help and solidarity among the population.

5.3.2.9 Interim Summary

Social interaction in Germany is characterised by a balance between individuality and community responsibility. Directness and reliability are the cornerstones of social interaction, and privacy is highly valued. At the same time, great value is placed on tolerance, justice and order, which often leads to a regulated and structured society. All in all, the outlined behaviours in an engineering context promote the finding of innovative solutions in product development and production process design.

5.4 Geographical and Location Impact: Japan's Special Situation

A strong impact on innovative or optimised solutions are geographical impacts. Within the direct comparison between Japan and Germany, the difference is obviously: Japan is located at the rupture zone of four tectonic plates of the earth crust:

- North American plate in the north,
- Eurasien plate in the west,
- Philippines plate in the south and
- Pacific plate in the east.

Each year, the drift is against each other within a range of some centimetres. The pacific plate moves under the Eurasien plate which leads to volcano eruption and many earthquakes. In total, Japan records around 7500 earthquakes every year, of which around 1500 are felt by humans. Regarding the strength, in an average in Japan happen 73 earthquakes per month (magnitude 4 or higher; thereof nine per month with magnitude 5 or higher; thereof 1.4 earthquakes with magnitude 6 or higher), cf. (Earthquake Research Committee et al. 2011).

However Germany is located in the centre of Europe, nowadays without any significant influences regarding earthquakes (the strength of earthquakes is on a very low level) and no influences regarding volcano eruption. Most earthquakes in Germany are very weak, with a magnitude of less than 3.0 on the Richter scale. Such earthquakes are often only registered by sensitive measuring devices and are usually barely perceptible to humans. On average, the magnitude of perceptible earthquakes in Germany is around 3.0 to 4.0. Earthquakes with a magnitude of over 4.0 are rare and usually only occur every few years.

This facts leads in Germany and Japan to total different strategies in case of safety and reliability engineering in many subjects, like facility engineering (e.g. buildings), mechanical engineering (e.g. railway systems) and or electromechanical engineering (e.g. consumer goods). Japanese risk- and reliability analysis considering always safety aspects based on earthquake-vibrations impact in comparison to German products. This is reflected e.g. in choose of construction, material, electric connections and shut-down strategies respectively product concepts.

6 Example Railway System: Comparison of German and Japanese Engineering

In this chapter, the outlined differences (impacts) between the industrialised nations of Germany and Japan in the four sub-areas of history, education system, society and geography are substantiated using the case study of the railway system and railway operations. The impact is compared in terms of the differences in the effectiveness of the long-distance transport systems of Deutsche Bahn (DB; 'Intercity Express (ICE)', 'Intercity (IC)', 'Eurocity (EC)') and the Japan Railway (JR) system ('Shinkansen'). The dimensions of punctuality of long-distance connections (Sect. 6.1), technology (Sect. 6.2) and operation with a focus on passenger boarding and alighting (Sect. 6.3) are analysed.

6.1 Comparison Parameter Punctuality

The easiest way to compare the effectiveness of long-distance transport systems in Germany and Japan from the user's point of view is to analyse punctuality. Firstly, long-distance transport for travellers is compared: Punctuality has the highest priority in Japan, explicitly for the *Shinkansen (high-speed train)*: The average delay is 1.6 min per train in 2023 (cf. Fig. 6), although it should be noted that this delay time also includes delays due to force majeure (natural events such as typhoons, earthquakes).

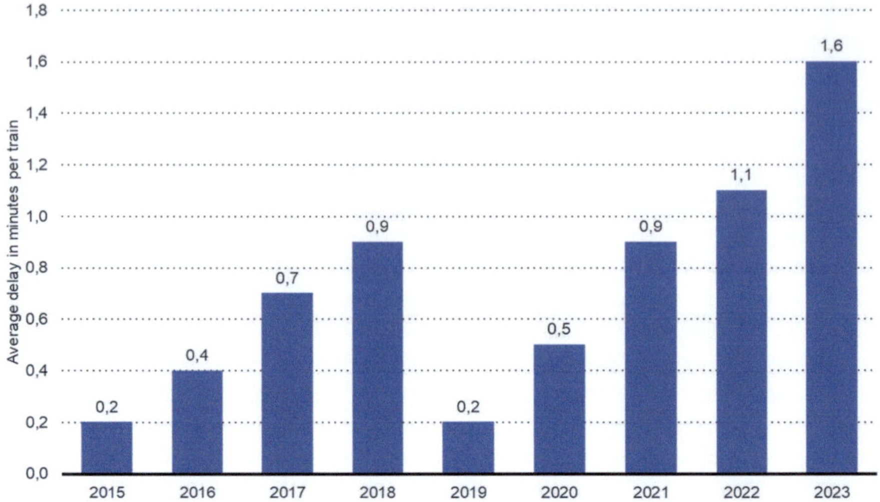

Fig. 6 Average delay of JR Central Tokaido Shinkansen in Japan FY 2015–2023 (Source: statista.com). Note: Japan; fiscal years 2015 to 2023; the company's fiscal years start on April 1 of the stated year and end on March 31 of the following year; including delay due to uncontrollable causes such as natural disasters, etc.

In contrast, Deutsche Bahn (Germany) defines punctuality in the *long-distance network* in such a way that delays of up to 6 min compared to the scheduled time are classified as punctual. In 2024, the punctuality rate for long-distance transport based on this definition was around 62.4%; cf. Fig. 7).

In Japan, 92% of *freight trains* reach their destination with a maximum delay of 30 min; cf. Fig. 8 (Source: Ministry of Land, Infrastructure, Transport and Tourism in Japan).

By contrast, 98,104 trains (41.6%) of 235,949 trains in cross-border rail freight transport at Deutsche Bahn (here: period from 1 January 2021 to 30 November 2021) reached their destination with a delay of more than one hour (source: statista.com).

6.2 Technology Long-Distance Transport Systems: Differences

The technology of the long-distance transport systems in Japan and Germany is compared on the basis of the following criteria:

- separate rail networks
- train frequency and organisation
- technical maintenance and precision
- responsiveness to disruptions

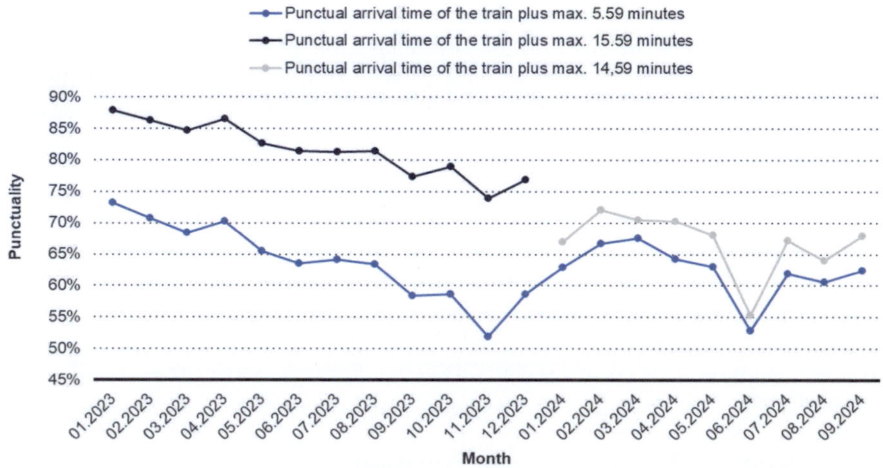

Fig. 7 Punctuality of Deutsche Bahn long-distance trains (EC, IC, ICE) until September 2024. *Source* statista.com

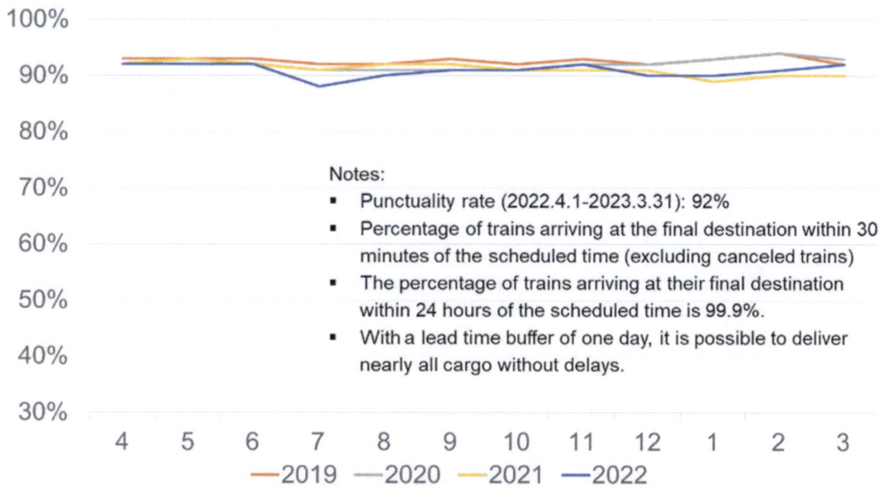

Fig. 8 Punctuality in freight transport on the Japan Railway (JR) in 2023. *Source* Ministry of Land, Infrastructure, Transport and Tourism in Japan

6.2.1 Separate Rail Networks

Japan: The Shinkansen trains use dedicated high-speed lines that are completely separated from the rest of the rail network (regional trains and freight transport). This avoids conflicts with slower trains and minimises delays. In the Japanese regional/freight network, goods trains are often operated outside peak hours or at night in order to minimise disruption to local traffic.

Germany: ICE and IC trains use the same rail network as regional and goods trains. This increases the susceptibility to delays, as all trains are affected in the event of bottlenecks or disruptions on a route. Priority regulations for long-distance traffic can only compensate for this to a limited extent, as all train types are affected in the event of signalling or points faults, for example.

6.2.2 Train Frequency and Organisation

Japan: Trains run at short intervals, sometimes every 3 to 10 min, especially in large cities and on high-speed lines. The system is highly organised and the train teams follow strict, coordinated procedures.

Germany: Although Germany also has a dense network, there are longer intervals on long-distance services, especially between cities away from major conurbations.

6.2.3 Technical Maintenance and Precision

Japan: Maintenance is a top priority in Japan. Trains and infrastructure are regularly inspected and maintained to ensure maximum precision.

Germany: Deutsche Bahn also undertakes regular maintenance work, but there are occasional bottlenecks in maintenance budgets and staffing, which can affect punctuality and availability.

6.2.4 Service Orientation and Customer Information

Japan: High service quality and customer-orientation are typical. Railway staff apologise even for minimal delays, and delay certificates are issued when the trains are delayed by more than 5 min.

Germany: Deutsche Bahn generally informs its customers via app and loudspeaker announcements. Apologies and information are available, but less emphasis is placed on minimal delays, as delays of less than 6 min are not defined as delays (cf. Sect. 6.1).

6.2.5 Reactivity in the Event of Malfunctions

Japan: In the event of disruptions, alternative measures and replacement journeys are organised immediately in order to normalise operations as quickly as possible.

Germany: Deutsche Bahn is also working to resolve disruptions quickly, but the tightly synchronised and mixed rail network often makes it difficult to return to normal operations quickly.

6.3 Operation of Long-Distance Trains: Passenger Boarding and Alighting Procedure

The efficiency of boarding and alighting on long-distance trains is a key factor that can influence the punctuality of a train.

In Japan, boarding and disembarking on trains is very well organised and takes place quickly and efficiently. In the following, the process of boarding and disembarking long-distance trains in Japan and Germany is compared using the following features:

- Markings on the platform
- Disembarking before boarding
- Railway staff as support
- Anticipatory waiting for reservations

6.3.1 Markings on the Platform

There are clear lines and positions on the Japanese platform: There are clear markings on each platform indicating where the doors of the train will open. Passengers know exactly where to queue, depending on the carriage or seat; cf. Fig. 9 right. Furthermore, Japanese passengers form disciplined queues at these marked points and wait until the train arrives and the doors open; cf. Fig. 10 right.

On a German platform, on the other hand, only an area is marked where the defined train waggon is expected, cf. Fig. 9 left (here: area 8). Accordingly, the queuing of passengers is disorganised and boarding is not efficient; cf. Fig. 10 left.

6.3.2 The Principle of Disembarking Before Boarding

In Japan, there is a clear rule that all passengers must get off before anyone gets on. This is supported by announcements and pictograms and monitored by railway staff. Japanese passengers are very disciplined in adhering to this rule, which speeds up the whole process.

In contrast, although the same rule applies in Germany, it is not fully implemented by travellers, which leads to delays in the boarding/disembarking process.

6.3.3 Railway Staff as Support

In large railway stations such as Tokyo, 'train pushers' (*oshiya*) or 'train pullers' (*hagashiya*) are used to support the boarding process during peak times. They help to ensure that trains depart on time by guiding passengers onto the train in an organised manner. In Germany, on the other hand, there are no staff to support the boarding/deboarding process. In Japan, travellers generally do not leave rubbish on the train (cf. also Sect. 5.3), so that trained cleaning staff can clean each train in a few minutes.

In contrast, cleaning trains in Germany is time-consuming, as travellers leave significantly more rubbish behind in comparison.

6.3.4 Anticipatory Waiting for Reservations

There are both reserved and unreserved carriages on Japanese high-speed lines, with reserved seats accounting for the vast majority. Passengers with reservations can queue at the marked positions for their carriage, which speeds up boarding considerably. Only passengers holding reservation tickets are allowed to sit in designated seating carriages. Passengers without reserved seats are not permitted to enter designated seating carriages, and wait at the lines of unreserved carriages as shown in Fig. 10 right (cf. also Fig. 9 right) to board in an organized way, whereby discipline also facilitates boarding here (cf. Sect. 5.3).

Fig. 9 Left side railway station Germany (Photo: D. Kruse, pixelio.de): No markings on platform, approximate indication of where the train wagon is expected to arrive (here: area number 8); Right side railway station Japan: markings on platform with information about train waggon, door number related to seat, boarding/disembarking

Fig. 10 Procedure boarding/disembarking in the long-distance transport system of Deutsche Bahn (DB; Germany) versus Japan Railway (JR; Japan)

In contrast, the majority of seats on German long-distance trains are not reserved, which leads to a disorganised, inefficient boarding process.

This combination of clear rules, disciplined behaviour and well-organised stations ensures that boarding and alighting on Japanese trains runs smoothly and efficiently. In contrast, the system in German railway stations is less regulated and the discipline of travellers is less pronounced, which favours disruptions in operations.

6.3.5 Interim Summary

The Japanese railway system is characterised by a highly efficient, dedicated network and uncompromising punctuality, while the German system, with a larger network and diversified operations, is more flexible but also more prone to disruption.

7 Summary

This paper shows essential aspects of the differences in German and Japanese reliability and safety engineering. Both industrialised nations are able to develop technical products at a high level in terms of functionality and reliability. However, the product development process ('the path to the goal') differs considerably due to the different developments of the industrialised nations: on the one hand, the clearly structured, serial processes of German engineering, on the other hand, the '*Suriawase*' principle of Japanese engineering. The approaches to avoiding technical risks also differ: Frontloading approach versus refinement of the '*Suriawase*' principle through multi-layered interaction. Furthermore, the technological solutions of Japanese engineering differ from German engineering:

German product functionality is often characterised by innovative solutions, Japanese products often show optimised solutions regarding the previous product generation. The reasons for this fundamental difference are manifold.

Base of operations are the influencing factors, which supports or restrains creative thinking and as a consequence, support product innovations or optimisations. Afterwards, the historical factors, social factors (society and interaction), educational system factors and geographical conditions respectively locations in Germany and Japan are analysed, which leads to the different technical reliability and safety engineering solutions in Japanese and German engineering as well as to a different field operation of technical products in Germany respectively Japan.

The identified differences in engineering and the different influences from history, society, education system and geography between the two countries are substantiated and analysed in detail using the example of long-distance transport systems (Deutsche Bahn, Germany and Japan Railway, Japan). On the one hand, design and logistical processes are outlined, and on the other hand, user behaviour in both countries is analysed. The Japanese railway system is characterised by a highly efficient, dedicated network and uncompromising punctuality, while the German system, with a larger network and diversified operations, is more flexible but also more prone to disruption. However, both innovative and optimised product design approaches have advantages and gives the chance to learn from each other.

References

Baumgart, F.: Between Reform and Reaction. Prussian school politics 1806–1859. Wissenschaftliche Buchgesellschaft, Darmstadt. (1990)

Bender, B. and Gericke, K.: Pahl/Beitz Konstruktionslehre: Methoden und Anwendung erfolgreicher Produktentwicklung. 9th ed., Springer, Imprint: Springer Vieweg. (2021)

Bracke, S.: Reliability engineering: data analytics, modeling, risk prediction. Berlin, Germany: Springer. (2024)

Bracke, S. and Inoue, M.: Reliability and safety engineering: The principles innovation and optimisation of German and Japanese product constructions. In ESREL 2018 Proceedings (June 17–21, 2018, Trondheim, Norway): S. Haugen, J. E. Vinnem, A. Barros, T. Kongsvik, C. v.

Gulijk: Safety and Reliability – Safe Societies in a Changing World. CRC Press, Pages: 221–226. (2018)

Braess, H.-H. and Seiffert, U. Vieweg Handbuch Kraftfahrzeugtechnik. 7th ed. Wiesbaden: Springer Fachmedien Wiesbaden. (2013)

Deutsches Institut für Normung e.V.: DIN ISO 15226 . Technische Produktdokumentation – Lebenszyklusmodell und Zuordnung von Dokumenten. Beuth Publisher. (2017)

Earthquake Research Committee: Supplementary Information to the Evaluation of Seismic Activity for January 2011. Earthquake Research Committee, 9. Februar 2011; release on 12. März 2011. (2011)

Givoni, M.: Development and Impact of the Modern High-speed Train: A Review. In: Transport Reviews. 26, Nr. 5, ISSN 0144-1647, S. 593–611. (2006)

Griffin, E.: A Short History of the British Industrial Revolution. London, Palgrave. (2010)

Inoue, M., Mogi, R., Nahm, Y.-E. & Ishikawa, H.: Design Support for "Suriawase": Japanese Way for Negotiation among Several Teams. In Dan Frey, Shuichi Fukuda and Georg Rock (eds.), Improving Complex Systems Today, Springer-Verlag, pp. 385–392. (2011)

Japan Statistical Yearbook 2012.: Statistics Bureau, Ministry of Internal Affairs and Communication. (2012)

Jehle, P., Naumann, R. and Schach, R.: Transrapid and Wheel-Rail-High speed trains: A comprehensive comparison of systems. (in German). Springer. (2006)

Morinosuke, K.: History of japanese external relations. Volume 1: Country opening to Meiji Restauration. (in German) Steiner, Wiesbaden. (1976)

Nussbaum, L.-F. and Roth, K.: Japan encyclopedia. Cambridge: Harvard University Press. (2005)

Pollard, S.: Peaceful Conquest. The Industrialisation of Europe 1760–1970, Oxford. (1981)

Sobek, II DK., Ward, AC., Liker, JK.: Toyota's principles of set-based concurrent engineering. Sloan Manage Review 40(2), 67–83. (1999)

Takeishi, A., Fujimoto, T.: Modularisation in the auto industry: interlinked multiple hierarchies of product, production and supplier systems. International Journal of Automotive Technology and Management 1(4):379–396. (2001)

Tomlinson, W. W.: The North Eastern Railway: Its rise and development. Andrew Reid and Company. (1915)

Methods for Analyzing Reliability Engineering: Probabilistics and Accelerated Testing in Early Phases of Product Emergence Process

Alicia Auer, Philipp Heß and Stefan Bracke

Contents

Abstract

The product emergence process of a product is significantly influenced by its reliability, which in turn determines both customer satisfaction and operating costs. It is therefore evident that reliability engineering plays a pivotal role in the development and optimization of products. This paper presents the application of reliability engineering methods using three representative case studies based on real test data, which have been enhanced by simulation. The data presented here originates from fatigue tests of shape memory actuators. In order to facilitate comprehension, an initial presentation will be made of the fundamentals of the shape memory effect, together with an account of the nature of shape memory actuators and the procedures employed in their testing. The initial case study entails an examination of the failure behavior exhibited by three distinct test channels of a test rig. The application of descriptive statistics (boxplots) and significance tests (Kruskal-Wallis test with Conover post-hoc analysis) enables the identification and interpretation of differences in the failure behavior of the channels. The second case study focuses on accelerated testing. The determination

A. Auer (✉) · P. Heß · S. Bracke
Chair of Reliability Engineering and Risk Analytics, University of Wuppertal, Wuppertal, Germany
e-mail: aauer@uni-wuppertal.de

S. Bracke and R. Pieper (eds.), *Safety Engineering*,
https://doi.org/10.1007/978-3-658-47473-7_2

of suitable acceleration levels is achieved through the analysis of different load levels. The investigation is conducted using the two-parameter Weibull distribution, including the calculation of confidence intervals for the model parameters and the determination of acceleration factors. In the third case study, a Sudden Death test is conducted in which candidate data is extrapolated using the Johnson rank correction in order to make reliable statements about the service life distribution. The methods presented illustrate the practical advantages of reliability engineering approaches in the analysis and optimization of system reliability, and offer valuable insights for product development and testing.

Keywords

Product emergence process · Reliability analysis · Accelerated testing · Weibull distribution · Sudden Death test · Kruskall-Wallis test · Shape memory alloy · Conover post hoc analysis · Shape memory actuator · Fatigue test

1 Introduction

The product life cycle of a technical product can be divided into distinct stages, including concept development, series development, production, and sales. The product complexity continues to increase due to the constant expansion of the range of functions and the increased utilization of materials and available installation space (Bracke 2024). The continual rise in expectations regarding components, systems and products is giving rise to new challenges and the necessity to devise innovative concepts. Technical reliability is of particular importance, as it is essential to guarantee the functionality, safe operation and reliability of technical products throughout their entire service life, which is becoming increasingly longer.

One method for evaluating the potential for failure at an early stage in the product development process is the utilization of distribution models, with the Weibull distribution being a particularly effective approach (Weibull 1951). The Weibull distribution is particularly well-suited for the assessment of damage cases and the analysis of field data, as the interpretation of the Weibull parameters enables the characterization of failure behavior (Bracke 2024, Pfeifer 2001). The typical failure behavior or failure rate of a technical product encompasses three main categories: early failure behavior, random failure behavior and wear failure behavior (Birolini 1999, Bracke 2024, Pfeifer 2001). The overall results are represented by a bathtub curve (Fig. 1). This demonstrates that there is a high failure rate at the outset of the utilization phase, given that products are typically not fully mature when they are launched onto the market. Subsequently, a stabilized process is initiated, which is followed by an increase in the frequency of runtime failure behavior.

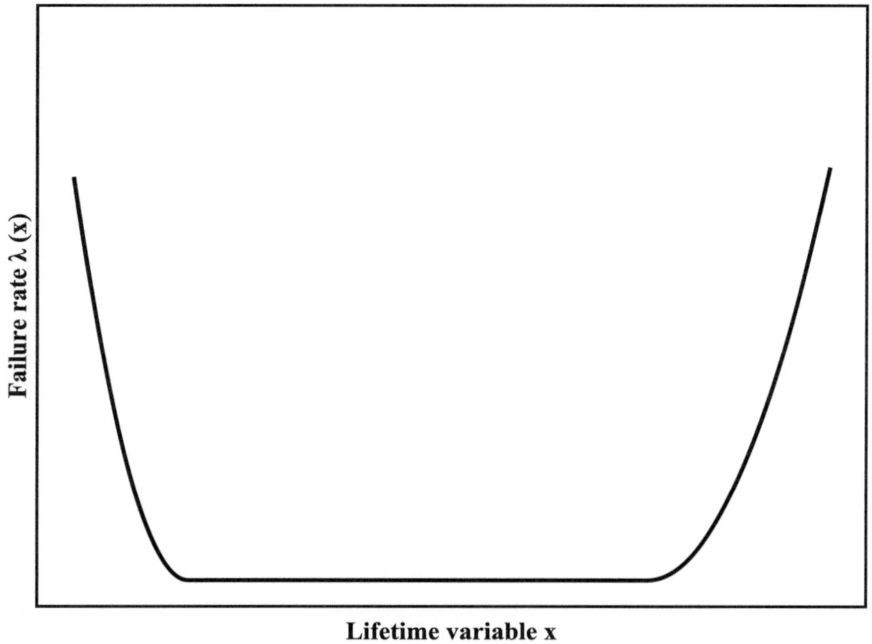

Fig. 1 Schematic representation of the failure rate $\lambda(x)$ of a component or product based on a lifetime variable x

2 Methodology

The following section provides an explanation of the methodologies employed in this article, with the objective of facilitating a more comprehensive understanding of the subject matter.

2.1 Weibull Distribution

The Weibull distribution is a statistical distribution model that has been widely employed in the field of reliability engineering (Weibull 1951). In this paper, the two-parameter form is utilized, as defined in Eq. 1 in the form of the cumulative distribution function. The parameters, in addition to the life span variable x, are the shape parameter b and the scale parameter T (characteristic life span).

$$F(x) = 1 - e^{-\left(\frac{x}{T}\right)^b} \tag{1}$$

The exponential-based distribution model is characterized by its remarkable flexibility and straightforward mathematical formulation. By varying parameter b, different failure rates can be described. A shape parameter b below 1 indicates an early failure behavior, while a shape parameter b around 1 gives a hint regarding a random failure behavior. Runtime failure behaviors are often characterized by a shape parameter b above 1. If the failure rates of these three failure behaviors are plotted graphically, the wellknown bathtub curve results, as illustrated in Fig. 1. The scale parameter T represents the failure probability of 63.3% (Bracke 2024).

The fitting of the distribution parameters can be conducted using maximum likelihood estimation (MLE), as proposed by Fisher (1912). An initial estimation of the failure probabilities can be achieved by mapping the sum probabilities (median ranks) (Bracke 2024). Eq. 2 demonstrates an approximation of the failure probability for rank i, taking into account the sample size n.

$$F(x_i) = \frac{i - 0.3}{n + 0.4} \tag{2}$$

When parameters are estimated on the basis of a sample, it is inevitable that uncertainties will emerge. In order to quantify these uncertainties, confidence intervals can be calculated based on the estimated parameters. In the case studies presented in this paper, the confidence interval for the shape parameter based on the approach proposed by Verband der Automobilindustrie e. V. (1976) is used. In accordance with the definition provided in Eq. 3, the calculation includes the estimated shape parameter $\hat{b}$, the sample size n and a constant k, which is dependent upon the selected confidence level γ. For the analyses presented in this paper, a confidence level of 0.95 has been chosen, resulting in $k = 2.0$.

$$\hat{b}\frac{1}{1 + \sqrt{\frac{k}{n}}} \leq b \leq \hat{b}\left(1 + \sqrt{\frac{k}{n}}\right) \tag{3}$$

The confidence interval for the characteristic life span T is calculated in accordance with the approach proposed by Bitter (1986). As shown in Eq. 4, this confidence interval is based on the χ^2-distribution, taking into account the estimated characteristic life span $\hat{T}$, the estimated shape parameter $\hat{b}$, the sample size n and the significance level α, where $\alpha = 1 - \gamma$.

$$\hat{T}\left(\frac{2n}{\chi^2_{2n,1-\frac{\alpha}{2}}}\right)^{\frac{1}{\hat{b}}} \leq T \leq \hat{T}\left(\frac{2n}{\chi^2_{2n,\frac{\alpha}{2}}}\right)^{\frac{1}{\hat{b}}} \tag{4}$$

In regard to the graphical representation of a fitted Weibull distribution model, the depiction in a Weibull probability paper is an appropriate choice. In a Weibull probability paper, the ordinate is divided in a double-logarithmic manner, while the abscissa is divided in a single-logarithmic manner. This allows for the visualization of a Weibull distribution model as a straight line, which enables an initial, simple analysis. In this representation, the shape parameter b corresponds to the slope of the straight line. For example, two Weibull distribution models with identical failure behavior (the same shape parameter b) would appear as parallel lines in a Weibull probability plot (Bracke 2024).

2.2 Kruskal-Wallis Test and Conover Post-Hoc Analysis

Kruskal-Wallis rank sum test. The Kruskal-Wallis rank sum test is a non-parametric statistical method that allows the comparison of mean parameters across multiple samples, with a minimum of three (Kruskal and Wallis 1952, Kruskal 1952). The null hypothesis of this test states that the mean values of the analyzed samples are identical, while the alternative hypothesis states that these differ in at least one. To perform the test, all values from all samples i (number k) are assigned ranks in ascending order of size, taking ties into account. The rank sum R_i is then determined for each sample, as illustrated in Eq. 5.

$$R_i = \sum_{j=1}^{n_i} R_{ij} \tag{5}$$

The (tie corrected) test statistic H is calculated using the summed rank sums of the samples, the total sample size (sum of all sample sizes) n, number of ties r and the number of identical values of a tie t_i in the case of the same size of all samples, in accordance with Eq. 6.

$$H = \frac{\frac{12}{n \cdot (n+1)} \cdot \sum_{i=1}^{k} \frac{R_i^2}{n_i} - 3 \cdot (n+1)}{1 - \frac{\sum_{i=1}^{r} (t_i^3 - t_i)}{n^3 - n}} \tag{6}$$

The test decision is made by means of a comparison with the χ^2-distribution according to Eq. 7 considering the number of samples k and the significance level α. The null hypothesis is rejected if the test statistic H exceeds the critical value.

$$H > \chi^2_{k-1;1-\alpha} \tag{7}$$

A detailed description of the Kruskal-Wallis test can be found in Bracke (2024).

Conover Post-hoc analysis. If a significant difference in the locations of the samples is detected using the Kruskal-Wallis test, a post-hoc analysis according to Conover and Iman (1979), Conover (1999) can be performed to determine which of the samples are different in terms of location. This non-parametric approach entails a comparison of the relative differences in the rank sums on a pairwise basis. The null hypothesis of equality between the samples i and j in their location is rejected if the condition set forth in Eq. 8 is met.

$$|\bar{R}_i - \bar{R}_j| > t_{1-\alpha/2;n-k} \cdot \sqrt{s^2 \cdot \left(\frac{n-1-H}{n-k}\right) \cdot \left(\frac{1}{n_i} + \frac{1}{n_j}\right)} \tag{8}$$

The Kruskal-Wallis test statistic H, the quantile of the t-distribution in dependency of the significance level α, and the total mean square s^2, as defined in Eq. 9, are included in the analysis.

$$s^2 = \frac{1}{n-1} \cdot \left(\sum R_i^2 - n \cdot \frac{(n+1)^2}{4} \right) \tag{9}$$

2.3 Sudden Death Testing and Johnson Rank Correction

In a classic component test, each component of the sample is tested until it fails. This approach has the advantage that all failure times can be used for modelling. However, it is disadvantageous in terms of the duration of the tests, which is lengthy and therefore requires a significant testing effort. One solution to these challenges is Sudden Death testing. This method involves dividing the parts from the sample into groups of equal size. If the first part in the group fails, the non-failed units (candidates) in this group are not tested further. The prerequisite for utilizing this methodology is that the identical load spectrum is applied to all constituent parts of a given group. To fit a suitable distribution model, the ranks of the known failure times must be corrected in accordance with the candidates.

One method for this is the rank correction according to Johnson (1964). In this approach, the candidates are considered by assigning mean, hypothetical rankings with reference to the failed units. The calculation of the average rank number r_j is shown in Eq. 10. The calculation includes the average rank r_{j-1} of the previous default time, the number of failures n, and the number of potential candidates n_j no longer available at the failure time t_j. The cumulative probability of t_j can be approximated with Eq. 11, based on Eq. 2 using the corrected ranks. The fit of the extrapolated distribution function is performed using MLE (Fisher 1912), taking into account the censoring of the data (Bracke 2024).

$$r_j = r_{j-1} + x_j \cdot \frac{1 + n - r_{j-1}}{1 + n - n_j} \tag{10}$$

$$F(t_j) = \frac{r_j - 0.3}{n + 0.4} \tag{11}$$

2.4 Accelerated Testing

The implementation of tests during the product development phase requires a considerable investment of time, as the tests are conducted under conditions that are representative of those that will prevail in subsequent operational use (Bertsche 2008,

Escobar and Meeker 1995, Härtler 2016, Nelson 2004). Furthermore, a considerable number of products possess a lengthy operational lifespan. In many cases, the tests would therefore require a significant investment of time, spanning months or even years, to fully analyse the degradation behavior of a product over its entire service life. The objective of accelerated testing is to reduce the necessary test duration by, for instance, applying higher loads selectively while maintaining the same failure behavior. There are various types of accelerated tests, which depend on the method or type of acceleration employed (Pascual et al. 2006, Pham 2006). The acceleration of a products degradation behavior can be achieved through a number of methods, including increasing the frequency of use, accelerating the ageing process, and increasing the product load. Test specimens are typically exposed to conditions that exceed normal operating limits, such as elevated temperatures, increased humidity, excessive loads, or higher stresses. This allows for the determination of data on durability or degradation behavior, enabling statements to be made on long-term reliability without the need for time-consuming real-time tests over months or years.

In the context of accelerated testing, it is of paramount importance that the failure mechanisms observed at elevated loads are identical to those that would occur under normal operating conditions, albeit in an accelerated form. The relationship between load and life expectancy can be expressed or described by various models, including the Arrhenius equation (for temperature-based acceleration), the Coffin-Manson model (for fatigue under mechanical load), and the Eyring model (influence of electrical voltage) (Escobar and Meeker 1995, Härtler 2016, Kremer and Bertsche 2018, Pascual et al. 2006). This allows engineers to extrapolate the service life of a product under typical conditions of use based on data obtained from accelerated tests.

The various types of accelerated tests are associated with particular failure mechanisms and defined stress factors. Highly Accelerated Life Testing (HALT) is frequently employed during the product development phase with the objective of identifying potential weaknesses by subjecting the product to extreme stresses (worst-case scenarios). In contrast, Highly Accelerated Stress Screening (HASS) is more commonly utilized in production environments. It employs less extreme stresses than HALT with the aim of detecting production-related weaknesses and defects that could lead to premature failure.

The implementation of accelerated testing methodologies offers a number of advantages, including the reduction of product validation timelines, the minimization of expenditure through the curtailment of requisite testing periods, and the capacity to discern and rectify design deficiencies or material weaknesses at an early stage of the product development process. Furthermore, accelerated testing facilitates the enhancement or assurance of long-term reliability by discerning and rectifying potential deficiencies before a product is released to the market. This can enhance overall safety and customer satisfaction. Nevertheless, accelerated testing is not without its own limitations and challenges. A fundamental requirement is that the failure mechanisms observed under high stress must precisely align with those that occur under normal operating conditions. Misinterpretation of test results can lead to inaccurate estimates of product life or reliability. Therefore, a comprehensive understanding of the underlying failure mechanisms and precise control of the test conditions is essential.

The utilization of accelerated testing is a prevalent practice across a multitude of industries, particularly in domains where product reliability is of paramount importance, such as aerospace, automotive, electronics and medical technology.

3 Fundamentals of Shape Memory Actuators

The shape memory effect (SME) describes the ability of shape memory alloys (SMA) to restore an originally trained or imprinted shape. This distinctive material behavior is based on a thermomechanical phase transformation between the two solid states of martensite and austenite, which enables deformations that can significantly exceed the elastic deformability or elongation of alloys (Elahinia 2015, Gümpel 2004, Lagoudas 2008, Lexcellent 2013, Rao et al. 2015). SMA, which are primarily composed of nickel-titanium (NiTi) and occasionally copper- or iron-based compounds, undergo structural transformations at the microscopic level that can be transferred to macroscopic movements. These transformations can be employed in actuators for compact, lightweight, and precise actuation (Gandhi and Thompson 1992, Hornbogen 1991, Peters 2010, Van Humbeeck 1991, 2001). SMA represent a class of intelligent materials or alloys that possess the distinctive property of retaining an original shape and reverting to this shape when specifically activated.

SMA are based on two fundamental phenomena: the SME and superelasticity. The SME enables shape memory alloys to revert to a predefined shape following a deformation caused by an increase in temperature above a defined threshold value, which is referred to as the transformation temperature. The alloy exists in two distinct phases (Fig. 2). The martensite phase, which exists at lower temperatures and is easily deformed, and the austenite phase, which is stable at higher temperatures and is the 'trained' form of the material. If the SMA is heated above the transformation temperature, this results in a phase transformation from martensite to austenite, which causes the SME (activation of shape memory actuator). The transformation between deformation and recovery can be repeated cyclically, whereby the alloy is subject to a degradation process. Superelasticity is a phenomenon observed in shape memory alloys above the austenite transformation temperature. This phenomenon enables the material to withstand considerable deformation under mechanical stress and return to its original shape without the need for thermal activation. This behavior is particularly advantageous in applications that require damping or energy absorption.

The most commonly utilized SMA is nickel-titanium (NiTi), due to its potential for repeated transformation, force generation and ductility. Copper-based alloys (Cu-Al-Ni, Cu-Zn-Al) are employed in applications that operate within specific temperature fields. However, they typically exhibit reduced fatigue resistance and durability compared to other materials. Iron-based alloys (Fe-Mn-Si) exhibit distinctive magnetic shape memory properties, rendering them suitable for utilization in magnetic actuators. SMA are manufactured in a range of forms, including wires, springs, strips and

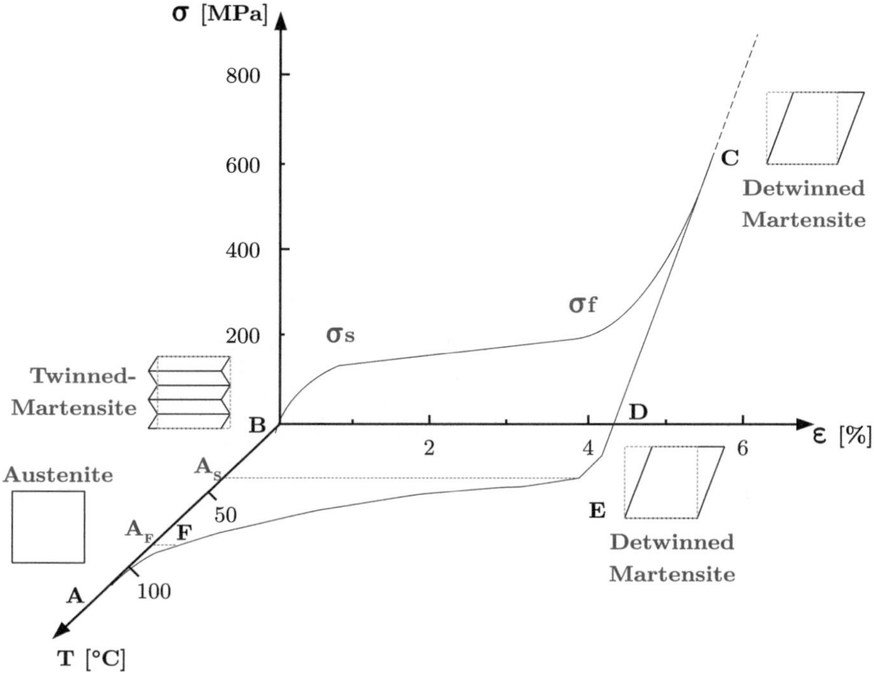

Fig. 2 Stress-strain-temperature diagram of shape memory alloys (cf. Heß 2024, Lagoudas 2008)

thin foils, in accordance with the specific requirements of the intended application, in terms of force, elongation or travel and response time.

The transformation process of SMA actuators is characterised by a cyclic transformation between martensite and austenite, which is controlled by an external heat input. When subjected to a heating process, the alloy undergoes contraction as it transforms into austenite, thereby generating a movement or force along the structure of the actuator. Upon cooling, the transformation is reversed, returning the actuator to the martensitic phase and resetting its position. This behavior can be designed for unidirectional actuation, where heating triggers a movement in one direction and cooling enables the return. Alternatively, dual wire configurations enable bidirectional actuation by alternately heating each wire. As a result, SMA actuators exhibit a dependence on temperature, which is in contrast to conventional actuators.

SMA actuators offer a number of significant advantages, including high energy density, compactness, silent operation and simple control mechanisms. As a result, they can be used in technical applications and systems where space constraints, weight and noise reduction are required. However, it should be noted that they also have certain limitations. The control behavior that can be achieved with SMA actuation is limited due to the influence of ambient temperature on heat, which is a fundamental principle of this technology (Fig. 3). Furthermore, temperature control can be challenging due to hysteresis, which may result in potential inaccuracies in the actuating movement. Additionally, SMA actuators demonstrate signs of fatigue

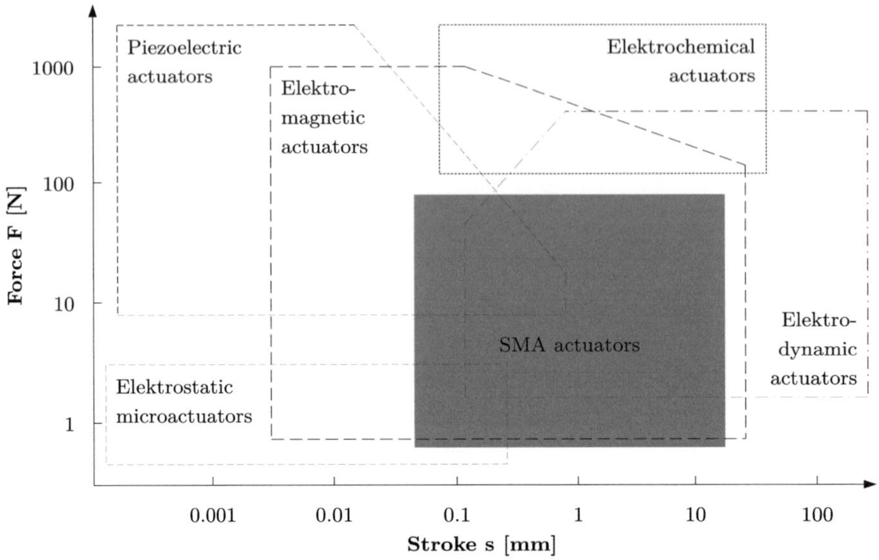

Fig. 3 Possible fields of application for different technologies of actuators (cf. Heß 2024, Heß and Bracke 2019, Heß and Bracke 2020a, b, c, 2021, Schiedeck 2009)

following repeated phase changes, leading to a gradual decline in performance over numerous cycles (Eggeler et al. 2004).

The versatility of SMA actuators makes them suitable for a diverse range of industries, including biomedical engineering, aerospace, and consumer electronics. Notable applications of SMA include medical devices, where they are used in stents, guidewires and surgical instruments due to their biocompatibility and flexibility; the aerospace and automotive industries, where they are used in extendable structures, valves and actuators for adaptive aerodynamics or fuel control systems; and consumer electronics, where SMA enable lightweight, compact actuation in camera lenses, foldable screens and haptic feedback mechanisms.

4 Case Studies

This section presents three case studies from the field of reliability engineering, which illustrate the application of typical methods and possible interpretations using both real and simulated data. All analyses presented in this paper were conducted using the R statistical computing environment (R Core Team 2024), with plots generated using the ggplot2 package (Wickham 2016) unless otherwise stated. The description of the test rig and the test procedure is based on Heß et al. (2024).

Test rig. Fig. 4 shows the test rig that was used for performing all fatigue tests of the case studies. The test rig includes three channels, allowing for simultaneous testing of three shape memory wires. Each channel comprises a fixed bearing and a

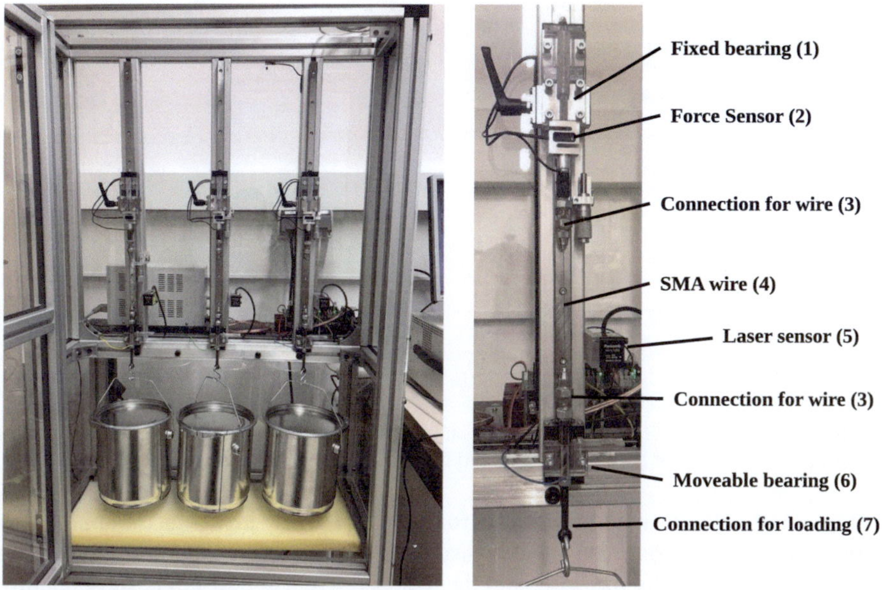

Fig. 4 Test rig for fatigue tests of SMA wires (Left); Detailed view of one of the three channels (Right) (Heß and Bracke 2020a, 2021, Heß et al. 2024, Theren et al. 2022a, b)

movable bearing, between which the wire is clamped. The connection technique is standardized according to VDI Guideline 2248 (Verein Deutscher Ingenieure e. V. 2017). Additionally, at the movable bearing, there is a connection for attaching various loads, which act as resetting elements. During the tests, the achieved displacement is measured by a laser sensor. Other parameters recorded during the tests include the current parameters, changes in weight force, and the ambient temperature within the test rig. The entire test rig has an enclosure to minimize the influence of the ambient temperature and convection.

The distance between the bearings can be adjusted for testing different wire diameters d and wire lengths l. The SMA wires tested and analyzed in this article have a total length of about 110 mm, which results in an effective length l of the wire between the two bearings of 90 mm. All SMA wires have the same pretension based on the applied load m. During a fatigue test, the SMA wires are subjected to cyclic loading, whereby they are heated above the austenite starting temperature by electrical activation, resulting in a lifting of the weight. Subsequently, a cooling phase is initiated, during which the weight once again elongates the wire. In addition to the aforementioned parameters, namely electrical voltage U, current I, ambient temperature T, force F and displacement or stroke s, the duration of the activation phase t_A and the cooling phase t_C are of importance. Furthermore, the mechanical load σ and electrical resistance R are crucial variables. The electrical resistance R can be calculated using Ohm's law. By relating the load to the wire crosssection and calculating the mechanical stress σ, it is possible to compare different diameters d and loads m.

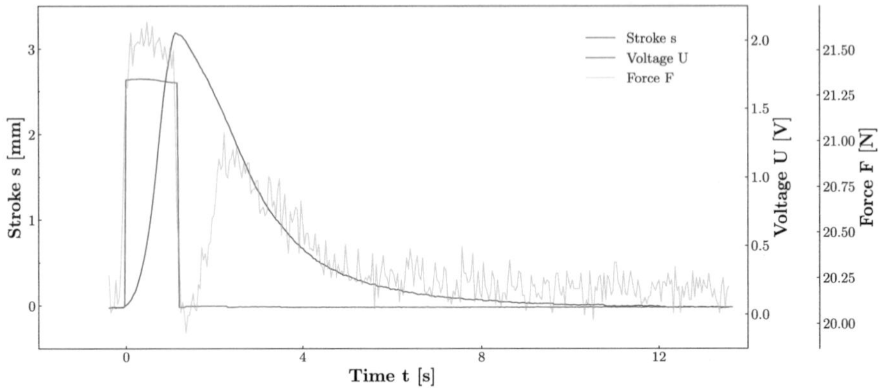

Fig. 5 Exemplary measurement series of the endurance test rig showing the measured variables stroke s [mm], voltage for activation U [V] and force F [N] over time t [s] (cf. Heß 2024)

Fig. 5 illustrates the measured values for one cycle. It is evident that the application of the voltage U (approximately 1.7 V) results in the SME being triggered. The weight is immediately lifted by the wire, and the distance reached (stroke) is subsequently measured and recorded by the distance sensor of the test rig. The maximum stroke of approximately 3.2 mm or the maximum strain of approximately 3.6 % is reached after two seconds. During the subsequent cooling phase, the weight elongates the wire to its original state (stretched). This allows the next cycle to be initiated directly, thereby maximising the usage rate. This increase or maximisation of the usage rate constitutes a simple setup for accelerated testing (Escobar and Meeker 2006, Meeker and Escobar 1998, Nelson 2004, Pascual et al. 2006).

Case study 1. The first case study is dedicated to a comparative analysis of the three test channels of the test rig, as detailed before. The primary objective is to ascertain whether the tested wires exhibit consistent failure behavior across all three test channels. This is a crucial aspect for subsequent analyses, including the potential implementation of a Sudden Death test with subsequent extrapolation of the candidates, cf. Sect. 2.3. This case study uses data from a real test series. In each of the three channels, 12 wires with a diameter of 0.3 mm were tested with a load of 300 MPa, a voltage of 2.7 V, an energization time of 1 s and a cooling time of 12 s until they failed. At first, the cycles until failure are presented in a box plot, differentiated by channel, as illustrated in Fig. 6.

While the failure cycles in channels 2 and 3 exhibit comparable characteristics with regard to both the mean value and dispersion, channel 1 displays a distinctive failure pattern. The mean value is markedly higher in channel 1 than in the other two channels, and the dispersion is considerably wider.

A Kruskal-Wallis rank sum test is conducted to ascertain the statistical significance of this anomaly. The null hypothesis posits that the mean values of the channels are identical. The Kruskal-Wallis rank sum test is conducted using the kruskal.test function (R Core Team 2024) with a significance level of $\alpha = 0.05$. The test result

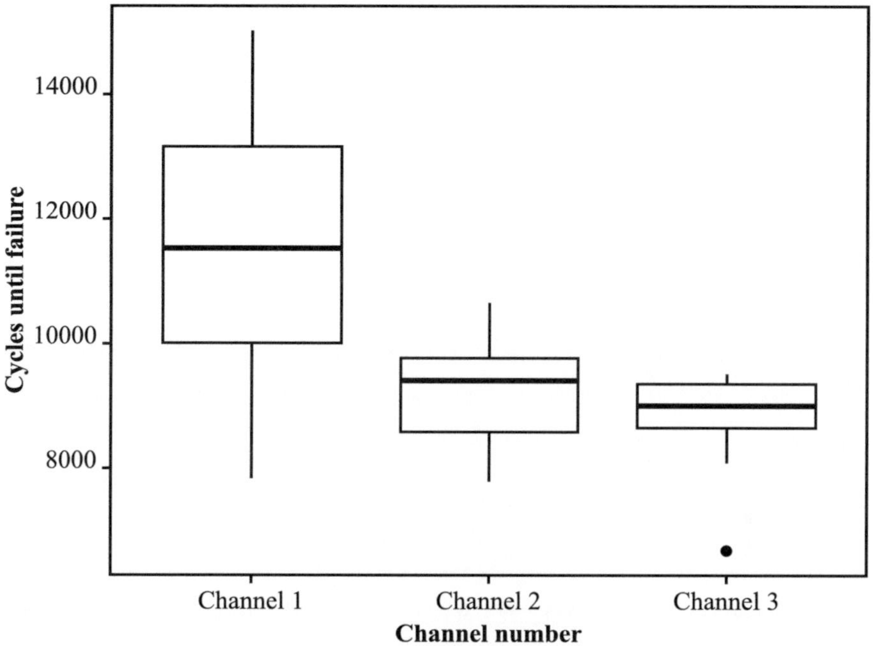

Fig. 6 Case Study 1: Boxplots cycles until failure per test channel

leads to a rejection of the null hypothesis, indicating that at least one channel differs from the others in the mean value (probability of 95%).

Subsequently, the ConoverTest function, as part of the DescTools package (Signorell 2024), is employed to facilitate pairwise comparisons of the channels in accordance with the Conover post-hoc analysis. In accordance with the outcome of the Kruskal-Wallis test, a significance level of $\alpha = 0.05$ is employed to ascertain whether the mean values of the two channels are comparable. The null hypothesis is rejected for the pairwise comparisons of channel 2 and 1, as well as channel 3 and 1. In these cases, there is a probability of 95% that there is a difference in the mean values of the channels under consideration, thereby confirming the assertion that the wires in channel 1 exhibit a disparate failure behavior than in channels 2 and 3. A reduction of the test effort through the use of e.g. a Sudden Death test would be inadmissible in this instance.

Case study 2. In the second case study, the results of an accelerated test are evaluated. For this purpose, data is analyzed from a real test series, in which three wires with a diameter of 0.28 mm were tested until failure under a load of 350 MPa with an energization time of 3 s and a cooling time of 40 s with five different currents (0.6 A, 0.8 A, 1.0 A, 1.2 A and 1.4 A). The data series of the different currents are synthetically expanded to a sample size of $n = 12$ using a random number generator and assuming a Weibull distribution model via the function rWeibull of the package ExtDist (Wu et al. 2023). A Weibull distribution model was fitted for each of the five data series using maximum likelihood estimation (MLE) using the function fitdist of

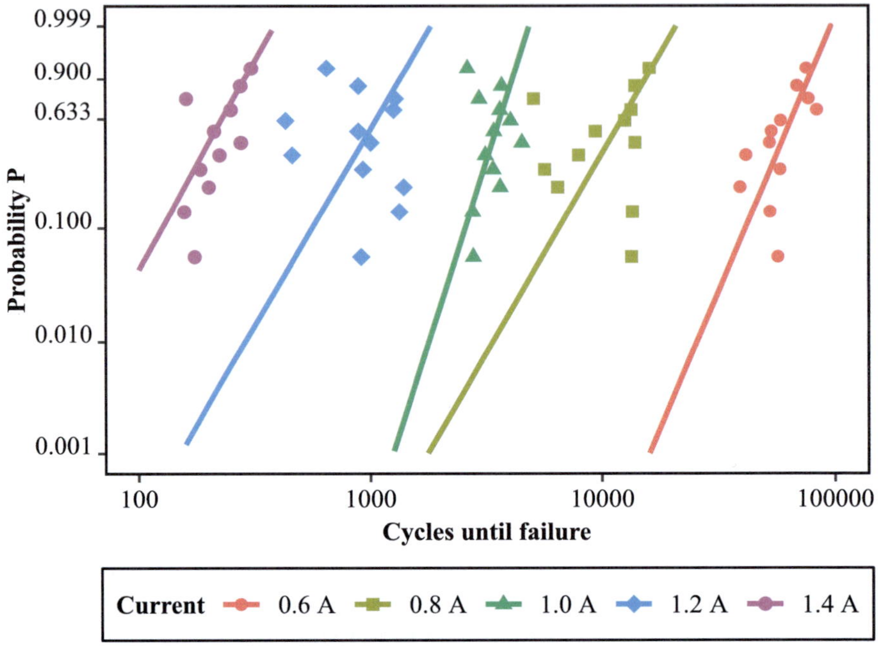

Fig. 7 Weibull distribution models of different currents as accelerated testing in case study 2. Weibull probability plot in double-logarithmic scale

the package fitdistrplus (Delignette-Muller and Dutang 2015). The resulting models are presented in Fig. 7 in the Weibull probability paper, together with the estimates of the failure probability over the median rank. The plot was created using the scales package (Wickham et al. 2016).

The distinctive characteristic lifetimes of the individual load levels of accelerated testing (currents) are readily discernible. It can be observed that an increase in current results in a reduction of the life time respectively the number of cycles until failure. Furthermore, it is evident that the curve for 1.0 A is steeper than the other curves, which, apart from that, are almost parallel. It seems reasonable to posit that the 1.0 A load level exhibits a distinct failure behavior. It should be noted that the Weibull fit only reflects the estimated failure probabilities with reasonable accuracy for the load levels 0.6 A, 1.0 A and 1.4 A. In the other two models, there are high deviations in some cases. One potential reason for this could be the relatively small sample size of 12 failures per data series. This must be taken into account as a source of uncertainty in the interpretation of the results.

In a quantitative analysis, the model parameters are compared with one another, with due consideration of their respective confidence intervals at the individual load levels of accelerated testing. The basis for this comparison is provided in Table 1.

A comparison of the characteristic lifetimes (scale parameter T) reveals a notable discrepancy across all load levels (currents), indicating that the test duration is sig-

Table 1 Parameters of Weibull distribution models of different currents as accelerated testing with confidence intervals (confidence level $\gamma = 0.95$) in case study 2

Current [A]	b_{LL}	b	b_{UL}	T_{LL}	T	T_{UL}
0.6	3.503	4.933	6.946	58,686	64,891	74,186
0.8	2.539	3.576	5.036	10,612	12,189	14,661
1.0	4.638	6.532	9.199	3337	3600	3983
1.2	2.526	3.557	5.009	921	1059	1275
1.4	2.641	3.720	5.238	202	231	276

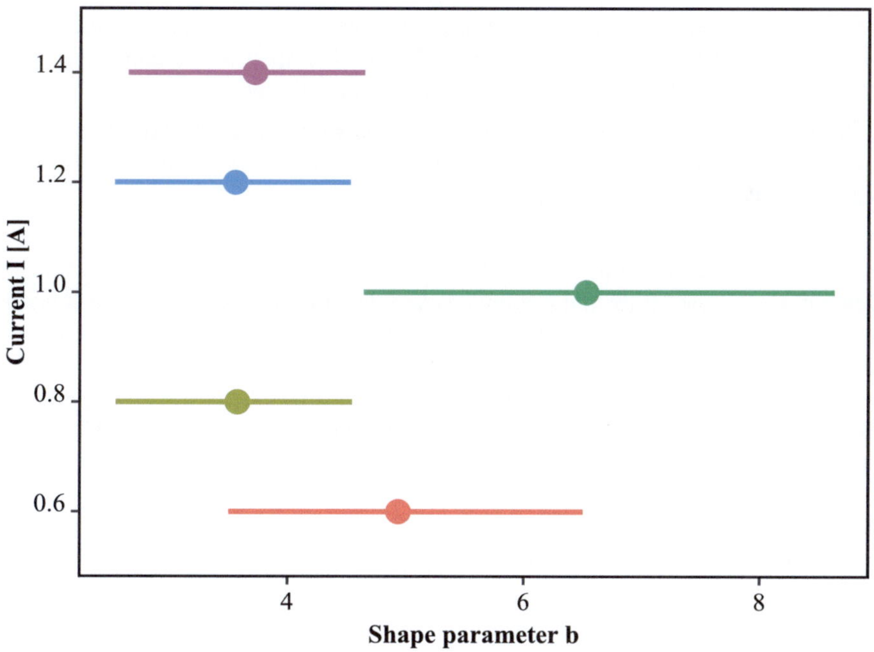

Fig. 8 Confidence intervals (confidence level $\gamma = 0.95$) of the shape parameter of Weibull distribution models of different currents as accelerated testing in case study 2

nificantly accelerated with increasing current. The comparison of shape parameters and their associated confidence intervals is a more challenging objective, which is why a graphical representation is also provided here for reference (see Fig. 8). For each load level of accelerated testing, the shape parameter is represented by a dot, and the respective confidence interval is represented by a line.

It is evident that the shape parameter of load level 1.0 A exhibits a notable divergence from those of load levels 0.8 A, 1.2 A, and 1.4 A. An overlap of the confidence intervals is observable only for load level 0.6 A. Consequently, load level 1.0 A is unsuitable for accelerated testing, as a different failure mechanism is presumably present here than in the other levels. The technical cause for this deviation in failure behavior cannot be identified on the basis of the available analyses.

Table 2 Acceleration factors based on characteristic life spans in relation to reference load level 0.6 A in case study 2

Current [A]	Scale parameter T	Acceleration factor
0.8	12,189	5.3
1.2	1059	61.3
1.4	231	280.6

In the final analysis of case study 2, Table 2 illustrates the potential acceleration factors of the test time for the load levels identified as suitable. Load level 0.6 A is selected as the reference point for the analysis. The reduction in the test duration is calculated using the quotient of the characteristic life span of the considered load level in relation to the characteristic life span of the reference.

It gets clear that the application of a higher current, specifically 0.8 A, 1.2 A, or 1.4 A, could significantly reduce the test time in the analyzed case study while maintaining the same failure mechanism.

Case study 3. The third case study delineates the implementation of a Sudden Death test as a case study and presents a candidate prognosis utilizing the Johnson rank correction. If the same load is applied in all test groups, the Sudden Death test can be used to reduce the required test scope, as discussed in Sect. 2.3. In accordance with the findings of the first case study, the data set utilized therein is employed once more. This data set comprises a real test series comprising three channels of 12 wires, each with a diameter of 0.3 mm, 300 MPa load, and 2.7 V voltage. A 1 s energization time and a 12 s cooling time are employed. The failure times of the first channel are simulated based on the values from channels 2 and 3, thereby ensuring that the basic requirement of a consistent load profile is met for each channel.

In order to demonstrate the feasibility of the candidate prognosis in this case study, the data set of the test series is subjected to an artificial pruning process. This signifies that the test is effectively terminated when the first wire in the three test channels fails. The remaining two wires are then considered as potential candidates. This process yields a censored data set.

In the extrapolation, the candidates are considered through the application of rank correction as described by Johnson. This approach entails the assignment of two candidates (wires) to each failed unit (wire) observed in the presented case study, with consideration given to the observed failure times. The implementation of the Weibull fit employs the maximum likelihood estimation (MLE) method, accounting for the presence of censoring. The programming is conducted using the weibulltools (Hensel and Barkemeyer 2023) and fitdistrplus (Delignette-Muller and Dutang 2015) packages.

Figure 9 illustrates the Weibull models of the censored (red) and extrapolated data (blue) in the Weibull probability plot. The red dots represent the approximation of the default probability via the median rank, while the blue dots represent the

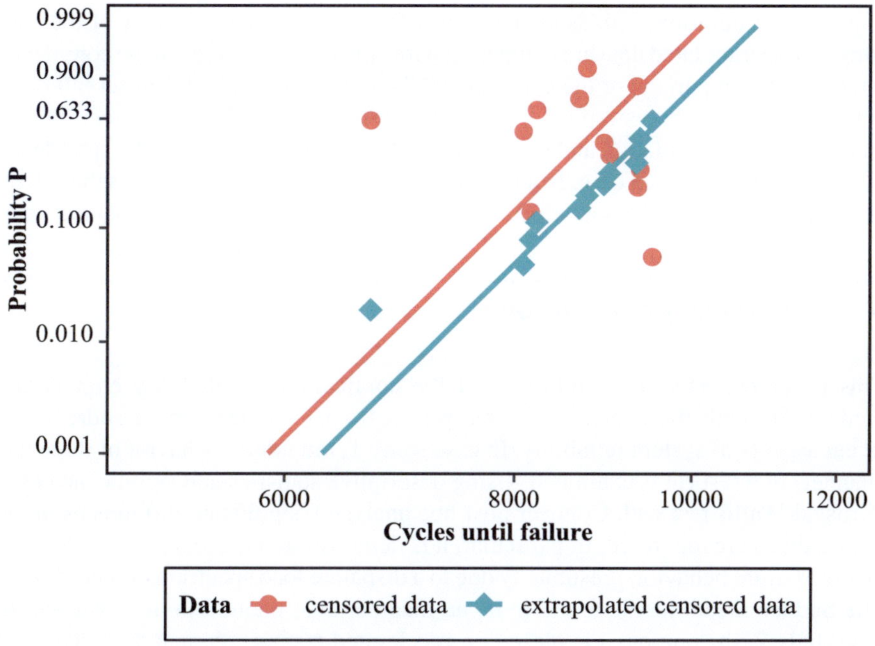

Fig. 9 Weibull distribution models of censored and extrapolated data in case study 3. Weibull probability plot in double-logarithmic scale

Table 3 Parameters of Weibull distribution models of censored and extrapolated data with confidence intervals (confidence level $\gamma = 0.95$) in case study 3

Data	b_{LL}	b	b_{UL}	T_{LL}	T	T_{UL}
Censored	11.528	16.235	22.862	8721	8991	9365
Extrapolated censored	13.140	16.237	20.064	9443	9620	9834
All	11.145	13.772	17.018	9150	9353	9598

extrapolation estimated using the Johnson rank correction. The double-logarithmic representation was generated with the scales package (Wickham et al. 2016).

The rank correction, as proposed by Johnson, is clearly discernible. The points were shifted downwards in parallel, as was the Weibull model. This indicates a very similar shape parameter with a higher characteristic life time. Table 3 documents the associated model parameters, including their confidence intervals. Additionally, the Weibull parameters of the entire test series (total database of case study 3) are provided for comparison with extrapolation based on the artificially pruned censored data.

Furthermore, the parallel shift of the Weibull line can be identified in the shape parameters of the censored and extrapolated data. These are nearly identical. The application of the Sudden Death test has the potential to significantly reduce the required test duration. This is evident when the characteristic lifetimes of the censored and extrapolated data are compared, with consideration given to the confidence intervals. A comparison of the extrapolated data with the original data set (all data) demonstrates that the extrapolation was successful. The model parameters (both b and T) of the extrapolation and the original data set are identical, and the confidence intervals overlap in each case. In addition, it is noteworthy that in all cases there is an extreme late failure behavior, which can be identified by the high shape parameter b.

5 Summary and Outlook

This paper presents an examination of the application of reliability engineering methods through the analysis of three practical case studies, which address different aspects of system reliability. In case study 1, the failure behavior of three test channels of a test rig is examined. Using descriptive statistics and significance tests (Kruskal-Wallis test with Conover post-hoc analysis), significant differences in the failure times are identified. In particular, it is demonstrated that channel 1 exhibits a distinct failure behavior, presumably due to a disparate load spectrum or mechanism. The Sudden Death test is found to be an inadmissable method in this instance. In case study 2, the emphasis is placed on accelerated testing. By analyzing five load levels using the two-parameter Weibull distribution, three suitable acceleration levels are identified. These allow a significant reduction in test duration up to a factor of 280 while maintaining the failure mechanism. Case study 3 examines the utilization of the Sudden Death test with Johnson-Rank correction. Here, a simulated scenario based on the data from case study 1 is employed to reduce the test duration by censoring after the initial failure. The extrapolation demonstrates a high level of concordance with the complete test series, both in terms of the model parameters and the characteristic service life.

The case studies demonstrate the importance of technical reliability, particularly in the context of SMA testing. The case studies illustrate the manner in which diverse methodologies can enhance the efficacy and import of reliability testing. The work presents exemplary reliability engineering tools, but offers only an insight into the comprehensive range of methods that can be employed. Other approaches, such as fault tree analyses (FTA), test process suitability and the use of discrete distribution models (e.g. binomial or hypergeometric distributions) can be employed to enhance future analyses. Furthermore, parametric tests, such as the t-test, or parameter-free methods, such as the Siegel-Tukey test, can be employed for specific research questions. Another area of potential focus is the integration of modern technologies, in particular artificial intelligence (AI). The application of AI-supported models and algorithms holds the potential to further optimize reliability analyses through the automation of data processing, pattern recognition and predictive analysis. Such developments offer the potential for more precise and efficient reliability testing in the future.

Acknowledgements The authors thank Marcel Schmidt for his valuable feedback, insightful suggestions, and careful proofreading, which significantly improved the quality of this paper.

References

Bertsche, B.: 2008, *Reliability in Automotive and Mechanical Engineering*, Springer Berlin Heidelberg, Berlin, Heidelberg.

Birolini, A.: 1999, *Reliability Engineering: Theory and Practice*, 3. edn, Springer, Berlin.

Bitter, P.: 1986, *Technische Zuverlässigkeit: Problematik, Mathematische Grundlagen, Untersuchungsmethoden, Anwendungen*, 3. edn, Messerschmidt-Bölkow-Blohm GmbH/Springer, Berlin/Heidelberg.

Bracke, S.: 2024, *Reliability Engineering: Data Analytics, Modeling, Risk Prediction*, Springer Berlin Heidelberg, Berlin, Heidelberg.

Conover, W. and Iman, R.: 1979, Multiple-comparisons procedures, *Technical Report* .

Conover, W. J.: 1999, *Practical Nonparametric Statistics*, 3. edn, Wiley.

Delignette-Muller, M. L. and Dutang, C.: 2015, Fitdistrplus: An R package for fitting distributions, *Journal of Statistical Software* **64**(4), 1–34.

Eggeler, G., Hornbogen, E., Yawny, A., Heckmann, A. and Wagner, M.: 2004, Structural and functional fatigue of NiTi shape memory alloys, *Materials Science and Engineering: A* **378**(1-2), 24–33.

Elahinia, M. H.: 2015, *Shape Memory Alloy Actuators: Design, Fabrication, and Experimental Evaluation*, John Wiley and Sons, Ltd, Chichester, West Sussex.

Escobar, L. A. and Meeker, W. Q.: 1995, Planning Accelerated Life Tests With Two or More Experimental Factors, *Technometrics : a journal of statistics for the physical, chemical, and engineering sciences* **37**(4), 411–427.

Escobar, L. A. and Meeker, W. Q.: 2006, A Review of Accelerated Test Models, *Statistical Science* **21**(4).

Fisher, R. A.: 1912, On an absolute criterion for fitting frequency curves, *Messenger Math.* **41**, 155–160.

Gandhi, M. V. and Thompson, B. S.: 1992, *Smart Materials and Structures*, 1st edn, Chapman & Hall, London ; New York.

Gümpel, P. (ed.): 2004, *Formgedächtnislegierungen: Einsatzmöglichkeiten in Maschinenbau, Medizintechnik und Aktuatorik ; mit 9 Tabellen*, number 655 in *Kontakt & Studium*, expert-Verl, Renningen.

Härtler, G.: 2016, *Statistik für Ausfalldaten*, Springer Berlin Heidelberg, Berlin, Heidelberg.

Hensel, T. and Barkemeyer, D.: 2023, Weibulltools: Statistical methods for life data analysis.

Heß, P.: 2024, *Technische Zuverlässigkeit von Formgedächtnisaktoren systematische Analyse des Degradationsverhaltens und Entwicklung eines Erprobungsprogramms*, Shaker Verlag, Düren.

Heß, P. and Bracke, S.: 2019, Reliability and degradation analysis of smart material actuators, *Proceedings of the 29th European Safety and Reliability Conference (ESREL)*, Research Publishing Services, pp. 1255–1261.

Heß, P. and Bracke, S.: 2020a, An Approach to Predict the Lifetime of Shape Memory Actuators based on Accelerated Testing Measurements, *Proceedings of the 30th European Safety and Reliability Conference and 15th Probabilistic Safety Assessment and Management Conference*, Research Publishing Services, pp. 1528–1535.

Heß, P. and Bracke, S.: 2020b, Smart material actuators as a contributor for IoT-based smart applications and systems: Analyzing prototype and process measurement data of shape memory actuators for reliability and risk prognosis, *Journal of Advanced Mechanical Design, Systems, and Manufacturing* **14**(2), JAMDSM0026–JAMDSM0026.

Heß, P. and Bracke, S.: 2020c, Zuverlässigkeitstechnik bei Formgedächtnisaktoren: Entwicklung von Prüfstandstechnik und Erprobungsprogramm, *in* R. H. Schmitt (ed.), *Potenziale Künstlicher*

Intelligenz für die Qualitätswissenschaft, Springer Berlin Heidelberg, Berlin, Heidelberg, pp. 39–56.
Heß, P. and Bracke, S.: 2021, Estimation Of The Remaining Lifetime Of Shape Memory Alloy Actuators During Prototype Testing: Analysis Of The Impact Of Different Currents, *Proceedings of the 31st European Safety and Reliability Conference (ESREL 2021)*, Research Publishing Services, pp. 3170–3177.
Heß, P., Schmelter, T., Kuhlenkötter, B. and Bracke, S.: 2024, Long-term reliability of shape memory wires: Analysis of the influence of different activation profiles on the degradation behavior and lifetime, *Proceedings of the ASME 2024 Conference on Smart Materials, Adaptive Structures and Intelligent Systems*, Atlanta (USA).
Hornbogen, E.: 1991, *Legierungen mit Formgedächtnis*, number 388 in *Vorträge / Rheinisch-Westfälische Akademie der Wissenschaften N, Natur-, Ingenieur- und Wirtschaftswissenschaften*, Westdt. Verl, Opladen.
Johnson, L. G.: 1964, *The Statistical Treatment of Fatigue Experiments*, Elsevier, Amsterdam u. a.
Kremer, A. and Bertsche, B.: 2018, Beschleunigte Zuverlässigkeitstests unter Berücksichtigung von Vertrauensbereichen, *Forschung im Ingenieurwesen* **82**(4), 395–409.
Kruskal, W. H.: 1952, A nonparametric test for the several sampling problem, *Annals of Mathematical Statistics* **23**(4), 525–540.
Kruskal, W. H. and Wallis, W. A.: 1952, Use of ranks in one-criterion variance analysis, *Journal of the American Statistical Association* **47**(260), 583–621.
Lagoudas, D. C.: 2008, *Shape Memory Alloys: Modeling and Engineering Applications*, Chemistry and Materials Science (Springer-11644; ZDB-2-CMS), Springer-Verlag US Springer e-books, Boston, MA.
Lexcellent, C.: 2013, *Shape-Memory Alloys Handbook*, Materials Science Series, ISTE ; Wiley, London, UK : Hoboken, NJ.
Meeker, W. Q. and Escobar, L. A.: 1998, *Statistical Methods for Reliability Data*, Wiley Series in Probability and Statistics. Applied Probability and Statistics Section, Wiley, New York.
Nelson, W.: 2004, *Accelerated Testing: Statistical Models, Test Plans and Data Analysis*, Wiley Series in Probability and Mathematical Statistics. Applied Probability and Statistics, Wiley-Interscience, Hoboken, NJ.
Pascual, F., Meeker, W. and Escobar, L.: 2006, Accelerated Life Test Models and Data Analysis, *in* H. Pham (ed.), *Springer Handbook of Engineering Statistics*, Springer London, London, pp. 397–426.
Peters, M. (ed.): 2010, *Titan und Titanlegierungen*, [3., völlig neu bearb. aufl.], 3. nachdr edn, Wiley-VCH, Weinheim.
Pfeifer, T.: 2001, *Qualitätsmanagement: Strategien, Methoden, Techniken*, 3., völlig überarb. und erw. aufl edn, Hanser, München Wien.
Pham, H. (ed.): 2006, *Springer Handbook of Engineering Statistics*, Springer London, London.
R Core Team: 2024, R: A language and environment for statistical computing.
Rao, A., Srinivasa, A. R. and Reddy, J. N.: 2015, *Design of Shape Memory Alloy (SMA) Actuators*, SpringerBriefs in Applied Sciences and Technology, Springer International Publishing, Cham.
Schiedeck, F.: 2009, *Entwicklung eines Modells für Formgedächtnisaktoren im geregelten dynamischen Betrieb*, number 03/2009 in *Berichte aus dem IDS / Institut für Dynamik und Schwingungen*, PHZ, Produktionstechnisches Zentrum, Garbsen.
Signorell, A.: 2024, DescTools: Tools for descriptive statistics.
Theren, B., Heß, P., Bracke, S. and Kuhlenkötter, B.: 2022a, The Development and Verification of a Simulation Model of Shape-Memory Alloy Wires for Strain Prediction, *Crystals* **12**(8), 1121.
Theren, B., Heß, P., Bracke, S. and Kuhlenkötter, B.: 2022b, Influence of the Phase Transformation Behaviour of NiTi Shape Memory Alloy Wires on the Predictability of Strain During Operation, *ASME 2022 Conference on Smart Materials, Adaptive Structures and Intelligent Systems*, American Society of Mechanical Engineers, Dearborn, Michigan, USA, p. V001T02A002.
Van Humbeeck, J.: 1991, Cycling Effects, Fatigue and Degradation of Shape Memory Alloys, *Le Journal de Physique IV* **01**(C4), C4–189–C4–197.

Van Humbeeck, J.: 2001, Shape Memory Alloys: A Material and a Technology, *Advanced Engineering Materials* **3**(11), 837.

Verband der Automobilindustrie e. V.: 1976, *VDA Band 3.2: Zuverlässigkeitssicherung Bei Automobilherstellern Und Lieferanten : Verfahren Und Beispiele*, Verband der Automobilindustrie e. V. (VDA)./Qualitätskontrolle in der Automobilindustrie, Berlin.

Verein Deutscher Ingenieure e. V.: 2017, VDI 2248 – produktentwicklung mit formgedächtnislegierungen (FGL).

Weibull, W.: 1951, A statistical distribution function of wide applicability, *Journal of Applied Mechanics* **18**, 293–297.

Wickham, H.: 2016, *Ggplot2: Elegant Graphics for Data Analysis*, Springer-Verlag, New York.

Wickham, H., Pedersen, T. and Seidel, D.: 2023, Scales: Scale functions for visualization.

Wu, H., Godfrey, A. J. R., Govindaraju, K. and Pirikahu, S.: 2023, ExtDist: Extending the range of functions for probability distributions.

Methods for Analyzing Reliability Engineering: AI-Based Production Capability Regarding Product Lifetime Reliability

Georgios Ioannou, Jannis Pietruschka, Saeideh Pourghasemian and Stefan Bracke

Contents

Abstract

Via the RAPEX rapid alert system for dangerous non-food products, a recall of biodegradable cutlery was initiated in 2024 due to a risk of injury from fragments breaking during use. The failure of such a safety-critical feature can be due to, for example, a design fault or a manufacturing defect. In the era of Industry 4.0, reliability engineering methods are increasingly being characterised and expanded by the application possibilities of artificial intelligence. Particularly within the product development process, opportunities arise to ensure product reliability for safety- and function-critical components. Especially for safety-critical features, it is essential to prove a consistent process capability, which can be achieved through

G. Ioannou (✉) · J. Pietruschka · S. Pourghasemian · S. Bracke
Chair of Reliability Engineering and Risk Analytics, University of Wuppertal, Wuppertal, Germany
e-mail: ioannou@uni-wuppertal.de

J. Pietruschka
e-mail: pietruschka@uni-wuppertal.de

S. Pourghasemian
e-mail: pourghasemian@uni-wuppertal.de

S. Bracke
e-mail: bracke@uni-wuppertal.de

S. Bracke and R. Pieper (eds.), *Safety Engineering*,
https://doi.org/10.1007/978-3-658-47473-7_3

inline inspection as a full inspection process. In this study, images of wooden forks were captured under constant boundary conditions in a test rig set up in a lab. These images serve as a training dataset for machine learning analysis, showing defects in the manufacturing process that could lead to user injury in application. Each fork image was analyzed for defects using Convolutional Neural Networks (CNNs). For the determination of an appropriate model, a comprehensive parameter study is performed. The approach of optimizing the algorithm results and identifying a reliable and reproducible CNN model is presented in detail. The evaluation of the results shows that the found model can detect the defective forks with an accuracy of 81.16%. Under these conditions, a transfer to real operation at an evaluation speed of 4.3 ms per object is possible to a certain extent.

Keywords

Reliability · Artificial intelligence · Machine learning · Convolutional neural networks · Condition monitoring · Product suitability · Optical inspection

1 Introduction

As part of a RAPEX (Rapid Exchange of Information System/Safety Gate) procedure for dangerous non-food products, a recall of wooden cutlery was initiated in 2024 due to the risk of injury it presented [4]. Fundamentally, the RAPEX procedure serves to disseminate potential hazards transparently and swiftly, making such warnings publicly accessible across the European Economic Area via an online platform. A key component of the procedure is that a potentially dangerous consumer item is reported EU-wide based on regulatory reporting obligations. Subsequently, a risk assessment is conducted according to established guidelines to evaluate the likelihood of injury or death resulting from the product defect, allowing for corresponding measures to be taken and liability or insurance claims to be addressed [10, 14]. When a serious risk is identified, typical actions include product recalls, which often fall at the expense of manufacturers or suppliers, leading to both economic and marketing repercussions. Since the product's entire lifecycle is considered in such an analysis, the technical reliability of products plays a crucial role. In the aforementioned recall of wooden cutlery, the issue was the failure of a safety-critical feature. During use, wooden splinters would break off, posing a swallowing hazard. Such splintering can occur due to deviations in the production process, which are exacerbated by inadequate quality control. To ensure 100% inspection of safety-critical features in the future production of wooden forks, this article examines the potential of artificial intelligence for automated inline inspection.

This article focuses on reliability during the manufacturing process and uses a case study to show how artificial intelligence can be used as a new technology in the optical inspection of functional and safety relevant surface topographies. Section 1 provides an introduction to the topic of technical reliability and why the usage of artificial intelligence is becoming increasingly relevant. The relevance of optical inspection during production is then demonstrated. Section 2 presents the basics of artificial intelligence, condition monitoring and optical inspection relevant to this article. Section 3 presents a case study of the production of wooden forks and our approach for an visual inline inspection of wooden forks during the production phase. The results of the case study are presented in Sect. 4. Section 5 concludes with a summary of this article and an outlook.

1.1 Reliability of Technical Products

Technical reliability is a key aspect in the development and manufacture of technical products, as it ensures that a product functions throughout its entire service life and fulfills the usage requirements. Reliability is the probability that a product will not fail in terms of its functional and safety-critical properties within a defined time-interval or an interval of other runtime related variables (cf. [8]). The development and manufacture of technical products is subject to dynamically increasing requirements, such as increasing product complexity, the reduction of development times and costs due to competition or the adaptation of laws and standards for product safety.

Product Development Process
Technical reliability plays an overriding role throughout the entire product development process. In Fig. 1 the product development process with its different phases is represented, which can also be divided into sub-phases. For the following section, also compare [8,42].

The product development process begins with *concept development*, which focuses on developing a concept for the product in preparation for series development. This includes the pre-development of the necessary technologies and functions, as well as the definition and confirmation of the concept feasibility. Once the concept development phase has been completed, the aim of the *series development* phase is to design and test the prototype and prepare the production process for the upcoming series production. Among other things, pre-series are produced under near-series conditions to validate the product and production process [8]. The *production* begins with a start-up phase in which production is initially ramped up to maximum capacity before moving on to series production. This phase lasts until the product is discontinued and is continuously adapted through market observation and product optimizations. In the final phase, the *distribution and field servicing*, the product is marketed and supported throughout the series. Market observation in the field plays a key role in product reliability, as it enables risk analysis and product

Fig. 1 Product development process, based on [8,13]

optimization for the current and subsequent generations, for example, on the basis of observed incidents of damage (cf. [8]).

Process Capability
Ensuring controllable and quality-capable manufacturing processes in the early phases of the product development process plays an important role for the technical reliability. This is the only way to ensure reliable manufacturing and thus the quality of the product over the entire series period [6]. In a *controlled process,* the distribution parameters of the measured values of a product feature (e.g. mean value and standard deviation of a component's weight) remain the same or only change to a known extent, so that reproducible results can be delivered. However, a controlled process can also lead to high reject rates if, for example, the component weight exceeds the specified component specification due to the known dispersion of the measured values. The process is then controlled, as the distribution of the component weight does not change, but is unsuitable due to the high reject rate [6]. In contrast to a controlled process, a *quality-capable process* complies with the given specification limits (e.g., component weight) after production and testing. This can already be achieved, for example, by a final inspection and subsequent sorting out of the existing scrap content. However, the distribution parameters can change compared to the controlled process [6,8]. The goal for a manufacturing process is that it is controlled and quality-capable. The process should be known with regard to its distribution parameters and at the same time comply with the specification limits, e.g. by centering the process and minimizing the dispersion of the measured values [6,8]. One possibility for monitoring production processes and supporting the assurance of controlled and quality-capable processes is the condition monitoring focused on in this work by means of an optical inline inspection.

1.2 Usage of Artificial Intelligence in Reliability Engineering

The use of artificial intelligence in reliability engineering has gained significant momentum, especially with the growing complexities of modern systems and the vast amounts of data generated during both the product development process and operational phases.

Increasing data volumes during the product development process
The increase in data volume during product development is primarily due to the growing complexity of products [8]. This complexity, characterized by a broader range of functionalities and advanced designs, has resulted in a substantial rise in data generated throughout the product lifecycle. This data is produced not only during manufacturing stages but also throughout operational phases and encompass diverse sources, including production data -such as process parameters and quality control metrics- as well as field data collected from the product's actual performance. Field data further includes operational metrics, damage reports (indicating timing,

symptoms, and causes), and load spectra, all of which are critical for assessing product reliability.

Given the scope and intricacy of this data volume, precise analysis is not feasible without advanced tools. Here, artificial intelligence (AI), with its capability to process and extract patterns from big data, plays a pivotal role in reliability engineering, enabling effective mapping, evaluation, and improvement of product reliability across all stages of development and usage [7,28].

Complex Causal Relationships
In addition to the aforementioned challenges, complex causal relationships pose significant obstacles in reliability engineering. These relationships often involve intricate dependencies among various factors, such as design parameters, manufacturing processes, and operating conditions, all of which collectively influence the overall reliability of a product. As product complexity increases, the need for modeling and predicting these complex interactions becomes increasingly important. Traditionally, these relationships were explored using physical models and expert judgment; however, the advent of AI and machine learning has opened new horizons in this domain. AI models, particularly those based on machine learning techniques, can identify patterns and causal relationships within large datasets, thereby facilitating more accurate predictions and risk assessments [28].

The usage of machine learning techniques is especially valuable in scenarios where the relationships between variables are non-linear or where interactions are too complex to be captured by traditional modeling methods. Probabilistic methods and data-driven models enable engineers to predict reliability more accurately by leveraging extensive datasets [7]. Furthermore, machine learning models can process sensor data collected during operation to identify signs of potential failures at early stages. This capability enhances predictive maintenance strategies and improves the overall resilience of systems.

1.3 Optical Inspection of Technical Products

As part of the series preparation of the production phase, the aim is to prove continuous "process-capable" production and to manufacture the products reliably (cf. section 1.1). Proof of the process capability can be provided, for instance, by inline testing of systems, component groups and at component level during series production, by final product testing of the overall product and by an external testing process by suppliers or laboratories. Testing is carried out on the basis of the features defined for the product or process. According to DIN 55350—Part 12 (DIN 1989), a distinction is made between qualitative and quantitative features (cf. [7]). Features to be tested are often function-critical and safety-critical. While the latter often have to be subjected to a full inspection (100% inspection), as a deviation from the specification can lead to a fault that is a human hazard, the function-critical feature inspection is also permissible for a part of the product population. Function-critical feature refer to the reliability of a product without directly endangering a person, so that destructive test procedures can also be carried out in random samples [7].

Optical inspection can be used in a variety of ways for quality control and feature testing. Depending on the feature, products are checked visually or with optical sensors for defects or deviations. Visual inspection is carried out by human labour, which strictly speaking does not represent 100% control through a partly subjective, cost-intensive and inconsistent perception, but is suitable for many types of feature inspection [5,41]. A sensor-based and objective but not universally applicable optical inspection can be carried out by machine vision, which uses cameras and image processing software for the automated detection of defects and deviations. This can be used, for example, to carry out geometric inspections, reference-based pattern recognition, validate surface topographie and position checks. These computer-aided processes are particularly suitable for inline inspection [5].

The rapid development of fields of application in which AI can be used is occurring at a rapid pace, with new applications being discovered constantly. This is due to the increasing availability of computer capacities, which is enabling the use of AI in ever more complex and larger data sets. In particular, the technological leaps in the further development of graphical processing units are enabling the increasingly diverse analysis of high-resolution images and videos with AI. For example, machine learning is already being used to check safety-critical features as part of computer vision (CV) for fault detection on printed circuit boards (PCB) [35,36], the inspection of weld seams [9] or aircraft parts for corrosion [29]. Others use CV to detect spot weld defects on batteries and sort them out before a possible explosion [44] or to detect production errors in the manufacture of plastic spoons [43].

2 Baseline

This chapter provides a comprehensive review of advanced methods for condition monitoring and data analysis aimed at improving and enhancing the health of industrial equipment. First, the concept of condition monitoring is introduced. Then, two approaches, supervised and unsupervised learning, are discussed, offering the capability to extract patterns and classify data for automated fault detection. Finally, optical inspection with a computer vision approach is introduced as a tool for advanced visual analysis.

2.1 Condition Monitoring

The health of a product refers to the degree of degradation or deviation from its standard operating conditions [32]. Defective products, if not detected immediately, can lead to increased costs and decreased efficiency in production processes, ultimately affecting overall productivity [11]. Accurate and real-time identifying failures aids industries implement effective quality control strategies, thereby maintaining their competitive advantage [46]. Production issues and errors are often caused by defects and malfunctions in machinery and production lines [24,39]. Therefore, condition monitoring and early detection of such defects are critical to ensuring optimal performance and preventing sudden breakdowns in modern industrial production [45].

Condition monitoring evaluates the operational status of a product by tracking key performance indicators for instance temperature, pressure, and vibration [30]. These indicators help identify undesirable trends and enable timely maintenance to prevent more significant failures. Condition monitoring can be performed in two main ways: continuous monitoring and periodic monitoring. In continuous monitoring, the product's health is constantly evaluated, providing real-time data for early detection of problems. While this method is highly effective, it can be costly. In contrast, periodic monitoring, which relies on scheduled inspections and processed data, can be a more cost-effective option [17].

Several methods for condition monitoring and fault detection have been proposed over the past decades. These methods are typically grouped into model-based, statistical, and artificial intelligence-based approaches [19]. Among these approaches, particularly artificial neural networks, are preferred due to their ability to handle complex, non-linear relationships and adapt to changing operational conditions, which are often challenging for traditional model-based and statistical methods [1]. Artificial neural networks can operate in both supervised and unsupervised learning modes [22].

2.2 Supervised and Unsupervised Learning

Supervised learning refers to the process where models are trained using labeled datasets to map inputs to corresponding outputs. This approach enables models to recognize patterns in the data and, as a result, predict or classify new data based on these learned patterns. Supervised learning is not a specific method, but rather a general framework applied to solve problems that require predictions or classifications based on labeled data. This approach plays a key role in many machine learning workflows where accurate predictions or classifications are essential [2].

As illustrated in Fig. 2a, the model is trained using labeled datasets and gradually learns to relate inputs to their correct outputs. Once trained, the model is capable of processing unseen data and providing accurate predictions or classifications. The output section of the figure also demonstrates how the model categorizes the data and determines the correct outcomes.

In contrast, unsupervised learning utilizes unlabeled data to uncover hidden patterns and structures within the data. Unlike supervised learning, which aims to predict specific outcomes, unsupervised learning autonomously identifies similarities and differences within the data, clustering them into groups of similar attributes. As shown in path (b) of Fig. 2, in unsupervised learning, the model processes the training data without labels. Instead of using labels to guide the learning process, the model automatically identifies patterns and clusters the data based on inherent similarities. This process is particularly effective for exploratory data analysis, clustering, customer segmentation, and anomaly detection [2,22].

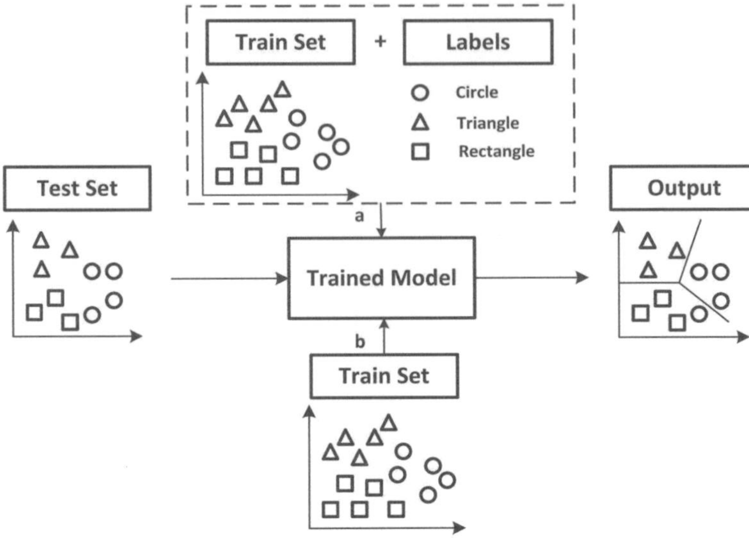

Fig. 2 A comparative illustration of supervised and unsupervised learning in machine learning. Path (**a**) follows supervised learning with labeled data, while path (**b**) shows unsupervised learning, where the model discovers patterns from unlabeled data

2.3 Computer Vision

Optical inspection and computer vision are increasingly recognized as vital components in condition monitoring, playing a crucial role in the assessment of machinery, equipment, and products. Optical inspection refers to the process of capturing images or videos of assets using advanced camera technologies. This method serves as the foundation for visual assessment, enabling the identification of defects and the evaluation of overall condition [31]. On the other hand, CV employs sophisticated algorithms to analyze the visual data obtained through optical inspection. By interpreting images and videos, CV systems can detect anomalies and provide insights into the operational health [18,31].

One of the significant advantages of vision-based condition monitoring is its ability to provide enhanced objectivity and reliability compared to traditional inspection methods. Automated analysis through computer vision can identify subtle defects that might be overlooked by human inspectors, thereby improving safety and reducing the likelihood of operational failures [18,37]. Additionally, this technology allows for remote, non-contact monitoring, which is particularly beneficial in hazardous environments where human intervention could pose risks. The implementation of such systems can lead to substantial cost savings by optimizing maintenance schedules and minimizing unplanned downtimes. At the core of advancements in status monitoring through CV lies deep learning. As a specialized branch of AI, deep learning is highly effective in analyzing complex visual data [46]. This concept is represented as a layered structure of related technologies, as shown in Fig. 3. As illustrated, AI forms the outermost layer, encompassing the broader field that includes

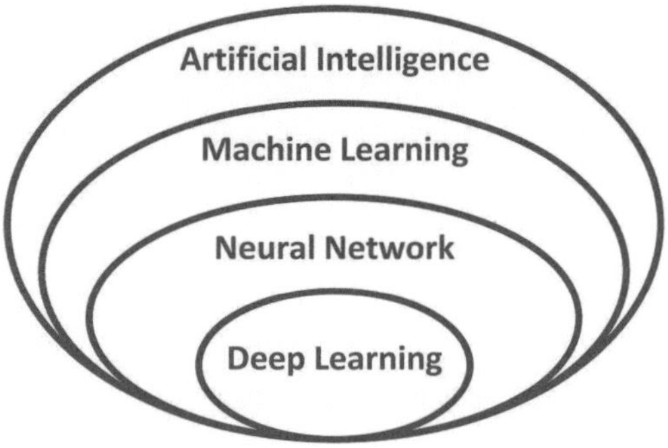

Fig. 3 Neural network-related fields

machine learning as a subset. Machine learning enables systems to learn from data without explicit programming. At the next level, we find neural networks, which are modeled on the human brain's structure and capable of learning patterns and features from data. Finally, deep learning is a specialized and advanced branch of neural networks that uses multi-layer architectures, such as convolutional neural networks, to analyze and extract complex features from visual data. This layered structure shows how each level of technology leads to a more specialized approach, ultimately enabling solutions for complex tasks like anomaly detection and precise status monitoring.

Artificial neural networks

Artificial neural networks stem from early attempts to simulate the learning and error-correction capabilities of biological nervous systems [22]. The fundamental concept behind these networks is the creation of a novel structure for processing information, composed of highly interconnected processing elements known as neurons. These neurons work in unison to solve problems. Each neuron amplifies its output by a specific weight coefficient and then transmits it to the input of other neurons via a nonlinear activation function. This process is illustrated in Fig. 4 (left). The structure of these networks consists of three primary components: 1) the input layer, which receives raw data from the environment and transmits it to the network for model training and testing, 2) the hidden (or intermediate) layers, which process information and identify complex patterns; these layers also facilitate connections between the input and output layers, and 3) the output layer, which generates the final result and delivers the model's output.

Learning in neural networks is conducted through supervised learning, primarily relying on iterative adjustments of the weights. To train these networks, a dataset of input and output samples is required, which includes examples of input data paired with output labels. One of the most common iterative methods in neural networks is the backpropagation of error. In this method, the gradient of the error function is calculated with respect to the network weights to optimally adjust them. This process occurs in two main phases: the feedforward phase and the backpropagation phase.

In the feedforward phase, input data moves through the layers and reaches the output layer, producing an initial result. Subsequently, in the backpropagation phase,

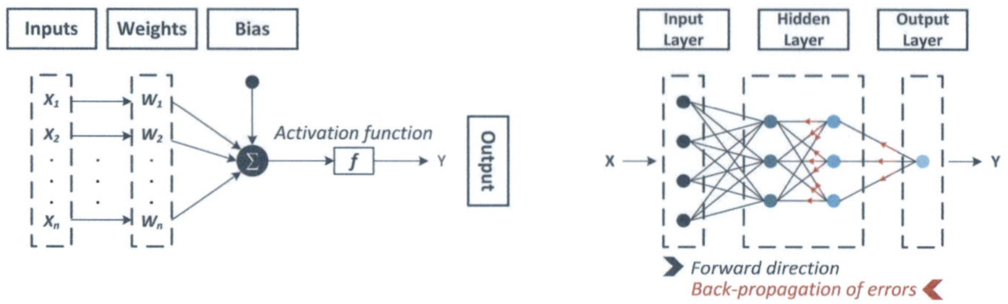

Fig. 4 (Left) Structure of the artificial neural network. (Right) Structure of the backpropagation error algorithm

the output error is calculated and propagated back toward the initial layers, allowing weights to be updated based on this error. This reverse movement of error information enables the network to efficiently compute the gradient at each layer. In Fig. 4, right, the directions of data flow and error propagation in the network are clearly depicted; distinct arrows indicate the data's movement during feedforward and backpropagation phases [25].

This method allows the network to effectively identify complex dependencies between inputs and outputs, achieving increased accuracy over time.

Convolutional neural networks
Convolutional neural networks are a specialized type of artificial neural networks designed primarily for processing grid-like structured data, such as images. These networks have gained widespread popularity in the field of computer vision due to their ability to autonomously and adaptively learn hierarchical spatial features from images [23]. The architecture of CNNs typically consists of several layers, with the most important being convolutional layers, pooling layers, and fully connected layers [12].

In the general structure of convolutional neural networks, as shown in Fig. 5, the input image is first passed through a convolutional layer. In this layer, convolution operations are performed on the image, where a set of learnable filters are applied. These filters help the network identify and extract key features such as edges, textures, and patterns in the image. The filters move across the input image, performing element-wise multiplication in different regions, creating a feature map that highlights the presence of specific features in the image [21,33]. After this, an activation function such as ReLU (Rectified Linear Unit) is typically applied to introduce nonlinearity into the model, which allows the network to better capture complex patterns in the data.

After the convolutional layers, pooling layers are applied to reduce the spatial dimensions of the feature maps. These layers preserve the most important information while making the data more manageable and enhancing the network's robustness to small shifts in feature positions. Pooling layers typically use smaller filters than the convolutional layers. Two common pooling methods are max pooling and average pooling, which select the maximum or average values from the filter's region,

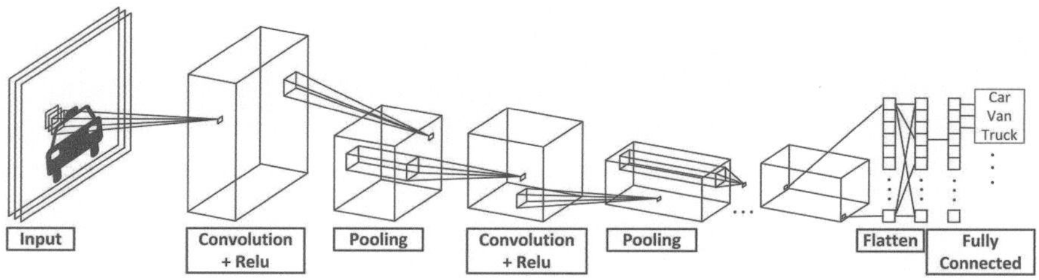

Fig. 5 Schematic architecture of a convolutional neural network with its different layers [33]

respectively. This operation is applied to each feature map independently, reducing computational load and improving the model's resistance to distortions and redundant information [12,33]. Together, convolutional and pooling layers form a key part of a convolutional neural network. In deeper networks, which consist of multiple layers, this architecture allows for the extraction of more complex features. Specifically, the layers closer to the input tend to extract low-level features such as edges and lines, while deeper layers focus on higher-level features like shapes and intricate patterns [12]. At the end of the CNN architecture, fully connected layers are used to transform the high-level feature maps produced by the convolutional and pooling layers into a single vector. This vector then passes through one or more fully connected layers to produce the final output of the model, such as class probabilities in a classification task [12,33].

To learn effectively, CNNs require large amounts of labeled data. During the training process, the network continuously updates its weights using backpropagation algorithms and gradient descent. The goal of this process is to minimize the error function, which measures the difference between the predicted output and the actual output. Once trained, CNNs can generalize well to new, unseen data, making them highly effective for tasks such as image recognition, object detection, and segmentation [21,27]. The success of CNNs is attributed to their ability to learn and extract hierarchical representations of features. Each successive layer identifies increasingly complex features, allowing the network to develop a deeper understanding of the data. This capability has led to significant advances in various fields, such as facial recognition, medical image analysis, and autonomous driving [38].

3 Case Study: Optical Inspection in the Manufacturing of Wooden Forks

This chapter presents the case study carried out, the optical inspection in the manufacturing of wooden forks. To this end, we first describe how wooden forks are manufactured and which manufacturing defects can occur, focusing on function- and safety-critical defects. This is followed by a presentation of the data and the AI model used in this case study.

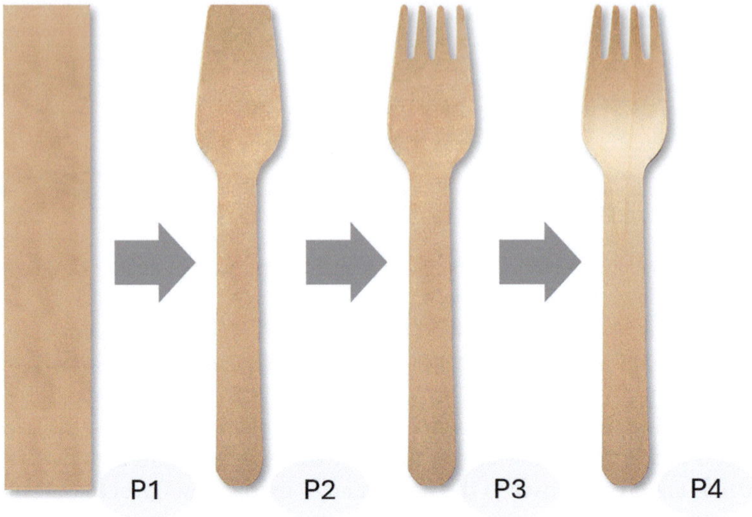

Fig. 6 Manufacturing states in the wooden fork production

3.1 Manufacturing Process of Wooden Forks

The manufacturing process of wooden forks differs in detail depending on the manufacturer and type of cutlery (e.g. type of wood, curvature, surface coating), but can basically be divided into the following main process steps based on [3,26,34,40], which are shown in Fig. 6. In the first step, smooth and clean wooden veneers are cut from the blocks of wood using a rotary process (also known as peeling), whereby the unclean and irregular bark is also removed (P1). In the next step, blanks are punched from these wood veneers, which provide the basic dimensions of the fork, and then milled into the correct shape. This work step is particularly relevant for the tines of a fork (P2 and P3). The milled forks must then be polished to minimize the risk of fraying, chipping and sharp edges. After intermediate drying, the forks are hot-pressed into the desired shape and coated with safe and food-grade materials to protect them from moisture and increase their durability (P4).

Wooden forks are generally considered safe products because they are made from natural materials and, as food-grade materials, they do not contain any harmful chemicals. This is additionally ensured by various regulations and laws that apply to the marketing of products in the EU. Examples include Regulation (EC) No. 1935/2004 and Regulation (EC) No. 2023/2006 [15,16], which apply to materials and products, or their manufacturing processes, that come into contact with food. Despite careful production, defects can occur in the manufacture of wooden forks that can be attributed to individual process steps and must be recognized by visual inspection. Some faults can be functionally and safety-critical, meaning that personal injury cannot be ruled out if the faulty forks reach the customer despite being inspected. Examples of faults that may occur are as follows:

- Brittle cutlery that breaks quickly during manufacture or use due to unsuitable quality wood, drying processes or other process steps. A fork could quickly splinter during use and unknowingly be swallowed with food by the customer.
- The wood may crack during drying or processing. These cracks make the fork susceptible to breakage.
- An uneven application of coating materials leave stains or an uneven surface, which can impair the functionality.
- Punching and milling processes can result in rough or even sharp points (e.g. fringes) during production, which can make the fork uncomfortable to hold or even cause injury during usage.
- Improper temperature treatment during hot pressing can also lead to uneven surface topographies and discoloration, which impair both aesthetics and functionality.

These examples show that the consideration of function- and safety-critical defects is relevant in the production of wooden forks, and that the controllability and quality capability of the manufacturing processes must be ensured. The latter can be supported by automated visual inspection, which can identify any rejects in order to minimize the risk of function and safety-critical failures for the customer.

3.2 Database

The quantity and quality of the data plays a decisive role in the successful development of machine learning models. A total of 673 images of wooden forks were taken for this case study. Of this database, 401 fork images were labeled manually as OK and 273 as defect. The images were taken with constant boundary conditions in an experimental setup illustrated in Fig. 7. In order to take comparable images of different forks, a test rig that provides consistent lighting conditions was designed and build up. The test rig is based on a design with white inside-walls, to keep ambient light outside and diffuse the light inside to prevent reflections. Two LED Spotlights were pointed directly on area of interest, because the degree of reflection from the recording area is low. The background consists of a matt dark black surface, which creates a strong contrast to the light-coloured forks and thus makes the edges more visible. In addition, the dark background is based on the real conditions of an assembly line, so that the models generated can be transferred to an inline inspection system or automatic inspection system (AOI). The position and rotation of the forks was also varied while taking the pictures. The photos are taken by an Olympus E-520 DSLR, equipped with an Olympus Zuiko Digital 14–42 mm f 1:3.5–5.6 lens and have a resolution of 3648 × 2736 pixel. The lens is placed normal to the forks at a distance of 25 cm. Further setting parameters of the camera are:

- Aperture value: F/9
- Exposure: 1/320 s.
- ISO sensitivity: ISO-100

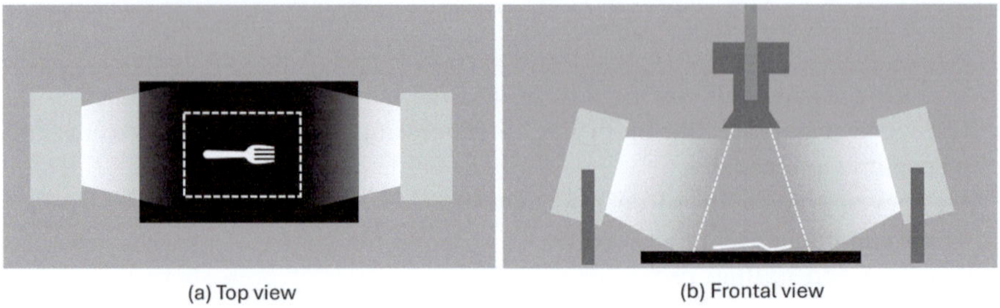

Fig. 7 Schematic illustration of the camera lighting system in the test rig setup. (**a**) Top view; (**b**) Front view

Fig. 8 Pictures of the synthetic generated defect types. (D1: Broken tines and handles; D2: Frayed edges; D3: Natural holes; D4: Burned surface)

- Focal length: 14 mm
- Colour representation: sRGB (3 channels)

Based on the defect types described in Sect. 3.1, four defects to be detected by the AI were generated. In Fig. 8 examples of synthetically generated reject forks are shown. D1, D2 and D4 represent the defects that can arise due to deviating manufacturing processes. While D1 and D2 show sharp and pointed edges that can injure people during use, D4 shows the result of improper temperature treatment during hot pressing. D3, on the other hand, shows the influence of using a natural product, whereby knots in the wood, for example, can result in an inhomogeneous structure or a pitted surface. The identification of these error patterns allows for the direct derivation of conclusions regarding the discrepancies observed in the manufacturing process. This, in turn, allows for the real-time modification of process parameters in order to rectify the identified issues.

3.3 Methology

In order to achieve optimal results for optical inline inspection on this basis, a number of problem-specific pre-processing steps and techniques are necessary before training and applying the CNN model. These include the preparation of the raw data, the selection and preparation of the architecture of the model to be used, the appropriate selection of the relevant evaluation metrics and finally the training step, where the resulting model is prepared for the operation in the field.

Data preparation

First, the raw image data is pre-processed to prepare it for input into the CNN model. The aim is to increase training speed and improve prediction accuracy. Individual data pre-processing steps have to be implemented regarding the data and situation. In the case of the underlying image data of wooden forks, the first step is to reduce the data due to the large storage size of the image data. This allows the algorithm to process more image data in shorter time, which has a significant impact on training speed. For each image, additional information about the camera system used is stored, which does not add additional information. This information can therefore be deleted. At the same time, the image data is stored in three channels as RGB images. Due to the way the images are captured (light brown wooden forks on a black background), the images consist largely of two color tones, which can be reduced to a third of the total size (RGB to gray values) with little loss of information, significantly increasing efficiency and training speed.

The third and final step in data preparation is to reduce the resolution and change the image format. A resolution of 3648×2736 pixels is very high compared to the level of detail and information content of the image data, so reducing the resolution can further reduce the processing and increase the training speed. At the same time, the reduction level should not be set too high so that relevant information is not lost and still recognizable for the algorithm. For this purpose, various reduction levels are tested in the training step and the reduction with the best results is retained.

The rectangular images are also cropped to a square format or padded to the square format, as the CNNs are designed for square image formats by default. In Fig. 9 the steps of image data reduction up to an exemplary reduction by a factor of 8 are shown. These data preparation steps result in a data reduction from 7.16 MB up to 2.31 KB, which significantly influences the training efficiency and allows the model to process much more data in shorter time.

After the images were processed in the first step, the data set is now prepared for training the CNN. With a total of 673 images, The dataset contains a challenging amount of data for training AI models such as CNNs. The training and subsequent application would produce poor results and be unsuitable for visual inspection. Therefore, the technique of data augmentation is used for further training, in which the data set is artificially expanded with the original data in order to increase the amount of available training data. Various techniques are used to enhance image data in a dataset, such as adding noise or varying the brightness to create sufficient differ-

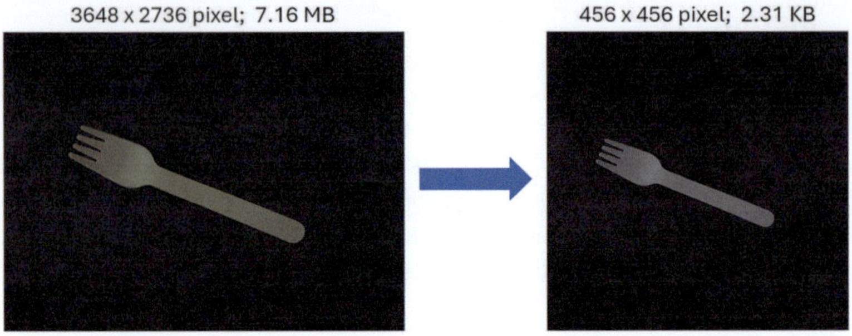

Fig. 9 Image of an exemplary wooden fork from the data set with information on resolution and data size. (Left) Before the data preparation; (Right) After the image preparation steps with an image reduction factor of 8

ences between the resulting images and the original images. For this case study, each original image is modified as follows:

- Random mirage (horizontal and vertical).
- Random rotation of the image up to a factor of 0.25.
- Adding a random noise level within a range factor of 0.25.
- Adding a random contrast level within a range factor of 0.02.

Each rotated image was post-processed with padding and cropping so that it has the same square format as the rest of the images. If a part of the wooden fork was cut off during augmentation, the corresponding image was removed from the data set to ensure complete images and therefore to prevent possible information loss. The data augmentation step can be repeated as often as required to generate the desired number of images in the dataset. As this is an artificial expansion of the dataset, the information gain is not as pronounced compared to an expansion with real data, as the same images are available and the differences between the images may be too small, especially with a minor modification. To ensure that too frequent augmentation does not lead to a reduction in efficiency and performance, various artificially enlarged datasets were tested when training the CNN.

Machine Learning based methods such as CNNs are particularly sensitive to different value ranges due to the use of gradient descent methods to iteratively adjust the weights in the neurons (cf. Sect. 2.3). Due to the different distances between the values, the weights can be adjusted to different degrees, which in the worst case can lead to convergence problems. In order to minimize the influence of these different step sizes and value ranges, it is common practice in data preparation to scale the data. One method that is also used here is Normalization. This involves scaling the pixel values from the value range [0, 255] to the value range [0, 1], whereby the structure of the values, such as the standard deviation, is retained.

Modelling
As mentioned above, the modelling process uses a CNN that is trained on the image data during the training step to learn the patterns and features and determine the correct class from the image data. The general structure of the used CNN is as follows: For each layer of the CNN, a convolutional layer with a learnable filter of size 3×3 and a Rectified Linear Unit (ReLu) activation function is used. The feature map resulting from the convolution operation is quartered in size for each layer duo to the usage of a max-pooling layer of size 2×2. A dropout layer is then added to the layer. Dropout layers generally cause randomly selected connections between layers to be interrupted. This reduces the information passed on to the next layer and forces the network to learn more robust and general features in the data. In the case of CNNs, randomly selected regions of the feature map are set to zero. The number of dropped connections is specified by a value, e.g. with a dropout value of 0.5, 50% of the connections are randomly dropped.

After the CNN layers, a flatten layer is used to convert the last feature maps before the flattening layer from a grid-like structure into a vector. This vector is then passed to one or more fully connected layers with the ReLu activation function to determine the final output—the error class of the input image. To classify the five different classes, the Softmax activation function in the output layer is used to convert the output into a value range between zero and one. The output then represents the respective class probabilities.

Evaluation metrics
Various evaluation metrics can be used to check the prediction quality and the training of machine learning methods, depending on the problem and the data basis. For a supervised classification problem with several classes (multiclass classification), the following metrics are used for evaluation in this case study:

- *Accuracy* describes the percentage of correctly classified images out of all images. For a multi-class classification, this is the number of correct predictions divided by the total number of predictions. The Accuracy can therefore be interpreted directly.
- *Categorical cross-entropy loss* (L_{CE}) with

$$L_{CE}(y, \hat{y}) = -\sum_{i=1}^{K} y_i * log(\hat{y_i}), \tag{1}$$

where y_i describes the true class and $\hat{y_i}$ the predicted class for every class i in K. This loss function is used for the optimization in the training step to measure the error between the true class and the predicted value or probability.

- *Recall* describes the sensitivity of the prognosis from the model by measuring how well the model recognizes actual positive cases. The recall is calculated as

$$Recall = \frac{TruePositives}{TruePositives + FalseNegatives} \tag{2}$$

- *Precision* measures how many of the predicted positive cases are actually positive. It therefore indicates the accuracy of the positive predictions and is calculated as

$$Precision = \frac{TruePositives}{FalsePositives + FalseNegatives} \tag{3}$$

Training step
In order to control the training of the model and achieve the best possible prediction results, the training is carried out as follows. First, the image data from the entire dataset is randomly divided into three categories: 80% training data, which is used directly to train the model; 10% validation data, which is used to monitor the progress of the training; and 10% test data, which is used after the training step to check the performance of the model on unseen data, and thus to estimate performance in the field. To ensure stable and reliable results, we used a 5-fold cross-validation to divide the data into 5 folds. Each model is trained 5 times, each time using a different fold for validation and testing, and the other 4 folds for training. The final performance of the model is the average of all 5 trainings.

From the data distribution shown, a stable model was first identified by manual analysis, which was used to investigate different configurations of image resolution and repeated application of data augmentation. This provided the basis for an efficient and effective training.

To find the optimum model for the objective of this case study, a comprehensive parameter study was then carried out in which a number of different values of the model parameters were tested. For each model, a random configuration of model parameters is selected from a specific range of values, with which the respective model is then trained using 5-fold cross-validation. The Table 1 shows the variable model parameters and the respective range of values. For the training, the established Adam optimizer [20] is used to solve the gradient descent problem. Each model was trained for a maximum of 100 epochs and was stopped early if the training process converged. As the classes are represented in the data set with different frequencies, there is a risk of class imbalance, especially if one class strongly outweighs the others in the number of samples. This can lead to the model preferentially learning the most frequent class and ignoring less frequently represented classes. To ensure that all classes are included equally in the learning process, the less frequent classes are assigned correspondingly higher weights to compensate for their low proportion in the dataset to ensure a better performance of the model.

Table 1 Parameters used for parameter tuning and their possible values

Parameter	Value
Number of CNN layers	1, 2, 3, 4
Number of filters per layer	5, 10, 16, 32
Number of fully connected layers	0, 1, 2, 3
Number of neurons per layer	2, 4, 8, 16, 32, 64
Learning rate	0.1, 0.01, 0.001, 0.0001
Batch size	4, 8, 16, 32, 64, 128, 256
Dropout rate	0.0, 0.25, 0.5

4 Results

The following chapter shows the results of modelling a CNN for an AI-based optical inspection carried out in the case study. We show the model achieving the best results based on the different evaluation metrics and discuss the prognosis results in the context of optical inspection.

Firstly, the preceding testing and analysis evaluating the different configurations for the image resolution and data augmentation showed that an 8-fold image reduction to 456 × 456 pixels and an 8-fold data augmentation with overall 5,939 images achieved the best results in terms of increasing training efficiency and improving performance. For image resolution, both a larger and a smaller image reduction resulted in poorer performance. For data augmentation, the performance increased with the number of augmentations performed until it converged at 8 times. With further augmentations, only the training time increased further; The dataset has thus reached its maximum information. Therefore, an 8-fold image reduction and an 8-fold data augmentation were used for the parameter study.

With the comprehensive parameter study, 300 different configurations for the model parameters were tested. In Table 2 the found parameter configuration for the best performing model are shown. A small CNN with a low number of layers and convolutional filters is considered as the best model. This is probably due to the low information content of the data. On the one hand, the images do not have a high level of detail, which means that the extraction of features and information in the convolution layer can be done with only a few layers. On the other hand, the case study was conducted with a very small dataset for the application of AI, which should be enlarged in the future to achieve better results.

The best model achieved an accuracy of 81.16% with a standard deviation of 1.16% and a categorical cross entropy loss of 0.9198 with a standard deviation of 0.057 on the test dataset. Precision and recall are in a similar range to accuracy with 81.79% and 81.34% respectively. A comparison between the true classes and the predicted classes in the confusion matrix is shown in Fig. 10. Most of the values, which are mostly the classes IO (OK) and nIO 1 (D1: Broken tines and handles),

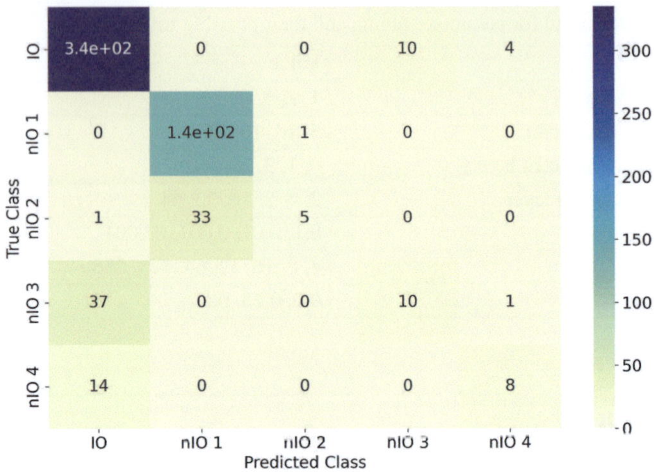

Fig. 10 Representative confusion matrix of the true and predicted classes of the test dataset

are on the main diagonal and therefore predicted correctly. The model struggled to predict the other defect classes. Classes nIO 3 (D3: Natural holes) and nIO 4 (D4: Burnt surface) were mostly interpreted as OK, while most of the class nIO 2 (D2: Frayed edges) images were predicted as class nIO 1, resulting in an improvable performance for these two classes and an overall accuracy of only 81.16%.

A representative training process of the best model is shown in Fig. 11. The growing gap between the training loss curve (blue) and the validation loss curve (yellow) over the epochs is an indicator of an overfitting model. The model starts to memorize the training data but cannot generalize the information in the images to predict unknown images, hence the almost perfect prediction on the training data and the slightly increasing loss after the low point at 12 epochs on the validation data. Either the model is too complex for the underlying classification problem, or more data is needed for training. The same gap is visible for the accuracy over the epochs. While the model achieved perfect accuracy on the training data, it stagnates at around 81% accuracy on the validation data.

Table 2 Parameter configuration for the best performing model

Parameter	Value
Number of CNN layers	1
Number of filters per layer	10
Number of fully connected layers	3
Number of neurons per layer	64
Learning rate	0.001
Batch size	256
Dropout rate	0.25

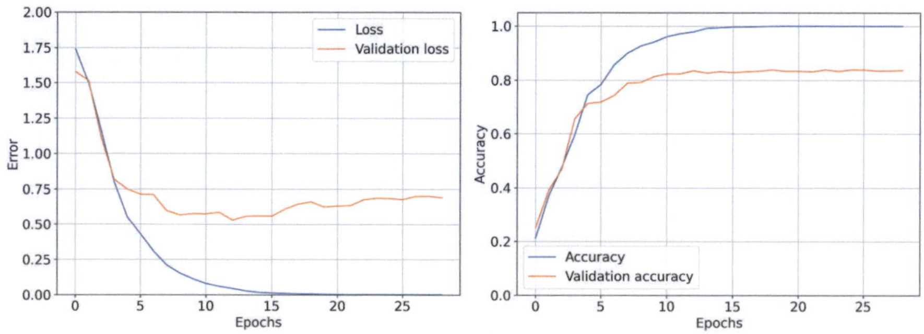

Fig. 11 Representative training process of the best model. The curves for the training data are in blue, those for the validation data are in yellow. (Left) The loss over the epochs; (Right) The accuracy over the epochs

Overall, the CNNs show positive results for the initial analyses in the case study, which could lead to significantly better results with further in-depth investigations, additional parameter studies and a larger amount of data. However, the overall model performance has to be further improved for an industrial application. The model found struggled particularly with predicting defects D3 and D4. A real-world application would require higher prediction accuracies (>99%), especially for functional and safety critical failures that could lead to human injury.

5 Summary and Outlook

This paper presents a comprehensive study on CNN-driven inline quality control for the reliable production of wooden forks. The proposed approach relies on a synthetically generated error spectrum that emulates defects occurring during production, thus ensuring safety-critical functionality for reliable consumer use. Using various augmentation methods, the number of image-based training samples was increased from 673 to 5,939 images, creating a dataset suitable for training CNNs. Through extensive hyperparameter tuning, a final model accuracy of 81.16% was achieved, allowing for the evaluation of each fork in approximately 4.3 ms. While this processing speed is suitable for real-time industrial application, the model's accuracy shows room for improvement. Specifically, naturally occurring defects such as knotholes are often misclassified as non-defective by the model, which are also challenging to replicate in a lab setting due to their high natural variability. Moreover, instances of "frayed edges" are occasionally misclassified as other defects, while further defects induced during production, such as "burned surfaces", remain challenging to detect. Consequently, the algorithm must be refined to attain a quality inspection performance that is on par with that of a human inspector. Expanding the dataset appears to be the most straightforward approach to improve the CNN model. The focus of dataset expansion should prioritize rarer defect types to establish a more balanced representation among defect classes. The overrepresented defect class of "broken tines and handles" can lead to a biased activation threshold in the Softmax function, causing less frequent defect classes to generate insufficient activation in the neuron, which may result in those defects going undetected. For this reason, training multiple models, each specialized in a specific defect class, could be beneficial. When these models are connected using an ensemble learning approach, they may achieve higher accuracy. Furthermore, the resulting binary decision of "this defect is present" or "this defect is not present" reduces complexity, potentially leading to a more reliable evaluation of the model.

Acknowledgements The authors would like to thank Nils Biermann, Luca Braun and Lauri Schütz for the professional exchange and valuable suggestions during the preparation of this work.

References

1. Agrawal, P., Jayaswal, P.: Diagnosis and classifications of bearing faults using artificial neural network and support vector machine. Journal of The Institution of Engineers (India): Series C **101**(1), 61–72 (2020). https://doi.org/10.1007/s40032-019-00519-9
2. Almuqati, M.T., Sidi, F., Rum, S.N.M., Zolkepli, M., Ishak, I.: Challenges in supervised and unsupervised learning: A comprehensive overview. International Journal on Advanced Science, Engineering and Information Technology **14**(4), 1449–1455 (2024). https://doi.org/10.18517/ijaseit.14.4.20191
3. Ancheng: Disposable wooden cutlery (manufacturer website) (2024). https://de.anchenggy.com/wooden-cutlery/
4. Baua: Datenbank: „Gefährliche Produkte in Deutschland: Besteck" (2024). https://www.baua.de/DE/Themen/Monitoring-Evaluation/Marktueberwachung-Produktsicherheit
5. Beyerer, J., Frese, C., Meyer, J.: Automatische Sichtprüfung: Grundlagen, Methoden und Praxis der Bildgewinnung und Bildauswertung, 3. auflage 2025 edn. Springer Berlin and Springer Vieweg, Berlin and Erscheinungsort nicht ermittelbar (2024). https://link.springer.com/978-3-662-69950-8
6. Bracke, S.: Einleitung: Prozessfähigkeitsanalysen bei der Herstellung komplexer Produkte. In: S. Bracke (ed.) Prozessfähigkeit bei der Herstellung komplexer technischer Produkte, pp. 1–3. Springer Berlin Heidelberg, Berlin, Heidelberg (2016). https://doi.org/10.1007/978-3-662-48214-8_1
7. Bracke, S.: Technische Zuverlässigkeit. Springer Berlin Heidelberg, Berlin, Heidelberg (2022). https://doi.org/10.1007/978-3-662-65015-8
8. Bracke, S.: Reliability Engineering. Springer Berlin Heidelberg, Berlin, Heidelberg (2024). https://doi.org/10.1007/978-3-662-67446-8
9. Buongiorno, D., Prunella, M., Grossi, S., Hussain, S.M., Rennola, A., Longo, N., Di Stefano, G., Bevilacqua, V., Brunetti, A.: Inline defective laser weld identification by processing thermal image sequences with machine and deep learning techniques. Applied Sciences **12**(13), 6455 (2022). https://doi.org/10.3390/app12136455
10. BVL: Für sichere verbraucherprodukte: Das RAPEX-Behördennetzwerk für gesundheitlichen Verbraucherschutz in Europa. https://www.bvl.bund.de/DE/Arbeitsbereiche/01_Lebensmittel/01_Aufgaben/04_Warn_und_In-formationssysteme/00_Warn_und_Informationssysteme/03_RAPEX/RAPEX_node.html
11. Carvajal Soto, J.A., Tavakolizadeh, F., Gyulai, D.: An online machine learning framework for early detection of product failures in an industry 4.0 context. International Journal of Computer Integrated Manufacturing **32**(4-5), 452–465 (2019). https://doi.org/10.1080/0951192X.2019.1571238

12. Casini, M., de Angelis, P., Chiavazzo, E., Bergamasco, L.: Current trends on the use of deep learning methods for image analysis in energy applications. Energy and AI **15**, 100,330 (2024). https://doi.org/10.1016/j.egyai.2023.100330
13. DIN ISO 15226:2017-03, Technische Produktdokumentation-Lebenszyklusmodell und Zuordnung von Dokumenten (iso_15226:1999). https://doi.org/10.31030/2603414
14. EU: Durchführungsbeschluss (eu) 2019/417 der Kommission – vom 8. November 2018 – zur Festlegung von Leitlinien für die Verwaltung des gemeinschaftlichen Systems zum raschen Informationsaustausch „RAPEX" gemäß Artikel 12 der Richtlinie 2001/95/eg über die allgemeine Produktsicherheit und für das dazugehörige Meldesystem – (bekannt gegeben unter Aktenzeichen c(2018) 7334) (2018)
15. European Union: Regulation (ec) no 1935/2004 of the european parliament and of the council of 27 october 2004 on materials and articles intended to come into contact with food: Regulation (ec) no 1935/2004 (2004). https://eur-lex.europa.eu/legal-content/DE/TXT/?uri=CELEX:32004R1935
16. European Union: Commission regulation (ec) no 2023/2006 of 22 december 2006 on good manufacturing practice for materials and articles intended to come into contact with food (2006). https://eur-lex.europa.eu/legal-content/EN/TXT/?uri=CELEX:32006R2023
17. Goldman, S.: Vibration spectrum analysis: a practical approach. Industrial Press Inc (1999)
18. Huang, X., Liu, Z., Zhang, X., Kang, J., Zhang, M., Guo, Y.: Surface damage detection for steel wire ropes using deep learning and computer vision techniques. Measurement **161**, 107,843 (2020). https://doi.org/10.1016/j.measurement.2020.107843
19. Jardine, A.K., Lin, D., Banjevic, D.: A review on machinery diagnostics and prognostics implementing condition-based maintenance. Mechanical Systems and Signal Processing **20**(7), 1483–1510 (2006). https://doi.org/10.1016/j.ymssp.2005.09.012
20. Kingma, D.P., Ba, J.: Adam: A method for stochastic optimization. https://doi.org/10.48550/arXiv.1412.6980
21. Krizhevsky, A., Sutskever, I., Hinton, G.E.: Imagenet classification with deep convolutional neural networks. Communications of the ACM **60**(6), 84–90 (2017). https://doi.org/10.1145/3065386
22. Kudelina, K., Vaimann, T., Asad, B., Rassõlkin, A., Kallaste, A., Demidova, G.: Trends and challenges in intelligent condition monitoring of electrical machines using machine learning. Applied Sciences **11**(6), 2761 (2021). https://doi.org/10.3390/app11062761
23. Lecun, Y., Bottou, L., Bengio, Y., Haffner, P.: Gradient-based learning applied to document recognition. Proceedings of the IEEE **86**(11), 2278–2324 (1998). https://doi.org/10.1109/5.726791
24. Li, J., Lu, Y., He, F., Miao, L.: Motion control of hydraulic actuators with nonlinear friction compensation: Applied to variable valve systems of diesel engine. ISA transactions **137**, 561–573 (2023). https://doi.org/10.1016/j.isatra.2023.01.037
25. Mohtasham Moein, M., Saradar, A., Rahmati, K., Ghasemzadeh Mousavinejad, S.H., Bristow, J., Aramali, V., Karakouzian, M.: Predictive models for concrete properties using machine learning and deep learning approaches: A review. Journal of Building Engineering **63**, 105,444 (2023). https://doi.org/10.1016/j.jobe.2022.105444
26. Newood: Wooden cutlery production line (manufacturer website) (2024). https://cnnewood.com/disposable-wooden-cutlery-making-machine/
27. O'Shea, K., Nash, R.: An introduction to convolutional neural networks. http://arxiv.org/pdf/1511.08458v2
28. Payette, M., Abdul-Nour, G.: Machine learning applications for reliability engineering: A review. Sustainability **15**(7), 6270 (2023). https://doi.org/10.3390/su15076270
29. Prakash, N., Nieberl, D., Mayer, M., Schuster, A.: Learning defects from aircraft ndt data. NDT & E International **138**, 102,885 (2023). https://doi.org/10.1016/j.ndteint.2023.102885
30. Ramakrishnan, A., Pecht, M.G.: A life consumption monitoring methodology for electronic systems. IEEE Transactions on Components and Packaging Technologies **26**(3), 625–634 (2003). https://doi.org/10.1109/TCAPT.2003.817654

31. Rani, A., Ortiz-Arroyo, D., Durdevic, P.: A survey of vision-based condition monitoring methods using deep learning: A synthetic fiber rope perspective. Engineering Applications of Artificial Intelligence **136**, 108,921 (2024). https://doi.org/10.1016/j.engappai.2024.108921
32. Rao, B.: Handbook of condition monitoring. Elsevier (1996)
33. Rawat, W., Wang, Z.: Deep convolutional neural networks for image classification: A comprehensive review. Neural computation **29**(9), 2352–2449 (2017). https://doi.org/10.1162/NECO_a_00990
34. Romiter Group: Disposable wooden forks production line (manufacturer website) (2017). https://www.cnwoodmachine.com/disposable-wooden-forks-production-line/
35. Rosebrock, C., Bracke, S. (eds.): Detection of defects on printed circuit boards using instance segmentation: Proceedings of the 31st European Safety and Reliability Conference; 19–23 September 2021, Angers, France. Research Publishing Services, Chennai (2021). https://www.rpsonline.com.sg/proceedings/9789811820168/index.html
36. Rosebrock, C., Grams, L., Bracke, S. (eds.): Fault detection in the manufacturing process of printed circuit boards using computer vision: Proceedings of the 30th European Safety and Reliability Conference and the 15th Probabilistic Safety Assessment and Management Conference, 01–05 November 2020, Venice, Italy. Research Publishing, Singapore (2020). https://www.rpsonline.com.sg/proceedings/esrel2020/
37. Schlanbusch, R., Oland, E., Bechhoefer, E.R.: Condition monitoring technologies for steel wire ropes—a review. International Journal of Prognostics and Health Management **8**(1) (2017)
38. Shen, D., Wu, G., Suk, H.I.: Deep learning in medical image analysis. Annual review of biomedical engineering **19**, 221–248 (2017). https://doi.org/10.1146/annurev-bioeng-071516-044442
39. Stavropoulos, G., Violos, J., Tsanakas, S., Leivadeas, A.: Enabling artificial intelligent virtual sensors in an iot environment. Sensors (Basel, Switzerland) **23**(3) (2023). https://doi.org/10.3390/s23031328
40. Wei, J.: Production technology of disposable wooden cutlery (internetblog) (2024). https://westshorecutlery.com/production-technology-of-disposable-wooden-cutlery/
41. Wennker, P.: Künstliche Intelligenz in der Praxis: Anwendungen in Unternehmen und Branchen: KI wettbewerbs- und zukunftsorientiert einsetzen. Springer Gabler, Wiesbaden, Germany and Heidelberg (2020)
42. Widmann, U., Weissinger, J., Breitling, T., Hackenberg, U., Wundram, K., Goß, S.: Produktentstehungsprozess. In: S. Pischinger, U. Seiffert (eds.) Vieweg Handbuch Kraftfahrzeugtechnik, pp. 1271–1382. Springer Fachmedien Wiesbaden, Wiesbaden (2021). https://doi.org/10.1007/978-3-658-25557-2_12
43. Yang, J., Qin, Y., Zhu, Z., Xu, X., Guan, D.: Design and fabrication of a high-efficiency defect inspection prototype for diary plastic cutlery based on machine-vision with improved deep leaning algorithm. Multimedia Tools and Applications (2024). https://doi.org/10.1007/s11042-024-19395-2
44. Yang, Y., Yang, R., Pan, L., Ma, J., Zhu, Y., Diao, T., Zhang, L.: A lightweight deep learning algorithm for inspection of laser welding defects on safety vent of power battery. Computers in Industry **123**, 103,306 (2020). https://doi.org/10.1016/j.compind.2020.103306
45. Zhang, W., Jia, M.P., Zhu, L., Yan, X.A.: Comprehensive overview on computational intelligence techniques for machinery condition monitoring and fault diagnosis. Chinese Journal of Mechanical Engineering **30**(4), 782–795 (2017). https://doi.org/10.1007/s10033-017-0150-0
46. Zhou, P., Zhou, G., Zhu, Z., Tang, C., He, Z., Li, W., Jiang, F.: Health monitoring for balancing tail ropes of a hoisting system using a convolutional neural network. Applied Sciences **8**(8), 1346 (2018). https://doi.org/10.3390/app8081346

System Reliability Engineering in the Age of Industry 4.0: Challenges and Innovations

Isabelle Müller, Zikai Zhang, Antoine Tordeux, Jannis Pietruschka, Stefan Bracke, Tim M. Julitz, Nadine Schlüter, Manuel Löwer and Nicola Fricke

Contents

Abstract

In the era of Industry 4.0, system reliability engineering faces both challenges and opportunities. On the one hand, the complexity of cyber-physical systems, the integration of novel numerical technologies, and the handling of large amounts of data pose new difficulties for ensuring system reliability. On the other hand, innovations such as AI-driven prognostics, digital twins, and IoT-enabled systems enable the implementation of new methodologies that are transforming reliability engineering. Condition-based monitoring and predictive maintenance are exam-

I. Müller · Z. Zhang · A. Tordeux (✉)
Chair for Traffic Safety and Reliability, University of Wuppertal, Gaußstraße 20, 42119 Wuppertal, Germany
e-mail: tordeux@uni-wuppertal.de

J. Pietruschka · S. Bracke
Chair for Reliability Engineering and Risk Analytics, University of Wuppertal, Gaußstraße 20, 42119 Wuppertal, Germany

T. M. Julitz · N. Schlüter · M. Löwer
Chair for Product Safety and Quality, University of Wuppertal, Gaußstraße 20, 42119 Wuppertal, Germany

N. Fricke
Chair for Sociotechnical Systems, University of Wuppertal, Gaußstraße 20, 42119 Wuppertal, Germany

S. Bracke and R. Pieper (eds.), *Safety Engineering*,
https://doi.org/10.1007/978-3-658-47473-7_4

ples of key advancements, leveraging real-time sensor data collection and AI to predict and prevent equipment failures. These approaches reduce failures and downtime, lower costs, and extend equipment lifespan and sustainability. However, it also brings challenges such as data management, integrating complexity, and the need for fast and accurate models and algorithms. Overall, the convergence of advanced technologies in Industry 4.0 requires a rethinking of reliability tasks, emphasising adaptability and real-time data processing. In this chapter, we propose to review recent innovations in the field, related methods and applications, as well as challenges and barriers that remain to be explored. In the red lane, we focus on smart manufacturing and automotive engineering applications with sensor-based monitoring and driver assistance systems.

Keywords

System Reliability Engineering · Industry 4.0 · Prognosis and Health Management · System Design · Simulation-based Analysis

1 Introduction

The recent advent of computational learning algorithms has revolutionised many disciplines [76], whether in engineering [3,72,149], the natural sciences [23,58,108, 129] or the social sciences [40,150,160]. This revolution, which began during the 1990s, has also profoundly impacted reliability engineering [2,39,161,163]. Rather than the emergence of new methods, the rise of Internet-of-Things (IoT), cyber-physical systems and Industry 4.0, along with the growing use of (intelligent) sensors, connected and autonomous systems, data storage, and advanced high-performance computational prognostics, are the key drivers of this change [31,47,75,137]. In fact, learning approaches have reached maturity and meet the need to address the reliability of increasingly complex systems.

Condition monitoring and predictive maintenance stand out as remarkable advances, using real-time data and AI to predict and prevent equipment failures [74,119,132]. These approaches reduce downtime, lower costs, and extend equipment lifespan, thereby improving system sustainability. Other recent progresses concern system design, e.g. fault-tolerant, fail-safe, fail-operational system structures and dynamic reconfiguration, especially for safety-critical systems [24,114]. In addition, computer-aided reliability analysis based on simulation, digital twins and real-time data assimilation allows the effects of degradation and other dynamic factors in mechanical and cyber-physical systems to be taken into account in real time [100,145,146]. Applications cover various fields, including (smart) manufacturing and industrial production [4,31,52], energy [39,89], robotics, transport and logistics [105], (smart) cities and healthcare [99]. The increasing complexity of systems and algorithms, the integration of new digital technologies and the management of large amounts of data generated in real time remains major challenges. Overall, the convergence of advanced technologies in Industry 4.0 requires rethinking reliability tasks with a focus on adaptability and real-time data processing.

The aim of this chapter is to review the latest innovations, applications, and relevant methodologies in the field, as well as the challenges that remain to be addressed. A red lane in the chapter is dedicated to smart manufacturing, automotive and traffic engineering applications, in particular sensor-based monitoring and driver assistance systems. The chapter is structured as follows. We identify and review in Sect. 2 three recent innovation domains in advanced system reliability engineering, namely real-time sensor-based reliability analysis in Sect. 2.1, robust system design in Sect. 2.2, and simulation-aided reliability analysis in Sect. 2.3. Then, we present some challenges and discuss barriers that remain to be explored in Sect. 3, covering difficulties related to safeguarding AI and real-time data processing in Sect. 3.1, system complexity in Sect. 3.2; and human-machine collaboration and sociotechnical systems in Sect. 3.3. Finally, in Sect. 4, we review related statistical methods and machine learning techniques for multivariate datasets. The methods include unsupervised learning approaches for data visualisation, dimension reduction and clustering in Sect. 4.1, and supervised learning methods for advanced predictive analytics, e.g., multivariate regression, classification, and machine learning techniques in Sect. 4.2.

2 Innovations

In this section, we review the most important recent advances in the three innovation areas of prognostics, system design and simulation-based reliability analysis (see Figs. 1 and 2). These research areas of reliability engineering have benefited from important innovations in recent decades. First, sensor-based monitoring based on machine learning techniques, supported by the Internet of Things and Industry 4.0, has shown significant progress in prognostics and health management, real-time condition monitoring, fault detection, self diagnosis and predictive maintenance. Then system design, which showed how advanced architectures can improve the system reliability through redundant configurations, fault-tolerance, fail-safe, fail-operational architectures, dynamic reconfiguration and automation. Finally important innovations have also been provided by simulation-based reliability analysis, e.g. using digital twins, accelerated life testing, or scenario-based simulation. Figure 2 shows the number of research articles over the last twenty years with related keywords in the Google Scholar engine. The trends are clearly more than linear for all the selected keywords. We can observe that the keywords *Digital Twin(s)*, *Internet-of-Things* and, to a lesser extent, *Prognostics* and *Fault Detection* are very topical. The keyword *Machine Learning* peaked about five years ago, although it is still the most cited keyword. *Fault Tolerant*, *Accelerated Life/Degradation*, *Reconfiguration*, *Fail-Safe/Operational*, and *Scenario-Based* are keywords that were established earlier in the 2000s.

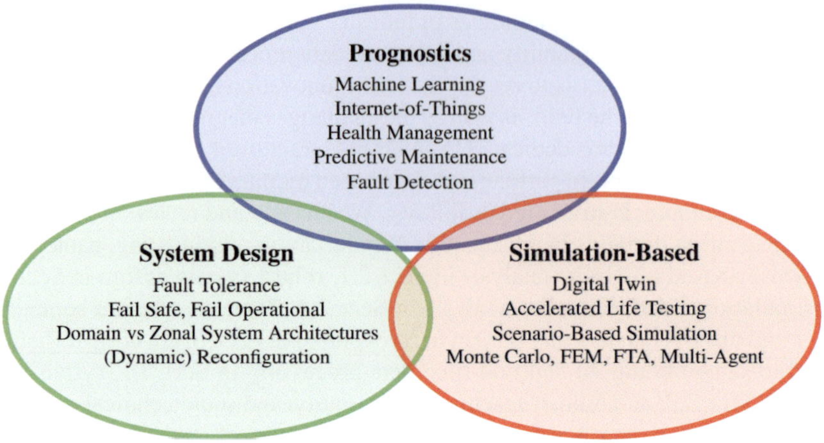

Fig. 1 Main sectors that contribute to advanced reliability engineering techniques during the last decades

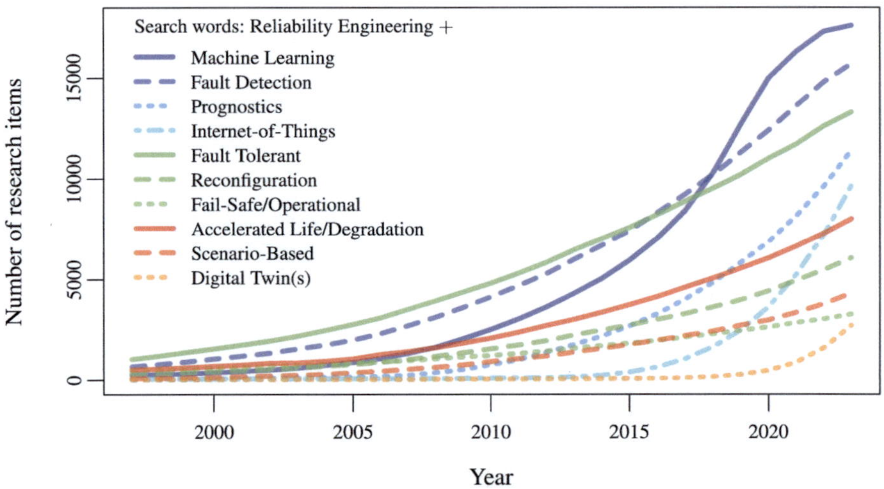

Fig. 2 Cumulative number of research articles from 1997 to 2023 with the tags 'Reliability Engineering' and related keywords in the Google Scholar engine (query made on 01.10.2024)

2.1 Prognostics

Sensor-based monitoring has emerged as an innovation in reliability engineering, driven by advances in data acquisition, real-time analysis, and integration with machine learning (ML) and AI technologies. IoT integration in the context of Industry 4.0 enables continuous or periodic monitoring of operating conditions in systems by using for example **Wireless Sensor Networks** (WSN). WSN collect vast amounts of data (i.e. vibration, temperature, pressure, etc.) remotely and decentralised. For (real-time) analysis purposes, IoT platforms provide access to this transmitted data, where it is processed to assess system health. The data acquisition is a crucial step in the process. Indeed, the labelled data enables machine learning algorithms to be trained and validated. Numerous open-access online data repositories exist for this purpose, such as Kaggle with specif keywords[1], NASA Prognostics Data Repository[2], UC Irvine Machine Learning Repository[3], etc.

A comprehensive framework that incorporates these innovations is **Prognostics and Health Management** (PHM). PHM consists of four main components:

1. Data Acquisition and Condition Monitoring,
2. Diagnostics,
3. Prognostics,
4. Health Management.

Within the PHM framework, data acquisition and condition monitoring integrates sensor-based monitoring with data analytics and machine learning techniques to monitor system health. Such a area relies on advanced sensor technology (e.g., WSN). Diagnostics aims to detect abnormal system behaviour, identify and isolate faults, and analyse the root causes of system failures, while Prognostics goes a step further by estimate the remaining useful life (RUL) and failure prediction based on current operational conditions. System maintenance strategy (i.e. predictive maintenance) and performance optimisation are enabled through the use of diagnostics and prognostics. Optimising maintenance schedules by using diagnostics and prognostics minimises unplanned downtime and maximises the operational life of (industrial) systems and performance optimisation, which is referred to Health Management. Many reviews exist on the topics, see, e.g., [31,47,74,75,119,132,137] and references therein.

2.1.1 Fault Detection and Diagnosis

In modern industrial applications where sensor networks are implemented, a large amount of process data is available. They contain important information about process dynamics and system health. In general, the diagnostics process integral to the **Fault Detection and Diagnosis** (FDD) framework is a series of transformations on system measurements [153], shown in Fig. 3. The FDD system uses these measurements from the process as an input. Then, in the next step features are derived from measurements through feature engineering or residual generation. Features act as health indicators of faulty (abnormal) process behaviour and provide a basis for detecting changes, which refer to fault symptoms. For example [170] provides an overview of the health indicator (HI) construction for rotating machinery. In the decision space, changes in the operating conditions are detected with threshold

---

[1] https://www.kaggle.com/datasets?search=predictive+maintenance.

[2] https://www.nasa.gov/intelligent-systems-division/discovery-and-systems-health/pcoe/pcoe-data-set-repository/.

[3] https://archive.ics.uci.edu/.

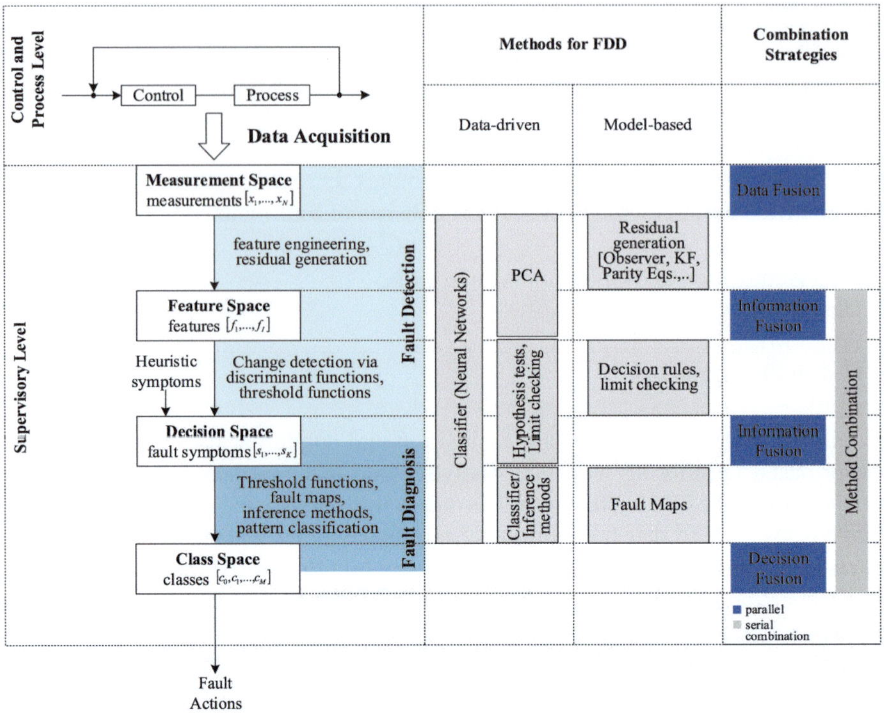

Fig. 3 Transformations within the FDD system, used data-driven and model-based methods and combination strategies for hybrid approaches [158]

or discriminant functions. Subsequently, the fault diagnosis constitutes a mapping between the set of fault symptoms and the set of fault classes. In order to ascertain the fault classes, including the normal operating condition, a variety of methods may be employed, such as pattern recognition, classifiers, inference methods or threshold functions. Afterwards the output of the diagnostics process, i.e. features and fault classes, may be used as an input for prognostics and health management [144, 172].

The methods employed in diagnostics can be broadly classified into two categories: data-driven and model-based, shown in Fig. 3. In addition, the work of [158] considered knowledge-based approaches. Data-driven methods extract information from historical process data and are generally divided into supervised (i.e., classification and regression techniques) and unsupervised methods (i.e., statistical models and clustering techniques). Supervised methods train algorithms to historical data with the objective of learning and distinguishing patterns within the high-dimensional space of process measurements or features. The patterns may then be used to perform fault detection and fault diagnosis, specifically to predict the fault symptoms or classes for new data. However, such methods require labelled training data, whereby for each input feature vector, the corresponding class label must be known [158].

Unsupervised methods do not require class labels and are therefore suitable for a wider range of applications where these are not available. Most unsupervised methods focus only on the first or second transformation step to extract features or fault symptoms [158]. Principle component analysis (PCA) is the most common approach to dimensionality reduction and feature extraction. Typical data-driven methods especially used for classification purposes are artificial neural networks (ANN), support vector machines (SVM), k-nearest-neighbour (kNN), decision trees, and Random Forest, see e.g. [5,69,123].

Model-based approaches rely on numerical reproduction in the form of mathematical equations of the real physical process. The mathematical equations are based on the knowledge of the input-output relationships of the process [67]. Due to these known input-output relationships, the model requires less process data compared to data-driven approaches. The operating state of the process is determined by comparing measurements of the real process state with the state predicted by the physical model. An inconsistent comparison indicates a faulty condition. Specific inconsistencies are referred to as residuals. The residuals correspond to the characteristics shown in the workflow Fig. 3. The subsequent fault diagnosis can be carried out using threshold logic functions, for example. Typical methods for FDD based on physical models are bond graphs, parity equations, state estimation (e.g., using Kalman filter), state observers or simple limit and trend checking, see e.g. [33,50,66,67,127].

Hybrid FDD approaches combine two or more established methodologies. Additionally, some approaches consider direct integration of a priori knowledge. Hybrid FDD methods may be combined parallel or serial through a variety of algorithms, see [158]. In the case of a data set comprising a significant amount of unlabelled data, the following hybrid methods may be employed to automatically expand the data set, for instance:

- **Semi-supervised**: A model is trained using a small, manually labelled data set, which is then applied to unseen, unlabelled data. Subsequently, the previously unlabelled data is incorporated into the training data set, with a label selected by the model itself. It is crucial that the base data set with the labelled data is selected to be representative of the use case, in order to facilitate the optimal training of the model. In an iterative process, the data labelled by the model is then added to the training data set, with the objective of optimising the model [21,92].
- **Activ Learning**: The model is trained in an iterative manner, whereby the most "uncertain" or informative data points are selected for human labelling. This approach entails that only the most necessary data is labelled manually, while the model assumes responsibility for the remaining labels automatically. This reduces the labelling effort, but with the potential for incorrect labelling of data points in which the model is certain [26,30].

2.1.2 Remaining Useful Life Prediction and Sampling Methods

Predictive maintenance strategies are generally based on real-time prediction of the **Remaining Useful Life** (RUL) of a system. These strategies allow for proactive maintenance planning, which helps avoid unexpected downtime and optimizes operational efficiency. As RUL predictions reach critical thresholds, maintenance interventions are initiated, enabling actions to be taken before failures occur. RUL prediction methods are widely applicable, particularly in industries that rely on battery and bearing maintenance, as reviewed in [55,60,65,74].

Common predictive maintenance algorithms can be categorised into *shallow* and *deep learning* methods, each providing unique advantages. In shallow learning, methods like *Logistic Regression* are computationally efficient and simple to implement, though they assume linear relationships, which limits their suitability in complex environments [55,65]. *Artificial Neural Networks* (ANNs) are also common in shallow learning; they store complex patterns and are robust to noise but may yield unpredictable results under certain conditions [90].

Other popular shallow learning techniques include *Support Vector Machines* (SVM), which perform well in high-dimensional data spaces, though they are less effective with large datasets. *Deep Belief Networks* (DBNs) require minimal labeled data and have quick training times but pose challenges in determining optimal network structures [60]. Deep learning approaches, such as *Convolutional Neural Networks* (CNNs) and *Recurrent Neural Networks* (RNNs), are highly effective for processing high-dimensional or sequential data, albeit with substantial computational requirements [74].

Convolutional Neural Networks excel in tasks like image recognition due to their automatic feature detection capabilities, although they require considerable computational resources [123]. Recurrent Neural Networks are particularly well-suited for time-series data because of their capacity to store sequential information; however, they are prone to the vanishing gradient problem, which affects prediction accuracy over long sequences [67]. *Hidden Markov Models* (HMMs) are also effective for capturing sequential dependencies, although they are unsupervised and necessitate large datasets for training [50,66].

Signal sampling and acquisition are foundational in digital technology, with applications from control systems to digital communication. Traditional sampling employs a uniform time-based cycle, known as *Riemann Sampling* (RS), which integrates along the time axis. This method is widely used for its simplicity in analysis and implementation [122,164]. In contrast, *Lebesgue Sampling* (LS) is an event-driven approach where data points are recorded only when a measurement transitions between states. This approach, which integrates along the y-axis (i.e., measurement value), reduces the number of data points and is ideal for systems with limited computational resources [122].

In RUL prediction, physical quantities indicating equipment health gradually change over time due to wear or aging. In early operational phases, signals typically remain stable with minimal variation, making continuous monitoring through RS both redundant and resource-intensive. In such cases, Lebesgue-based sampling techniques offer a novel solution, especially when RUL prediction is constrained by computational limits [122]. By focusing on state changes rather than continuous time-based measurements, LS-based sampling in hybrid models can reduce data volume while maintaining predictive accuracy, making it well-suited for edge devices with limited processing capacity.

Table 1 Comparison of predictive maintenance algorithms based on shallow and deep learning approaches [55,60,65,74]

Algorithm	Advantages	Limitations
Logistic Regression	Simple, efficient	Assumes linearity
Artificial Neural Network (ANN)	Robust to noise, stores patterns	Unpredictable behavior
Support Vector Machine (SVM)	Effective in high dimensions	Not suitable for large datasets
Deep Belief Network (DBN)	Quick training on small datasets	Network structure complexity
Convolutional Neural Network (CNN)	Excels in feature detection	Computationally demanding
Recurrent Neural Network (RNN)	Effective for time-series data	Suffers from gradient vanishing
Hidden Markov Model (HMM)	Captures sequential dependencies	Requires large datasets
Cluster Visualization	No need for labeled data	Lacks standardised parameter tuning

Hybrid models combining Lebesgue Sampling with traditional data-driven or model-based methods benefit from event-driven data acquisition, allowing for efficient resource allocation without compromising RUL prediction quality. This sampling adaptation enhances the hybrid model's ability to manage large datasets more efficiently, particularly under constrained computational conditions, contributing significantly to predictive maintenance solutions.

As shown in Table 1, each algorithm presents distinct strengths and limitations, and their selection depends on the system's specific data and operational requirements. These predictive maintenance approaches, particularly with the integration of Lebesgue Sampling in hybrid models, contribute significantly to enabling real-time, data-driven maintenance solutions across various industrial applications.

2.2 System Design

The chapter discusses fault-tolerant system design in safety-critical and highly automated environments. It provides an overview of key strategies and architectural principles needed to ensure system reliability, availability and safety under failure conditions. Section 2.2.1 introduces core fault tolerance concepts, including *fail-safe*, *fail-degradation*, and *fail-operational strategies*. Section 2.2.2 explores redundancy techniques and their impact on system reliability and availability. Section 2.2.3 outlines different system architectures, highlighting *domain* and *zonal approaches*. Finally, Sect. 2.2.4 focuses on *dynamic reconfiguration*, emphasizing real-time adaptation to maintain system functionality.

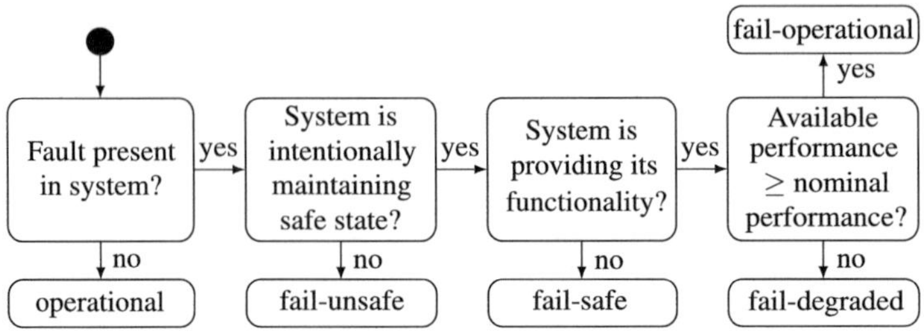

Fig. 4 Comparison of fail-safe, fail-operational, and fail-degradation strategies [140]

2.2.1 Fail-Safe, Fail-Degradation and Fail-Operational Concepts

Fault tolerance is a critical aspect of modern system design, especially in safety-critical and highly automated environments [68, 84, 130, 131]. As systems become more complex, integrating numerous electronic components and software subsystems, the likelihood of failures increases. Faults in one component can cascade through interconnected systems, leading to catastrophic consequences if not properly managed. For example, in autonomous vehicles, the failure of a single sensor or control module could potentially lead to dangerous driving conditions if not mitigated. Therefore, fault tolerance ensures that the system can continue to operate safely and effectively, even when individual components fail. To achieve fault tolerance, three key concepts: Fail-safe, fail-operational, and fail-degradation provide distinct approaches to managing failures and ensuring system reliability and availability [140]. To further clarify these strategies, Fig. 4 is illustrating how fail-safe, fail-operational, and fail-degradation approaches differ from each other and how they work together to create a fault-tolerant system.

Fail-Safe systems transition into a predefined safe state when a failure occurs [154]. This safe state is designed to minimize risks and prevent any further harm. In automotive systems, this often means ceasing functionality to avoid dangerous situations, such as in braking systems where mechanical fallback mechanisms may be activated upon failure of electronic components. Although the system stops functioning, fail-safe strategies ensure that safety is prioritised over continued operation. The focus is on bringing the system to a state that poses no unreasonable risk, even if this means halting operations entirely.

Fail-Operational systems maintain their full functionality even after a fault has occurred [157]. This is crucial in applications where uninterrupted operation is essential, such as in autonomous driving or aerospace systems. Fail-operational systems are designed to continue providing their specified functionality without performance degradation. For instance, in autonomous vehicles, even if one sensor fails, the vehicle must continue to operate using redundant sensors to complete its mission or reach a safe state. Fail-operational strategies ensure that critical systems can continue functioning without a loss in performance, enhancing both system availability and safety.

Fail-Degradation provides a middle ground, allowing systems to continue operating with reduced performance in the event of a failure [166]. Unlike fail-operational systems, which maintain full functionality, fail-degradation allows for impaired functionality while still ensuring safety. In the automotive context, a fail-degradation strategy could involve reducing the speed of an autonomous vehicle if its primary navigation system fails, allowing it to safely continue driving at a reduced capability or move to a safer location. This concept is especially useful when full operational redundancy is too expensive or unnecessary. Fail-degradation increases both reliability and availability by ensuring that the system can still function, albeit at a lower capacity, instead of shutting down entirely. Fail-degradation can also be part of fail operational concepts [157].

2.2.2 Redundancy and Its Role in Fault Tolerant Systems

Redundancy is fundamental to achieve fault tolerant capabilities in safety-critical systems [71]. By introducing multiple components or subsystems that can take over in case of failure, redundancy ensures that the system remains operational or enters a safe state when faults occur. There are different strategies for implementing redundancy, often through **m-out-of-n** systems, where n components are in operation, and the system continues functioning as long as at least m components are working [70, 152].

In redundant systems, **NooN** represents a *serial configuration*, where all N components must function correctly for the system to operate. This configuration offers high reliability but is vulnerable, as any single failure causes the entire system to fail. On the other hand, **1ooN** represents a *parallel configuration*, where the system can function as long as at least one of the N components works.

Reliability focuses on the probability that the system will operate without failure over a specified time. For a redundant system with n identical components, each with reliability $R_i(t)$, the overall system reliability $R_s(t)$ for an *m-out-of-n* configuration is given by:

$$R_s(t) = \sum_{k=m}^{n} \binom{n}{k} R_i^k(t) \, (1 - R_i(t))^{n-k} \tag{1}$$

This formula calculates the probability that at least m components are functioning correctly, ensuring system operation. Redundancy in *fail-safe* systems guarantees that, upon failure of one or more components, the system transitions into a *safe state*. The goal in such systems is not to continue operation, but rather to prevent unsafe conditions by shutting down or moving to a controlled, safe condition.

Availability, on the other hand, is crucial for *fail-operational* systems, where the system must remain functional even after failures. Availability is a measure of the system's ability to perform its intended function at any given time, factoring in maintenance and repair periods. In *fail-operational* systems, maintaining availability is often more critical than ensuring perfect reliability, as the system needs to continue functioning despite component failures. The *availability* of a redundant system is influenced by both its reliability and its ability to be repaired or reconfigured, and is asymptotically calculated as:

$$A_s = \frac{MTTF}{MTTF + MTTR}, \tag{2}$$

where $MTTF$ is the *mean time to failure*, and $MTTR$ is the *mean time to repair*. In this context, redundancy allows the system to continue operating while repairs are made, ensuring minimal downtime. The effectiveness of repair or reconfiguration in fault-tolerant systems is directly influenced by the ability to separate or diversify system elements [126, 138].

Separation refers to the physical isolation of redundant subsystems, ensuring that a failure in one part of the system does not affect others. For example, in aircraft systems, placing critical sensors or control units in different parts of the vehicle prevents damage in one area, such as from fire or impact, from compromising the entire system. This isolation allows faulty components to be repaired or replaced while the rest of the system remains operational, thus maintaining availability [126, 138].

Diversification, on the other hand, involves using different technologies, designs, or suppliers for redundant components to reduce the risk of common-cause failures. In a fail-operational system, diversity ensures that if one component fails due to a specific failure mode or environmental condition, another component with a different design or tolerance can continue functioning. For instance, in autonomous vehicles, using a mix of sensors like lidar, radar, and cameras ensures that the system can maintain functionality even if one type of sensor is compromised by conditions like fog or rain [126, 138].

In *fail-degradation* systems, redundancy enables the system to degrade gracefully in response to failures, continuing to function at a reduced capacity rather than completely ceasing operation [140]. For instance, if several sensors fail in an autonomous vehicle, the remaining sensors may allow the vehicle to continue driving, although with reduced accuracy or slower speeds. This kind of redundancy ensures that the system can maintain essential functions and prevent complete shutdown, balancing *reliability* and *availability*.

2.2.3 System Architectures

The concepts of redundancy, separation, and diversification are foundational to designing robust electronic control units (ECUs), particularly in safety-critical systems such as automotive and aerospace applications [138]. In practice, ECUs use redundancy strategies to maintain operation in the face of failures. For instance, a 1ooN configuration may be employed for non-critical systems where basic functionality must continue despite component failures, while a 2oo2 or 2oo3 configuration could be used in safety-critical systems that require more stringent fault-tolerance measures [84]. By deploying multiple ECUs in parallel, the system can isolate faults or switch to a backup ECU if necessary [87]. Two different fail-operational ECU architectures are shown in Fig. 11 in Appendix 1.

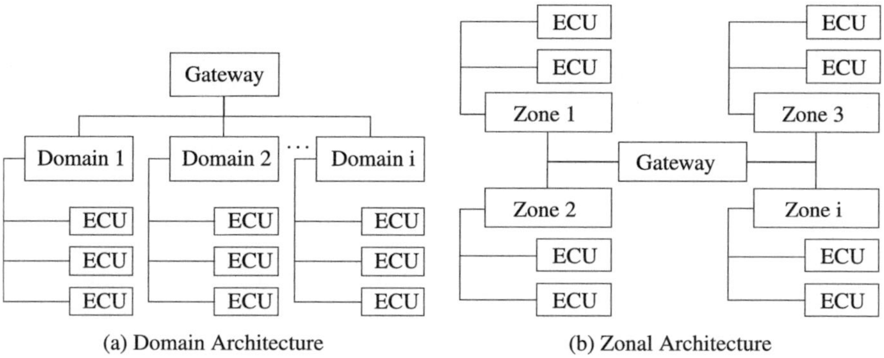

Fig. 5 Electronic control unit (ECU) architectures at the system level

ECU redundancy schemes can be integrated into broader architectures at the system level, which are commonly organised into either domain architectures or zonal architectures [87, 125]. These system architectures dictate how ECUs are distributed, communicate, and handle redundancy across the entire System. The two approaches are illustrated in Fig. 5.

In a *domain architecture*, ECUs are organised based on their functional roles, with each domain representing a specific set of functions. For example vehicle, the powertrain domain manages functions related to the engine, transmission and drivetrain. The body control domain oversees lighting, doors, windows, and climate control. The infotainment domain handles multimedia, navigation, and user interfaces and the safety domain manages systems such as airbags, ABS (anti-lock braking system), and electronic stability control. In this architecture, each domain operates somewhat independently, with ECUs dedicated to specific tasks within that domain. Redundancy strategies like 2oo3 TMR or 2oo2 DFS can be applied within each domain to ensure critical functions like powertrain or safety can continue to operate even in the event of a failure. The advantage of a domain architecture is that it allows for easier management and development, as each domain is functionally isolated, allowing for specialization and optimization of ECUs within that domain. However, a significant drawback is that communication between different domains, such as powertrain and safety, requires more complex wiring and can introduce latency. Physical separation between ECUs in different domains may also limit fault tolerance across domains, as sharing resources in failure conditions is more difficult.

A *zonal architecture*, on the other hand, organizes ECUs based on their physical location in the vehicle, dividing the system into zones. For instance, the front zone of a vehicle would include ECUs handling headlights, front sensors, and climate control for the front passengers, while the rear zone would include ECUs for tail lights, rearview cameras, and rear passenger controls. Each zone houses ECUs that manage multiple functions from different domains. For example, an ECU in the front zone could handle both body control functions like headlights and safety functions like front sensors. In this approach, redundancy strategies like 2oo2 DFS or 2oo3 TMR can be applied within or across zones to improve fault tolerance and ensure continuous operation. One of the key advantages of a zonal architecture is that it reduces wiring complexity, as ECUs in the same zone can communicate and share resources locally, improving efficiency. Real-time communication between different systems within a zone is also faster, as there is less distance and latency in the connections. Moreover, zonal architectures allow for better scalability, as new zones or systems can be added without redesigning the entire vehicle architecture. However, this approach makes it more challenging to specialize ECUs for specific functions, since each ECU within a zone may need to handle multiple tasks across different domains, increasing the complexity of the system. Additionally, ensuring redundancy and fault tolerance for critical systems that span across zones can introduce challenges in managing inter-zone communication and resource sharing.

2.2.4 Dynamic Reconfiguration in Automated Systems

As systems become increasingly automated, dynamic reconfiguration-the ability to adapt the configuration of electronic control units (ECUs) in real time-plays a critical role in ensuring continuous operation [114,139]. This capability is particularly important in fail-operational systems, where safety and availability are paramount. Dynamic reconfiguration allows the system to respond to faults or failures by redistributing tasks or switching over to backup resources.

In highly automated systems, where there is no human fallback option, dynamic reconfiguration becomes crucial. As, e.g. autonomous vehicles, takes full control over critical functions, the system must have the ability to adapt in real time to any failures or faults [62]. This is especially important in fail-operational systems, where the vehicle must continue functioning even after a component failure, without relying on human intervention. The shift towards fail-operational systems requires transitioning from static E/E-architectures to more dynamic solutions that can handle real-time reconfiguration [114]. Current Electric/Electronic (E/E) architectures typically implement a fail-safe mode, where the driver must intervene in case of system failure. However, as the industry moves towards higher levels of automation (SAE levels 4 and 5), necessitating the development of fail-operational architectures that can dynamically reconfigure to ensure the system remains operational without human intervention. Dynamic reconfiguration is realised through both hardware and software adaptations, allowing ECUs to switch tasks or functions between redundant components. For example, in a fail-operational system for brake and steering control, multiple ECUs work together, and if one ECU fails, another ECU-previously in a hot standby mode-takes over the failed ECU's tasks [114].

Dynamic reconfiguration can be effectively modeled using **Markov processes**, which are suited to describe probabilistic transitions between system states based on failures, reconfigurations, or repairs [35,59]. An example of a Markov transition diagram of a 2-out-of-3 System with failure rates λ_i and repair rates μ_i can be found in Appendix 2, Fig. 12.

In such a system, the state of each ECU evolves based on transition probabilities that account for different operational conditions. The probability of transitioning from one state i (e.g., fully operational) to another state j (e.g., reconfiguration) at a time t is given by:

$$p_{ij}(t) = P(\text{transition from state } i \text{ to state } j \text{ at time } t) \tag{3}$$

This transition probability captures how likely the system is to reconfigure itself in response to detected failures.

The **reconfiguration time** is a critical factor, defined as the time it takes to detect a fault and switch control to the standby ECU. According to [114], the **fault reaction time** (FRT), which includes both fault detection and reconfiguration time, must be shorter than the **failure tolerance time interval** (FTTI) for the system to maintain safe operation. For example, in a braking system, if the FTTI is 160 ms, the reconfiguration process must occur within that time to prevent hazardous events, as outlined by ISO 26262 standards.

To further quantify this, the system's **transition rate matrix** Q describes the rate of transitioning between states, where q_{ij} represents the rate at which the system moves from state i to state j. For instance, the transition from normal operation to a reconfigured state (following ECU failure) can be described as:

$$q_{ij} = \lambda_{ij} e^{-\lambda_{ij} t} \tag{4}$$

where λ_{ij} is the rate of failure or reconfiguration. The system must then balance between fail-operational and fail-safe states based on real-time conditions.

In practice, dynamic reconfiguration ensures that critical systems like braking and steering remain operational, even in highly automated vehicles. [114] describes a three-ECU system for braking and steering, where the third ECU serves as a backup and is dynamically activated in the event of a failure in one of the primary ECUs. This approach provides both redundancy and flexibility, allowing the system to reconfigure in real time while adhering to strict timing requirements, as outlined in the failure detection Time and reconfiguration time tables provided in [114]. For example, if a nominal steering function is lost, the system must detect the failure within 30 ms and reconfigure within 100 ms to maintain safety and operational integrity.

By leveraging real-time monitoring, predictive maintenance, and Markov-based probabilistic modeling, the system can anticipate and mitigate failures before they escalate into critical issues. This aligns with the fail-operational concepts we have previously discussed, ensuring that the system continues to function even when individual components experience faults.

2.3 Simulation-Based Reliability Analysis

In reliability engineering, simulation techniques are critical tools for modelling, analysing, and improving system reliability [117, 121]. They allow engineers to predict the performance and failure behaviour of complex systems under different operating conditions, without the need for time-consuming or expensive physical testing. Simulation analysis is particularly relevant for safety-critical systems involving humans, where experimentation and testing is not possible for safety, ethical or cost reasons. A typical example is automated driving. In fact, the probability of a traffic accident per kilometre driven is extremely low, while testing and experimenting new systems in real traffic situations is difficult and expensive [81]. Generally speaking, simulation-aided analysis can help optimise maintenance schedules, component designs, and system architectures for maximum reliability or test different scenarios in a *what-if* analysis. Simulation techniques are especially relevant for analysing the reliability of dynamic mechanical systems that exhibit wear and fatigue effects, or more generally systems where the failure rates are time-dependent [37,48,117,121]. Indeed, such systems cannot be modelled using Markovian processes and are in general not analytically tractable. Simulations can also handle the complexity of real world systems with multiple interacting components, for instance by identifying critical components, where redesigns can enhance reliability [52,106]. Simulation techniques are also very useful for scheduling of complex maintenance strategies for repairable systems [34,42,151]. They can also incorporate sensor data and be used as digital twins for real-time monitoring of the system health [9,14,64,98,100].

Simulation techniques in reliability engineering showing particular innovations during the last decades include

- **Monte Carlo Simulation** (MCS), one of the most widely used simulation techniques in reliability engineering (see the reviews [2,116,136]). Monte Carlo techniques are particularly well suited to discrete-event simulation. MCSs use random sampling to simulate the behaviour (typically failure times) of complex systems, especially when analytical solutions are difficult or impossible. For example, they can address the structural reliability of multi-component systems including degradation processes [96,101] or with interdependent failure rates [13,79].
- **Fault Tree Analysis** (FTA) Simulation, a top-down, deductive failure analysis tool used to identify potential causes of system-level failures. FTA methods can be used to quantify the likelihood of different failure scenarios and to identify critical components [133]. FTA is particularly useful in safety-critical systems, such as aerospace or nuclear power, where understanding the root cause of potential failures is important [1,104,106], see also the bibliographic review [165].
- **Finite Element Method** (FEM), a numerical technique widely used in reliability engineering to analyse and predict the behaviour of physical systems described by (nonlinear) partial differential equations. FEM is especially relevant when dealing with complex structures, materials, and loading conditions [41,48,88]. FEM allows engineers through numerical schemes to break down complex systems into smaller, simpler parts (finite elements) to study stress, strain, heat distribution, vibration, and other physical phenomena that can affect the reliability of a system.
- **Multi-Agent Systems** (MAS), to analyse the reliability of complex systems where multiple autonomous components (or subsystems) must work together to maintain system reliability. MAS can dynamically adjust to changes, detect failures, and coordinate repairs or optimisations in a distributed and autonomous manner. They are generally based on a reinforcement learning processes for which model parameters are adjusted through feedback reward mechanisms [168]. Typical appli-

cations concern robotics and automated (smart) manufacturing [112,115,141]. MAS are especially suited for modelling emergent collective behaviours such as cascading failures [97,162].

2.3.1 Digital Twin

In reliability engineering, a digital twin is a dynamic, virtual model of a physical system that uses real-time data, simulation, and analytics to predict system performance and behaviour. The digital twin concept is becoming increasingly important as it enables more accurate, detailed, and real-time analysis of system reliability, including failure prediction, maintenance scheduling, and optimisation of system performance, see the reviews [9,145,146]. A digital twin is created using data from sensors, historical information, and models of the physical system, often differential models. This virtual model is continuously updated with real-time data from the physical counterpart, allowing engineers to monitor, analyse, and optimise the physical system in real time. In the context of reliability engineering, digital twins are particularly useful for:

- Monitoring system health and performance in real time [98],
- Predicting failures using predictive analytics [14,100],
- Optimising maintenance strategies (e.g., predictive maintenance) [42,151],
- Improving reliability by simulating and testing different configurations, conditions, scenarios or repair strategies [9,64,124].

Applications of digital twins in reliability engineering cover many areas, including aerospace and aviation, manufacturing, energy and power systems, automotive industry, smart cities and infrastructure, etc., see the reviews [16,64,100]. Typical examples are in automotive engineering for accident analysis and reconstruction of traffic accidents. The modelling of the vehicle structure and the deformation due to impacts and collisions can be performed at low cost using continuous physical models and FEM simulation analysis [44,113]. Another example of applications is the field of automated vehicles and advanced driver assistance systems. Here, the extremely low probability of accidents in traffic and the obvious safety concerns prevent testing and experimentation under real conditions and are generally carried out using numerical simulation, see, e.g., the review [36].

2.3.2 Accelerated Life Testing

For safety-critical technical products or systems, the term *reliability* is defined as a characteristic or behavioural trait that indicates the extent to which a specific function is performed in a reliable manner within a given time interval [17]. To determine the exact lifespan for purposes such as maintenance intervals, service policy or optimal operating conditions, it is essential to conduct targeted tests that accurately reflect actual failure behavior. Nevertheless, given that the anticipated product lifespan can span several years, accelerated life testing is employed to facilitate the assessment of the planned service life within a shorter time frame.

As part of the Design of Experiments (DoE), targeted stresses and loads are simulated by adjusting individual or multiple parameters in combination, thereby accelerating the ageing process or degradation. The challenge in simulating degradation or failure behavior lies in replicating or representing the causative factors of damage responsible in reality at an earlier point in time, despite the accelerating factors [135, 156]. A basic distinction is made here between qualitative High Accelerated Testing (HALT) and quantitative Accelerated Testing (ALT). While HALT aims to provoke failure causes as quickly as possible through overloading, ALT is conducted strictly within specified product specification limits, making it transferable to real-world conditions. Although sensor data can be collected during HALT, it only allows for qualitative assessment, whereas ALT can be used for quantitative modelling.

While [61] categorises the possible forms of accelerated life testing in more detail, [156] and [17] provides a methodological overview of the quantitative modelling types depending on the degradation or damage event. Specific applications include, for example, modelling the complex degradation behaviour of corroding lead anodes (see reference [29]), or deriving an acceleration factor (scuffing factor) relevant for the maintenance forecast of products from the construction industry that are designed to last longer than 19 years [45].

2.3.3 Scenario-Based Simulation

Scenario-based simulation is a widely used approach for evaluating the reliability of complex systems, such as those encountered in safety-critical domains [121, 143]. These systems are typically characterised by numerous interacting components and dynamic behaviour, leading to high-dimensional problems that are often NP-hard. The complexity of such systems can escalate rapidly, making it infeasible to derive general solutions analytically. Consequently, it is essential to constrain the analysis space and focus on identifying specific safety-critical scenarios. This process requires domain expertise to determine which scenarios pose significant risks to the system. For example, in the context of autonomous vehicles, understanding critical traffic situations such as lane changes or interactions at intersections is fundamental for ensuring system safety. Identifying these scenarios enables a targeted evaluation, facilitating more effective safety assessments [142, 143].

In [57], the authors categorise scenario methods into three main groups. **Intuitive Logics** is a qualitative approach, introduced by [80], relies on expert judgment to develop scenarios without using mathematical models. It focuses on generating diverse future environments to explore strategic alternatives and assess potential impacts, making it suitable for exploratory long-term planning. **Cross-Impact Analysis** uses probabilistic modeling to analyze interdependencies between influencing factors. Starting from initial probabilities for each factor, it computes how changes in one factor affect the probabilities of others using a cross-impact matrix. Scenarios are then generated based on the highest combined probabilities, making it a mathematically rigorous approach for identifying the most likely future developments. **Consistency-Based Approaches** evaluate the pairwise consistency between different projections to create scenarios that are logically coherent. By scoring how well different assumptions fit together, it identifies consistent sets of scenarios, ensuring internal alignment and plausibility, but without emphasizing likelihood of occurrence.

3 Challenges and Development Perspectives

Although sensor-based monitoring, system design and simulation-based reliability analysis have provided important advances and innovations in reliability engineering over the last decades, many challenges remain to be addressed. In this section we suggest some of the more important ones.

3.1 Trustworthy AI and Real-Time Data Processing

Ensuring that machine learning systems perform as expected, minimising risks and operating reliably under different conditions is crucial in advanced reliability methods. This aspect is commonly referred to as *Trustworthy* or *Safeguarding AI* in engineering [25,82,118]. With the increasing integration of machine learning algorithms into safety-critical systems (e.g., transportation or healthcare) and industrial automation, failures can lead to catastrophic outcomes. Therefore, it is necessary to quantify confidence and uncertainty in AI predictions and to explore the robustness of the algorithms in different situations, and especially rare ones. Safeguarding AI in reliability engineering includes various strategies and frameworks to ensure safe and consistent performance [146]. The following are some recently developed methods and concepts.

- **Robust and Redundant AI** Key considerations for safeguarding AI are robustness, ensuring that AI systems can handle unexpected inputs or disturbances without failing, and redundancy, when incorporating multiple AI algorithms or fallback mechanisms to ensure that if one system fails, another can take over. These aspects remain challenging. They can be addressed through Monte Carlo simulation, fuzzy logic and ensemble learning methods [103,136], and in system design through redundant, fail-safe, fail-operational or fault-tolerant architectures that drive AI systems to safe or operational states in case of failure.
- **Explainable and Transparent AI** The explainability and transparency are also key considerations, by building AI models that are interpretable so that engineers can understand why a certain decision was made, and by ensuring that the development, training, and operation of AI systems are documented [118]. This is especially important in safety-critical applications. Indeed, as AI models become more complex, particularly with the use of ensemble learning or deep learning, their decisions become harder to explain and trust in safety-critical applications.

A typical application concern the driving automation where AI algorithms for vehicle perception and motion planning quickly turn out to be black boxes whose operation is difficult to interpret. More research is needed into explainable AI (XAI) techniques to ensure that human operators can understand and trust AI-based systems in high-stakes environments [8,25,110].

- **AI Model Validation and Verification** Model validation is an important task of AI modules, by ensuring that the AI model performs well under various operational conditions and uncertainties [73]. Model verification, by checking that the AI system adheres to its specified requirements and does not deviate from expected behavior, even in dynamic or unpredictable environments, is also fundamental for trustworthy AI [36,94].
- **Handling of Rare Events** Another challenge is the handling of rare events. Indeed, AI systems often struggle with rare or outlier events, especially in reliability engineering where catastrophic failures are rare but highly impactful [2,134]. Developing models that can reliably predict rare events is still a challenge. Methods like extreme value theory, enhanced Monte Carlo simulations, or specialised outlier detection algorithms are needed to better handle these edge cases.
- **Scalability and real-time processing of AI Models** The scalability of AI models is also especially challenging. In fact, many AI models, such as networks or ensemble methods, are difficult to scale for large, complex systems with many interacting components [11,118,171]. There is an important need for more efficient algorithms and computational techniques to handle the large-scale, high-dimensional systems found in industries like aerospace, healthcare, and energy. Another challenge is to make AI model processing operational in real time. This challenge is particularly relevant for smart manufacturing, autonomous vehicles and other safety-critical dynamic systems.

3.2 Addressing System Complexity

Dealing with system complexity is an important task in reliability engineering. In fact, the operation of complex systems often consists of combination of different modules and components interacting with each other. The system architecture may also be complex and include state and time dependent mechanisms, for instance for redundant, safe tolerant or fail safe designs. To address the reliability of these systems, the complexity of the system must be assessed, taking into account the system architecture. This task is not straightforward, especially when the system modules and components are interconnected. The interdependence of the component failures, such as *common cause failures* and *cascading failures* makes the reliability analysis much more complex. Furthermore, the failure interdependency can drastically affect the system reliability [85], for example for the perception or motion planning of autonomous vehicles [56,79,86]. Indeed, dependent components require the system to be considered as a whole (i.e., in a multidimensional framework), since the components cannot be regarded individually. Several models have been developed for this purpose, including lifetime distribution models [83,111,169], system state models [93,109], failure interaction models [22,46], failure propagation models [32], see also the review [167] and references therein. However, the task remains challenging and is and currently under active development [46,79].

Another challenge in addressing system complexity relates to the diverse technologies involved in system reliability, particularly for cyber-physical systems. The combination of IoT, AI, robotics and digital control can lead to unforeseen failure modes that traditional reliability methods may not be able to address. Another challenging component of complexity is the collaboration between humans and machines in sociotechnical systems. This aspect is explored below.

3.3 Sociotechnical Systems

A sociotechnical system can be defined as a system consisting of two interacting components: a technical component and a social component (see, e.g., [148]). The concept emphasises that both components of an overall system have to be addressed when changes are introduced in one part - due to the complex interaction structure (see, e.g., [155]). Two aspects are mainly relevant in sociotechnical theory:

1. Interactions between technical and social subsystems determine the overall system success and
2. optimising either social or technical components without considering the other part or the interaction can lead to unintended, non-linear relationships diminishing overall system performance [155].

The ideas stem from original coal mining workplace studies, where technological innovations meant to improve working situations and increase performance, introduced unintended side-effects of changing social structures among the workers [147]. Outcomes were high fluctuation and sick-leaves instead of performance increases [95]. The theory has implications for any change in case of complex systems. Taking a look at the changes of industry 4.0, several aspects related to sociotechnical theory have to be taken into account.

First of all, the increasing digitalisation and interconnectedness through e.g. cyber-physical systems increases the amount of available and relevant information and secondly autonomous and intelligent processes operating in real time, as well as interactions with virtual reality displays create new strain on workers [54]. Automation for example, poses the challenge of changing the role of the workers from active controllers to monitors which in turn leads to degradation in skills that might be necessary when abnormal situations occur [10]. This -along with the notion, that humans are not well suited for prolonged vigilance tasks- has already been formulated in the *ironies of automation* [10]. The idea of compensating this challenge through training and getting the human to adapt to the new system challenges will not produce satisfactory outcomes. Instead, the sociotechnical systems approach shall be applied: before introducing technological innovations simultaneous estimation of social changes and influences on the human and possible countermeasures shall be addressed.

4 Statistical Learning Methods

In this section we review statistical, learning and simulation methods for advanced reliability techniques of complex systems discussed in Sects. 2 and 3. The methods include unsupervised techniques for data visualisation and clustering in Sect. 4.1. They also cover supervised techniques for prognostics, such as classification, regression and machine learning algorithms in Sect. 4.2. See also [17] for in-depth reliability engineering basics.

4.1 Unsupervised Learning Methods

We first consider a multivariate dataset X consisting of $n \in \{1, 2 \dots\}$ observations of $p \in \{1, 2 \dots\}$ variables (features)

$$X = \begin{bmatrix} x_1^1 & x_1^2 & \dots & x_1^p \\ x_2^1 & x_2^2 & \dots & x_2^p \\ \vdots & \vdots & & \vdots \\ x_n^1 & x_n^2 & \dots & x_n^p \end{bmatrix} \in \mathbb{R}^{n \times p}, \qquad \left| \begin{array}{ll} x_i = (x_i^1, \dots, x_i^p) \in \mathbb{R}^p, & i = 1, \dots, n \\ x^j = (x_1^j, \dots, x_n^j)^\top \in \mathbb{R}^n, & j = 1, \dots, p \end{array} \right. \tag{5}$$

We assume that the variables $(x^1, \dots, x^p)$ are dependent, i.e., the multivariate sample X has a structure that we aim to explore. In general, the data are sensor measurements of various physical quantities such as temperature, pressure, vibration, speed (e.g. rotation), noise, etc. The values describe the state of the system, e.g., a machine.

4.1.1 Principal Component Analysis

There exist many different graphical methods to visualise multivariate datasets, including parallel plots, radar charts, Andrews' plot [6] or again Chernoff's face [28]. However, the dimension of the dataset, i.e. the number of features p observed in the system, can be large, making it difficult to visualise the global structure of the data. In addition, several features may be highly correlated (e.g. noise with rotation speed) and the dimension of the dataset can be reduced without losing a significant amount of information.

Principal component analysis (PCA) can be traced back to the pioneering work of Karl Pearson and, later, Harold Hotelling [63]. The idea is to use a proper orthogonal decomposition of the data, the principal components, that maximises the amount of information. In this way, the first principal component is the best representation of the data in one dimension in terms of variability, the first two components the best representation in two dimensions, and so on. If the original multivariate dataset is highly correlated, the last components contain almost no information and can be

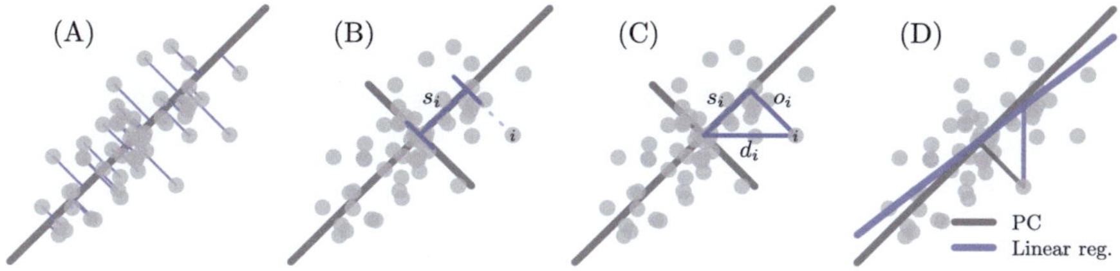

Fig. 6 Principles of the principal components (example in two dimensions). (**A**): orthogonal projections of the data. (**B**): maximisation of the information on the axe $\sum_i s_i^2$. (**C**): As the total variability $\sum_i d_i^2 = \sum_i s_i^2 + \sum_i o_i^2$ is conserved, maximising $\sum_i s_i^2$ leads to minimising the squared orthogonal distances $\sum_i o_i^2$. (**D**): Principal component and linear regression generally do not match

neglected. Therefore, the dimensionality of the dataset can be reduced by using only the first components.

Principal components are orthogonal linear combinations of the original variables. Geometrically, we maximise the variability of projections of the data (see Fig. 6). Thanks to the Pythagorean theorem, the total variability is conserved and maximising the variability on the axes resumes to minimising the sum of the squared orthogonal distances to the axes. This differs from the technique used in linear regression, which minimises the squared difference between a linear model and the data. For this reason, PCA is sometimes referred to as orthogonal regression.

Analytically, the principal components can be determined using basic linear algebra. Let $\tilde{X}$ be the standard score transformation of the data with mean zero and variance one over the p features and $\Gamma = \frac{1}{n}\tilde{X}^\top\tilde{X} \in \mathbb{R}^{p\times p}$ the variance/covariance matrix, i.e., $\Gamma_{j,k} = \text{covar}(\tilde{x}^j, \tilde{x}^k)$. Then, thanks to the diagonal formulation $\Gamma = P^{-1}DP$, where $D \in \mathbb{R}^{p\times p}$ is the diagonal matrix of the ordered eigenvalues of Γ and $P \in \mathbb{R}^{p\times p}$ is the matrix formed by the associated eigenvectors, the principal components are directly the columns of the matrix $\tilde{X}P \in \mathbb{R}^{n\times p}$. An example of application of PCA on measurements of banknotes is given in Appendix 3[4].

4.1.2 Clustering Techniques

Clustering techniques are unsupervised data mining methods designed to explore multivariate datasets and to identify patterns and clusters. In principle, the characteristics of the observations in a given cluster are more similar than those of the observations in the other clusters. Clustering techniques can be distinguished according to the following criteria

- Monitored and unmonitored clustering, whether the number of clusters is known or unknown
- Strict clustering, where each observation belongs to exactly one cluster

[4] The principal component analysis of banknote measurements can also be computed online at https://colab.research.google.com/drive/1D-Z-lWD_Z74Ts88EQfEjc2W1BLZBdqNw?usp=sharing.

- Clustering with outliers, where some observations can also belong to no cluster
- Overlapping clustering, where observations can belong to more than one cluster
- Fuzzy clustering, where each observation belongs to each cluster to some extent
- Hierarchical clustering, where observations of a child cluster also belong to the parent cluster
- Centroid clustering, where clusters are represented by a centroid (mean value)

Generally speaking, non-parametric clustering are empirical methods based on distance metrics between the observations or kernels. Parametric clustering rely on multivariate probabilistic distributions, typically multivariate normal distributions.

K-means Clustering

K-means algorithm in a non-parametric monitored strict clustering technique where a cluster is represented by a centroid. The terminology *K-means* dates from the late 1960s [102] although the method has been used since the 1950s. Consider $(x_1, \dots, x_n) \in \mathbb{R}^{n,p}$ observations and a partition of clusters $S = \{S_1, \dots, S_K\}$ where $\mu_1, \dots, \mu_K$ are the means on each cluster. Then K-means clustering is the partition that minimises the inter-cluster variability:

$$S_{K\text{-means}} = \arg\min_S \sum_{k=1}^{K} \sum_{i \in S_k} \|x_i - \mu_k\|^2. \tag{6}$$

In two dimensions, the K-means partition corresponds to a Voronoi diagram. Note that since the total variability is constant for any clustering, minimising the inter-cluster variability is equivalent to maximising the intra-cluster variability.

The K-means problem is computationally NP-hard difficult. However, efficient heuristic algorithms quickly converge to a local optimum. For instance, the **Lloyd-Forgy algorithm** is an iterative process consisting of three steps:

1. initialisation: Choose K random mean values $u_1^1, \dots, u_K^1 \in \mathbb{R}^p$.
2. Assignment: Each data object is assigned to the cluster with the smallest variance increment.

$$S_k^t = \left\{x_i : \left\|x_i - u_k^t\right\|^2 \le \left\|x_i - u_{k*}^t\right\|^2 \text{ for all } k^* = 1, \dots, K\right\},\ k = 1, \dots, K. \tag{7}$$

3. Update: Recalculate the centres of the clusters:

$$u_k^{t+1} = \frac{1}{|S_k^t|} \sum_{i \in S_k^t} x_i, \qquad k = 1, \dots, K. \tag{8}$$

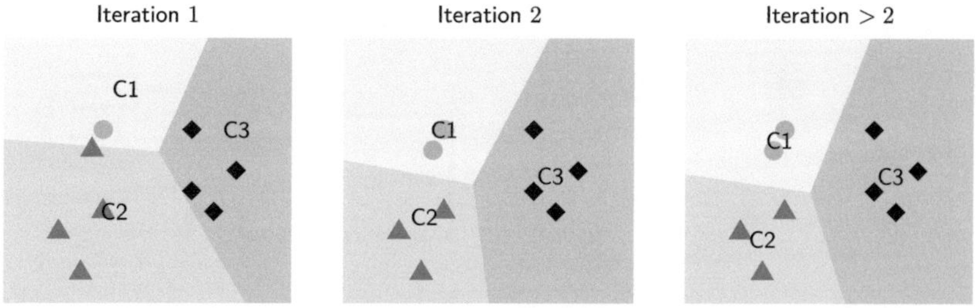

Fig. 7 Iteration of the Lloyd-Forgy algorithm for the K-means clustering with $n = 9$ synthetic bivariate observations. The algorithm converges in three iterations

Steps 2–3 are repeated until the assignments no longer change. Figure 7 shows an illustrative example with $n = 9$ synthetic bivariate observations where the Lloyd-Forgy algorithm converges in three iterations. The number of iterations before convergence depends on the initial mean values chosen. In this example, the achieved local optimum is also the global optimum.

Hierarchical Clustering

Hierarchical methods are unmonitored clustering based on tree representations. The top of the tree is a single cluster with all observations while each observation is a cluster in the bottom of the tree (the leaves). Hierarchical methods are based on a distance metric (e.g., Euclidean, Manhattan, Mahalanobis, etc.), and a linkage between two clusters (centroid distance, min/max distance, etc.) from which we can build an objective function. In general, the problem of finding the optimal hierarchical clustering, i.e., the one that minimises a global objective function is NP-hard. This means that as the size of the data grows, finding the exact optimal hierarchical clustering for large datasets becomes computationally infeasible because the number of possible hierarchical clustering grows exponentially with the observation number.

Agglomerative and divisive methods are efficient heuristic algorithms that produce good hierarchical clustering in practice, but they do not guarantee the globally optimal solution due to the NP-hardness of the problem. Agglomerative clustering methods (ACM) are bottom-up approaches, in contrast to top-down divisive methods. In agglomerative methods, each observation is initially a cluster. ACM consists of repeatedly merging the two closest clusters until the number of clusters is greater than one. A dendrogram is a common graphical representation of hierarchical clustering. Dendograms are trees with the linkage distances in the x coordinate (branches) and observations in the y coordinate (leaves). The section of the dendogram allows the number of clusters and the members of each cluster to be determined. A basic ACM with Euclidean distance and centroid linkage for seven synthetic bivariate observations and its dendogram representation is given in Fig. 8. Here the shape of the dendogram suggests a configuration with two clusters: $\{A, B, C\}$ and $\{D, E, F, G\}$, or with three: $\{A\}$, $\{B, C\}$ and $\{D, E, F, G\}$, as shown in the figure.

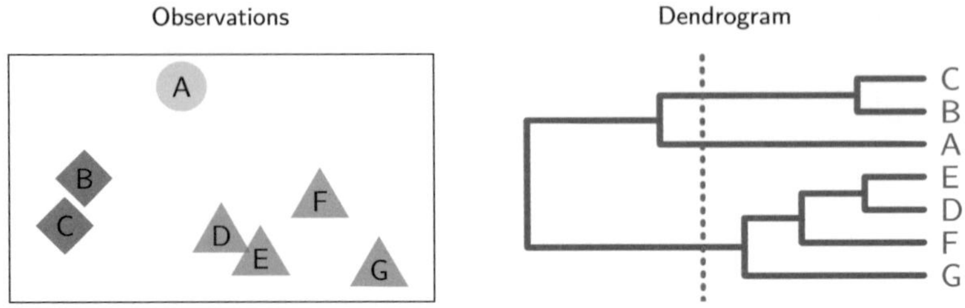

Fig. 8 Seven bivariate observations (left panel) and their dendogram representation from an agglomerative clustering method with Euclidean distance and centroid linkage, with a section for three clusters: $\{A\}$, $\{B, C\}$ and $\{D, E, F, G\}$

Density-Based Clustering

Density-based clustering techniques are non-parametric methods based on empirical data probability density functions, typically using kernel density estimates. These algorithms are designed to identify regions where data points are densely packed and separate them from regions of lower density including noise and outliers. Density-based clustering problems are considered NP-hard in the general case, especially when the goal is to find the globally optimal clustering. However, heuristic methods offer practical solutions without necessarily finding the globally optimal clustering. Popular density-based non-parametric clustering heuristics were developed in the late 1990s, including mean shift [27], DBSCAN (Density-Based Spatial Clustering of Applications with Noise) [43] or OPTICS (Ordering Points to Identify the Clustering Structure) [7] algorithms.

The mean shift algorithm iteratively shifts points towards the mode (peak) of a density function until convergence [27]. Clusters form around these modes. A bandwidth parameter controls the size of the window used to shift points. Similarly, data points are clustered based on the density of points in a neighbourhood with the DBSCAN algorithm [43]. This technique is controlled by two parameters: the neighbourhood search radius and the minimum number of points required to form a dense region (i.e., a cluster). The OPTICS algorithm is an extension of DBSCAN where the radius for neighbourhood radius is adaptive [7]. Oppositely to DBSCAN, OPTICS allows detecting meaningful clusters in data of varying density.

Unlike K-means and, to a lesser extent depending on the metric and linkage used, hierarchical algorithms which mainly identify circular clusters, density-based clustering methods are designed to find clusters of arbitrary shape, including non-convex clusters[5]. In addition, density-based clustering algorithms do not require the number of clusters to be defined. However, density-based models are generally sensitive to parameter values, especially the bandwidth and clustering threshold.

[5] A comparison of K-means, AHM and mean-shift clustering algorithms on synthetic bivariate datasets including circular and non-convex clusters can be computed online at https://colab.research.google.com/drive/1vVzTL_Cf4kUUnbSpAZ4vVmjPKlgAljsC?usp=sharing.

Parametric Clustering

Parametric clustering refers to a type of clustering where the data is modelled using a fixed number of parameters, such as the number of clusters or parameters related to the shape of the clusters. Often the data are assumed to follow a particular mixture distribution model, typically multivariate Gaussian distributions. The clusters may have a particular shape, like spherical or ellipsoidal, depending on the parametric model. The main objective is to fit the parameters of the model to the data in a way that best describes the underlying clusters.

For a Gaussian mixture model with multivariate normal probability density function f, the parameters $\theta = (\mu_k, \sigma_k, \pi_k, k = 1, \ldots, K)$ are, besides the number of clusters K, the expected values μ_k and standard deviations σ_k as well as the proportions of observations per cluster π_k. The likelihood of the model is

$$\mathscr{L}_\theta(X) = \prod_{i=1}^{n} \sum_{k=1}^{K} \pi_k f(x_i, \mu_k, \sigma_k). \tag{9}$$

Local maxima can be obtained using the expectation-maximisation heuristic [107], variational Bayesian methods, or randomised algorithms such as Gibbs sampling. In addition, the number of clusters can be selected using information criteria such as AIC, BIC or likelihood ratio.

4.2 Supervised Learning Methods

In the context of supervised learning methods the multivariate dataset reads

$$X = \begin{bmatrix} x_1^1 & x_1^2 & \ldots & x_1^p \\ x_2^1 & x_2^2 & \ldots & x_2^p \\ \vdots & \vdots & & \vdots \\ x_n^1 & x_n^2 & \ldots & x_n^p \end{bmatrix} \in \mathbb{R}^{n \times p}, \qquad Y = \begin{bmatrix} y_1 \\ y_2 \\ \vdots \\ y_n \end{bmatrix} \in \mathbb{R}^n. \tag{10}$$

The observations X are the set of explanatory variables (sometimes called independent variables, predictor variables, regressors, covariates, features, or input variables), while Y is the variable to be explained (sometimes called the dependent variable, response variable, predicted variable, or output variable), which we assume to be univariate for simplicity. Typically, the data are sensor measurements of various physical quantities such as temperature, pressure, vibration, speed (e.g. rotation), noise, etc. of a machine, and Y is the operational state of the machine or its remaining useful life.

Supervised learning methods are algorithms that aim to predict Y from X. To this end, the parameters of the algorithm are inferred from the data, hence the learning process. Classification methods refer to the case where Y is discrete or categorical, e.g. a class such as the qualitative state of the machine. While regression methods refer to the prediction of continuous numeric quantities, typically remaining useful life.

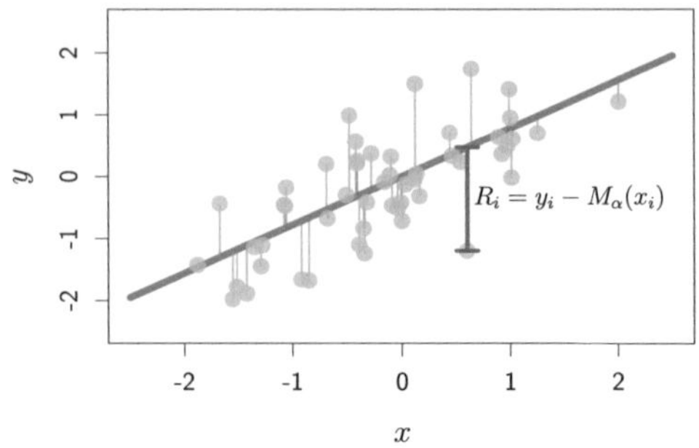

Fig. 9 Illustrative scheme of a univariate linear regression model. The parameters α are calibrated by minimising the squared difference between the model $M_\alpha(x_i)$ and the observation y_i. This corresponds to minimising the variability of the residuals $R_i = y_i - M_\alpha(x_i)$

Machine learning methods refer to (nonlinear) algorithms that are flexible enough to be used for both classification and continuous prediction.

4.2.1 Multivariate Regression Techniques

We focus first on regression methods where the variable to be predicted is numerical and continuous. The algorithms are models given by

$$y_i = M_\alpha(x_i) + \sigma\varepsilon_i, \tag{11}$$

where α is a set of parameters (coefficients) to be calibrated and ε_i is a centred random variable with amplitude σ corresponding to the unexplained part of the models. For instance, a linear model reads

$$M_\alpha(x_i) = \alpha_0 + \alpha_1 x_i^1 + \cdots + \alpha_n x_i^n. \tag{12}$$

The objective is to calibrate the parameters in order to fit the data as well as possible, within the limits of the model's shape. In regressive models, the coefficients are estimated by minimising the variability of the residuals $R_i(\alpha) = y_i - M_\alpha(x_i)$, i.e. the difference between the model and the observation (see Fig. 9). This is

$$\tilde{\alpha} = \arg\max_\alpha \sum_{i=1}^{n} \left(y_i - M_\alpha(x_i)\right)^2. \tag{13}$$

For linear models, it is easy to check that

$$\tilde{\alpha} = \left(X^\top X\right)^{-1} X^\top Y, \tag{14}$$

while the unbiased estimate of the noise amplitude is given by

$$\tilde{\sigma}^2 = \frac{1}{n-p-1} \sum_{i=1}^{n} R_i^2(\tilde{\alpha}). \tag{15}$$

For example, for a univariate linear model $M(x) = ax + b$, $a, b \in \mathbb{R}$, the least squares estimate is given by $\tilde{a} = \text{covar}(x, y)/\text{var}(x)$ and $\tilde{b} = \bar{y} - \tilde{a}\bar{x}$. This least squares estimate maximises the likelihood if the residuals are independent and identically normally distributed. It is in general a good estimate for compact distributions, but is sensitive to extreme values and outliers. Fisher tests can be used to assess the influence of a single feature, a group of features, or the linear model as a whole. Nonlinear models can be handled by using invertible nonlinear transformations. For instance, linear regression of a log/log transformation $\hat{x} = \log(x)$ and $\hat{y} = \log(y)$ leads to the exponential model

$$M_\alpha(x) = e^{\alpha_0} \cdot (x^1)^{\alpha_1} \dots (x^p)^{\alpha_p}. \tag{16}$$

The logarithmic model

$$M_\alpha(x) = \log(\alpha_0 + \alpha_1 x^1 + \dots + \alpha_p x^p) \tag{17}$$

can be treated with the exponential transformation $\hat{y} = e^y$. More generally, regressions for differentiable nonlinear models are based on linearisation and iterative heuristics, such as the Gauß-Newton algorithm, which converge to local optima.

The quality of the fit can be assessed by partitioning the total variability SST into the variability explained by the model SSM and the variability of the residuals SSR, where

$$SST = \sum_{i=1}^{n} (y_i - \bar{y})^2, \quad SSM = \sum_{i=1}^{n} (\bar{M} - M_{\tilde{\alpha}}(x_i))^2, \quad SSR = \sum_{i=1}^{n} (y_i - M_{\tilde{\alpha}}(x_i))^2.$$

Indeed, we have $SST = SSM + SSR$ when the residuals are centred. Therefore, minimising the variability of the residuals leads back to maximising the variability of the model. In addition, the proportion of variability explained

$$R^2 = \frac{SSM}{SST} = 1 - \frac{SSR}{SST}, \tag{18}$$

known as the coefficient of determination, can be used to quantify the quality of the fit. We have $R^2 \in [0, 1]$ for centred residuals. The coefficient of determination is equal to one if the total variance is recovered by the model, while it is equal to zero if the variability of the model is equal to zero. Note that for univariate data, $R^2 = \text{cor}^2(x, y)$.

4.2.2 Classification Techniques

In this section, we focus on statistical classification methods where the variable to be predicted is discrete.

Logistic Regression

The logistic regression is a binary classification algorithm for which $Y \in \{0, 1\}^n$. The first mention of the logistical model can be traced back to the work of Berkson in 1944 [12]. In the logistic model, the membership of an observation to a class is quantified with the help of the sigmoid (logistic) function

$$\tilde{y}_i = \pi_i(\alpha) = \frac{1}{1 + \exp\big(- (\alpha_0 + \alpha_1 x_i^1 + \ldots + \alpha_p x_i^p)\big)}. \tag{19}$$

This quantity lies between 0 and 1 with a relatively high transition speed: the logistic function tends quickly to zero if $\alpha_0 + \alpha_1 x_i^1 + \ldots + \alpha_p x_i^p \ll 0$ while it fast tends to one if $\alpha_0 + \alpha_1 x_i^1 + \ldots + \alpha_p x_i^p \gg 0$. The scalars $(\alpha_0, \alpha_1, \ldots, \alpha_p)$ are the $p + 1$ parameters (coefficients) of the logistic model to be estimated and interpreted. Note that binary variables can be generalised to categorical variables when there are more than two possible values, and the binary logistic regression can be generalised to multinomial logistic regression.

The parameters of the binary logistic regression can be estimated by maximising the likelihood

$$\mathscr{L}(\alpha) = \prod_{i=1}^{n} \pi_i(\alpha)^{y_i} (1 - \pi_i(\alpha))^{1-y_i} \tag{20}$$

using heuristics, typically gradient methods. In practice, regularisation is often added to the optimisation problem to prevent overfitting, especially when dealing with high dimensional data. The parameters (coefficients) in logistic regression reflect the log-odds of the outcome as a function of the input features. The intercept α_0 represents the log-odds of the outcome when all the input features are zero. If α_0 is large and positive, it indicates that the predicted probability of the outcome is high when all feature values are zero, and vice versa. For a positive coefficient α_j, the feature x^j increases the likelihood of the outcome $y = 1$. The higher the value of the feature, the higher the probability of the positive outcome. The opposite happens for negative coefficients. To make the coefficients easier to interpret, they are often exponentiated to convert them from log-odds to odds ratios.

Linear Discriminant Analysis

Linear discriminant analysis consists of a linear transformation of the explanatory variables to maximise the inter-class variability or, equivalently, to maximise the intra-class variability. The principle is similar to that of the principal component analysis. The optimal linear transformation can be obtained explicitly by matrix calculus, using the variance/covariance matrix. Linear discriminant analysis deals with binary output data with two classes and can be extended to multi-class analysis [120]. In fine, the binary classification resumes to a linear discriminant

$$D(\alpha) = \alpha_0 + \alpha_1 x^1 + \cdots + \alpha_p x^p. \tag{21}$$

In the standard score transformation of the data, the belonging of an observation to one of the two classes is given by the sign of the discriminant. As in logistic regression, the coefficient estimates allow the influence of a given feature on the class to be interpreted. Note that the linear discriminant analysis is much more sensitive to extreme values and outliers than the logistic regression.

Naive Bayes Algorithm

Unlike previous logistic and discriminant classifiers, the naive Bayes algorithm is directly designed for multi-class problems. The classifier is based on the Bayes theorem

$$\mathbb{P}(Y|X) = \frac{\mathbb{P}(Y)\mathbb{P}(X|Y)}{\mathbb{P}(X)}, \qquad \mathbb{P}(X) > 0, \tag{22}$$

where Y is the (latent) variable to be predicted and X the (observable) explanatory variables. A natural estimation for Y is the value that maximise its probability $\mathbb{P}(Y|X)$ conditionally to the observations. In (22), $\mathbb{P}(X)$ is a constant that does not depend on Y while $\mathbb{P}(Y)$ can be estimated using empirical means $\tilde{p}_k$. The objective is therefore to estimate the conditional probability $\mathbb{P}(X|Y)$. In the naive Bayes algorithm, the features are assumed to be independent and normally distribution so that the prediction is given by

$$\tilde{y}_i = \arg\max_k \; \tilde{p}_k \prod_{j=1}^{p} f(x_i^j, \mu_j^k, \sigma_j^k) \tag{23}$$

with f the normal probability density function. The $2Kp$ parameters (μ_j^k, σ_j^k) can be estimated by maximum likelihood, i.e., by least square estimates over the classes as the distributions are assumed normal and independent. Similarly to the linear regression, the Bayes classifier is sensitive to extreme values.

Decision Tree

Logistic regression, linear discriminant analysis and naive Bayes classifier are linear algorithms providing limited classification for groups whose shape does not allow differentiation by hyperplanes. In contrast, decision trees are based on recursive binary partition rules that can recover nonlinear features. The goal is to split the data in a way that results in the most homogeneous subgroups (in terms of the target variable). As Bayes classifier, they are natural designed for multi-class problem and categorical data. Furthermore, they are easy to interpret and visualise using dendograms, especially for small trees, and don't assume any specific data distribution, making them flexible for various tasks. Decision trees also naturally rank the importance of features based on how often and effectively they are used to split the data. However, they are due to splitting iteration computationally expensive, can lead to overfitting (that can be mitigated by pruning the tree or setting stopping criteria) and can prove unstable (i.e. sensitive to particular observations).

4.2.3 Machine Learning Techniques

Machine learning techniques are statistical algorithms that can learn from data and generalise to make predictions. However, in contrast to the classical statistical models described above, machine learning algorithms are almost not constrained by the shape of the model [20, 128]. In fact, they contain many more coefficients, giving them great malleability and allowing them to approximate a wide variety of relationships, including complex nonlinear and categorical ones. The learning mechanisms shape the algorithms through the data, empirically identifying the underlying mechanisms at work. Unlike traditional statistical approaches, the predictive power comes from the data rather than the shape of the model. In this sense, machine learning approaches are not based on a theory used to formulate a model. They have no modelling bias and can be applied to a wide range of problems, including classification and regression. However, although their predictive ability can be much more accurate than classical statistical models, they do not allow the influence of the features to be understood directly by interpreting the parameter values, as can be done with models. For this reason, they are often referred to in the literature as *black boxes*.

The learning processes that allow the data to be fitted and the coefficients to be calibrated are fundamental to machine learning algorithms. This is the model training phase, formulated as a multidimensional optimisation problem. The objective is to minimise a prediction error function, which can be a squared error or an entropy function. The size of the problem, i.e. the number of coefficients to be calibrated, is very large (hundreds, even thousands if not more), so training is performed systematically using iterative heuristics, typically backpropagation methods, which compute the gradient of a loss function with respect to the network weights. Tuning the complexity of the algorithms and their hyperparametrisation is, in the absence of theory, an open problem that needs to be specified according to the input data. Hyperparameters are, for example, the number of layers and neurons in artificial neural networks, the number of support vectors in support vector machines or the number of decision trees in random forests. Although theoretical tools exist for defining the optimal complexity of algorithms, such as the Vapnik-Chervonenkis dimension, they are often difficult to apply. In practice, determining the optimal complexity of algorithms is usually determined empirically by partitioning the data using cross-validation techniques, where part of the data is used to train the algorithms while the remaining part is used to test the predictions on new data. These operations can be repeated by randomly partitioning the data in training and testing to quantify the uncertainties of the prediction in so-called bootstrapping techniques. If the training error function decreases systematically as the complexity of the algorithms increases, the test error decreases before reaching a minimum and then increases. The minimum corresponds to optimal complexity, the decreasing part to insufficient complexity, known as underfitting, and the increasing part to overparametrisation, known as overfitting, which prevents generalisation.

Artificial Neural Network

Inspired by the way a brain works, artificial neural networks (ANN) are networks of fully connected neurons organised in layers (see Fig. 10, left panel). The first artificial neural network algorithms can be traced back to the pioneering work of Lettvin et al. on the eye and brain of a frog [91]. Each neuron is a activation function (e.g., a sigmoid) on a linear combination of the input, calibrated by weighting coefficients.

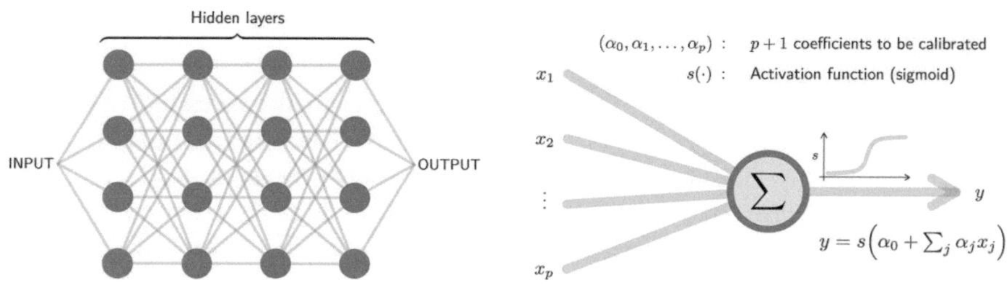

Fig. 10 Scheme of an artificial feed-forward neural network. Left panel: Layer structure of fully connected neurons. Right panel: Activation operation of a neuron

Since neurons are fully connected, the coefficient number is the dimension of the input plus one times the number of neurons. Using a logistic activation function, a single neuron network is simply a logistic model.

The malleability of the algorithm arises through the superposition of many layers and neurons. In fact, any relationship can approximated using a neural network, provided the network is sufficiently complex. ANNs have found applications in many disciplines, including regression analysis and classification. Different types of architecture are used in practice, depending on the discipline. The simplest networks are feed-forward networks in which information propagates in one direction only. Deep neural networks Recursive approaches include loops that can operate at different scales, as in long-short term memory (LSTM) networks. These approaches are well adapted to the analysis of time series. Convolutional neural networks consist of a succession of subsampling of overlapping inputs and convolution, which is particularly well suited to image processing. In addition, networks are said to be deep when the number of layers (and neurons) is large.

Support Vector Machine

Artifical neural networks consists of a tangle of nonlinear activation functions where the phase of training requires using approximation heuristics. In contrast, support vector machines (SVM) rely on combinations of optimal linear hyperplane in a high dimensional space, making them analytically tractable. The optimisation is carried out by maximising margins between the closest data points of different classes (i.e., the hyperplane and support vectors). It focuses on the points that are hardest to predict. This makes SVMs to be resilient to noisy data, extreme values and outliers unlike classical linear statistical models such as linear discriminant, naive Bayes classifier, or linear regression.

As ANNs, SVMs can be used for both classification and regression [38], although the algorithm was initially designed for classification [15]. Indeed, the used of multiple hyperplane make it possible the fitting of various relationships, whether nonlinear or categorical. In addition, similarly to nonlinear regression models, the input space can be rescaled using nonlinear transformations. This allows the algorithms to fit maximum-margin hyperplane in a transformed feature space that is nonlinear in the original input space and so addressing complex data structures. Another approach to treat data that is not linearly separable, is to use a technique called the kernel trick projecting the data into higher-dimensional space, where it becomes linearly separable [15]. Common kernels include polynomial, Gaussian and radial basis function kernels. SVMs are especially effective in high-dimensional spaces and when the number of dimensions exceeds the number of samples. They are also memory efficient, as it uses only the support vectors. However, SVMs can be less effective with large datasets because training complexity can become high. In addition choosing the right kernel and hyperparameters (e.g., number of support vectors) can be tricky.

Ensemble Learning
Ensemble methods in machine learning are techniques that combine the predictions of multiple algorithms (often referred to as "weak learners" or "base learners") to produce a more accurate and robust prediction than any single algorithm. The key idea behind ensemble methods is that by leveraging the diversity of algorithms, the ensemble can correct for the weaknesses of individual models and improve overall performance. Indeed different models may capture different patterns or aspects of the data, and combining them tends to compensate the weakness and reduce errors. The main types of ensemble methods include bagging methods (bootstrap aggregating), stacking (or stacked generalisation) and boosting.

- **Bagging methods** [18] reduce variance by training multiple instances of the same algorithm on different random subsets (bootstrapped samples) of the training data, and then averaging their predictions. Random forest is a popular bagging algorithm that trains multiple decision trees and averages their results (for regression) or votes on them (for classification).
- **Stacking methods** [19, 159] involve training several basic algorithms and then using a *meta-heuristic* to combine their predictions. The meta-heuristic learns how best to aggregate the outputs of the base models. A common setup is to use random forests, SVMs, and neural networks as base models, and then train a linear regression or another model to combine their outputs.
- **Boosting methods** aim to reduce bias by training algorithms sequentially, focusing on the errors made by the previous algorithms, and progressively improving the prediction. AdaBoost (Adaptive Boosting) [51] and Gradient Boosting [53] are common boosting algorithms. In AdaBoost, misclassified data points are given higher weights in each iteration, while Gradient Boosting iteratively optimises the loss function.

Ensemble learning algorithms are particularly suited to the analysis of complex multivariate datasets. They are widely used in both academic research and industrial applications. Ensemble methods, such as bagging, can help reduce overfitting by averaging out the biases of individual models, leading to a more general and robust prediction algorithm. Boosting techniques can reduce bias by focusing on correcting the errors made by previous algorithms. These two strategies help in balancing the trade-off between bias and variance in machine learning models. However, training multiple algorithms requires more computational resources and time than training a single algorithm. In addition, ensemble methods, combining many algorithms, are difficult to interpret.

Applications of different algorithms for the prediction of pedestrian speeds in single-file movement is given for illustration in Appendix 4[6].

Appendix

Appendix 1: Examples of Fail-Operational M-out-of-N Architectures

We detail two examples of fail-operational M-out-of-N electronic control unit architectures in this appendix: a *2-out-of-2 Dual Fail Safe* (DFS) architecture and a *Tri-modular 2-ouf-of-3* (TMR) architecture.

The *2-out-of-3 (TMR)* architecture involves a redundancy scheme where three independent processing paths receive sensor information from the same sources (e.g., camera, radar, lidar), see Fig. 11b . A majority decision is made between the three independent processing paths, and the system continues to operate as long as at least two of the three paths remain functional, ensuring fail-operational behavior. However, a drawback of the 2oo3 architecture is that all three processing paths typically reside within the same ECU. This makes the system more vulnerable to common-cause failures, such as hardware defects or environmental impacts affecting the entire ECU.

In contrast, the *2-out-of-2 Dual Fail Safe (DFS)* architecture consists of two independent fail-safe subsystems distributed across two separate ECUs, see Fig. 11a . These subsystems communicate with each other in isolation, and in the event of a fault, one ECU can take over the functionality of the other. A key advantage of the 2oo2 architecture is the greater ability to implement separation and diversification, as the two ECUs can be physically isolated and use different technologies, which significantly reduces the risk of common-cause failures. However, a drawback is that once one ECU fails, the system is no longer fail-operational and can only maintain fail-safe behavior, requiring restoration of the faulty ECU for full functionality.

Appendix 2: Markov Transition Diagram of a 2-out-of-3 System

We present in this appendix the Markov transition diagram of a 2-out-of-3 System with 4 operational modes, see Fig. 12.

---

[6] Multi-algorithm predictions of pedestrian speeds can be computed online at https://colab.research.google.com/drive/1zGHJQbosu-gZPB8HUaf0GVHSJb_hdcbG?usp=sharing
Additionally, multi-algorithm predictions for other datasets including discrete lane-changing maneuvers on highways and RUL aircraft engine forecasting can be computed online at https://colab.research.google.com/drive/1R_p1oJfP_PRke9eYXxRCpknQXTiUuh8W?usp=sharing
https://colab.research.google.com/drive/1KSbi3Lms9hSTx-pnceS0SjltV1XTm7ZU?usp=sharing.

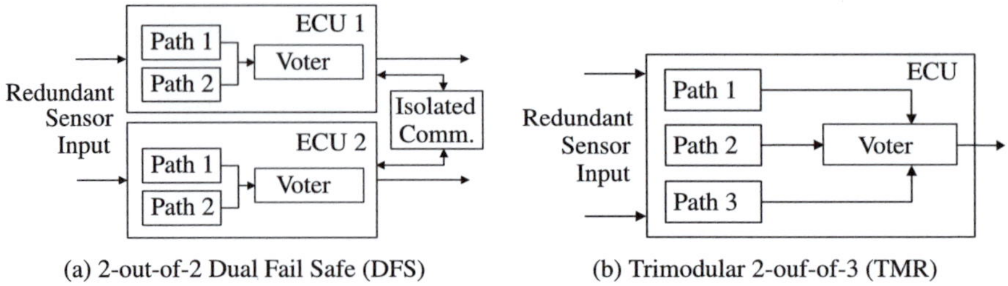

Fig. 11 Fail-Operational M-out-of-N Electronic Control Unit (ECU) Architectures [77,78]

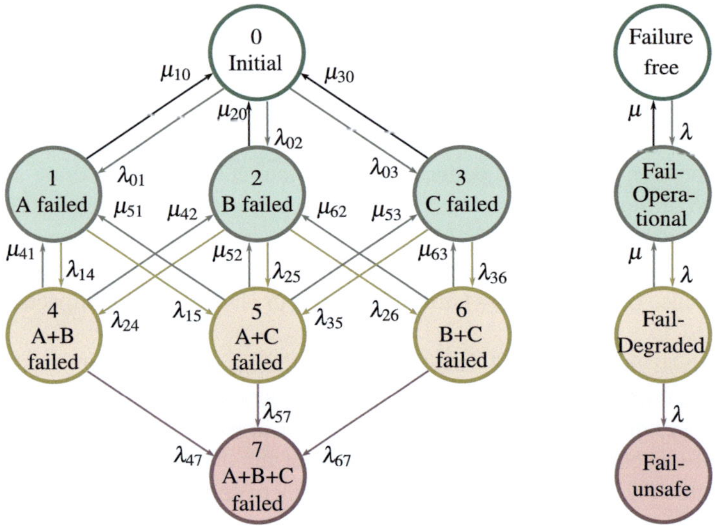

Fig. 12 Markov transition diagram of a 2-out-of-3 System with 4 operational modes [59]

Appendix 3: Principal Component Analysis of Banknote Measurements

As an illustrative example, we analyse the principal components of a dataset of six measurements made on 100 genuine and 100 counterfeit old Swiss 1000 franc banknotes [49] (see Fig. 13)[7]. The height measurements $x^2, \ldots, x^5$ are all positively correlated while the banknote width x^1 is more independent. In addition, the diagonal measurement x^6 is negatively correlated with all the height measurements $x^2, \ldots, x^5$ and slightly positively correlated with the banknote width x^1.

The pairwise scatter plots of the six measurements are shown in Fig. 14. The blue circles are the genuine banknotes, while the grey triangles are the counterfeits. Two clusters of observations corresponding to the genuine and counterfeit notes

[7] The principal component analysis of the banknote measurements can be computed online at https://colab.research.google.com/drive/1D-Z-lWD_Z74Ts88EQfEjc2W1BLZBdqNw?usp=sharing.

Fig. 13 Multivariate dataset of six measurements made on 100 genuine and 100 counterfeit old Swiss 1000 franc banknotes [49]. Left panel: six measurements made. Top right panel: Boxplots of the normalised measurements. Bottom right panel: Correlation table of the measurements

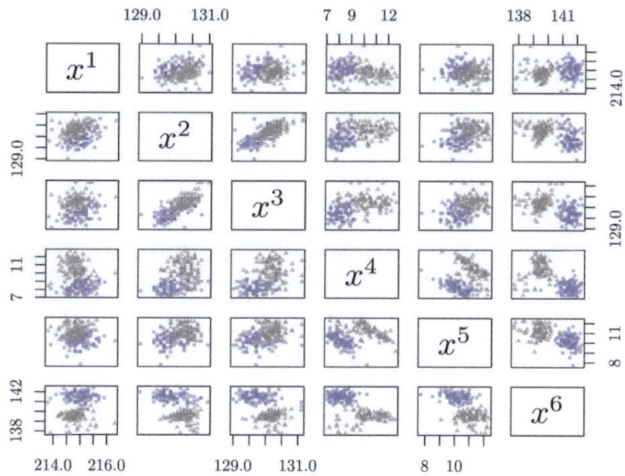

Fig. 14 Pairwise scatterplots of the six measurements on the notes. The blue circles are the genuine banknotes while the grey triangles are the counterfeits

can be identified, particularly using the diagonal measurement x^6. However, it is not straightforward to determine whether a given banknote is genuine or counterfeit from the scatter plots. Figure 15 presents some synthetic representations of the datasets using a parallel plot, a radar chart and Andrews' curves, while Fig. 16 shows Chernoff's faces of twenty genuine and twenty counterfeit banknotes, specifically in the parallel plot and the radar chart. Different features can be observed for genuine and counterfeit banknotes. However, again, it is not easy to determine from these plots whether a particular banknote is genuine or counterfeit.

Figure 17 shows the principal component analysis of the banknote dataset. The first two principal components contain about 90% of the data variability (top left panel). Therefore, we can reasonably resume the data in two dimensions using the

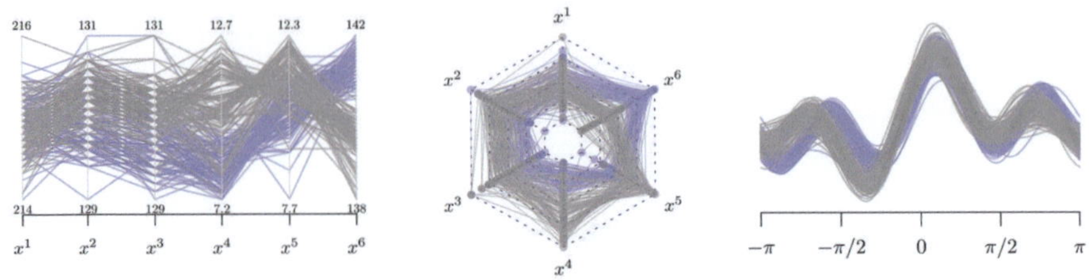

Fig. 15 Parallel-plots (left panel), radar charts (middle panel) and Andrews' curves (right panel) for the six measurements on the notes. The blue curves are the genuine banknotes while the grey curves are the counterfeits

Fig. 16 Examples of Chernoff's faces for the six measurements on the notes. The twenty faces on the left are genuine banknotes, while the twenty grey faces on the right are counterfeits

first two components. The correlation table with the original data shows that both first components are correlated with all measurements (bottom left panel). The first component is negatively correlated with all height measures $x^2, \ldots, x^5$ and positively correlated with the banknote width x^1 and the diagonal x^6. This component characterises the fact that a banknote is wide rather than long. The second component is mainly negatively correlated with the height of the lower edge x^5 and positively correlated with the height of the upper edge x^4 and the diagonal x^6. This component rather focuses on the position of the central drawing. Notwithstanding these interpretations, the scatter plot of the first two components (right panel) clearly shows two groups consisting of the genuine banknotes (blue circles) and the counterfeits (grey triangles). Most of the variability in the data is specific to the genuine and counterfeit banknotes, and the representation over the first two components captures this directly.

Appendix 4: Benchmarking Pedestrian Speed Prediction Algorithms

Pedestrian trajectory prediction is of great interest, especially for autonomous vehicles. In this appendix we compare the performance of different pedestrian speed prediction algorithms. The data come from a single file experiment where pedes-

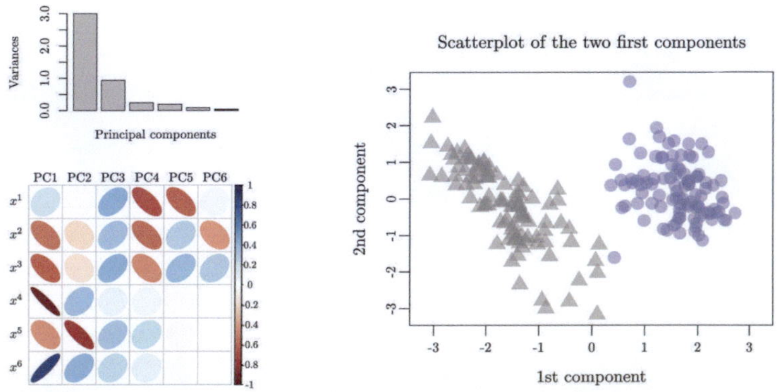

Fig. 17 Principal component analysis of the banknote dataset. Top left panels: variance per component—the first two components contain most of the variability. Bottom left panel: Correlation table with the original data - the first two components are correlated with all the six measurements. Right panel: scatter plot of the first two components, clearly showing two groups consisting of the genuine banknotes (blue circles) and the counterfeits (grey triangles)

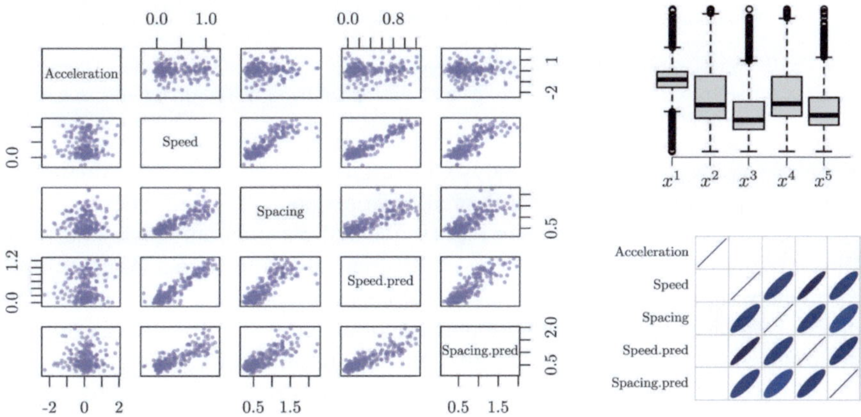

Fig. 18 Acceleration (in m/s^2), speed (in m/s), spacing (m) and speed and spacing of the neighbour in front of pedestrian in a single-file experiment. Left panel: scatterplots of the features. Top right panel: Boxplots of the normalised measurements. Bottom right panel: Correlation table

trians walk in a raw on a periodic geometry without overtaking possibility[8]. The explanatory variables are the acceleration, the spacing to the predecessor and its speed, and the spacing of the predecessor with a total of six thousand observations. The predicted variable is the speed of the pedestrian. A summary of the data including scatterplot, boxplot and correlation table is given in Fig. 18. Speed is correlated with all explanatory variables except acceleration. In addition, we can identify a nonlinear relationship between speed and distance.

[8] Some details on the data are available at https://ped.fz-juelich.de/da/doku.php?id=corridor2.

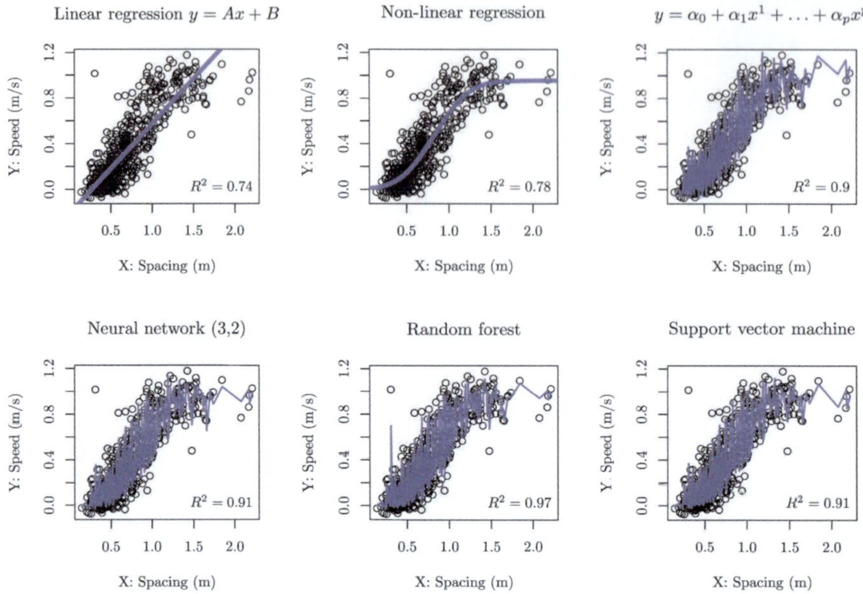

Fig. 19 Prediction of the speed with different algorithms. For univariate algorithms (linear and nonlinear regressions in top left and middle panels), the speed is regressed on the distance to the neighbour in front only

A cross-validated benchmark analysis of different pedestrian speed prediction algorithms is performed[9]. The testing prediction results and the R^2 metric quantifying the goodness of fit are shown in Fig. 19. We first consider univariate linear and nonlinear regression before considering multivariate algorithms, including multiple linear regression, a two-layer artificial neural network with three and two neurons respectively, a random forest and a support vector machine. Multivariate algorithms clearly outperform univariate algorithms. Among the multivariate algorithms, the random forest performs best, followed by multiple linear regression, artificial neural network and support vector machine which perform similarly.

References

1. M. Abdelghany, W. Ahmad, and S. Tahar. Event tree reliability analysis of safety-critical systems using theorem proving. *IEEE Systems Journal*, 16(2):2899–2910, 2021.
2. S. S. Afshari, F. Enayatollahi, X. Xu, and X. Liang. Machine learning-based methods in structural reliability analysis: A review. *Reliability Engineering & System Safety*, 219:108223, 2022.

[9] Benchmarking of pedestrian speed prediction algorithm can be computed online at https://colab.research.google.com/drive/1zGHJQbosu-gZPB8HUaf0GVHSJb_hdcbG?usp=sharing.

3. S. S. Afshari, C. Zhao, X. Zhuang, and X. Liang. Deep learning-based methods in structural reliability analysis: A review. *Measurement Science and Technology*, 34(7):072001, 2023.
4. J. Aldrini, I. Chihi, and L. Sidhom. Fault diagnosis and self-healing for smart manufacturing: a review. *Journal of Intelligent Manufacturing*, 35(6):2441–2473, 2024.
5. M. Aly. Survey on multiclass classification methods. *Neural Netw*, 19(1–9):2, 2005.
6. D. F. Andrews. Plots of high-dimensional data. *Biometrics*, pages 125–136, 1972.
7. M. Ankerst, M. M. Breunig, H.-P. Kriegel, and J. Sander. OPTICS: Ordering points to identify the clustering structure. *ACM Sigmod record*, 28(2):49–60, 1999.
8. A. B. Arrieta, N. Díaz-Rodríguez, J. Del Ser, A. Bennetot, S. Tabik, A. Barbado, S. García, S. Gil-López, D. Molina, R. Benjamins, et al. Explainable artificial intelligence (xai): Concepts, taxonomies, opportunities and challenges toward responsible ai. *Information fusion*, 58:82–115, 2020.
9. M. F. Bado, D. Tonelli, F. Poli, D. Zonta, and J. R. Casas. Digital twin for civil engineering systems: An exploratory review for distributed sensing updating. *Sensors*, 22(9):3168, 2022.
10. L. Bainbridge. Ironies of automation. In *Analysis, design and evaluation of man–machine systems*, pages 129–135. Elsevier, 1983.
11. M. R. Belgaum, Z. Alansari, S. Musa, M. M. Alam, and M. Mazliham. Role of artificial intelligence in cloud computing, iot and sdn: Reliability and scalability issues. *International Journal of Electrical and Computer Engineering*, 11(5):4458, 2021.
12. J. Berkson. Application of the logistic function to bio-assay. *Journal of the American statistical association*, 39(227):357–365, 1944.
13. L. Bian, G. Wang, and P. Liu. Reliability analysis for multi-component systems with interdependent competing failure processes. *Applied Mathematical Modelling*, 94:446–459, 2021.
14. J. Bofill, M. Abisado, J. Villaverde, and G. A. Sampedro. Exploring digital twin-based fault monitoring: Challenges and opportunities. *Sensors*, 23(16):7087, 2023.
15. B. E. Boser, I. M. Guyon, and V. N. Vapnik. A training algorithm for optimal margin classifiers. In *Proceedings of the fifth annual workshop on Computational learning theory*, pages 144–152, 1992.
16. D. M. Botín-Sanabria, A.-S. Mihaita, R. E. Peimbert-García, M. A. Ramírez-Moreno, R. A. Ramírez-Mendoza, and J. d. J. Lozoya-Santos. Digital twin technology challenges and applications: A comprehensive review. *Remote Sensing*, 14(6):1335, 2022.
17. S. Bracke. *Reliability Engineering*. Springer, 2024.
18. L. Breiman. Bagging predictors. *Machine learning*, 24:123–140, 1996.
19. L. Breiman. Stacked regressions. *Machine learning*, 24:49–64, 1996.
20. L. Breiman. Statistical modeling: The two cultures (with comments and a rejoinder by the author). *Statistical science*, 16(3):199–231, 2001.
21. D. Brüggemann, M. Hinz, and S. Bracke. An application of semi-supervised learning to sparsely labelled data. In M. C. Leva, E. Patelli, L. Podofillini, and S. Wilson, editors, *Book of Extended Abstracts for the 32nd European Safety and Reliability Conference*, pages 799–806, Singapore, 2022. Research Publishing Services.
22. B. Cai, Y. Liu, Z. Liu, X. Tian, X. Dong, and S. Yu. Using bayesian networks in reliability evaluation for subsea blowout preventer control system. *Reliability Engineering & System Safety*, 108:32–41, 2012.
23. G. Carleo, I. Cirac, K. Cranmer, L. Daudet, M. Schuld, N. Tishby, L. Vogt-Maranto, and L. Zdeborová. Machine learning and the physical sciences. *Reviews of Modern Physics*, 91(4):045002, 2019.
24. J. P. Cerrolaza, R. Obermaisser, J. Abella, F. J. Cazorla, K. Grüttner, I. Agirre, H. Ahmadian, and I. Allende. Multi-core devices for safety-critical systems: A survey. *ACM Computing Surveys (CSUR)*, 53(4):1–38, 2020.
25. V. Chamola, V. Hassija, A. R. Sulthana, D. Ghosh, D. Dhingra, and B. Sikdar. A review of trustworthy and explainable artificial intelligence (xai). *IEEe Access*, 2023.
26. B. Chen. *Active Learning*. Synthesis Lectures on Artificial Intelligence and Machine Learning. Springer International Publishing and Imprint Springer, Cham, 1st ed. 2012 edition, 2012.

27. Y. Cheng. Mean shift, mode seeking, and clustering. *IEEE transactions on pattern analysis and machine intelligence*, 17(8):790–799, 1995.
28. H. Chernoff. The use of faces to represent points in k-dimensional space graphically. *Journal of the American statistical Association*, 68(342):361–368, 1973.
29. Christoph Rosebrock. *Beitrag zur Prognose der Restnutzungsdauer von komplex beanspruchten Werkzeugen auf der Grundlage einer empirischen Fallstudie zur Korrosion von Bleianoden*. Shaker Verlag, 2023.
30. Christoph Rosebrock and Stefan Bracke, editors. *DETECTION OF DEFECTS ON PRINTED CIRCUIT BOARDS USING INSTANCE SEGMENTATION: Proceedings of the 31st European Safety and Reliability Conference (ESREL 2021)*. Research Publishing Services, Chennai, 2021.
31. Z. M. Çınar, A. Abdussalam Nuhu, Q. Zeeshan, O. Korhan, M. Asmael, and B. Safaei. Machine learning in predictive maintenance towards sustainable smart manufacturing in industry 4.0. *Sustainability*, 12(19):8211, 2020.
32. V. da Costa Bueno and I. M. do Carmo. Active redundancy allocation for a k-out-of-n: F system of dependent components. *European Journal of Operational Research*, 176(2):1041–1051, 2007.
33. A. Dardanelli, S. Corbetta, I. Boniolo, S. M. Savaresi, and S. Bittanti. Model-based kalman filtering approaches for frequency tracking. *IFAC Proceedings Volumes*, 43(10):37–42, 2010.
34. B. De Jonge and P. A. Scarf. A review on maintenance optimization. *European journal of operational research*, 285(3):805–824, 2020.
35. A. Djoudi, R. Zitouni, N. Zangar, and L. George. Lora network reconfiguration with markov decision process and fuzzy c-means clustering. *Computer Communications*, 196:129–140, 2022.
36. R. Donà and B. Ciuffo. Virtual testing of automated driving systems. a survey on validation methods. *IEEE Access*, 10:24349–24367, 2022.
37. W. Dong, A. R. Nejad, T. Moan, and Z. Gao. Structural reliability analysis of contact fatigue design of gears in wind turbine drivetrains. *Journal of Loss Prevention in the Process Industries*, 65:104115, 2020.
38. H. Drucker, C. J. Burges, L. Kaufman, A. Smola, and V. Vapnik. Support vector regression machines. *Advances in neural information processing systems*, 9, 1996.
39. L. Duchesne, E. Karangelos, and L. Wehenkel. Recent developments in machine learning for energy systems reliability management. *Proceedings of the IEEE*, 108(9):1656–1676, 2020.
40. A. Edelmann, T. Wolff, D. Montagne, and C. A. Bail. Computational social science and sociology. *Annual Review of Sociology*, 46(1):61–81, 2020.
41. S. Ereiz, I. Duvnjak, and J. F. Jiménez-Alonso. Review of finite element model updating methods for structural applications. In *Structures*, volume 41, pages 684–723. Elsevier, 2022.
42. I. Errandonea, S. Beltrán, and S. Arrizabalaga. Digital twin for maintenance: A literature review. *Computers in Industry*, 123:103316, 2020.
43. M. Ester, H.-P. Kriegel, J. Sander, X. Xu, et al. A density-based algorithm for discovering clusters in large spatial databases with noise. In *kdd*, volume 96, pages 226–231, 1996.
44. S. Evtiukov, E. Golov, and G. Ginzburg. Finite element method for reconstruction of road traffic accidents. *Transportation research procedia*, 36:157–165, 2018.
45. Fabian Hartwig, Marcin Hinz, Severin Dahms, and Stefan Bracke, editors. *Auslegung eines Accelerated Life Tests zur Abbildung der Langzeit-Zuverlässigkeit von Produkten im Prototypenstadium: Bericht zur GQW-Jahrestagung 2022 in Chemnitz*. Springer Vieweg, Wiesbaden and Heidelberg, 2024.
46. R. Fang and X. Li. On allocating one active redundancy to coherent systems with dependent and heterogeneous components' lifetimes. *Naval Research Logistics (NRL)*, 63(4):335–345, 2016.
47. C. Ferreira and G. Gonçalves. Remaining useful life prediction and challenges: A literature review on the use of machine learning methods. *Journal of Manufacturing Systems*, 63:550–562, 2022.

48. M. Feyzi, K. Fallahnezhad, M. Taylor, and R. Hashemi. A review on the finite element simulation of fretting wear and corrosion in the taper junction of hip replacement implants. *Computers in Biology and Medicine*, 130:104196, 2021.
49. B. Flury. *Multivariate statistics a practical approach*. Chapman and Hall, London ; New York, 1988.
50. P. M. Frank and X. Ding. Survey of robust residual generation and evaluation methods in observer-based fault detection systems. *Journal of process control*, 7(6):403–424, 1997.
51. Y. Freund and R. E. Schapire. A desicion-theoretic generalization of on-line learning and an application to boosting. In *European conference on computational learning theory*, pages 23–37. Springer, 1995.
52. J. Friederich and S. Lazarova-Molnar. Reliability assessment of manufacturing systems: A comprehensive overview, challenges and opportunities. *Journal of Manufacturing Systems*, 72:38–58, 2024.
53. J. H. Friedman. Greedy function approximation: a gradient boosting machine. *Annals of statistics*, pages 1189–1232, 2001.
54. M. Frost and S. Sandrock. Neue Belastungsarten in der Arbeitswelt 4.0. Zahlen, Daten, Fakten. Technical report, ifaa—Institut für angewandte Arbeitswissenschaft, 2018.
55. M.-F. Ge, Y. Liu, X. Jiang, and J. Liu. A review on state of health estimations and remaining useful life prognostics of lithium-ion batteries. *Measurement*, 174:109057, 2021.
56. H. Gottschalk, M. Rottmann, and M. Saltagic. Does redundancy in ai perception systems help to test for super-human automated driving performance? *Deep Neural Networks and Data for Automated Driving: Robustness, Uncertainty Quantification, and Insights Towards Safety*, pages 81–106, 2022.
57. I. Gräßler, P. Scholle, and H. Thiele. Scenario technique. In S. Vajna, editor, *Integrated Design Engineering*, pages 615–645. Springer International Publishing, Cham, 2020.
58. J. G. Greener, S. M. Kandathil, L. Moffat, and D. T. Jones. A guide to machine learning for biologists. *Nature Reviews Molecular Cell Biology*, 23(1):40–55, 2022.
59. I. Häring, Y. Satsrisakul, J. Finger, G. Vogelbacher, C. Köpke, F. Höflinger, and P. Gelhausen. Advanced markov modeling and simulation for safety analysis of autonomous driving functions up to sae 5 for development, approval and main inspection. In M. C. Leva, E. Patelli, L. Podofillini, and S. Wilson, editors, *Proceedings of the 32nd European Safety and Reliability Conference*, pages 104–111. Research Publishing Services, 2022.
60. S. A. Hasib, S. Islam, R. K. Chakrabortty, M. J. Ryan, D. K. Saha, M. H. Ahamed, S. I. Moyeen, S. K. Das, M. F. Ali, M. R. Islam, et al. A comprehensive review of available battery datasets, rul prediction approaches, and advanced battery management. *Ieee Access*, 9:86166–86193, 2021.
61. M. Hinz, P. Temminghoff, and S. Bracke, editors. *Beitrag zur multivariaten Analyse von Qualitäts- und Zuverlässigkeitsdaten in der Produkt-Erprobungsphase am Beispiel der Fahrzeugtechnik: Bericht zur GQW-Jahrestagung 2014 in Chemnitz*. Shaker Verlag, 2014.
62. J. M. Horcas, J. Monteil, M. Bouroche, M. Pinto, L. Fuentes, and S. Clarke. Context–dependent reconfiguration of autonomous vehicles in mixed traffic. *Journal of Software: Evolution and Process*, 30(4), 2018.
63. H. Hotelling. Analysis of a complex of statistical variables into principal components. *Journal of educational psychology*, 24(6):417, 1933.
64. W. Hu, T. Zhang, X. Deng, Z. Liu, and J. Tan. Digital twin: A state-of-the-art review of its enabling technologies, applications and challenges. *Journal of Intelligent Manufacturing and Special Equipment*, 2(1):1–34, 2021.
65. X. Hu, L. Xu, X. Lin, and M. Pecht. Battery lifetime prognostics. *Joule*, 4(2):310–346, 2020.
66. R. Isermann. Fault diagnosis of machines via parameter estimation and knowledge processing—tutorial paper. *Automatica*, 29(4):815–835, 1993.
67. R. Isermann. *Fault-Diagnosis Systems: An Introduction from Fault Detection to Fault Tolerance*. Springer Science & Business Media, 2006.

68. T. Ishigooka, S. Honda, and H. Takada. Cost-effective redundancy approach for fail-operational autonomous driving system. In *2018 IEEE 21st International Symposium on Real-Time Distributed Computing (ISORC)*, pages 107–115. IEEE, 2018.
69. M. M. Islam and J.-M. Kim. Reliable multiple combined fault diagnosis of bearings using heterogeneous feature models and multiclass support vector machines. *Reliability Engineering & System Safety*, 184:55–66, 2019.
70. N. Ivanova and V. Vishnevsky. Application of k-out-of-n:g system and machine learning techniques on reliability analysis of tethered unmanned aerial vehicle. In A. Dudin, A. Nazarov, and A. Moiseev, editors, *Information Technologies and Mathematical Modelling. Queueing Theory and Applications*, volume 1605 of *Communications in Computer and Information Science*, pages 117–130. Springer International Publishing, Cham, 2022.
71. M. J. and V. G. Bayesian estimation of linear/circular consecutive k-out-of-n: F system reliability. *International Journal of Performability Engineering*, 16(10):1509, 2020.
72. R. H. Jhaveri, A. Revathi, K. Ramana, R. Raut, and R. K. Dhanaraj. A review on machine learning strategies for real-world engineering applications. *Mobile Information Systems*, 2022(1):1833507, 2022.
73. X. Jiang and S. Mahadevan. Bayesian risk-based decision method for model validation under uncertainty. *Reliability Engineering & System Safety*, 92(6):707–718, 2007.
74. P. Jieyang, A. Kimmig, W. Dongkun, Z. Niu, F. Zhi, W. Jiahai, X. Liu, and J. Ovtcharova. A systematic review of data-driven approaches to fault diagnosis and early warning. *Journal of Intelligent Manufacturing*, 34(8):3277–3304, 2023.
75. J. J. M. Jimenez, S. Schwartz, R. Vingerhoeds, B. Grabot, and M. Salaün. Towards multi-model approaches to predictive maintenance: A systematic literature survey on diagnostics and prognostics. *Journal of manufacturing systems*, 56:539–557, 2020.
76. M. I. Jordan and T. M. Mitchell. Machine learning: Trends, perspectives, and prospects. *Science*, 349(6245):255–260, 2015.
77. T. M. Julitz, A. Tordeux, and M. Löwer. Reliability of fault-tolerant system architectures for automated driving systems. In M. C. Leva, E. Patelli, L. Podofillini, and S. Wilson, editors, *Proceedings of the 32nd European Safety and Reliability Conference*, pages 120–127, Singapore, 2022. Research Publishing Services.
78. T. M. Julitz, A. Tordeux, and M. Löwer. Computer-aided design of fault-tolerant hardware architectures for autonomous driving systems. *Proceedings of the Design Society*, 3:1047–1056, 2023.
79. T. M. Julitz, A. Tordeux, N. Schlüter, and M. Löwer. Reliability of redundant M-Out-Of-N architectures with dependent components: Comprehensible approach with monte carlo simulation. In K. Kołowrocki and E. Dąbrowska, editors, *Advances in Reliability, Safety and Secruity. ESREL 2024 Monograph Book Series*, volume 4, pages 97–106. Polish Safety and Reliability Association, Gydnia, 2024.
80. H. Kahn and A. J. Wiener. The next thirty-three years: A framework for speculation. *Daedalus, Vol. 96*, 96(3):705–732, 1967.
81. N. Kalra and S. M. Paddock. Driving to safety: How many miles of driving would it take to demonstrate autonomous vehicle reliability? *Transportation Research Part A: Policy and Practice*, 94:182–193, 2016.
82. D. Kaur, S. Uslu, K. J. Rittichier, and A. Durresi. Trustworthy artificial intelligence: a review. *ACM computing surveys (CSUR)*, 55(2):1–38, 2022.
83. M. Kelkinnama and M. Asadi. Optimal redundancy allocation in coherent systems with heterogeneous dependent components. *Journal of Applied Probability*, 59(4):1144–1177, 2022.
84. A. Kohn, M. Kabmeyer, R. Schneider, A. Roger, C. Stellwag, and A. Herkersdorf. Fail-operational in safety-related automotive multi-core systems. In *10th IEEE International Symposium on Industrial Embedded Systems (SIES)*, pages 1–4. IEEE, 2015.
85. S. Kotz, C. D. Lai, and M. Xie. On the effect of redundancy for systems with dependent components. *IIE Transactions*, 35(12):1103–1110, 2003.
86. K. Kowol., M. Rottmann., S. Bracke., and H. Gottschalk. Yodar: Uncertainty-based sensor fusion for vehicle detection with camera and radar sensors. In *Proceedings of the 13th Interna-*

tional Conference on Agents and Artificial Intelligence—Volume 2: ICAART, pages 177–186, 2021.

87. R. Krishnamoorthy, B. Chokkalingam, and J. L. Munda. Design of fault-tolerant automotive gateway architecture using mc9s12xdp512 microcontroller device. *Energies*, 16(16):5923, 2023.
88. J. Kudela and R. Matousek. Recent advances and applications of surrogate models for finite element method computations: a review. *Soft Computing*, 26(24):13709–13733, 2022.
89. S. Kumar, R. Saket, D. K. Dheer, J. B. Holm-Nielsen, and P. Sanjeevikumar. Reliability enhancement of electrical power system including impacts of renewable energy sources: a comprehensive review. *IET Generation, Transmission & Distribution*, 14(10):1799–1815, 2020.
90. Y. Lei, N. Li, L. Guo, N. Li, T. Yan, and J. Lin. Machinery health prognostics: A systematic review from data acquisition to rul prediction. *Mechanical systems and signal processing*, 104:799–834, 2018.
91. J. Y. Lettvin, H. R. Maturana, W. S. McCulloch, and W. H. Pitts. What the frog's eye tells the frog's brain. *Proceedings of the IRE*, 47(11):1940–1951, 1959.
92. M. C. Leva, E. Patelli, L. Podofillini, and S. Wilson, editors. *Book of Extended Abstracts for the 32nd European Safety and Reliability Conference*, Singapore, 2022. Research Publishing Services.
93. J. Li, Y. Huang, Y.-F. Li, and S. Wang. Redundancy allocation under state-dependent distributional uncertainty of component lifetimes. *Production and Operations Management*, 32(3):930–950, 2023.
94. X. Li, P. Ye, J. Li, Z. Liu, L. Cao, and F.-Y. Wang. From features engineering to scenarios engineering for trustworthy AI: I&I, C&C, and V&V. *IEEE Intelligent Systems*, 37(4):18–26, 2022.
95. K. Liggieri and O. Müller. *Mensch-Maschine-Interaktion. Handbuch zu Geschichte, Kultur und Ethik*. Springer, 2019.
96. Y.-H. Lin, Y.-F. Li, and E. Zio. A reliability assessment framework for systems with degradation dependency by combining binary decision diagrams and monte carlo simulation. *IEEE Transactions on Systems, Man, and Cybernetics: Systems*, 46(11):1556–1564, 2015.
97. G. Liu, V. Pandey, C. Somarakis, and N. Motee. Risk of cascading failures in multi-agent rendezvous with communication time delay. In *2022 American Control Conference (ACC)*, pages 2172–2177. IEEE, 2022.
98. H. Liu, M. Xia, D. Williams, J. Sun, and H. Yan. Digital twin-driven machine condition monitoring: A literature review. *Journal of Sensors*, 2022(1):6129995, 2022.
99. H.-C. Liu, L.-J. Zhang, Y.-J. Ping, and L. Wang. Failure mode and effects analysis for proactive healthcare risk evaluation: a systematic literature review. *Journal of evaluation in clinical practice*, 26(4):1320–1337, 2020.
100. M. Liu, S. Fang, H. Dong, and C. Xu. Review of digital twin about concepts, technologies, and industrial applications. *Journal of manufacturing systems*, 58:346–361, 2021.
101. H. Lyu, H. Qu, H. Xie, Y. Zhang, and M. Pecht. Reliability analysis of the multi-state system with nonlinear degradation model under markov environment. *Reliability Engineering & System Safety*, 238:109411, 2023.
102. J. Macqueen. Some methods for classification and analysis of multivariate observations. In *Proceedings of 5-th Berkeley Symposium on Mathematical Statistics and Probability/University of California Press*, pages 281–297, 1967.
103. Y. A. Mahmood, A. Ahmadi, A. K. Verma, A. Srividya, and U. Kumar. Fuzzy fault tree analysis: a review of concept and application. *International Journal of System Assurance Engineering and Management*, 4:19–32, 2013.
104. M. R. Mamdikar, V. Kumar, and P. Singh. Dynamic reliability analysis framework using fault tree and dynamic bayesian network: A case study of npp. *Nuclear Engineering and Technology*, 54(4):1213–1220, 2022.

105. N. V. Martyushev, B. V. Malozyomov, S. N. Sorokova, E. A. Efremenkov, D. V. Valuev, and M. Qi. Review models and methods for determining and predicting the reliability of technical systems and transport. *Mathematics*, 11(15):3317, 2023.
106. A. Maurya and D. Kumar. Reliability of safety-critical systems: A state-of-the-art review. *Quality and Reliability Engineering International*, 36(7):2547–2568, 2020.
107. X.-L. Meng and D. Van Dyk. The em algorithm-an old folk-song sung to a fast new tune. *Journal of the Royal Statistical Society Series B: Statistical Methodology*, 59(3):511–567, 1997.
108. M. Meuwly. Machine learning for chemical reactions. *Chemical Reviews*, 121(16):10218–10239, 2021.
109. J. Mi, Y.-F. Li, W. Peng, and H.-Z. Huang. Reliability analysis of complex multi-state system with common cause failure based on evidential networks. *Reliability Engineering & System Safety*, 174:71–81, 2018.
110. S. Narteni, M. Ferretti, V. Orani, I. Vaccari, E. Cambiaso, and M. Mongelli. From explainable to reliable artificial intelligence. In *International Cross-Domain Conference for Machine Learning and Knowledge Extraction*, pages 255–273. Springer, 2021.
111. J. Navarro and P. Fernández-Martínez. Redundancy in systems with heterogeneous dependent components. *European Journal of Operational Research*, 290(2):766–778, 2021.
112. M. R. Nazabadi, S. E. Najafi, A. Mohaghar, and F. Movahedi Sobhani. The joint policy of production, maintenance, and product quality in a multi-machine production system by reinforcement learning and agent-based modeling. *International journal of research in industrial engineering*, 13(1):71–87, 2024.
113. G. Noorsumar, S. Rogovchenko, K. G. Robbersmyr, and D. Vysochinskiy. Mathematical models for assessment of vehicle crashworthiness: a review. *International journal of crashworthiness*, 27(5):1545–1559, 2022.
114. F. Oszwald, J. Becker, P. Obergfell, and M. Traub. Dynamic reconfiguration for real-time automotive embedded systems in fail-operational context. In *2018 IEEE International Parallel and Distributed Processing Symposium Workshops (IPDPSW)*, pages 206–209. IEEE, 2018.
115. V. O'Neill and B. Soh. Improving fault tolerance and reliability of heterogeneous multi-agent iot systems using intelligence transfer. *Electronics*, 11(17):2724, 2022.
116. Z. Qiu, R. Huang, X. Wang, and W. Qi. Structural reliability analysis and reliability-based design optimization: Recent advances. *Science China Physics, Mechanics and Astronomy*, 56:1611–1618, 2013.
117. M. Ram. *Modeling and simulation based analysis in reliability engineering*, volume 40. CRC Press, 2019.
118. S. B. Ramezani, L. Cummins, B. Killen, R. Carley, A. Amirlatifi, S. Rahimi, M. Seale, and L. Bian. Scalability, explainability and performance of data-driven algorithms in predicting the remaining useful life: A comprehensive review. *Ieee Access*, 11:41741–41769, 2023.
119. Y. Ran, X. Zhou, P. Lin, Y. Wen, and R. Deng. A survey of predictive maintenance: Systems, purposes and approaches. *arXiv preprint* arXiv:1912.07383, 2019.
120. C. R. Rao. The utilization of multiple measurements in problems of biological classification. *Journal of the Royal Statistical Society. Series B (Methodological)*, 10(2):159–203, 1948.
121. M. Rao and V. Nakain. Review of simulation approaches in reliability and availability modeling. *International Journal of Performability Engineering*, 12(4):369, 2016.
122. C. Reyes, F. Jaramillo, B. Zhang, C. Kulkarni, and M. Orchard. Just-in-time point prediction using a computationally-efficient lebesgue-sampling-based prognostic method: Application to battery end-of-discharge prediction. *Annual Conference of the PHM Society*, 10(1), 2018.
123. R. F. Ribeiro Junior, I. A. dos Santos Areias, M. M. Campos, C. E. Teixeira, L. E. B. da Silva, and G. F. Gomes. Fault detection and diagnosis in electric motors using convolution neural network and short-time fourier transform. *Journal of Vibration Engineering & Technologies*, 10(7):2531–2542, 2022.
124. C. Rosebrock and S. Bracke. Simulation of anode corrosion in the electrowinning process of non-ferrous metals. In *2020 Annual Reliability and Maintainability Symposium (RAMS)*, pages 1–6. IEEE, 2020.

125. M. Rumez, D. Grimm, R. Kriesten, and E. Sax. An overview of automotive service-oriented architectures and implications for security countermeasures. *IEEE Access*, 8:221852–221870, 2020.
126. S. Safari, M. Ansari, H. Khdr, P. Gohari-Nazari, S. Yari-Karin, A. Yeganeh-Khaksar, S. Hessabi, A. Ejlali, and J. Henkel. A survey of fault-tolerance techniques for embedded systems from the perspective of power, energy, and thermal issues. *IEEE Access*, 10:12229–12251, 2022.
127. A. K. Samantaray, K. Medjaher, B. O. Bouamama, M. Staroswiecki, and G. Dauphin-Tanguy. Diagnostic bond graphs for online fault detection and isolation. *Simulation Modelling Practice and Theory*, 14(3):237–262, 2006.
128. G. Saporta. Models for understanding versus models for prediction. In *COMPSTAT 2008: Proceedings in computational statistics*, pages 315–322. Springer, 2008.
129. N. Sapoval, A. Aghazadeh, M. G. Nute, D. A. Antunes, A. Balaji, R. Baraniuk, C. Barberan, R. Dannenfelser, C. Dun, M. Edrisi, et al. Current progress and open challenges for applying deep learning across the biosciences. *Nature Communications*, 13(1):1728, 2022.
130. B. Sari. *Fail-operational Safety Architecture for ADAS/AD Systems and a Model-driven Approach for Dependent Failure Analysis*. Springer Fachmedien Wiesbaden, Wiesbaden, 2020.
131. T. Schmid, S. Schraufstetter, S. Wagner, and D. Hellhake. A safety argumentation for fail-operational automotive systems in compliance with iso 26262. In *2019 4th International Conference on System Reliability and Safety (ICSRS)*, pages 484–493. IEEE, 2019.
132. S. Schwendemann, Z. Amjad, and A. Sikora. A survey of machine-learning techniques for condition monitoring and predictive maintenance of bearings in grinding machines. *Computers in Industry*, 125:103380, 2021.
133. X. Shi and A. M. Bazzi. Fault tree reliability analysis of a micro-grid using monte carlo simulations. In *2015 IEEE Power and Energy Conference at Illinois (PECI)*, pages 1–5. IEEE, 2015.
134. C. Shyalika, R. Wickramarachchi, and A. Sheth. A comprehensive survey on rare event prediction. *arXiv preprint* arXiv:2309.11356, 2023.
135. K. Siebertz, D. van Bebber, and T. Hochkirchen. *Statistische Versuchsplanung: Design of Experiments (DoE)*. VDI-Buch. Springer Vieweg, Berlin and Heidelberg, 2. auflage edition, 2017.
136. C. Song and R. Kawai. Monte carlo and variance reduction methods for structural reliability analysis: A comprehensive review. *Probabilistic Engineering Mechanics*, 73:103479, 2023.
137. M. L. H. Souza, C. A. da Costa, G. de Oliveira Ramos, and R. da Rosa Righi. A survey on decision-making based on system reliability in the context of industry 4.0. *Journal of Manufacturing Systems*, 56:133–156, 2020.
138. R. Stetter. *Fault-Tolerant Design and Control of Automated Vehicles and Processes*, volume 201. Springer International Publishing, Cham, 2020.
139. H. Stoll, D. Grimm, M. Schindewolf, M. Brodatzki, and E. Sax. Dynamic reconfiguration of automotive architectures using a novel plug-and-play approach. In *2021 IEEE Intelligent Vehicles Symposium Workshops (IV Workshops)*, pages 70–75. IEEE, 2021.
140. T. Stolte, S. Ackermann, R. Graubohm, I. Jatzkowski, B. Klamann, H. Winner, and M. Maurer. Taxonomy to unify fault tolerance regimes for automotive systems: Defining fail-operational, fail-degraded, and fail-safe. *IEEE Transactions on Intelligent Vehicles*, 7(2):251–262, 2022.
141. J. Su, J. Huang, S. Adams, Q. Chang, and P. A. Beling. Deep multi-agent reinforcement learning for multi-level preventive maintenance in manufacturing systems. *Expert systems with applications*, 192:116323, 2022.
142. J. Sun, H. Zhang, H. Zhou, R. Yu, and Y. Tian. Scenario-based test automation for highly automated vehicles: A review and paving the way for systematic safety assurance. *IEEE transactions on intelligent transportation systems*, 23(9):14088–14103, 2021.
143. S. Tang, Z. Zhang, Y. Zhang, J. Zhou, Y. Guo, S. Liu, S. Guo, Y.-F. Li, L. Ma, Y. Xue, et al. A survey on automated driving system testing: Landscapes and trends. *ACM Transactions on Software Engineering and Methodology*, 32(5):1–62, 2023.

144. P. Tchakoua, R. Wamkeue, T. A. Tameghe, and G. Ekemb. A review of concepts and methods for wind turbines condition monitoring. In *2013 world congress on computer and information technology (WCCIT)*, pages 1–9. IEEE, 2013.
145. A. Thelen, X. Zhang, O. Fink, Y. Lu, S. Ghosh, B. D. Youn, M. D. Todd, S. Mahadevan, C. Hu, and Z. Hu. A comprehensive review of digital twin-part 1: modeling and twinning enabling technologies. *Structural and Multidisciplinary Optimization*, 65(12):354, 2022.
146. A. Thelen, X. Zhang, O. Fink, Y. Lu, S. Ghosh, B. D. Youn, M. D. Todd, S. Mahadevan, C. Hu, and Z. Hu. A comprehensive review of digital twin-part 2: roles of uncertainty quantification and optimization, a battery digital twin, and perspectives. *Structural and multidisciplinary optimization*, 66(1):1, 2023.
147. E. L. Trist and K. W. Bamforth. Some social and psychological consequences of the longwall method of coal-getting: An examination of the psychological situation and defences of a work group in relation to the social structure and technological content of the work system. *Human Relations*, 4(1):3–38, 1951.
148. E. Ulich. *Arbeitspsychologie*. Vdf, Hochschulverlag AG an der ETH Zürich, 2001.
149. S. R. Vadyala, S. N. Betgeri, J. C. Matthews, and E. Matthews. A review of physics-based machine learning in civil engineering. *Results in Engineering*, 13:100316, 2022.
150. J. Vamathevan, D. Clark, P. Czodrowski, I. Dunham, E. Ferran, G. Lee, B. Li, A. Madabhushi, P. Shah, M. Spitzer, et al. Applications of machine learning in drug discovery and development. *Nature reviews Drug discovery*, 18(6):463–477, 2019.
151. R. van Dinter, B. Tekinerdogan, and C. Catal. Predictive maintenance using digital twins: A systematic literature review. *Information and Software Technology*, 151:107008, 2022.
152. A. J. C. van Gemund and G. L. Reijns. Reliability analysis of k-out-of-n systems with single cold standby using pearson distributions. *IEEE Transactions on Reliability*, 61(2):526–532, 2012.
153. V. Venkatasubramanian, R. Rengaswamy, K. Yin, and S. N. Kavuri. A review of process fault detection and diagnosis: Part i: Quantitative model-based methods. *Computers & chemical engineering*, 27(3):293–311, 2003.
154. S. vom Dorff, B. Boddeker, M. Kneissl, and M. Franzle. A fail-safe architecture for automated driving. In *2020 Design, Automation & Test in Europe Conference & Exhibition (DATE)*, pages 828–833. IEEE, 2020.
155. G. H. Walker, N. A. Stanton, P. M. Salmon, and D. P. Jenkins. A review of sociotechnical systems theory: a classic concept for new command and control paradigms. *Theoretical Issues in Ergonomics Science*, 9(6):479–499, 2008.
156. Wayne Nelson. *Accelerated Testing: Statistical Models, Test Plans, and Data Analysis*. John Wiley & Sons, Inc., 1990.
157. P. Weiss, A. Weichslgartner, F. Reimann, and S. Steinhorst. Fail-operational automotive software design using agent-based graceful degradation. In *2020 Design, Automation & Test in Europe Conference & Exhibition (DATE)*, pages 1169–1174. IEEE, 2020.
158. Y. Wilhelm, P. Reimann, W. Gauchel, and B. Mitschang. Overview on hybrid approaches to fault detection and diagnosis: Combining data-driven, physics-based and knowledge-based models. *Procedia Cirp*, 99:278–283, 2021.
159. D. H. Wolpert. Stacked generalization. *Neural networks*, 5(2):241–259, 1992.
160. R. Wu, D. Kang, Y. Chen, and C. Chen. Assessing academic impacts of machine learning applications on a social science: Bibliometric evidence from economics. *Journal of Informetrics*, 17(3):101436, 2023.
161. L. Xing. Reliability in internet of things: Current status and future perspectives. *IEEE Internet of Things Journal*, 7(8):6704–6721, 2020.
162. B. Xu, G. Bai, Y. Fang, J. Tao, et al. Failure analysis of unmanned autonomous swarm considering cascading effects. *Journal of Systems Engineering and Electronics*, 33(3):759–770, 2022.
163. Z. Xu and J. H. Saleh. Machine learning for reliability engineering and safety applications: Review of current status and future opportunities. *Reliability Engineering & System Safety*, 211:107530, 2021.

164. W. Yan, B. Zhang, G. Zhao, S. Tang, G. Niu, and X. Wang. A Battery Management System With a Lebesgue-Sampling-Based Extended Kalman Filter. *IEEE Transactions on Industrial Electronics*, 66(4):3227–3236, Apr. 2019.
165. M. Yazdi, J. Mohammadpour, H. Li, H.-Z. Huang, E. Zarei, R. G. Pirbalouti, and S. Adumene. Fault tree analysis improvements: A bibliometric analysis and literature review. *Quality and Reliability Engineering International*, 39(5):1639–1659, 2023.
166. J. Yu, S. Zheng, H. Pham, and T. Chen. Reliability modeling of multi–state degraded repairable systems and its applications to automotive systems. *Quality and Reliability Engineering International*, 34(3):459–474, 2018.
167. Z. Zeng, A. Barros, and D. Coit. Dependent failure behavior modeling for risk and reliability: A systematic and critical literature review. *Reliability Engineering & System Safety*, page 109515, 2023.
168. K. Zhang, Z. Yang, and T. Başar. Multi-agent reinforcement learning: A selective overview of theories and algorithms. *Handbook of reinforcement learning and control*, pages 321–384, 2021.
169. Y. Zhang, E. Amini-Seresht, and W. Ding. Component and system active redundancies for coherent systems with dependent components. *Applied Stochastic Models in Business and Industry*, 33(4):409–421, 2017.
170. H. Zhou, X. Huang, G. Wen, Z. Lei, S. Dong, P. Zhang, and X. Chen. Construction of health indicators for condition monitoring of rotating machinery: A review of the research. *Expert Systems with Applications*, 203:117297, 2022.
171. E. Zio. Some challenges and opportunities in reliability engineering. *IEEE Transactions on Reliability*, 65(4):1769–1782, 2016.
172. E. Zio. Prognostics and health management (phm): Where are we and where do we (need to) go in theory and practice. *Reliability Engineering & System Safety*, 218:108119, 2022.

Modelling Vehicle and Pedestrian Collective Dynamics: Challenges and Advances

Antoine Tordeux, Cécile Appert-Rolland, Alexandre Nicolas, Armin Seyfried and Denis Ullmo

Contents

Abstract

In our urbanised societies, the management and regulation of traffic and pedestrian flows is of considerable interest for public safety, economic development, and the conservation of the environment. However, modelling and controlling the collective dynamics of vehicles and pedestrians raises several challenges. Not only are the individual entities self-propelled and hard to describe, but their complex

A. Tordeux (✉)
Chair for Traffic Safety and Reliability, University of Wuppertal, Wuppertal, Germany
e-mail: tordeux@uni-wuppertal.de

C. Appert-Rolland
Laboratory of the Physics of the two Infinities Irène Joliot-Curie, CNRS/IN2P3 & University Paris-Saclay, Orsay, France

A. Nicolas
Institut Lumière Matière, CNRS & University Lyon 1, Villeurbanne CEDEX, France

A. Seyfried
Chair for Computer Simulation for Fire Safety and Pedestrian Traffic, University of Wuppertal, Wuppertal, Germany

Civil Safety Research (IAS-7), Forschungszentrum Jülich, Jülich, Germany

D. Ullmo
Laboratory of Theoretical Physics and Statistical Models, University Paris-Saclay, Orsay CEDEX, France

S. Bracke and R. Pieper (eds.), *Safety Engineering*,
https://doi.org/10.1007/978-3-658-47473-7_5

nonlinear physical and social interactions make the multi-agent problem of crowd flow and traffic even more involved. In this chapter, we purport to review the suitability and limitations of classical modelling approaches through four examples of collective behaviour: stop-and-go waves in traffic flow, lane formation, long-term avoidance behaviour, and load balancing in pedestrian dynamics. While stop-and-go dynamics and lane formation can both be addressed by basic reactive models (at least to some extent), the latter two require anticipation and/or coordination at the level of the group. The results highlight the limitations of classical force-based models, but also the need for long-term anticipation mechanisms and multiscale modelling approaches. In response, we review new developments and modelling concepts.

Keywords

Pedestrian Dynamics · Traffic Flow · Collective Dynamics · Force-Based Model · Anticipation-Based Model · Socio-Psychological Concept

1 Introduction

Our societies are becoming increasingly urbanised. Managing traffic flow, mixed urban traffic, and pedestrian dynamics is of considerable interest for public safety, economic development, the conservation of the environment, and the quality of everyday life. In particular, crowd management in large buildings, such as train stations or stadiums, or during large outdoor events, such as festivals or concerts, has become the focus of much attention [36]. However, controlling traffic flow and pedestrian dynamics is not straightforward. In fact, pedestrian crowding and traffic flow are collective behaviours of many-body systems subject to complex social and physical interactions. In practice, intelligent traffic systems and crowd management are generally designed based on simulations and computational optimisation techniques. Simulations rely on pedestrian and traffic flow modelling approaches, and the demand for reliable and realistic models is getting more and more pressing. However, much of the collective drivers' and pedestrians' behaviour remains poorly understood. For example, traffic flow stabilisation is still a challenge, especially in the context of driving automation [44], while the notion of anticipation for a pedestrian is still somewhat fuzzy.

In congested conditions, pedestrian and road traffic flows essentially describe collective dynamics and coordinated behaviours. These coordinated dynamics can result from a form of collective intelligence, for example, when pedestrian counterflows self-organise into lanes to optimise flow [47] (although this phenomenon is also observed in passive particle systems such as colloids with different directions of motion), when crowds clear the way to facilitate the passage of an *intruder* [11,93], for example an ambulance (e.g., in the video [2]), or when pedestrians spread out over different possible exits to balance the load on each exit during evacuation [136]. That being said, collective dynamics are not always beneficial and can even be strongly detrimental. Consider the stop-and-go waves (accordion traffic) commonly observed on congested motorways, which negatively impact traffic safety and performance and heighten the environmental harm caused by traffic because of the increased gas consumption due to repeated accelerations and braking. For all these problems, modelling, understanding and controlling collective dynamics starting from the local interactions of the agents from a complex systems perspective is a challenging task [9].

1.1 Modelling Vehicle and Pedestrian Collective Dynamics

The modelling of vehicle and pedestrian collective dynamics is generally divided into three different levels: strategic, tactical, and operational. This hierarchy of levels, that originally finds its roots in the military field, was introduced to the modelling of driver behaviour in the 1970s by Michon [83] and was transferred to pedestrian dynamics in the early 2000s [54]. At the strategic level, the modeller is concerned with the largest scales, those of activity choices and scheduling (departure time planning). The focus of the tactical level is on smaller scales; it includes manoeuvre planning, the choice of a path, or the choice of an exit in the context of an evacuation. Operational modelling describes local motion in interaction with neighbours and obstacles. The state of the art generally treats the strategic, tactical and operational modelling scales separately. However, we argue in this review that these classifications are not entirely distinct from one another, but could be strongly interlaced, especially the operational and tactical modelling scales. This leads to conceptual problems in modelling these systems. Among operational models, three categories are classically distinguished: microscopic, mesoscopic, and macroscopic approaches [19,46]. Microscopic models deal with the individual motion of the agents in Lagrangian coordinates, while mesoscopic and macroscopic approaches aggregate the dynamics in Eulerian coordinates using probabilistic agent distribution functions and averaged quantities such as flow and density, respectively. Among these approaches, first-order models posit that the agent's speed *instantaneously* adapts to the environment, typically using proactive models based on anticipation and optimisation, whereas force-based models are second-order approaches that are formally similar to Newton's equation and thus incorporate into the dynamics 'inertia', i.e., gradual adjustments of an agent's velocity to the current situation, along with possible mechanical forces.

Microscopic operational models reproduce the outcome of competition between conflicting goals. Agents are driven towards a destination defined at the tactical and strategic modelling levels, but at the same time they are intent on avoiding collisions with other entities and unfavourable regions in space. The resulting interactions between entities need not obey Newton's third law of motion, where action equals reaction, because vehicles and pedestrians self-propel in a direction set by their own decisions, rather than the resultant of mechanical forces. In addition, interactions are generally strongly anisotropic, the presence of neighbours or obstacles being much more dominant when located in the direction of motion, and nonlinear, through short-range interaction models or finite dynamic capacities. The analysis of such systems can be cumbersome, particularly when considering nonlinear anisotropic interaction models. In addition, many situations require explicit anticipation on the long-term and on-the-fly decisions about collective tactics, resulting in dynamic choices at different modelling scales.

1.2 Objectives and Layout of the Chapter

In this chapter, we purport to review the suitability of vehicle and pedestrian models to describe phenomena that seem to pertain to coordinated collective dynamics. Specifically, we highlight that current models fail to consider goals beyond the operational movement of a single pedestrian and represent a collective goal of a group or subgroup. These are typical limitations of classical reactive approaches with pairwise interaction, motivating the need for collective prediction and proactive models that exhibit long-term anticipation and coordination with complex interaction schemes coupling operational and tactical scales, among other modelling concepts. We study four examples of collective phenomena: 1) stop-and-go waves in traffic flow, 2) lane formation for pedestrian counterflow, 3) long-term anticipation in crowd crossed by an intruder, and 4) load balancing between multiple exits in the context of an evacuation. While both stop-and-go and lane formation collective phenomena can be addressed by basic reactive models, the latter require proactive control, optimisation, and collective coordination. The chapter is organised as follows. Section 2 provides a concise overview of the state-of-the-art in operational vehicle and pedestrian models in one and two spatial dimensions. The focus is on continuous approaches using ordinary and delay differential equation systems. In Sect. 3, we review the four collective behaviours of stop-and-go waves, lane formation, coordinated obstacle avoidance, and load balancing in light of current modelling approaches and concepts. Finally, in Sect. 4, we discuss some challenges and limitations of current models and propose some development perspectives and research directions.

2 Vehicle and Pedestrian Models

We start by briefly reviewing continuous vehicle and pedestrian operational models using differential systems. The review includes car-following in one dimension in Sect. 2.1 and pedestrian models in two dimensions in Sect. 2.2, covering force-based reactive models and anticipation-based proactive approaches. We refer to the books [27, 79, 106, 127] for more in-depth reviews of the topic, including mesoscopic and macroscopic modelling approaches.

2.1 Car-Following Models

Car-following models are totally asymmetric interaction models, where the dynamics of a vehicle n depends only on the nearest vehicle ahead $n+1$ (see Fig. 1).

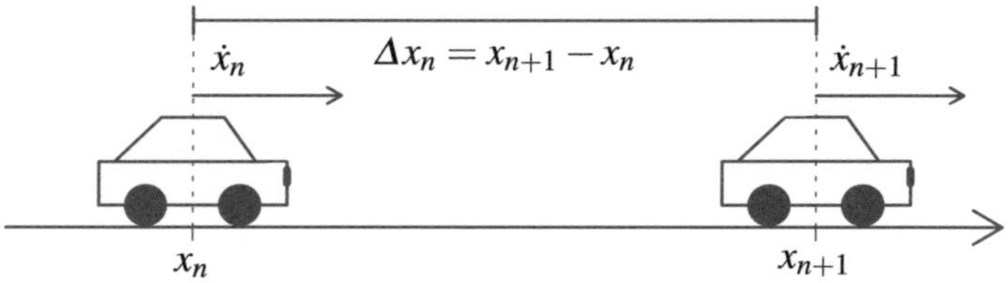

Fig. 1 Main variables for a car-following situation in one dimension. The dynamics of the vehicle n are totally asymmetric. They solely depend on the distance and speed of the predecessor $n + 1$

With $x_n(t) \in \mathbb{R}$ the position of the vehicle n at time t, $\Delta x_n(t) = x_{n+1}(t) - x_n(t)$ the distance to the predecessor and $\Delta \dot{x}_n(t) = \dot{x}_{n+1}(t) - \dot{x}_n(t)$ the speed difference, ordinary car-following models are of the form

$$\dot{x}_n(t) = V(\Delta x_n(t), \Delta \dot{x}_n(t)), \tag{1}$$

$$\ddot{x}_n(t) = A(\Delta x_n(t), \Delta \dot{x}_n(t), \dot{x}_n(t)). \tag{2}$$

Here (1) is a first-order speed model while (2) is a second-order acceleration model. Both models admit for any distance d at least one equilibrium speed $v(d)$ such that $V(d, 0) = v(d)$ and $A(d, 0, v(d)) = 0$.

Car-following models have gone through two major periods of development. The first dates back to their inception in the 1950s and the craze that followed. The second, more recent, dates back to the 1990s and is still in use today.

2.1.1 Pioneering Follow-the-Leader Models (1950–1970)

The first studies on car-following behaviour can be traced back to the 1950s and the pioneering work by Reuschel on the dynamics of road vehicles [102]. In this first car-following model, the forerunner of a long series of works, the speed is directly proportional to distance to the vehicle in front

$$\dot{x}_n(t) = \frac{1}{T}\big(\Delta x_n(t) - \ell\big), \qquad T > 0, \;\; \ell \geq 0, \tag{3}$$

where T is the time gap and ℓ the vehicle size. For this first-order linear model, all the states of the form $(d, v(d) = (d - \ell)/T)$ are systematically stable equilibrium states. The model proposed by Pipes a few years later is also linear, but of the second-order

$$\ddot{x}_n(t) = \lambda \Delta \dot{x}_n(t), \qquad \lambda > 0, \tag{4}$$

where λ is a speed relaxation time. In this model, there is no regulation of distance as a function of velocity, and any uniform speed configuration is systematically stable. The end of the 1950s saw the fruitful development of delayed linear models and collective stability analysis [17,51,65]. Nonlinear models date back to early 1960s with, e.g., the general *Stimulus-Response* follow-the-leader model by Gazis et al.

[41]

$$\ddot{x}_n(t+\tau) = \lambda \dot{x}_{n+1}^{\alpha}(t+\tau)\frac{\Delta\dot{x}_n(t)}{\Delta x_n^{\beta}(t)}, \qquad \tau, \lambda, \alpha, \beta > 0, \tag{5}$$

where α and β are some exponent parameters, and the first-order exponential delayed model of Newell [90]

$$\dot{x}_n(t+\tau) = v_0\left(1 - \exp\left(-\frac{\Delta x_n(t) - \ell}{T v_0}\right)\right), \quad v_0, T > 0, \ \ell \geq 0. \tag{6}$$

This first series of models highlighted three main characteristic aspects of car-following behaviour:

1. the need for nonlinearity in models so as to account for the two regimes of (free and congested) traffic
2. the effect of reaction and relaxation times on traffic stability – it became clear that the models must be sufficiently reactive to achieve traffic stability
3. the possibility to identify parameters such as the maximum speed (or desired speed) v_0 for free traffic, and the reaction time τ and the time gap T that prevail in congested traffic.

In fact, in congested traffic, the speed of the vehicles is assumed to be proportional to the spacing ahead, the time gap T being the inverse of the proportionality factor. It corresponds to the time before collision between a vehicle and its predecessor if the latter were to stop abruptly while the speed of the follower remains constant. The *constant-time-gap-policy* is a following strategy that was initiated in the *California Code* in the US [17] and physically related to safety distances in the early 1980s [42]. The policy is now widely accepted for modelling driver behaviour and also for Adaptive Cruise Control (ACC) systems (see the pioneering work [116]) that is recommended in ISO standards (such as the ISO15622 with a time gap gap ranging from 0.8 to 2.2 s [55] or again 1.8 s for the German ADAC guidelines).

2.1.2 Modern Car-Following Models (1990-Today)

A second spate of nonlinear models emerged in the 1990s. In parallel, the terminology *follow-the-leader* was on the wane, in favour of *car-following* (see [24, Fig. 2]). This second spate is often referred to as *Optimal Velocity* (OV) models. The concept of *Optimal Velocity* comes from the systematic use of a nonlinear equilibrium speed function depending on the distance to the preceding vehicle, a notion initially introduced in the pioneering models by Reuschel [102], Chandler [17] and Newell [90]. Note that the terminology *Optimal Velocity* can be misleading, as there is no optimisation process here. That said, it has made its way into the literature. In modern OV models, the OV function $V : \mathbb{R} \mapsto \mathbb{R}_+$ can take the form of a sigmoidal arc-tangent as in the seminal *Optimal Velocity* model of Bando et al. [6]

$$\ddot{x}_n(t) = \frac{1}{\tau}\big[V\big(\Delta x_n(t)\big) - \dot{x}_n(t)\big], \qquad \tau > 0, \ \ V \in C^1(\mathbb{R}, \mathbb{R}_+), \tag{7}$$

or the *Full Velocity Difference* (FVD) model by Jiang et al. [57]

$$\ddot{x}_n(t) = \frac{1}{\tau_1}\big[V\big(\Delta x_n(t)\big) - \dot{x}_n(t)\big] + \frac{\Delta\dot{x}_n(t)}{\tau_2}, \quad \tau_1, \tau_2 > 0, \quad V \in C^1(\mathbb{R}, \mathbb{R}_+) \tag{8}$$

including two relaxation rates: τ_1 is associated with the relaxation to the distance-dependent OV, whereas the speed difference with the predecessor relaxes over a characteristic time τ_2. This double dependence is not surprising; it is sensible to assume that car-following behaviour depends not only on the distance to the preceding car, but also on their speed difference. Nowadays all modern car-following models depend on both of these variables and also on individual speeds, as *inter alia* in the famous *Intelligent Driver* (ID) model by Treiber et al. [126]

$$\begin{cases} \ddot{x}_n(t) = a\left[1 - \left(\frac{\dot{x}_n(t)}{v_0}\right)^4 - \left(\frac{f(\dot{x}_{n+1}(t), \dot{x}_n(t))}{\Delta x_n(t) - \ell}\right)^2\right], \\ f(v, v_1) = \ell + Tv + v\frac{v - v_1}{2\sqrt{ab}} + \Delta\sqrt{v/v_0}, \end{cases} \tag{9}$$

where $a, b, v_0, T > 0$ and $\ell \geq 0$ are the model parameters, with a a maximum acceleration and b a comfortable deceleration.

The emergence of new nonlinear models in the 1990s and early 2000s notably followed from simulation results showing realistic stop-and-go waves in stationary states for fine-tuned parameter settings [5,6,57]. The mathematical analysis of the models yielded collective stability conditions, typically $2\tau < 1/V'(d) =: T$ for the OV model [6]. In other words, the reaction time τ must be sufficiently small compared to the time gap T, that is, the drivers must be sufficiently reactive. Such findings opened the door to the modelling and control of traffic jam and stop-and-go dynamics [96,139]. A more recent revival comes from the advent of driving automation and the need to develop stable models [22,44]. For example, the Adaptive Time Gap model [122], where the time gap $T_n(t) = \big[\Delta x_n(t) - \ell\big]/\dot{x}_n(t)$, which is the distance gap divided by the velocity, is relaxed to the desired time gap parameter T, i.e., $\dot{T}_n(t) = \frac{1}{\tau}[T - T_n(t)]$, or again in Newtonian notation

$$\dot{x}_n(t) = \frac{1}{T_n(t)}\Big[\frac{\dot{x}_n(t)}{\tau}(T_n(t) - T) + \Delta\dot{x}_n(t)\Big], \qquad \tau, T > 0, \tag{10}$$

is unconditionally stable for all $\tau, T > 0$.

2.2 Pedestrian Models

Turning to pedestrians, the world gains a new geometric dimension, as pedestrians do not evolve on predefined tracks. Thus, pedestrian models are two-dimensional models that describe the dynamics of the pedestrian position $\vec{x}_n(t) \in \mathbb{R}^2$ according to the positions and velocities of its neighbours. In contrast to car-following models which have nearest neighbour interaction schemes in one dimension, the modelled

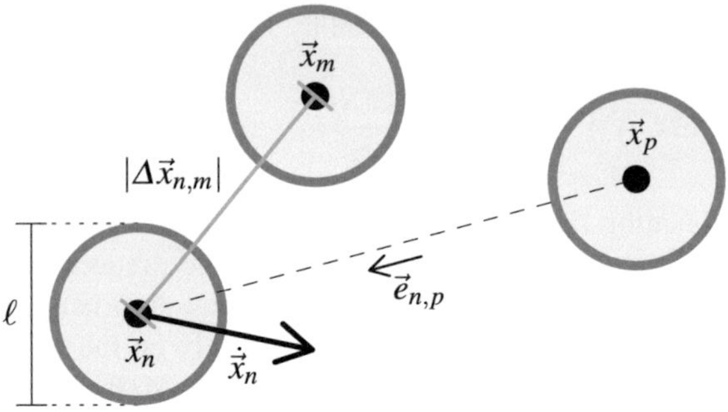

Fig. 2 Main variables of pedestrian dynamics in two dimensions. $d_{n,m}$ is the distance from pedestrian n to pedestrian m, while $\vec{e}_{n,p}$ is the unit vector joining pedestrian p to pedestrian n

pedestrians typically interact with all their close neighbours in two dimensions. The interactions may depend on distance, velocity difference, and also safety related variables such as the time-to-collision (TTC). Pioneering models are based on a superposition of pairwise interaction forces with the neighbours, the so-called *social forces* [50], while more recent approaches rely on collision avoidance, anticipation, and optimisation concepts. Pedestrians are generally represented as disks, as in Fig. 2, although more realistic descriptions use ellipses or even skeleton models for the pedestrian shapes.

2.2.1 Force-Based Models

The concept of a *social force field* can be traced back to the seminal works of Lewin in the social psychology of group dynamics [20, 68, 69]. *Force-based* pedestrian models date back to the pioneering work of Hirai and Tarui in the 1970s on the simulation of crowd behaviour in panic [52] where the dynamics of a pedestrian result from the superposition of different pairwise forces, including *attraction and repulsion forces* with the neighbours and *velocity alignment force*. The concept of *social force* for pedestrian dynamics was later democratised from the 1990s with the work of Helbing and co-authors [48, 50]. Around the same time, swarming-type models such as the *Boids model* by Reynolds [103] and *self-propelled particle model* by Vicsek and co-authors [134], or, later, Cucker-Smale and other swarming and flock models [15, 30] have been developed with interaction mechanisms based on averages with neighbours and similar principles of pairwise interaction superposition. However, in contrast to swarming models, space is not isotropic in pedestrian models: agents experience a drift towards their desired destinations in addition to the interaction field.

Minimal force-based models are given by the second-order dynamics

$$\ddot{\vec{x}}_n(t) = \frac{1}{\tau}\big[\vec{v}_0 - \dot{\vec{x}}_n(t)\big] + \sum_{m \neq n} \omega\big(\phi_{n,m}(t)\big)\,\nabla_{\vec{x}}\mathscr{U}\Big[\Delta\vec{x}_{n,m}(t),\,\Delta\dot{\vec{x}}_{n,m}(t)\Big], \qquad (11)$$

where $\tau > 0$ is a reaction (relaxation) time, $\vec{v}_0 \in \mathbb{R}^2$ is the *desired velocity* coming from a tactical modelling level, the social force is the gradient in space of the interaction potential $\mathscr{U} \in C^2(\mathbb{R}^4, \mathbb{R}^2)$, while $\omega \in C^1(\mathbb{R}^2, \mathbb{R})$ is an anisotropic factor based on the bearing angle $\phi_{n,m}$ that provides more weight to the pedestrians in the direction of motion. For example, in the pioneering *Social Force* (SF) model [50], the interaction force is derived from an exponential repulsive distance-based interaction potential and a piecewise constant anisotropic factor

$$\mathscr{U}_{SF}(\Delta\vec{x}) = AB\exp(-|\Delta\vec{x}|/B), \quad \omega_{SF}(\phi) = \begin{cases} 1 \text{ if } |\phi| < \kappa, \\ c \text{ otherwise,} \end{cases} \tag{12}$$

where $A, B > 0$ are the repulsion range and the characteristic distance, respectively, while $0 < c \ll 1$. Due to the exponential potential, the interaction of the social force model is short-ranged.

In the *Centrifugal Force* (CF) model [144] and the *Generalised Centrifugal Force* (GCF) model [21], the range of interaction decays algebraically. The repulsive force for the CF model is

$$\nabla_{\vec{x}}\mathscr{U}_{CF}(\Delta\vec{x}, \tilde{\Delta\dot{\vec{x}}}) = \beta\frac{|\tilde{\Delta\dot{\vec{x}}}|^2}{|\Delta\vec{x}|}\frac{\Delta\vec{x}}{|\Delta\vec{x}|}, \tag{13}$$

while the force for the GCF model reads

$$\nabla_{\vec{x}}\mathscr{U}_{GCF}(\Delta\vec{x}, \tilde{\Delta\dot{\vec{x}}}) = \beta\frac{\left[\eta\vec{v}_0 + \tilde{\Delta\dot{\vec{x}}}\right]^2}{|\Delta\vec{x}| + r(\dot{\vec{x}})}\frac{\Delta\vec{x}}{|\Delta\vec{x}|}, \tag{14}$$

where $\beta, \eta > 0$ are parameters, $r \in C(\mathbb{R}^2, R_+)$ models a velocity-dependent pedestrian shape, e.g. an ellipse [21], while $\tilde{\Delta\dot{\vec{x}}}$ is the relative velocity projected on the direction to the neighbouring pedestrians. Unlike the SF model, the CF and GCF models also rely on the relative velocity to the neighbours. Nowadays, many extensions of the SF model include relative velocity terms and other velocity-dependent mechanisms; see the review [18] and references therein. For instance, in [98], the SF model extension consists of some respect areas in the vicinity of the pedestrians that force them to stop. Similarly, but in a dynamic way, in [145] the SF model is extended with explicit collision prediction. In another recent approach, the interaction force is anisotropic to take into account for a preferred crossing direction [125].

2.2.2 Velocity-Based Models

A stream of models developed subsequently to the force-based models are the velocity-based models. In contrast to their Newton-like forebears, velocity-based models instantly regulate velocity at the first order. They are inherently free of inertia and allow strict control of body exclusion—an aspect that is difficult to manage with a force-based model [20, 113]. Velocity-based models are initially borrowed from the robotics. They generally rely on velocity optimisation processes under collision constraints over given anticipation times.

Such developments can be traced back to the pioneering work of Fiorini et al. using purely geometric *Velocity Obstacle* (VO) models [37]. The model is based on linear extrapolations of the pedestrian trajectories to determine potential collision sets. Geometrically, these sets assume the form of so called collision cones under the hypothesis of constant pedestrian velocities. The cone is empty if no collision is expected. The velocity dynamics of agent n is then obtained by minimising the deviation from the desired velocity under the exclusion constraint of the time-dependent collision cones $\mathscr{C}_n^m(t)$ with other agents m

$$\dot{\vec{x}}_n(t) = \underset{\vec{v}_n \notin \cup_{m \neq n} \mathscr{C}_n^m(t)}{\arg\min} \|\vec{v}_0 - \vec{v}_n\|^2 \qquad (15)$$

where $\vec{v}_0$ is the desired velocity. Since the collision cone depends on the current velocities of the agents, the velocity model is implicit. In practice, the implicit system is solved numerically using semi-implicit schemes combining an explicit Euler solver for the velocity with an implicit Euler scheme for the positions of the agents. Initially, velocity obstacle models were developed for controlling robots in an environment including moving obstacles and automated vehicles driving on a highway [37]. First applications for pedestrian dynamics date back to the late 2000s [97,131].

Generally speaking, velocity obstacle models can be formulated as an optimisation problem under constraints [133]

$$\dot{\vec{x}}_n(t) = \underset{\vec{v}_n \notin \cup_{m \neq n} \mathscr{C}_n^m(t)}{\arg\min} C_{\vec{x}(t)}^{\vec{v}_0}(\vec{v}_n) \qquad (16)$$

where $C_{\vec{x}}^{\vec{v}_0}$ is a cost function depending on the desired velocity and also the positions and velocities of the neighbours. In the collision-free model developed in [97], the cost function includes components associated with speed variations, distance to desired speed, and orientation changes. In the *Reciprocal Velocity Obstacle* (RVO) model [131], the cost is the sum of a quadratic deviation from the desired velocity as in the original velocity obstacle model plus a repulsive potential based on the time to collision. The cost function is a quadratic deviation from the desired speed as in the initial VO model (15) in the *Optimal Reciprocal Collision Avoidance* (ORCA) model [130], while its extension PORCA model [75] includes an additional velocity-dependent term that prevents the system from freezing. No collision cones are used in the *Linear Trajectory Avoidance* model [99]. Instead, the optimisation is done by exponentially penalising the minimum distance under linear extrapolation to avoid collisions, in addition to a quadratic deviation from the desired speed as in the original VO model (15).

A general collision-free velocity-based modelling framework was proposed in the 2000s by Maury and Venel [80,81]. Its principle also rests on velocity optimisation under collision-free constraints. However, unlike the models described above where the pedestrian velocities are optimised individually, the optimisation is performed centrally over all the pedestrians. The pedestrians are considered as disks of diameter ℓ. The set of admissible configurations of the system excluding overlapping configurations is given by

$$Q = \{\vec{x},\ D_{m,n} = |\Delta\vec{x}_{n,m}| - \ell \geq 0,\ \ \forall m \neq n\}. \tag{17}$$

Then the set of admissible velocities constrains the system to remain in the admissible configurations

$$C_{\vec{x}} = \{\dot{\vec{x}},\ D_{m,n} = 0 \Rightarrow \nabla D_{m,n} \cdot \dot{\vec{x}}_n \geq 0,\ \ \forall m \neq n\}. \tag{18}$$

If a contact occurs, i.e., $D_{m,n} = 0$, then the admissible velocity enforces the motion away as $\nabla D_{m,n} \cdot \dot{\vec{x}}_n \geq 0$ and so the distance increases. This formally requires the system to remain within the admissible configurations. Finally, similar to velocity obstacle models (15), the velocities are computed using the orthogonal projection of the desired velocity onto the admissible velocities

$$\dot{\vec{x}}(t) = P_{C_{\vec{x}(t)}}\vec{v}_0. \tag{19}$$

The use of orthogonal projection reproduces dynamics where pedestrians try to reach their destination as quickly as possible under collision-free constraints, as in an emergency evacuation.

2.2.3 Hybrid Models

Many pedestrian models are hybrid approaches combining different concepts and mechanisms from first- and second-order models such as social force, optimal velocity (OV), or optimisation and collision avoidance techniques. For example, some extensions of force-based models are inspired by swarming approaches [48,66], where the desired velocity $\vec{v}_0$ in (11) is mitigated with the average velocity of the neighbouring pedestrians using

$$\bar{\vec{v}}_0(t) = p\vec{v}_0 + (1-p)\langle\dot{x}_m(t)\rangle_n, \qquad p \in [0, 1], \tag{20}$$

to address mass collective behaviour. In the same vein, the pedestrian velocity is relaxed to a combination of the desired velocity and a superposition of distance-based OV models [87,88], generalising concepts of car-following models [5,6,57] to two dimensions

$$\ddot{\vec{x}}_n(t) = \frac{1}{\tau}\left[\left[\vec{v}_0 + \sum_{m\neq n} f\big(|\Delta\vec{x}_{n,m}(t)|\big)\omega\big(\phi_{n,m}(t)\big)\vec{e}_{n,m}(t)\right] - \dot{\vec{x}}_n(t)\right], \tag{21}$$

where $f(x) = \alpha\big(a + \operatorname{atan}\beta(x-b)\big)$ is the OV function, $\omega(x) = 1 + \cos(x)$ is an anisotropic vision field factor, and $\vec{e}_{n,m}(t)$ is the unit vector directed from n to m.

Another hybrid two-dimensional OV pedestrian model relies on first-order dynamics [120]. As in the modelling framework introduced by Maury and Venel in [80,81], this model constrains pedestrian velocities to remain in admissible collision-free configurations [120]. However, the speed is regulated according to the distance to the next pedestrian in the direction of motion using an OV function, while the direction results from a superposition of distance-based repulsion with the neighbours as in force-based models. An extension of this collision-free velocity-based model relaxes the direction, making it a second-order model [143]. In the *Gradient Navigation* model introduced in [31], the speed is relaxed to the OV function, while the direction results from a the desired velocity and a superposition of distance-based repulsion with the neighbours as in force-based models

$$\begin{cases} \dot{\vec{x}}_n(t) = w_n(t)\big[C_1\vec{v}_0 + C_2 \sum_{m\neq n} \nabla \mathscr{U}_{n,m}\big], \\ \dot{w}_n(t) = \frac{1}{\tau}\big[V(\rho_n) - w_n(t)\big], \end{cases} \tag{22}$$

where C_1 and C_2 are two normalisation constants while $V(\rho_n)$ is the OV depending on the local density ρ_n. A similar approach is given in [86] with a heuristic model without social interaction and where, as in the collision-free velocity-based model [120], the OV depends on the distance to the nearest pedestrian in front.

The concept of anticipatory collision avoidance in force-based models introduced in [145] is deepened in a recent hybrid force-based model [58], where the interaction force is derived directly from the potential

$$\mathscr{U}_{TTC}(\tau_{n,m}) = \frac{a}{\tau_{n,m}^2}\exp(-\tau_{n,m}/\tau_0), \qquad a, \tau_0 > 0, \tag{23}$$

based on the time-to-collision (TTC)

$$\tau_{n,m} = \begin{cases} \frac{\Delta\vec{x}_{n,m}\Delta\dot{\vec{x}}_{n,m} - \sqrt{D}}{\Delta\dot{\vec{x}}_{n,m}^2} & \text{if } D > 0, \\ \infty & \text{otherwise,} \end{cases} \qquad D = \big(\Delta\vec{x}_{n,m}\cdot\Delta\dot{\vec{x}}_{n,m}\big)^2 - |\Delta\dot{\vec{x}}_{n,m}|^2(|\Delta\vec{x}_{n,m}|^2 - \ell^2). \tag{24}$$

The TTC $\tau_{n,m}$ is the time before a collision occurs between the n-th and m-th pedestrians (i.e., $|\Delta\dot{\vec{x}}_{n,m}| < \ell$), assuming that their velocities are constant. In [74], the force is based on similar collision anticipation mechanisms. A general hybrid modelling framework first introduced in the early 2000s [53] assumes pedestrian dynamics based on a *decisional layer*, resulting from an optimal control problem, and a *mechanical layer* modelling physical exclusion force and other inertia phenomena

$$\vec{\tilde{v}}_0^{\,n}(t) = \underset{\vec{v}_n\in\mathbb{R}^2}{\arg\min}\ \mathscr{E}_{\vec{x}(t)}^{\vec{v}_0}(\vec{v}_n), \qquad \ddot{\vec{x}}_n(t) = \frac{1}{\tau}\Big[\vec{\tilde{v}}_0^{\,n}(t) - \dot{\vec{x}}_n(t)\Big] + F_n^{\mathrm{M}}(t), \tag{25}$$

where $\mathscr{E}$ is an energy function involving several concepts, while $F_n^{\mathrm{M}}(t)$ are the mechanical force on the n-th pedestrian at time t. In the *ANticipatory Dynamic Algorithms* (ANDA) [34], the energy function consists of four main components:

1. *Target energy*, which models the desired destination and velocity from a tactical modelling level,

2. *Speed energy* including bio-mechanical costs for walking speed and inertia,
3. *Repulsion energy* in the form of a superposition of pairwise nonlinear distance-based repulsive potentials with the neighbours,
4. *Anticipation energy* based on the TTC potential (23)–(24) introduced in [58].

3 Modelling Collective Dynamics

Traffic and pedestrian flows describe many coordinated collective dynamics [10, 16,19,46] A typical example in traffic flow is stop-and-go dynamics on highways where jam waves resulting from instability phenomena [96,139], also called *jamitons* [38], propagate upstream in the flow without any apparent external cause. Typical examples for pedestrian dynamics are lane formation for counterflow, jamming and clogging at bottlenecks, oscillation and flow intermittency for counterflow at bottlenecks, or stripe formation for crossing flow [10,47]. Most of the aforementioned models describe these phenomena, with realistic features upon fine-tuning the parameters. However, conventional models fall short of replicating more complex situations involving long-term anticipation (on the order of several seconds) or socio-psychological aspects. A typical example is the formation of a clear lane in pedestrian crowds to facilitate the passage of a vehicle, typically an ambulance [2], which relies on early anticipation. Another example in the load balancing of pedestrians over several exits in case of normal evacuation. Here, the distance between the exits requires pedestrians to anticipate for long periods of time. These different collective phenomena and related modelling approaches are discussed in more detail in the following sections.

3.1 Stop-and-Go Waves in Traffic Flow

In this section, we focus on stop-and-go dynamics in congested traffic and show how collective linear stability analysis can be used to address this phenomenon.

3.1.1 Empirics of Stop-and-Go Waves in Traffic Flow

Stop-and-go dynamics are currently observed in road traffic flow. They are also found in single-file pedestrian movements and in bicycle flow [147]. In fact, stop-and-go and accordion traffic now commonly denotes congested traffic. In addition to its scientific interest, the jerky dynamics of congested traffic is a major safety concern. It also has a negative impact on the environment. Indeed, the acceleration and deceleration phases of stop-and-go traffic dynamics cause an increase in fuel consumption and pollutant emissions [71].

The terminology of *stop-and-go waves* can be traced back to the late 1960s [32]. A related word is *phantom jam* [114], as the waves seem to appear for no apparent reason. This was demonstrated in the late 2000s with an experiment involving 22 vehicles on a 230-m circuit [114] (see Fig. 3). Starting from a homogeneous config-

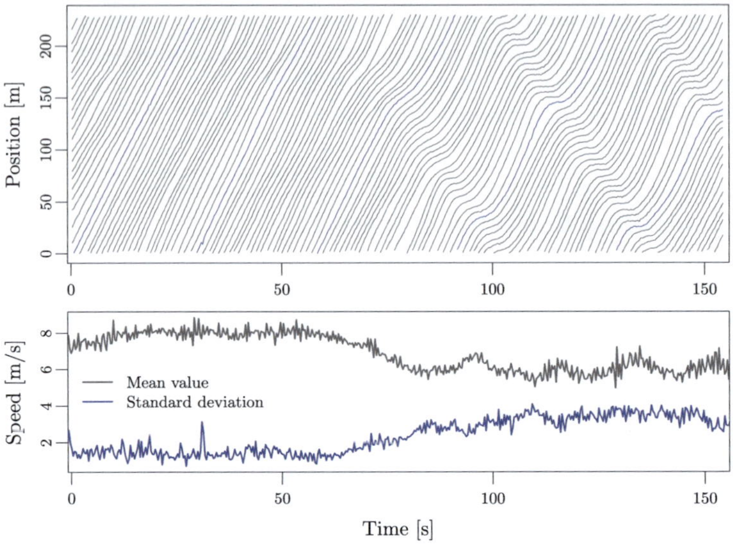

Fig. 3 Trajectories in the experiment of Sugiyama et al. with 22 vehicles on a circuit of length 230 m starting from an uniform configuration [114]. After a while, a stop-and-go wave appears, which causes a decrease in the average speed and an increase in the speed standard deviation

uration with equal distance gaps, a stop-and-go wave emerged after a while, with jerky vehicular dynamics not caused by any infrastructure peculiarity. This experiment shed new light on the formation of waves in traffic flow, and has helped to popularise it. A more recent experiment under similar conditions has shown that a single vehicle with particularly stable dynamics allows the waves to dissipate. This field of research remains particularly active (see Sect. 4 below).

3.1.2 Collective Stability Analysis

The formation of stop-and-go waves and phase separation from laminar traffic to oscillatory dynamics was theorised in the 2010s, mostly using linear stability analysis [96, 127, 139]. Local stability analysis considers a platoon of vehicles following a leader. Stability conditions are usually derived using Laplace transform and transfer function. Global (or string or collective) stability analysis considers a flow of vehicles with periodic boundary conditions. In general, global stability conditions are more restrictive than local ones, since global analysis takes into account advective or convective perturbations that vanish locally. The stability conditions for infinite systems are even more restrictive. The stability of finite periodic systems can be derived using matrix spectral analysis, while for infinite systems an exponential ansatz is generally used. The latter is equivalent to performing a Laplace transformation in time and a discrete Fourier transformation in space [96].

To be more specific, consider the linearisation of the car-following model around the homogeneous equilibrium solution

$$\ddot{\tilde{x}}_n(t) = a\Delta\tilde{x}_n(t)) + b\dot{\tilde{x}}_n(t) + c\dot{\tilde{x}}_{n+1}(t) \tag{26}$$

and N vehicles with periodic boundary conditions on a circuit of lane L (i.e., the N-th vehicle interacts with the first vehicle and its spacing is $\Delta x_N(t) = x_N(t) - x_1(t) + L$). The matrix system being circulant, it is easy to check that the characteristic equation of the dynamics reads

$$\lambda^2 + \lambda(b + c\iota_k) + a(1 - \iota_k) = 0, \quad \iota_k = \exp\big(\mathrm{i}2\pi k/N\big), \ k = 0, \ldots, N-1. \quad (27)$$

Here ι_k is the kth root of the unit, which appears thanks to the circulatory property of the linear system. The equation is a second order complex polynomial or again a real polynomial of order $2N$. Here $\lambda = 0$ is a solution for $k = 0$ resulting from the periodic boundary where the sum of the spacings $\sum_{n=1}^{N} \Delta x_n(t) = L$ is conserved at any time, while the real part of the remaining eigenvalues are strictly negative (i.e., the system is linearly stable) if $a > 0$, $b < 0$, and $b^2 - c^2 > 2a$, see, e.g., [123], [127, Chap. 15] and general stability conditions for complex polynomials in [39, Th. 3.2] which are a generalisation of the Hurwitz conditions. The conditions are sufficient for any finite system and become exact for infinite systems. For example, the linear stability condition is $0 < 2\tau < 1/V'(L/N)$ for the OV model (7), $0 < 2\tau_1\tau_2/\big(2\tau_1 + \tau_2\big) < 1/V'(L/N)$ for the FVD model (8), while the ATG model (10) is systematically linearly stable provided that $\tau, T > 0$, hence its prevalence for automated driving systems [61].

3.2 Lane Formation in Pedestrian Dynamics

In this section, we turn to the collective dynamics of lane formation for counterflow of pedestrians moving in opposite directions and the class of models that describe this phenomenon.

3.2.1 Empirical Observations in Pedestrian and Other Particle Systems

Counterflows, where two groups of pedestrians move in opposite directions, often lead to the formation of lanes of pedestrians walking in the same direction. Lane formation in pedestrian dynamics has been observed since the 1960s [89,95] and is studied in experiments (see Fig. 4 and [14]). It may also be related to the stripes formed at intersections between pedestrian streams [4]. In fact, the phenomenon of lane formation is widely encountered in discrete systems in which entities move in two opposite directions [33]. For example, lane formation occurs in passive matter, e.g., driven mixtures of oppositely charged colloids [135] or driven plasmas [105,115], and in active matter, such as in microswimmers [64], bacterial populations [109], or social insects such as ants [26]. Nowadays, laning phenomenon is recognised as a fundamental collective property of particle and self-avoiding agent systems driven by a mixture of two potentials [28,101].

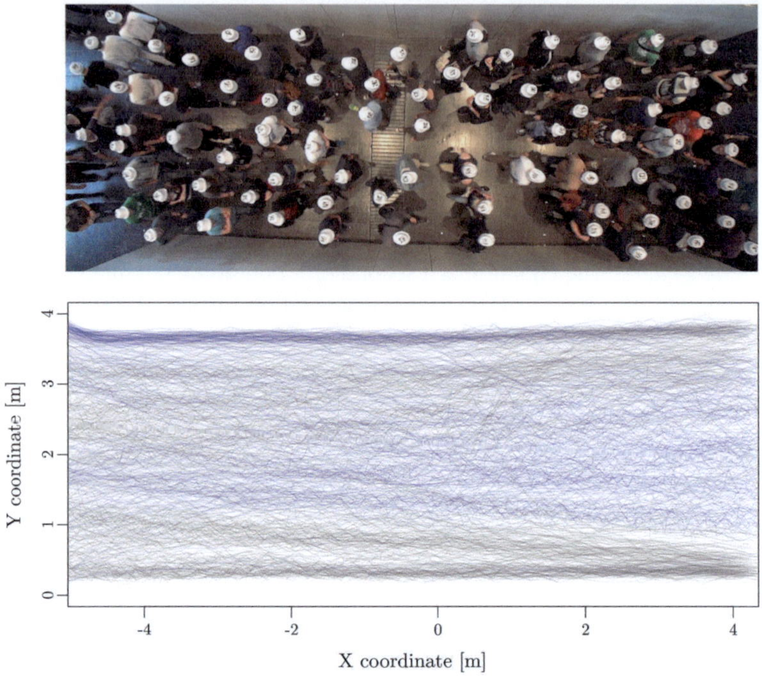

Fig. 4 Trajectories of pedestrians in a counterflow experiment [14]. The blue trajectories are pedestrians moving from right to left, while the grey trajectories are pedestrians moving from left to right. We can see the formation of lanes moving in opposite directions

3.2.2 Modelling Lane Formation

Lane formation in pedestrian and other mixed-particle models is difficult to analyse mathematically and is usually studied by simulation or using continuous appraoches (see, e.g., [4]). In fact, most of two-dimensional pedestrian models are able to reproduce laning in counterflow. Pioneering simulation results were obtained in the 2000s using the social force model [48], basic cellular automata [12], or, a few years later, velocity obstacle models [45]. The formation of lanes can also be explained by socio-psychological factors of group membership and social identity [117, 138]. In [60], lane formation is even observed for mixed flows moving in the same direction but at different speeds. However, some experiments with pedestrians walking at different speeds did not provide empirical evidence of lane formation [40].

In fact, the occurrence of lane formation is observed also in systems of inanimate particles. For example in [115], the volume exclusion and collisions due to the motion in different directions in driven binary complex plasma, combined with joint pushing of particles moving in the same direction, induce a segregation into lanes. Thus, any minimal models for pedestrian dynamics representing volume exclusion, such as force models with repulsive interactions, are able to reproduce lane formation. This example indicates that lane formation itself does not need social or psychological explanations, even if the steering process of humans in such a situation is based on the impressive cognitive and biomechanical abilities of the human body. These aspects are discussed in more detail in [111].

In general, lane formation in pedestrian counterflow depends on density. At low densities, it may be advantageous to avoid rather than follow, thus limiting lane formation. Conversely, at high densities, exclusion constraints may prevent the formation of lanes. However, there is no empirical evidence for a transition, although experiments have shown that counterflow can give rise to gridlock at densities greater than 3.5 persons per square metre [146]. In the models, these density thresholds above which lanes no longer emerge depend on the form of the model in addition to the initial configuration [121, 142]. They remain an open rearch question. Another factor that leads to counterflow gridlock is the introduction of noise into pedestrian motion models. This phenomenon is known in the literature as *freezing by heating* [49]. Even in simple deterministic models, a transition to lane formation occurs as the driven force increases relative to the interacting repulsive forces [124].

3.3 Far-Sighted Anticipation in Local Navigation

Rather than lanes arising from oppositely-directed flows, let us now inspect the response of a static crowd to a moving 'obstacle'. For this purpose, we focus on an experiment presented in [93] that highlights the long-term anticipatory abilities of pedestrians to avoid an intruder, and how mean-field games can be used to model this type of behaviour.

3.3.1 Empirical Evidence for Long-Term Anticipation

That pedestrians or cars plan their motion tens of minutes or even hours ahead should not come as a surprise. Unlike physical particles, their local navigation is steered towards a target destination set at larger modelling scales. The existence of a desired direction, imparted to agents at the higher tactical level, makes space anisotropic for them (i.e., not all directions are equivalent), irrespective of their interactions with the rest of the crowd, in contrast to boids [103] or simple swarming models [134] for which directional preferences only result from interactions.

What may be less obvious is that even the local, operational interactions with their neighbours and the built environment may involve some degree of planning. Direct visualisation of the simulation output of simple force-based models and the frequent near-collisions they imply evince that distance-based interaction forces are not sufficient to render realistic collision avoidance. Taking into account the relative velocities of the agents is thus necessary. Evaluating distances via this relative velocity, i.e., in terms of the imminence of a collision risk, gives rise to the well-known notion of anticipated times to collision (TTC). Experimental observations (collected in real life or in virtual reality) as well as empirical data show the central role played by the TTC for collision avoidance and the spatial arrangement of crowds [25, 58, 82, 100]. Interactions between people may start at a TTC of several seconds, when they are more than half a dozen meters away from each other.

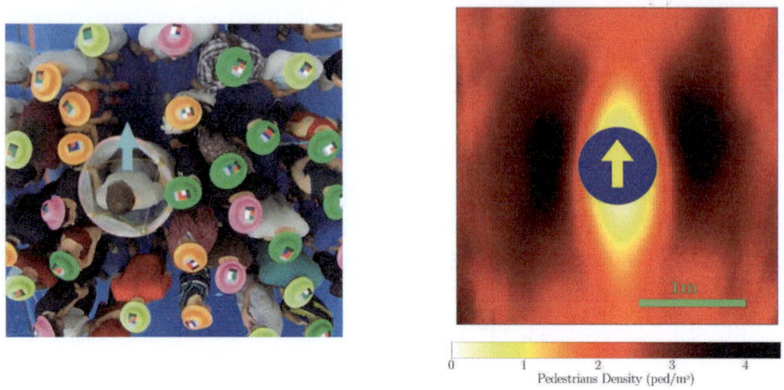

Fig. 5 Experiment [93] in which a dense crowd is crossed by an cylindrical intruder (Left). Pedestrians anticipate the passage of the cylinder as seen from the void in front of the cylinder, and temporarily accept the discomfort of the dense regions on both sides of the cylinder (Right, from [13]), because they know it will not last

Interestingly, anticipation may even need to be pushed beyond the nearest collision to describe operational pedestrian dynamics [136]. In particular, a crowd's response to an intruder's crossing [93] (by transverse moves opening a depleted tunnel ahead of the intruder, see Fig. 5) appears to elude state-of-the-practice models based (or not) on TTC [11, 13]. The observed response could actually be understood by assuming that pedestrians anticipate much beyond the next collision, and actually beyond the passage of the moving obstacle, as we shall see.

3.3.2 Mean-Field Games

Situations where pedestrians have to compete for space, such as the one shown in Fig. 5, can be modelled using game theory. It is assumed that each pedestrian chooses her strategy—her preferred velocity—in order to minimize some cost integrated over the whole future. Each pedestrian will then follow her preferred velocity up to some noise, namely $\dot{\vec{X}}_t = \vec{v}_0(t) + \sigma\xi(t)$, with ξ a δ-correlated normalized white noise. A possible cost is

$$
\begin{aligned}
c[\vec{v}_0(\cdot)](\vec{x}, t) = \mathbb{E}\Bigg[\int_t^T \left[\frac{\mu}{2}(\vec{v}_0(\tau))^2 - g\rho(\vec{X}_\tau, \tau) - U_0(\vec{X}_\tau, \tau)\right] e^{\gamma(t-\tau)} d\tau \\
+ e^{\gamma(t-T)} c_T(\vec{X}_T)\Bigg]_{|\vec{X}_t = x},
\end{aligned}
\tag{28}
$$

where the expectation value is taken on the noise of the Langevin equation. This cost includes some penalty for high velocities—see the square term $(\vec{v}_0)^2$—and, with $g < 0$, for high pedestrian densities ρ. Besides, the potential $U_0 < 0$, assumed large, prevents pedestrians to occupy the space under obstacles—for example here the cylinder. The last term c_T is a terminal cost which may be useful for example if pedestrians have to reach a given location. The integral is taken over a long future (T

is supposed to be large), but the exponential term allows to bound the anticipation to a typical timescale $1/\gamma$.

Note that the cost does not depend on the precise pedestrian, but is a function of $\vec{x}$. This means that all pedestrians at the same location at the same time will have the same strategy. The interaction between pedestrians is entirely contained in the density term in the cost, meaning that in this model, pedestrians do not react to the precise location of each other, but rather to an ensemble averaged density that they foresee, in the spirit of a mean-field assumption. Indeed, if all individuals had to be taken into account, the problem would become overwhelmingly complex.

The OV $\vec{v}_0^*(t)$, that is the velocity that optimized the cost Eq. (28), can then be computed introducing the *value function* $u(\vec{x}, t) = \min_{\vec{v}_0(\cdot)} c(\vec{x}, t)$, and making use of Bellman's optimality principle, which states that if we want to optimize the strategy along its whole history up to time T, we need to optimize it between any current time τ and T. This allows one to construct the history of u backward from the final time T, and u is shown to obey the so-called *Hamilton–Jacobi–Bellman* (HJB) equation

$$\begin{cases} \partial_t u = -\frac{\sigma^2}{2} \Delta u + \frac{1}{2\mu} (\vec{\nabla} u)^2 + \gamma u + g\rho(\vec{x}, t) + U_0(\vec{x}, t) \, . \\ u(\vec{x}, t = T) = c_T(\vec{x}) \end{cases} \tag{29}$$

From $u(\vec{x}, t)$, the OV is then obtained as $\vec{v}_0^*(t) = -\vec{\nabla} u/\mu$. This equation must be supplemented by the evolution of the density, given that all pedestrians evolve according to the aforementioned *Langevin* equation, with their preferred velocity being the one yielding the minimum cost u. The associated equation, forward in time, reads

$$\begin{cases} \partial_t \rho = \frac{\sigma^2}{2} \Delta \rho + \frac{1}{\mu} \nabla \cdot (\rho \nabla u) \\ \rho(\vec{x}, t = 0) = \rho_0(\vec{x}) \end{cases} \tag{30}$$

Solving the two coupled equations (29)–(30) can be done iteratively. Besides the equation system can be simplified in some cases, for example when a stationary state is reached.

This framework allows to reproduce the anticipative behaviour of pedestrians, in particular in the case of the incoming cylinder of Fig. 5. An underlying assumption is that pedestrians can predict the future density, relying on a repeated experience. In case where pedestrians have only a limited knowledge of what is happening—for example because they cannot see some parts of the scene—it can be useful to adjust their anticipation horizon $1/\gamma$ [13].

3.4 Load Balancing in Pedestrian Evacuation

Another experiment that reveals collective dynamics and a form of long-term anticipation behaviour is the evacuation of pedestrians under normal conditions in a geometry with multiple exits [136]. We can observe that the pedestrians spread over the

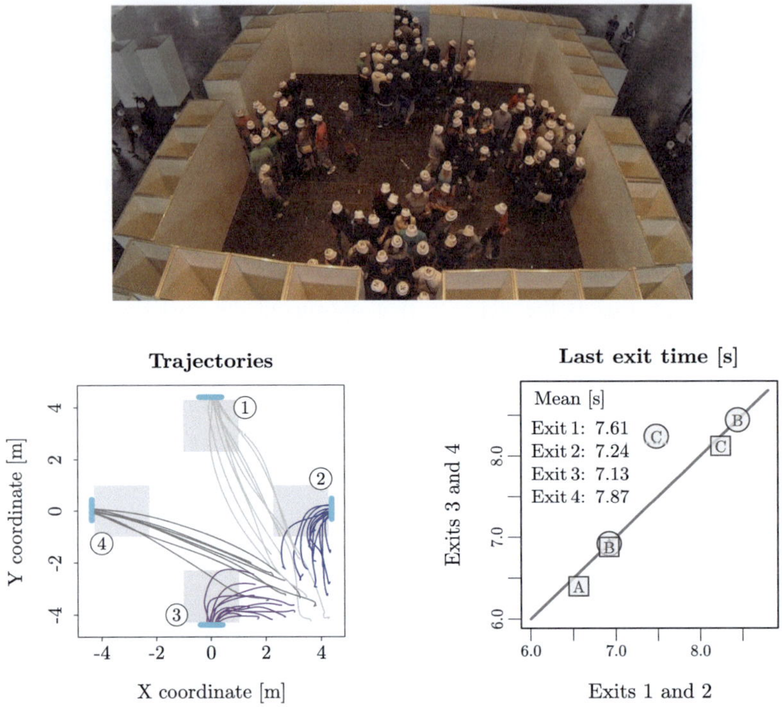

Fig. 6 Pedestrian evacuation experiment with four possible exits [136]. Three runs were repeated with pedestrians starting at the bottom right of the geometry (see the trajectories in the lower left panel for the run A). The distribution of the pedestrians over the four exits shows a load balancing with similar last exit times (see the lower right panel—the circles correspond to exits 2 and 3 while the squares correspond to exits 1 and 4; the letter identifies the run)

different exits to balance the load and optimise the final evacuation time, whether for rectangular, curved or square geometries, as in Fig. 6. As the exits are several metres apart, load balancing requires the agent to anticipate for a few seconds to select the exit and move accordingly. This typically exceeds the short-term anticipation of the operational model and requires the inclusion of some long-term tactical decision models. This involves operational/tactical coupling in modelling the dynamics, which may operate on different time scales.

The tactical choice of an exit is a problem of differential games, because the problem is dynamic and the choice of an exit by a pedestrian depends on the choices of his neighbours. In addition, static distance-based shortest path algorithms prove ineffective, as the choice of exit depends on the current level of congestion on the path to the exits. Real-time pedestrian exit choices can be modelled by centralised optimisation algorithms using, for example, a *Hamilton–Jacobi–Bellman* equation as in a mean field game [54], non-cooperative individual behavioural rules [59, 72], or learning processes based on previous experience [29]. In general, individual estimation of pedestrian travel time can be computationally intensive. However, heuristic tactical approaches based on local density levels prove to provide realistic behaviour [59, 136]. In any case, the modelling approach requires the tactical exit choice to be dynamic and the operational and tactical modelling scales to be coupled.

4 Conclusion and Challenges

In summary, we have presented the state-of-the-art in pedestrian and traffic modelling. However, many challenges and research directions for modelling the collective dynamics of vehicles and pedestrians still need to be addressed. In conclusion, we now present a few of them. These include the stabilisation of road traffic flow, the development of realistic pedestrian behaviour models, the consideration of socio-psychological aspects, or the development of generic order parameters for the characterisation of collective dynamics.

4.1 Stabilisation of Stop-and-Go Waves

Stop-and-go dynamics are currently observed in conventional road traffic flow. In addition, several recent experiments [22, 44, 70, 76, 84] show that adaptive cruise control (ACC) driver assistance systems present some stability issues, partly due to large response times of the systems [77]. In the context of driving automation, there is an important need to develop stable ACC systems that are robust to delays, latency, and stochastic perturbations in the dynamics. Indeed, many factors can disrupt the stability of the flow and lead to collective oscillations.

First experiments in the late 2010s show that stable system allows the wave to be dissipated [112]. However, this requires the system to maintain a very large time gap, which limits flow performance. Connected platooning systems, known as cooperative adaptive cruise control (CACC) systems, enable efficient automated flows, particularly for truck platooning [85, 137]. However, they rely on communication systems and are not autonomous. Various compensator modules for lag and delay in the dynamics, typically using feedback control or Smith predictor, as well as filters for noise attenuation are being developed [8, 62, 141]. However, the topic remains challenging. Recently, an experiment was conducted in the United States with up to 100 connected and autonomous vehicles to stabilise traffic flows [67], including reinforcement learning techniques to damp oscillations [56][1].

4.2 Developing Realistic Microscopic Pedestrian Models

The practical demand for better pedestrian dynamics models that would superside existing software suggests that one should strive for as much realism in these models as possible, hence greater sophistication. A seemingly natural conclusion would

[1] See also https://circles-consortium.github.io/.

be to try and refine the description of agents, notably by better accounting for the psychological processes at play, for the motivation experienced by people and their emotions. Indeed, using the pedestrian flow at a bottleneck as an illustration, it has been shown that the crowd's level of motivation impacts how densely people pack in front of the bottleneck [1], which in turn largely controls the flow rate at not-too-high pressure [92]. In this vein, Thalmann and Musse had thus prompted the idea that the interplay between perception and emotion determines an agent's behaviour, and should thus be accounted for [119].

Nonetheless, Corbetta et al.'s ability to reproduce the statistics of U-turns on a staircase platform with a simple scalar potential energy description [23], irrespective of the great variety of decisions that led to such U-turns in practice, demonstrates that the dynamical effect of complex psychological factors can be captured at a high level without delving into the substance of these factors.

In short, the required degree of realism obviously depends on the practical purpose, and in particular whether the consequences of the motivation experienced by people (in terms of desired speed, direction, etc.) can be taken as an input to the model, or need to be explained in greater depth. That being said, for any practical application, we believe that state-of-the-practice models can hardly ignore two long-overlooked factors: (i) the presence of social groups (of friends, co-workers, family, etc.), which implies a unique relative arrangement of people in space (taking up more lateral space on a corridor than would be predicted for single individuals), (ii) the non-uniformity of space as perceived by pedestrians in real settings, which sharply contrasts with the uniform space idealised in controlled experiments designed to single out intrinsic features of crowd dynamics.

4.3 Including Socio-Psychological Concepts

In many situations, a natural scientific perspective on the dynamics of pedestrians is not sufficient, but must be supplemented by sociological or psychological concepts, such as, for instance, group behaviour, social identity [107, 138] or social norms [110]. This is illustrated below with two examples.

In some situations, one can observe different behaviour of individuals which can change dynamically. In [1, 111] this can be observed in the time distance plots, Fig. 6 in [1] where the paths show that some people push their way to the front while others fall back. A more detailed analysis is provided in [129] where psychologist developed an individual annotation of the behaviour of people in a crowd. These behavioural data allow to study the space-time dynamics of pushing behaviour like the relation of distance to the entrance and the probability of pushing, see [128]. The authors showed, for example, that the probability to observe pushing behaviour increases when people approach their goal, in this experiment the entrance to a concert.

To reproduce such a space-time dynamic of behaviour, state-of-the-art models need layers that model behavioural changes. Furthermore, a critical look at classic movement models in this respect makes it clear that formulations of the interaction of the particles are mostly designed to model a mixture of behaviours, e.g., a force model is able to describe the transfer of forces when people push and is also used to model avoidance manoeuvres when crossing other people. Here, the models quickly reach their limits when it comes to realistically representing the diversity of behaviours in a single simulation.

A further example is the behavioural diversity and dynamic as described in [110]. The authors study a entrance situation with different motivation, where queuing but also pushing could be observed. An ethogram is created to document the diversity of behaviour and it is shown how pedestrians change their behaviour with time, see Fig. 7. The formation of a queue and its meaning as a social rule leads to a collective phenomenon for which many people work together towards a common goal. Pairwise interaction schemes such as the superposition mechanisms in force-based models corresponding to weighted averages are too simplistic and cannot consider the collective aim to form a queue. This also applies to common velocity models where the interaction with the neighbour is helpful to model the avoidance of collision but are not usable to consider the formation of an order by queuing. Furthermore, current models are unable to take into account the triggers for behavioural changes documented in [110, Fig. 7].

To describe collective interaction concepts from psychology and sociology should be integrated in modelling of the movement of agents. These could be sociological concepts such as norms and scripts [110] but also psychological concepts such as perception [140], cognition or the influence of leadership [73].

Pairwise interaction schemes, such as the superposition mechanisms in force-based models or single interaction with the neighbour with the shortest time-to-collision in certain velocity-based models, are too simplistic to account for complex behavioural changes. In fact, such mechanisms rely on intricate collective interaction schemes based on sociological and psychological cognitive aspects, and complex threshold mechanisms [43, 148]. Other aspects rely on safety components, especially for vulnerable users such as pedestrians and cyclists in mixed urban traffic [91]. The modelling of safety behaviours is based on pedestrian risk assessment that can affect the collective dynamics of a crowd and mixed traffic flow [108, 149]. In general, considerable efforts are still needed to identify the sociological, psychological and safety aspects involved and to model mechanisms capable of taking them into account [1, 3, 7, 118].

4.4 Developing Generic Order Parameters

Another important challenge is to quantify and characterise the collective dynamics of vehicles and pedestrians using aggregated order parameters. In general, specific order parameters are defined for each of the collective dynamics. For example, stop-and-go waves can be characterised using variability indicators of vehicle speeds or gaps, while lane formation can be determined by counting the number of pedestrians in front with the same direction [94]. However, the formulation of generic order parameters remains challenging.

Lately, a modelling paradigm for particle, vehicle and pedestrian dynamics based on port-Hamiltonian systems has been developed [35,63,78,104,124]. Port-Hamiltonian systems (PHS) are a recent but already well-established modelling approach for nonlinear physical systems [132]. They decompose the dynamics into between distance-based repulsion corresponding to the Hamiltonian part with energy conservation, and some relaxation to a tactical desired velocity corresponding to the dissipative part of the system and to the input ports. Here, the Hamiltonian can be used as a generic order parameter characterising the collective dynamics [124]. In addition, the Hamiltonian can be used as a Lyapunov function to address the system stability thanks to the dissipativity inequality without the need to linearise the system. The Hamiltonian in PHS is a promising modelling approach to quantify collective dynamics in a generic way. However, the subject remains challenging and is still under development.

References

1. J. Adrian, A. Seyfried, and A. Sieben. Crowds in front of bottlenecks at entrances from the perspective of physics and social psychology. *Journal of the Royal Society Interface*, 17(165):20190871, 2020.
2. Ambulance crossing a crow during the Lyon's Festival of Lights 2022. Place des Terreaux, Lyon, France, 2022-12-10, https://www.youtube.com/watch?v=1zqpJRnAqsM, Last accessed on 2024-10-09.
3. R. E. Amini, K. Yang, and C. Antoniou. Development of a conflict risk evaluation model to assess pedestrian safety in interaction with vehicles. *Accident Analysis & Prevention*, 175:106773, 2022.
4. K. A. Bacik, B. S. Bacik, and T. Rogers. Lane nucleation in complex active flows. *Science*, 379(6635):923–928, 2023.
5. M. Bando, K. Hasebe, K. Nakanishi, and A. Nakayama. Analysis of optimal velocity model with explicit delay. *Physical Review E*, 58(5):5429, 1998.
6. M. Bando, K. Hasebe, A. Nakayama, A. Shibata, and Y. Sugiyama. Dynamical model of traffic congestion and numerical simulation. *Physical review E*, 51(2):1035, 1995.
7. M. Beermann and A. Sieben. The connection between stress, density, and speed in crowds. *Scientific Reports*, 13(1):13626, 2023.
8. N. Bekiaris-Liberis and A. I. Delis. Pde-based feedback control of freeway traffic flow via time-gap manipulation of ACC-equipped vehicles. *IEEE Transactions on Control Systems Technology*, 29(1):461–469, 2020.
9. N. Bellomo, B. Piccoli, and A. Tosin. Modeling crowd dynamics from a complex system viewpoint. *Mathematical models and methods in applied sciences*, 22(supp02):1230004, 2012.
10. M. Boltes, J. Zhang, A. Tordeux, A. Schadschneider, and A. Seyfried. Empirical results of pedestrian and evacuation dynamics. *Encyclopedia of complexity and systems science*, 16:1–29, 2018.
11. T. Bonnemain, M. Butano, T. Bonnet, I. Echeverría-Huarte, A. Seguin, A. Nicolas, C. Appert-Rolland, and D. Ullmo. Pedestrians in static crowds are not grains, but game players. *Physical Review E*, 107(2):024612, 2023.
12. C. Burstedde, K. Klauck, A. Schadschneider, and J. Zittartz. Simulation of pedestrian dynamics using a two-dimensional cellular automaton. *Physica A: Statistical Mechanics and its Applications*, 295(3-4):507–525, 2001.

13. M. Butano, C. Appert-Rolland, and D. Ullmo. Discounted mean-field game model of a dense static crowd with variable information crossed by an intruder. *SciPost Phys.*, 16:104, 2024.
14. S. Cao, A. Seyfried, J. Zhang, S. Holl, and W. Song. Fundamental diagrams for multidirectional pedestrian flows. *Journal of Statistical Mechanics: Theory and Experiment*, 2017(3):033404, 2017.
15. J. A. Carrillo, M. Fornasier, G. Toscani, and F. Vecil. Particle, kinetic, and hydrodynamic models of swarming. *Mathematical modeling of collective behavior in socio-economic and life sciences*, pages 297–336, 2010.
16. C. Castellano, S. Fortunato, and V. Loreto. Statistical physics of social dynamics. *Reviews of modern physics*, 81(2):591–646, 2009.
17. R. E. Chandler, R. Herman, and E. W. Montroll. Traffic dynamics: studies in car following. *Operations research*, 6(2):165–184, 1958.
18. X. Chen, M. Treiber, V. Kanagaraj, and H. Li. Social force models for pedestrian traffic–state of the art. *Transport reviews*, 38(5):625–653, 2018.
19. D. Chowdhury, L. Santen, and A. Schadschneider. Statistical physics of vehicular traffic and some related systems. *Physics Reports*, 329(4-6):199–329, 2000.
20. M. Chraibi, U. Kemloh, A. Schadschneider, and A. Seyfried. Force-based models of pedestrian dynamics. *Networks and Heterogeneous Media*, 6(3):425–442, 2011.
21. M. Chraibi, A. Seyfried, and A. Schadschneider. Generalized centrifugal-force model for pedestrian dynamics. *Physical Review E—Statistical, Nonlinear, and Soft Matter Physics*, 82(4):046111, 2010.
22. B. Ciuffo, K. Mattas, M. Makridis, G. Albano, A. Anesiadou, Y. He, S. Josvai, D. Komnos, M. Pataki, S. Vass, et al. Requiem on the positive effects of commercial adaptive cruise control on motorway traffic and recommendations for future automated driving systems. *Transportation research part C: emerging technologies*, 130:103305, 2021.
23. A. Corbetta, C.-m. Lee, R. Benzi, A. Muntean, and F. Toschi. Fluctuations around mean walking behaviors in diluted pedestrian flows. *Physical Review E*, 95(3):032316, 2017.
24. J. Cordes, M. Chraibi, A. Tordeux, and A. Schadschneider. Single-file pedestrian dynamics: a review of agent-following models. *Crowd Dynamics, Volume 4: Analytics and Human Factors in Crowd Modeling*, pages 143–178, 2023.
25. J. Cordes, A. Schadschneider, and A. Nicolas. Dimensionless numbers reveal distinct regimes in the structure and dynamics of pedestrian crowds. *PNAS nexus*, 3(4):pgae120, 2024.
26. I. D. Couzin and N. R. Franks. Self-organized lane formation and optimized traffic flow in army ants. *Proceedings of the Royal Society of London. Series B: Biological Sciences*, 270(1511):139–146, 2003.
27. E. Cristiani, B. Piccoli, and A. Tosin. *Multiscale modeling of pedestrian dynamics*, volume 12. Springer, 2014.
28. J. Cristín, V. Méndez, and D. Campos. General scaling in bidirectional flows of self-avoiding agents. *Scientific reports*, 9(1):18488, 2019.
29. L. Crociani and G. Lämmel. Multidestination pedestrian flows in equilibrium: a cellular automaton-based approach. *Computer-Aided Civil and Infrastructure Engineering*, 31(6):432–448, 2016.
30. F. Cucker and S. Smale. Emergent behavior in flocks. *IEEE Transactions on automatic control*, 52(5):852–862, 2007.
31. F. Dietrich and G. Köster. Gradient navigation model for pedestrian dynamics. *Physical Review E*, 89(6):062801, 2014.
32. L. Duckstein. Control of traffic in tunnels to maximize flow. *Highway Research Record*, 154, 1967.
33. J. Dzubiella, G. Hoffmann, and H. Löwen. Lane formation in colloidal mixtures driven by an external field. *Physical Review E*, 65(2):021402, 2002.
34. I. Echeverría-Huarte and A. Nicolas. Body and mind: Decoding the dynamics of pedestrians and the effect of smartphone distraction by coupling mechanical and decisional processes. *Transportation research part C: emerging technologies*, 157:104365, 2023.

35. M. Ehrhardt, T. Kruse, and A. Tordeux. The collective dynamics of a stochastic port-Hamiltonian self-driven agent model in one dimension. *ESAIM: Mathematical Modelling and Numerical Analysis*, 58(2):515–544, 2024.
36. C. Feliciani, K. Shimura, and K. Nishinari. *Introduction to crowd management: Managing crowds in the digital era: Theory and Practice*. Springer Nature, 2022.
37. P. Fiorini and Z. Shiller. Motion planning in dynamic environments using velocity obstacles. *The international journal of robotics research*, 17(7):760–772, 1998.
38. M. R. Flynn, A. R. Kasimov, J.-C. Nave, R. R. Rosales, and B. Seibold. Self-sustained nonlinear waves in traffic flow. *Physical Review E—Statistical, Nonlinear, and Soft Matter Physics*, 79(5):056113, 2009.
39. E. Frank. On the zeros of polynomials with complex coefficients. *Bulletin of the American Mathematical Society*, 52:144–157, 1946.
40. A. Fujita, C. Feliciani, D. Yanagisawa, and K. Nishinari. Traffic flow in a crowd of pedestrians walking at different speeds. *Physical Review E*, 99(6):062307, 2019.
41. D. C. Gazis, R. Herman, and R. W. Rothery. Nonlinear follow-the-leader models of traffic flow. *Operations research*, 9(4):545–567, 1961.
42. P. G. Gipps. A behavioural car-following model for computer simulation. *Transportation research part B: methodological*, 15(2):105–111, 1981.
43. A. Gregorj, Z. Yücel, F. Zanlungo, C. Feliciani, and T. Kanda. Social aspects of collision avoidance: a detailed analysis of two-person groups and individual pedestrians. *Scientific reports*, 13(1):5756, 2023.
44. G. Gunter, D. Gloudemans, R. E. Stern, S. McQuade, R. Bhadani, M. Bunting, M. L. Delle Monache, R. Lysecky, B. Seibold, J. Sprinkle, et al. Are commercially implemented adaptive cruise control systems string stable? *IEEE Transactions on Intelligent Transportation Systems*,22(11):6992–7003, 2020.
45. S. J. Guy, J. Chhugani, S. Curtis, P. Dubey, M. C. Lin, and D. Manocha. Pledestrians: A least-effort approach to crowd simulation. In *Symposium on computer animation*, pages 119–128, 2010.
46. D. Helbing. Traffic and related self-driven many-particle systems. *Reviews of modern physics*, 73(4):1067, 2001.
47. D. Helbing, L. Buzna, A. Johansson, and T. Werner. Self-organized pedestrian crowd dynamics: Experiments, simulations, and design solutions. *Transportation science*, 39(1):1–24, 2005.
48. D. Helbing, I. Farkas, and T. Vicsek. Simulating dynamical features of escape panic. *Nature*, 407(6803):487–490, 2000.
49. D. Helbing, I. J. Farkas, and T. Vicsek. Freezing by heating in a driven mesoscopic system. *Physical Review Letters*, 84(6):1240, 2000.
50. D. Helbing and P. Molnar. Social force model for pedestrian dynamics. *Physical review E*, 51(5):4282, 1995.
51. R. Herman, E. W. Montroll, R. B. Potts, and R. W. Rothery. Traffic dynamics: analysis of stability in car following. *Operations research*, 7(1):86–106, 1959.
52. K. Hirai and K. Tarui. A simulation of the behavior of a crowd in panic. In *Proceedings of the 1975 International Conference on Cybernetics and Society*, pages 409–411, 1975.
53. S. Hoogendoorn and P. HL Bovy. Simulation of pedestrian flows by optimal control and differential games. *Optimal control applications and methods*, 24(3):153–172, 2003.
54. S. P. Hoogendoorn and P. H. Bovy. Pedestrian route-choice and activity scheduling theory and models. *Transportation Research Part B: Methodological*, 38(2):169–190, 2004.
55. ISO15622:2018. Intelligent transport systems—adaptive cruise control systems—performance requirements and test procedures, 2018.
56. K. Jang, N. Lichtlé, E. Vinitsky, A. Shah, M. Bunting, M. Nice, B. Piccoli,B. Seibold, D. B. Work, M. L. D. Monache, et al. Reinforcement learning based oscillation dampening: Scaling up single-agent RL algorithms to a 100 AV highway field operational test. arXiv preprint arXiv:2402.17050, 2024.

57. R. Jiang, Q. Wu, and Z. Zhu. Full velocity difference model for a car-following theory. *Physical Review E*, 64(1):017101, 2001.
58. I. Karamouzas, B. Skinner, and S. J. Guy. Universal power law governing pedestrian interactions. *Physical Review Letters*, 113(23):238701, 2014.
59. A. U. Kemloh Wagoum, A. Seyfried, and S. Holl. Modeling the dynamic route choice of pedestrians to assess the criticality of building evacuation. *Advances in Complex Systems*, 15(07):1250029, 2012.
60. B. Khelfa, R. Korbmacher, A. Schadschneider, and A. Tordeux. Heterogeneity-induced lane and band formation in self-driven particle systems. *Scientific reports*, 12(1):4768, 2022.
61. P. Khound, P. Will, A. Tordeux, and F. Gronwald. Extending the adaptive time gap car-following model to enhance local and string stability for adaptive cruise control systems. *Journal of Intelligent Transportation Systems*, 27(1):36–56,2023.
62. P. Khound, P. Will, A. Tordeux, and F. Gronwald. Unified framework for over-damped string stable adaptive cruise control systems. *Transportation research part C: emerging technologies*,148:104039, 2023.
63. S. Knorn, A. Donaire, J. C. Agüero, and R. H. Middleton. Passivity-based control for multi-vehicle systems subject to string constraints. *Automatica*, 50(12):3224–3230, 2014.
64. F. Kogler and S. H. Klapp. Lane formation in a system of dipolar microswimmers. *Europhysics Letters*, 110(1):10004, 2015.
65. E. Kometani and T. Sasaki. On the stability of traffic flow (report-i). *Journal of the Operations Research Society of Japan*, 2(1):11–26, 1958.
66. T. I. Lakoba, D. J. Kaup, and N. M. Finkelstein. Modifications of the Helbing-Molnar-Farkas-Vicsek social force model for pedestrian evolution. *Simulation*, 81(5):339–352, 2005.
67. J. W. Lee, H. Wang, K. Jang, A. Hayat, M. Bunting, A. Alanqary, W. Barbour, Z. Fu, X. Gong, G. Gunter, et al. Traffic control via connected and automated vehicles: An open-road field experiment with 100 CAVs. *arXiv preprint* arXiv:2402.17043, 2024.
68. K. Lewin. *Field theory in social science: selected theoretical papers*. Tavistock Publications in collaboration with Routledge & K.Paul, 1952.
69. K. Lewin. *The Conceptual Representation and the Measurement of Psychological Forces*. Contributions to psychological theory. Duke University Press, 1968.
70. T. Li, D. Chen, H. Zhou, J. Laval, and Y. Xie. Car-following behavior characteristics of adaptive cruise control vehicles based on empirical experiments. *Transportation research part B: methodological*, 147:67–91, 2021.
71. X. Li, J. Cui, S. An, and M. Parsafard. Stop-and-go traffic analysis: Theoretical properties, environmental impacts and oscillation mitigation. *Transportation Research Part B: Methodological*, 70:319–339, 2014.
72. S. M. Lo, H.-C. Huang, P. Wang, and K. Yuen. A game theory based exit selection model for evacuation. *Fire Safety Journal*, 41(5):364–369, 2006.
73. M. Lombardi, W. H. Warren, and M. di Bernardo. Nonverbal leadership emergence in walking groups. *Scientific reports*, 10(1):18948, 2020.
74. Y.-X. Lü, Z.-X. Wu, and J.-Y. Guan. Pedestrian dynamics with mechanisms of anticipation and attraction. *Physical Review Research*, 2(4):043250, 2020.
75. Y. Luo, P. Cai, A. Bera, D. Hsu, W. S. Lee, and D. Manocha. Porca: Modeling and planning for autonomous driving among many pedestrians. *IEEE Robotics and Automation Letters*, 3(4):3418–3425, 2018.
76. M. Makridis, K. Mattas, A. Anesiadou, and B. Ciuffo. OpenACC. An open database of car-following experiments to study the properties of commercial ACC systems. *Transportation research part C: emerging technologies*,125:103047, 2021.
77. M. Makridis, K. Mattas, and B. Ciuffo. Response time and time headway of an adaptive cruise control. An empirical characterization and potential impacts on road capacity. *IEEE transactions on intelligent transportation systems*,21(4):1677–1686, 2019.
78. I. Matei, C. Mavridis, J. S. Baras, and M. Zhenirovskyy. Inferring particle interaction physical models and their dynamical properties. In *2019 IEEE 58th Conference on Decision and Control (CDC)*, pages 4615–4621. IEEE, 2019.

79. B. Maury and S. Faure. *Crowds in equations: an introduction to the microscopic modeling of crowds*. World Scientific, 2018.
80. B. Maury and J. Venel. A mathematical framework for a crowd motion model. *Comptes Rendus Mathematique*, 346(23-24):1245–1250, 2008.
81. B. Maury and J. Venel. A discrete contact model for crowd motion. *ESAIM: Mathematical Modelling and Numerical Analysis*, 45(1):145–168, 2011.
82. L. A. Meerhoff, J. Bruneau, A. Vu, A.-H. Olivier, and J. Pettré. Guided by gaze: Prioritization strategy when navigating through a virtual crowd can be assessed through gaze activity. *Acta psychologica*, 190:248–257, 2018.
83. J. A. Michon. A critical view of driver behavior models: what do we know, what should we do? In *Human behavior and traffic safety*, pages 485–524. Springer, 1985.
84. V. Milanés and S. E. Shladover. Modeling cooperative and autonomous adaptive cruise control dynamic responses using experimental data. *Transportation Research Part C: Emerging Technologies*,48:285–300, 2014.
85. V. Milanés, S. E. Shladover, J. Spring, C. Nowakowski, H. Kawazoe, and M. Nakamura. Cooperative adaptive cruise control in real traffic situations. *IEEE Transactions on intelligent transportation systems*, 15(1):296–305, 2013.
86. M. Moussaïd, D. Helbing, and G. Theraulaz. How simple rules determine pedestrian behavior and crowd disasters. *Proceedings of the National Academy of Sciences*,108(17):6884–6888, 2011.
87. A. Nakayama, K. Hasebe, and Y. Sugiyama. Instability of pedestrian flow and phase structure in a two-dimensional optimal velocity model. *Physical Review E-Statistical, Nonlinear, and Soft Matter Physics*, 71(3):036121, 2005.
88. A. Nakayama, K. Hasebe, and Y. Sugiyama. Effect of attractive interaction on instability of pedestrian flow in a two-dimensional optimal velocity model. *Physical Review E-Statistical, Nonlinear, and Soft Matter Physics*, 77(1):016105, 2008.
89. F. P. Navin and R. J. Wheeler. Pedestrian flow characteristics. *Traffic Engineering, Inst Traffic Engr*, 39, 1969.
90. G. F. Newell. Nonlinear effects in the dynamics of car following. *Operations research*, 9(2):209–229, 1961.
91. Y. Ni, M. Wang, J. Sun, and K. Li. Evaluation of pedestrian safety at intersections: A theoretical framework based on pedestrian-vehicle interaction patterns. *Accident Analysis & Prevention*, 96:118–129, 2016.
92. A. Nicolas, S. Bouzat, and M. N. Kuperman. Pedestrian flows through a narrow doorway: Effect of individual behaviours on the global flow and microscopic dynamics. *Transportation Research Part B: Methodological*, 99:30–43, 2017.
93. A. Nicolas, M. Kuperman, S. Ibañez, S. Bouzat, and C. Appert-Rolland. Mechanical response of dense pedestrian crowds to the crossing of intruders. *Scientific reports*, 9(1):105, 2019.
94. S. Nowak and A. Schadschneider. Quantitative analysis of pedestrian counter flow in a cellular automaton model. *Physical Review E-Statistical, Nonlinear, and Soft Matter Physics*, 85(6):066128, 2012.
95. D. Oeding. *Verkehrsbelastung und Dimensionierung von Gehwegen und anderen Anlagen des Fußgängerverkehrs*. Der Bundesminister f. Verkehr, Abt. Strassenbau, 1963.
96. G. Orosz, R. E. Wilson, and G. Stépán. Traffic jams: dynamics and control, 2010.
97. S. Paris, J. Pettré, and S. Donikian. Pedestrian reactive navigation for crowd simulation: a predictive approach. In *Computer Graphics Forum*, volume 26(3), pages 665–674. Wiley Online Library, 2007.
98. D. R. Parisi, M. Gilman, and H. Moldovan. A modification of the social force model can reproduce experimental data of pedestrian flows in normal conditions. *Physica A: Statistical Mechanics and its Applications*, 388(17):3600–3608, 2009.
99. S. Pellegrini, A. Ess, K. Schindler, and L. Van Gool. You'll never walk alone: Modeling social behavior for multi-target tracking. In *2009 IEEE 12th international conference on computer vision*, pages 261–268. IEEE, 2009.

100. L. M. Pfaff and M. E. Cinelli. Avoidance behaviours of young adults during a head-on collision course with an approaching person. *Experimental brain research*, 236:3169–3179, 2018.
101. C. Reichhardt, J. Thibault, S. Papanikolaou, and C. Reichhardt. Laning and clustering transitions in driven binary active matter systems. *Physical Review E*, 98(2):022603, 2018.
102. A. Reuschel. Fahrzeugbewegungen in der kolonne. *Osterreichisches Ingenieur Archiv*, 4:193–215, 1950.
103. C. W. Reynolds. Flocks, herds and schools: A distributed behavioral model. In *Proceedings of the 14th annual conference on Computergraphics and interactive techniques*, pages 25–34, 1987.
104. B. Rüdiger, A. Tordeux, and B. E. Ugurcan. Stability analysis of a stochastic port-Hamiltonian car-following model. *Journal of Physics A: Mathematical and Theoretical*, 57(29):295203, 2024.
105. U. Sarma, S. Baruah, and R. Ganesh. Lane formation in driven pair-ion plasmas. *Physics of Plasmas*, 27(1), 2020.
106. A. Schadschneider, D. Chowdhury, and K. Nishinari. *Stochastic transport in complex systems: from molecules tovehicles*. Elsevier, 2010.
107. M. J. Seitz, A. Templeton, J. Drury, G. Köster, and A. Philippides. Parsimony versus reductionism: how can crowd psychology be introduced into computer simulation? *Review of General Psychology*, 21(1):95–102, 2017.
108. X. Shen and P. Raksincharoensak. Pedestrian-aware statistical risk assessment. *IEEE Transactions on Intelligent Transportation Systems*, 23(7):7910–7918, 2021.
109. T. Shimaya and K. A. Takeuchi. Lane formation and critical coarsening in a model of bacterial competition. *Physical Review E*, 99(4):042403, 2019.
110. A. Sieben and T. Postmes. Behavioral repertoires in crowds: An observational approach.
111. A. Sieben, J. Schumann, and A. Seyfried. Collective phenomena in crowds-where pedestrian dynamics need social psychology. *PLoS one*, 12(6):e0177328, 2017.
112. R. E. Stern, S. Cui, M. L. Delle Monache, R. Bhadani, M. Bunting, M. Churchill, N. Hamilton, H. Pohlmann, F. Wu, B. Piccoli, et al. Dissipation of stop-and-go waves via control of autonomous vehicles: Field experiments. *Transportation Research Part C: Emerging Technologies*, 89:205–221, 2018.
113. I. M. Sticco, G. A. Frank, and C. O. Dorso. Effects of the body force on the pedestrian and the evacuation dynamics. *Safety science*, 129:104829, 2020.
114. Y. Sugiyama, M. Fukui, M. Kikuchi, K. Hasebe, A. Nakayama, K. Nishinari, S.-i.Tadaki, and S. Yukawa. Traffic jams without bottlenecks-experimental evidence for the physical mechanism of the formation of a jam. *New journal of physics*, 10(3):033001, 2008.
115. K. Sütterlin, A. Wysocki, A. Ivlev, C. Räth, H. Thomas, M. Rubin-Zuzic, W. Goedheer, V. Fortov, A. Lipaev, V. Molotkov, et al. Dynamics of lane formation in driven binary complex plasmas. *Physical review letters*, 102(8):085003, 2009.
116. D. Swaroop, J. K. Hedrick, C. Chien, and P. Ioannou. A comparision of spacing and headway control laws for automatically controlled vehicles1. *Vehicle system dynamics*, 23(1):597–625, 1994.
117. A. Templeton, J. Drury, and A. Philippides. Placing large group relations into pedestrian dynamics: Psychological crowds in counter flow. *Collective Dynamics*, 4:1–22, 2020.
118. A. Templeton and F. Neville. Modeling collective behaviour: insights and applications from crowd psychology. *Crowd Dynamics, Volume 2: Theory, Models, and Applications*, pages 55–81, 2020.
119. D. Thalmann and S. R. Musse. *Crowd simulation*. Springer Science & Business Media, 2012.
120. A. Tordeux, M. Chraibi, and A. Seyfried. Collision-free speed model for pedestrian dynamics. In *Traffic and Granular Flow'15*, pages 225–232. Springer, 2016.
121. A. Tordeux, G. Lämmel, F. S. Hänseler, and B. Steffen. A mesoscopic model for large-scale simulation of pedestrian dynamics. *Transportation research part C: emerging technologies*, 93:128–147, 2018.
122. A. Tordeux, S. Lassarre, and M. Roussignol. An adaptive time gap car-following model. *Transportation research part B: methodological*, 44(8–9):1115–1131, 2010.

123. A. Tordeux, M. Roussignol, and S. Lassarre. Linear stability analysis of first-order delayed car-following models on a ring. *Physical Review E-Statistical, Nonlinear, and Soft Matter Physics*, 86(3):036207, 2012.
124. A. Tordeux and C. Totzeck. Multi-scale description of pedestrian collective dynamics with port-Hamiltonian systems. *Networks and Heterogeneous Media*, 18(2):906–929, 2023.
125. C. Totzeck. An anisotropic interaction model with collision avoidance. *Kinetic and Related Models*, 13(6):1219–1242, 2020.
126. M. Treiber, A. Hennecke, and D. Helbing. Congested traffic states in empirical observations and microscopic simulations. *Physical review E*, 62(2):1805, 2000.
127. M. Treiber and A. Kesting. *Traffic Flow Dynamics*. Springer, 2013.
128. E. Uesten, J. Schumann, and A. Sieben. Exploring the dynamic relationship between pushing behavior and crowd dynamics. *Collective Dynamics*, 8:1–29, 2023.
129. E. Üsten, H. Lügering, and A. Sieben. Pushing and non-pushing forward motion in crowds: A systematic psychological observation method for rating individual behavior in pedestrian dynamics. *Collective dynamics*, 7:1–16, 2022.
130. J. Van Den Berg, S. J. Guy, M. Lin, and D. Manocha. Reciprocal n-body collision avoidance. In *Robotics Research: The 14th International Symposium ISRR*, pages 3–19. Springer, 2011.
131. J. Van den Berg, M. Lin, and D. Manocha. Reciprocal velocity obstacles for real-time multi-agent navigation. In *2008 IEEE international conference on robotics and automation*, pages 1928–1935. IEEE, 2008.
132. A. Van Der Schaft. Port-Hamiltonian systems: an introductory survey. In *International congress of mathematicians*, pages 1339–1365. European Mathematical Society Publishing House (EMS Ph), 2006.
133. W. van Toll, F. Grzeskowiak, A. L. Gandía, J. Amirian, F. Berton,J. Bruneau, B. C. Daniel, A. Jovane, and J. Pettré. Generalized microscopic crowd simulation using costs in velocity space. In *Symposium on Interactive 3D Graphics and Games*, pages 1–9, 2020.
134. T. Vicsek, A. Czirók, E. Ben-Jacob, I. Cohen, and O. Shochet. Novel type of phase transition in a system of self-driven particles. *Physical Review Letters*, 75(6):1226, 1995.
135. T. Vissers, A. Wysocki, M. Rex, H. Löwen, C. P. Royall, A. Imhof, andA. van Blaaderen. Lane formation in driven mixtures of oppositely charged colloids. *Soft Matter*, 7(6):2352–2356, 2011.
136. A. K. Wagoum, A. Tordeux, and W. Liao. Understanding human queuing behaviour at exits: an empirical study. *Royal Society open science*, 4(1):160896, 2017.
137. Z. Wang, G. Wu, P. Hao, K. Boriboonsomsin, and M. Barth. Developing a platoon-wide eco-cooperative adaptive cruise control (CACC) system. In *2017 ieee intelligent vehicles symposium (iv)*, pages1256–1261. IEEE, 2017.
138. N. Wijermans and A. Templeton. Towards more realism in pedestrian behaviour models: First steps and considerations in formalising social identity. In *Advances in Social Simulation: Proceedings of the 16th Social Simulation Conference, 20–24 September 2021*, pages 53–64. Springer, 2022.
139. R. E. Wilson and J. A. Ward. Car-following models: fifty years of linear stability analysis—a mathematical perspective. *Transportation Planning and Technology*, 34(1):3–18, 2011.
140. T. D. Wirth, G. C. Dachner, K. W. Rio, and W. H. Warren. Is the neighborhood of interaction in human crowds metric,topological, or visual? *PNAS nexus*, 2(5):pgad118, 2023.
141. H. Xing, J. Ploeg, and H. Nijmeijer. Compensation of communication delays in a cooperative ACC system. *IEEE Transactions on Vehicular Technology*, 69(2):1177–1189, 2020.
142. Q. Xu, M. Chraibi, and A. Seyfried. Anticipation in a velocity-based model for pedestrian dynamics. *Transportation research part C: emerging technologies*, 133:103464, 2021.
143. Q. Xu, M. Chraibi, A. Tordeux, and J. Zhang. Generalized collision-free velocity model for pedestrian dynamics. *Physica A: Statistical Mechanics and its Applications*, 535:122521, 2019.
144. W. Yu, R. Chen, L.-Y. Dong, and S. Dai. Centrifugal force model for pedestrian dynamics. *Physical Review E-Statistical, Nonlinear, and Soft Matter Physics*, 72(2):026112, 2005.

145. F. Zanlungo, T. Ikeda, and T. Kanda. Social force model with explicit collision prediction. *Europhysics Letters*, 93(6):68005, 2011.
146. J. Zhang, W. Klingsch, A. Schadschneider, and A. Seyfried. Ordering in bidirectional pedestrian flows and its influence on the fundamental diagram. *Journal of Statistical Mechanics: Theory and Experiment*, 2012(02):P02002, 2012.
147. J. Zhang, W. Mehner, S. Holl, M. Boltes, E. Andresen, A. Schadschneider, and A. Seyfried. Universal flow-density relation of single-file bicycle, pedestrian and car motion. *Physics Letters A*, 378(44):3274–3277, 2014.
148. J. Zhang, Q. Qu, and X.-B. Chen. A review on collective behavior modeling and simulation: building a link between cognitive psychology and physical action. *Applied Intelligence*, 53(21):25954–25983, 2023.
149. L. Zhang, K. Yuan, H. Chu, Y. Huang, H. Ding, J. Yuan, and H. Chen. Pedestrian collision risk assessment based on state estimation and motion prediction. *IEEE Transactions on Vehicular Technology*, 71(1):98–111, 2021.

Product Development and Material Science

Development of Microstructured High Performance Materials Based on Finite Element Simulations

Swantje Bargmann and Jana Wilmers

Contents

Abstract

Microstructured materials, engineered with precise architectures at the microscale, exhibit unique properties and performance characteristics that extend beyond their chemical composition. Inspired by natural materials, these materials leverage advanced manufacturing techniques such as 3D printing to achieve complex geometries that enhance mechanical, thermal, optical, and electrical properties. Applications span aerospace, biomedical, electronics, and energy industries, where

S. Bargmann (✉)
Chair of Solid Mechanics, School of Mechanical and Safety Engineering,
University of Wuppertal, Wuppertal, Germany
e-mail: bargmann@uni-wuppertal.de

S. Bargmann
Wuppertal Center for Smart Materials, University of Wuppertal,
Wuppertal, Germany

J. Wilmers
School of Mechanical Engineering,
Stralsund University of Applied Sciences, Stralsund, Germany

S. Bracke and R. Pieper (eds.), *Safety Engineering*,
https://doi.org/10.1007/978-3-658-47473-7_6

they provide lightweight, durable, and multifunctional solutions. Microstructured materials also promote sustainability by optimizing material usage and efficiency.

Keywords

High performance materials · Microstructured materials · Finite element modeling

1 Introduction

Microstructured materials represent an extraordinary class of materials exhibiting unique material properties that arise from their carefully designed microstructures, rather than solely from their chemical composition. By carefully designing the arrangement and interaction of structural elements at the microscale, scientists and engineers can achieve a wide range of mechanical, thermal, optical, and electrical properties tailored for specific applications.

Biological materials such as bone, nacre, and dental enamel display extraordinary properties due to their hierarchical structures [Wegst and Ashby 2004, Fratzl and Weinkamer 2007]. For instance, dental enamel, the hardest substance in the human body, combines high strength with exceptional fracture toughness through its sophisticated arrangement of hydroxyapatite crystals. Inspired by such natural materials, researchers develop tailor-made microstructured high performance materials that aim to replicate and enhance these desirable properties for various technological applications.

Applications of microstructured materials are vast and diverse, spanning multiple industries including aerospace, biomedical, electronics, and energy. In the aerospace industry, for example, microstructured materials are used to develop lightweight and durable components that can withstand extreme conditions. In the biomedical field, these materials are engineered for implants and prosthetics that mimic the mechanical properties of natural tissues, promoting better integration and performance. Furthermore, in electronics, microstructured materials enable the fabrication of flexible and stretchable devices that can conform to various shapes and surfaces, opening new possibilities for wearable technology and advanced sensors.

The development of microstructured materials also aligns with the principles of sustainability and resource efficiency. By optimizing material usage and enhancing performance through microstructural design, these materials contribute to the creation of more efficient and environmentally friendly products. Additionally, the ability to tailor properties at the microscale allows for the development of multifunctional materials that can perform multiple roles, reducing the need for multiple separate components and further improving overall system efficiency.

Microstructured materials represent a significant advancement in material science, offering unparalleled opportunities to design and engineer materials with tailored properties for specific applications. Through innovative manufacturing techniques and a deep understanding of structure–property relationships, researchers continue to push the boundaries of what is possible with microstructured materials, driving progress across a wide range of industries and contributing to the development of sustainable and high-performance technologies.

At the Chair of Solid Mechanics, we investigate microstructured natural materials (e.g., dental enamel [Bargmann et al. 2013; Ma et al., 2016a; Ma et al., 2016b; Scheider et al. 2015; Scheider et al. 2016; Wilmers and Bargmann 2020; Wilmers et al. 2024], enameloid [Wilmers et al. 2021], limpet teeth [Oh et al. 2022; Wurmshuber et al. 2023], nacre [Peng et al. 2022], skeletal muscles [Lamsfuss and Bargmann 2021; Lamsfuss and Bargmann 2022; Lamsfuss and Bargmann 2023], skin [A. McBride et al. 2016], turtle shells [Alheit et al. 2020, Alheit et al. 2021, Alheit et al. 2022]) as well as engineering materials such as metals [Okulov et al. 2021; Peng and Bargmann 2021], polymers [Raza et al. 2019], metallic glasses [Sopu et al. 2016] and composites [Griffiths et al. 2020; Kowol et al. 2022; Schneider et al. 2016; Wilmers and Bargmann 2018] with computer simulations. Hereby, we develop mathematical models precisely describing the material behavior and advanced finite element codes.

In the following, we exemplarily present our research of damage-tolerant biological role model materials and their synthetic high-performance counterparts.

2 Biological Role Models for Damage Tolerant Composites: Dental Enamel and Nacre

2.1 Dental Enamel

The development of damage tolerant composites hinges on achieving a balance between high strength and exceptional fracture toughness as these two properties are critical for structural applications. Traditionally, materials that excel in strength tend to be brittle, while those that are tough are often soft. However, dental enamel defies this trade-off with remarkable success. Composed of brittle mineral crystals organized in a complex hierarchical microstructure, enamel demonstrates both high stiffness and excellent toughness.

Dental enamel's unique microstructure, primarily consisting of hydroxyapatite crystals interspersed with low amounts of organic proteins, contributes to its impressive mechanical properties. Its hierarchical organization (Fig. 1) extends across multiple scales, from the nanoscale arrangement of crystals to the microscale configuration of enamel rods and interrod regions. Such a microstructure allows to absorb and dissipate energy, and can guide and re-direct cracks, thereby preventing catastrophic failure and enhancing durability.

Different species exhibit a variety of structural adaptations in their dental enamel, optimizing not only fracture toughness but also hardness and abrasion resistance [Rensberger and Pfretzschner 1992, Ziscovici et al. 2014, Weng et al. 2016, see Fig. 2]. These adaptations reflect a sophisticated interplay between structure and property, which nature has perfected over millions of years. Despite the limited range of constituents available in enamel, these natural materials

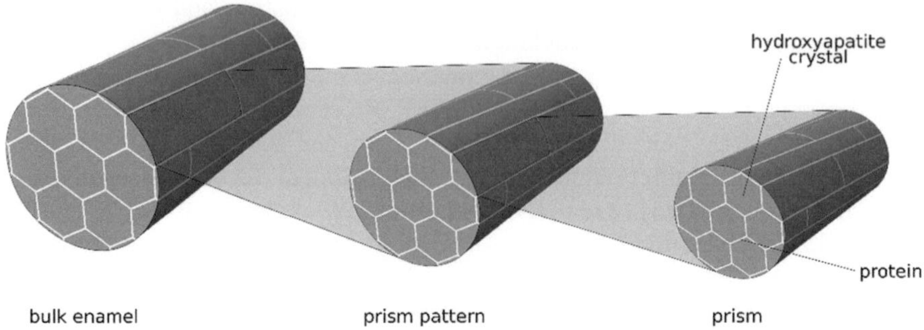

Fig. 1 Dental enamel exhibits high fracture toughness and stiffness due to a complex hierarchical and graded microstructure, optimally organized from nano- to macroscale. Bovine and human dental enamel, for example, consists of three levels of hierarchy: the first hierarchy level are so-called prisms, rods of thousands of crystallites. Complex patterns of these prisms and the interprismatic matrix form the second hierarchical level, commonly referred to as enamel types. On the third level, different complex groupings of enamel types form the bulk enamel. In the simplest enamel, radial enamel, prisms are straight and parallel, but most mammalian enamel exhibits far more complex patterns in which prisms are oriented in sharp angles to each other

consistently outperform their synthetic counterparts due to their intricate structure–property relationships.

Micromechanical modeling of dental enamel offers a pathway to decipher these relationships systematically [Ma et al. 2016a, Pro et al. 2019, Wilmers et al. 2024]. By employing advanced computational techniques such as finite element modeling, the material behavior is studied under various loading conditions, gaining insights into the stress distribution, deformation, and failure mechanisms at the microstructural level. This understanding is crucial for formulating design principles that could guide the development of advanced synthetic materials, mimicking the high strength and toughness of natural enamel [Renteria et al. 2021, Yu and Zhu 2024]. Furthermore, the integration of safety engineering and sustainability considerations into the development of dental materials ensures that these innovations are not only effective but also safe and environmentally friendly.

2.2 Nacre

Nacre, commonly known as mother of pearl, is a natural composite material with high damage tolerance that forms the innermost layer of the shells of some mollusks and the outermost coating of pearls. It is renowned for its iridescent, lustrous appearance, which is highly valued in jewelry and decorative arts. This visually striking material is not only admired for its aesthetic appeal but also for its remarkable combination of strength and resilience, which arises from its complex, natural architecture.

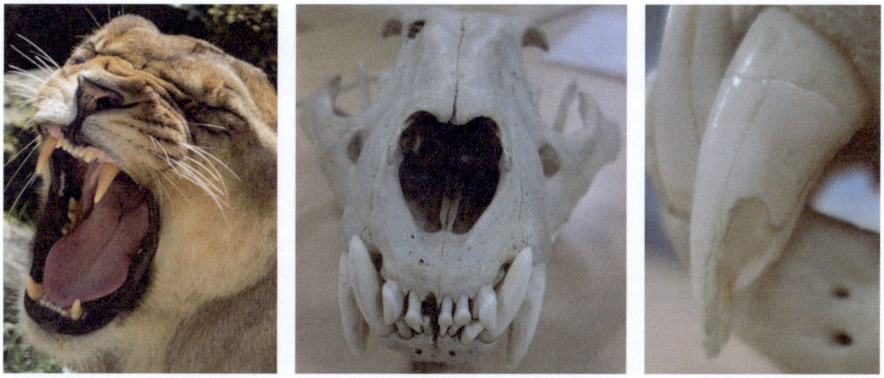

Fig. 2 Left: Female lion at Zoo Wuppertal, Germany. Middle: Skull of a female lion. Right: fractured dental enamel in the upper left fang of the lion. Skull located at Zoo Wuppertal, Germany

Nacre exhibits exceptional mechanical properties, particularly in terms of strength, toughness, and damage resistance [Menig et al. 2000, Song et al. 2003]. Despite being composed primarily of a brittle mineral, nacre is surprisingly resilient, with a toughness that far exceeds that of its individual components. This toughness is a result of its ability to absorb and dissipate energy, making nacre highly resistant to cracking and fracture. Its unique structure allows it to endure mechanical stress, distributing forces in a way that prevents catastrophic failure, which is why nacre is often studied as a model for developing strong, durable materials in engineering.

The remarkable mechanical properties of nacre are attributed to its highly organized "brick-and-mortar" microstructure [Barthelat et al. 2007]. The microstructure is composed of microscopic aragonite platelets, which are calcium carbonate crystals, arranged in a staggered, layered pattern resembling bricks in a wall. These platelets are only a few micrometers in size and are surrounded by a thin layer of organic matrix, often referred to as the "mortar". The matrix is primarily composed of proteins and polysaccharides, which serve as a binding material that holds the aragonite platelets together. The alternating layers of stiff platelets and flexible organic matrix give nacre its unique combination of rigidity and flexibility, allowing it to resist deformation while dissipating energy.

The intricate microstructure of nacre is directly linked to its superior mechanical properties, particularly its damage resistance. The brick-and-mortar arrangement allows nacre to effectively stop cracks from propagating through the material. When a crack begins to form, the organic matrix between the aragonite platelets absorbs the energy, forcing the crack to deviate and follow a tortuous path rather than a straight line. This crack deflection mechanism dissipates energy and prevents the crack from spreading rapidly, which is why nacre can sustain considerable damage without failing. Additionally, the layered structure enables nacre to undergo minor deformations, which further contributes to its toughness by distributing stress across the material. This synergy between structure and function is a key reason why nacre is an inspiration for the development of advanced, damage-resistant materials (Fig. 3).

3 Finite Element Modeling of Composites with High Damage Tolerance

3.1 Computational Generation of the Microstructure

Representative Volume Elements (RVEs) [Hill, 1963; Ostoja-Starzewski, 2006; Gitman et al. 2007; Bargmann et al. 2018] are crucial in computational modeling, particularly in the field of materials science and engineering. Mathematically, they constitute the smallest volume over which a material's properties can be averaged to represent the entire material in a larger-scale simulation. The need for RVEs arises from the complexity and heterogeneity of microstructured materials. The materials microstructural features such as the spatial distribution of phases, grains or inclusions are reproduced within the RVE. This makes it possible to model a material and its structure–property relationships with all its intricacies.

RVEs are necessary to bridge the gap between the microscopic details and the macroscopic properties of materials. By carefully selecting an RVE that is representative of the entire material, one can perform detailed simulations on this

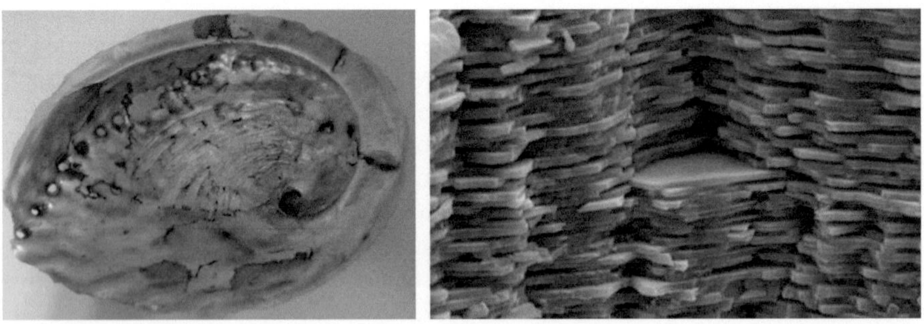

Fig. 3 Left: Image of an Haliotis midae (Midas Ear Abalone) shell found in Atlantic Ocean (South Africa). Nacre, the innermost layer of the shell, exhibiting iridescent optical properties as well as a high damage tolerance. Right: Microstructure of nacre, showing the 'brick-and-mortar' arrangement where aragonite platelets ('bricks') are embedded within an organic matrix ('mortar'), creating a staggered, layered structure that contributes to its remarkable mechanical properties. (Picture: F. Heinemann at German Wikipedia (available in the public domain, Wikimedia Commons))

smaller domain and then use the results to predict the overall behavior of the material. This approach allows for capturing essential microstructural characteristics without the computational expense of simulating every detail at the full scale.

In essence, RVEs enable the effective use of computational resources by reducing the simulated problem size while still capturing the necessary material behaviors. This is especially important in multiscale modeling, where different scales of the material's structure and behavior must be linked together. Without RVEs, complex materials or structures could not be simulated with sufficient accuracy, making RVEs a cornerstone of modern computational modeling.

Computationally generating RVEs is challenging due to the need for the RVE to accurately capture the complex and diverse microstructural features of a material while also being statistically representative of the whole material. This involves not only ensuring that the RVE encompasses all relevant phases, inclusions, grain boundaries, and defects but also that it does so in a way that reflects their actual distribution, orientation, and interaction within the larger material. Achieving this requires sophisticated algorithms and significant computational resources, as the RVE must balance detail and representativeness. Additionally, the size of the RVE must be large enough to capture all necessary features but small enough to remain computationally feasible, making the process of generating RVEs a delicate and complex task.

To effectively study structural variations in materials, a fully automatized coding system for the generation of RVEs with structural variations is essential. Such a set-up allows for the systematic and efficient exploration of a wide range of microstructural configurations, enabling us to assess how different structural features—such as grain size, shape, orientation, and phase distribution—affect the material's overall properties. Automation is crucial because it eliminates the need for manual intervention in the generation process, ensuring consistency, reducing human error, and significantly speeding up the analysis. By leveraging advanced algorithms, this approach generates a large number of RVEs with controlled variations, facilitating a comprehensive understanding of the relationship between microstructure and material behavior, and ultimately leading to better material design and optimization.

The process of RVE generation is exemplarily explained for nacre which is composed of alternating layers of thin aragonite platelets ($CaCO_3$), making up about 95% of its volume, and an organic matrix, which accounts for the remaining 5% [Katti et al. 2006; Barthelat et al. 2007]. These platelets and the matrix are organized into a "brick-and-mortar" structure. The platelets, which are pseudo-hexagonal in shape with diameters ranging from 5 to 15 μm, form planar layers approximately 0.4 μm thick [Barthelat et al. 2007]. In this arrangement, the centers of the platelets in successive layers are offset, creating a brick-laying pattern. The organic matrix surrounds the platelets within each layer and separates the layers, similar to mortar between bricks. This brick-and-mortar architecture is key to nacre's remarkable strength and toughness [Menig et al. 2000].

The generation of 3D microstructures is parameterized and automated to enable investigations across a wide range of RVE geometries. First, 2D sketches are generated for each layer of the final microstructure in Matlab. To create polygonal cells (which will later represent the platelets) with variation between layers, multiple Voronoi tessellations are generated. A Voronoi tessellation divides a domain into convex, polygonal regions based on the distances between specified points, often referred to as germs, seeds, or nucleation points, which are defined before the generation process begins. A periodic Voronoi tessellation is created by replicating the original domain and its germs around the RVE, followed by the computation of the Voronoi tessellation over the entire area (Fig. 4). To incorporate matrix material between the platelets, each polygon is shrunk, creating a gap (representing the matrix) between the original cells and their shrunken counterparts.

The geometrical information for each layer is stored and transferred to Abaqus/CAE using its Python interface to generate the 3D RVE. The layers are created sequentially by extruding the 2D structures into 3D layers and generating the matrix phase. This process is repeated until a cubic RVE is formed (Fig. 5).

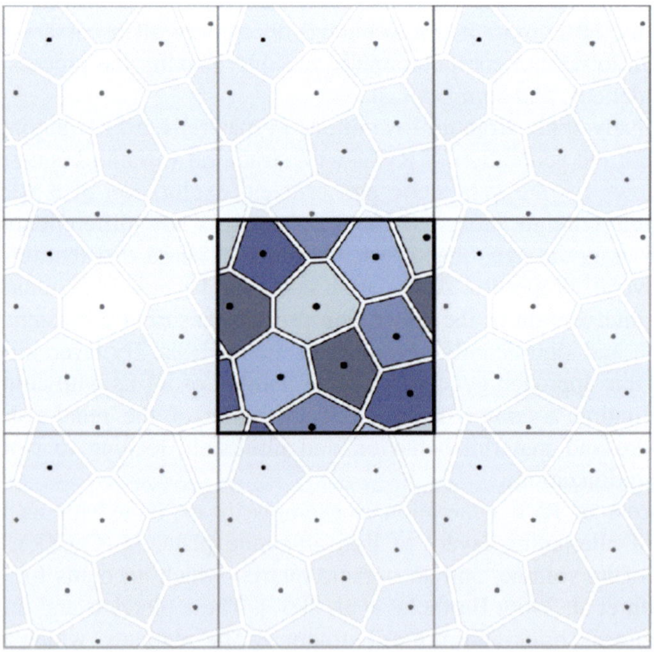

Fig. 4 Ten seeds are distributed, initially placed on a grid and then randomized, and subsequently copied to the eight adjacent RVEs. A periodic structure within the RVE is created by generating a Voronoi diagram based on these seeds. The resulting polygons are then shrunk to create spaces between them, which are later filled with matrix material. Figure from [Leuter et al. 2023] (an open access article under the CC BY license (http://creativecommons.org/licenses/by/4.0))

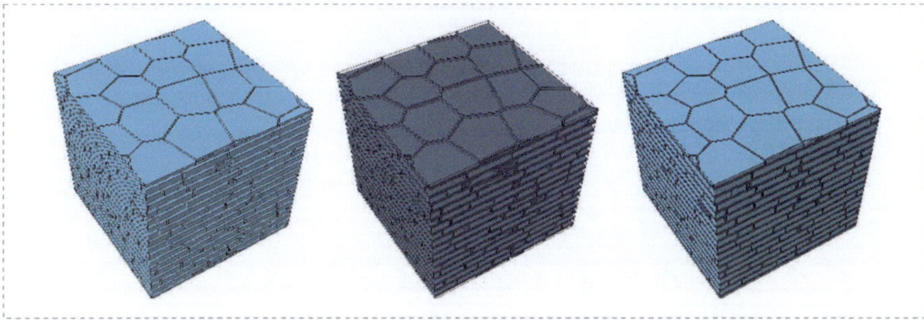

Fig. 5 Left: platelets. Middle: matrix material between platelets and layers. Right: 3D cubic RVE mimicking the brick-and-mortar like microstructure of nacre

3.2 Finite Element Simulations of Nacre

Micromechanical continuum mechanics models are essential for understanding the material behavior because they provide a framework to predict how materials deteriorate under stress and eventually fail. This insight is critical for designing materials and structures that can better withstand extreme conditions, optimize their durability, and ensure safety in applications ranging from aerospace to civil engineering.

Finite element simulations building on the principles of continuum mechanics provide a powerful computational tool to model and analyze complex microstructural processes in materials. These simulations allow for the detailed examination of how materials behave under various loading conditions, considering the effects of geometry, material properties, and boundary conditions. By discretizing a material or structure into finite elements, these simulations can track the evolution of stress at a very fine scale, enabling the prediction of stress concentrations, the influence of microstructural features, and the overall material behavior. This level of detail is invaluable for optimizing material design, improving structural integrity, and developing strategies to mitigate failure in critical applications.

Nacre's geometry is modeled as a composite with a periodic brick-and-mortar microstructure, excluding any tablet surface irregularities (Fig. 6a). In line with [Barthelat et al. 2007], the aragonite tablets are represented as a transversely isotropic elastic material, while the organic matrix, composed of a biopolymer, is treated as an isotropic linear elastic material. To study the effective elastic behavior of nacre, we analyze an elementary volume of the periodic microstructure under periodic boundary conditions using computational homogenization. This approach allows us to capture loading in any direction by imposing macrostrain on the elementary volume, enabling us to investigate how the loading direction influences nacre's mechanical performance.

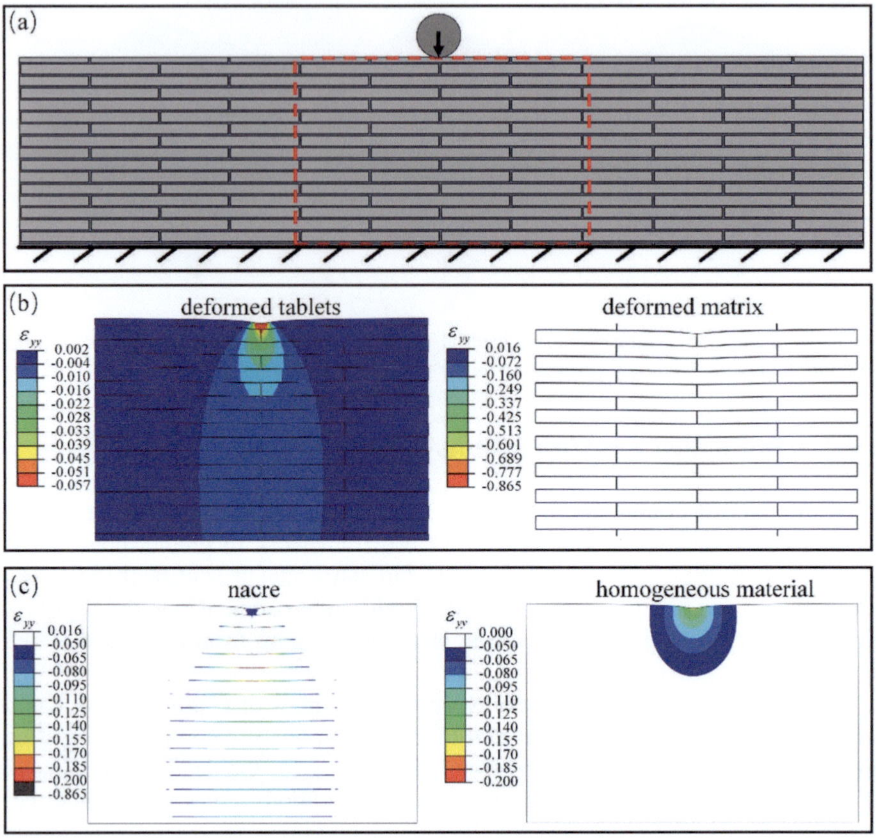

Fig. 6 **a** Sketch of a finite element model of a numerical indentation test performed on microstructure of nacre. **b** deformed platelets and deformed matrix at the end of the indentation test. **c** left: plot of normal strain in the center region. right: plot of normal strain for a homogeneous platelet-type material

Finite Element simulations are performed using the commercial finite element software package Abaqus/Standard, with Python scripts written to automate all simulation procedures. The tablets and matrix are meshed using 4-node bilinear plane strain elements (CPE4) and the rigid circular indenter is modeled as a rigid surface meshed with two-dimensional rigid elements (R2D2). Periodic boundary conditions are applied to the elementary volume using the multi-point constraint method in Abaqus. The built-in surface-to-surface contact algorithm is employed with hard contact and frictionless options to model the interaction between the indenter and the nacre RVE.

Our finite element simulations (Fig. 6b–c) show that in microstructured composites like nacre, microscopic deformation is heterogeneous even when subjected to uniform macroscopic deformation. In the organic matrix, the microscopic averaged strain components significantly exceed the applied macroscopic strain. The exceptional mechanical performance of nacre relies on the high deformability of the organic matrix. The organic matrix is observed to be strongly bonded to the platelets, enduring large strains before failure. Additionally, nacre exhibits high indentation resistance due to its layered structure, with the deformation under the indenter primarily absorbed by the organic matrix. See [Peng et al. 2022] for further experimental as well as numerical results.

3.3 Finite Element Simulations of Nacre-Inspired Synthetic Composites

Auxeticity refers to a unique property of materials that exhibit a negative Poisson's ratio, meaning that when stretched, they expand laterally rather than contracting, and when compressed, they contract laterally instead of expanding. This unusual behavior results in enhanced mechanical properties such as increased energy absorption, toughness, and resistance to fracture. Researchers develop bio-mimicking, tailor-made materials to replicate and enhance these desirable properties found in nature, aiming to create advanced materials with superior performance in specific applications. Nacre, with its intricate brick-and-mortar microstructure and high toughness, serves as an excellent role model for developing auxetic high-performance composites. By mimicking nacre's structure and the way it dissipates energy and resists damage, materials can be engineered that not only exhibit auxetic behavior but also possess enhanced strength, toughness, and durability, making them ideal for a wide range of advanced engineering applications.

In [Peng and Bargmann 2024], we designed synthetic 3D interlocking brick-and-mortar composites inspired by nacre with auxetic material behavior. By conducting finite element simulations and computational homogenization, we systematically explored how different geometrical parameters and base material properties influence the composites' effective elastic behavior. The findings reveal that these composites can achieve adjustable negative Poisson's ratios by precisely tailoring their microstructural design. The auxetic behavior is primarily driven by a sliding mechanism, where adjacent bricks with wavy surfaces slide relative to one another. With their unique combination of auxetic properties and high fracture toughness, these composites are highly suitable for a wide range of engineering applications, especially where robust auxetic materials are needed (Figs. 7 and 8).

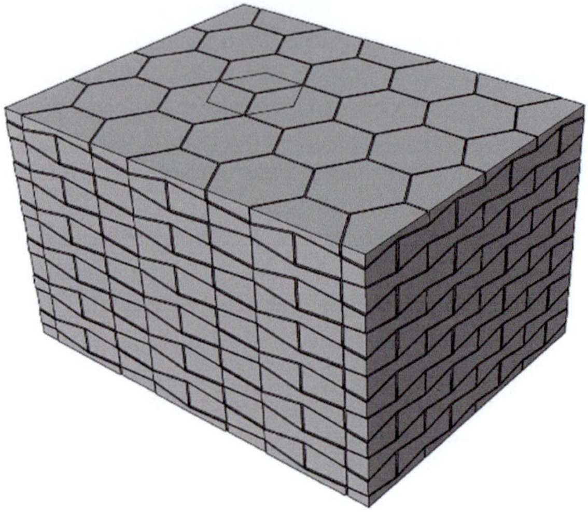

Fig. 7 Full nacre-mimicking microstructure based on the unit cell depicted in the middle

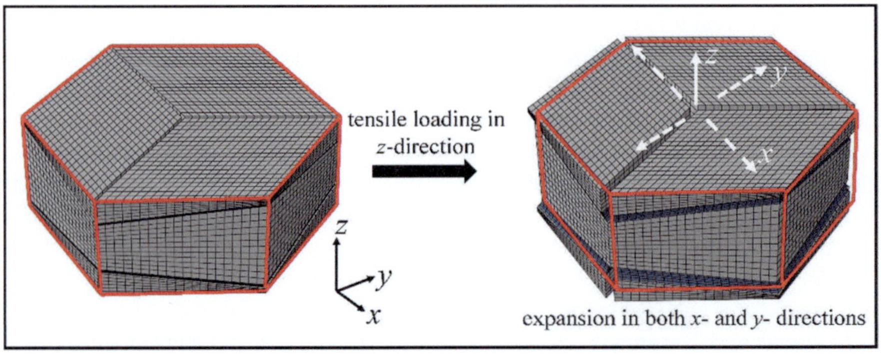

Fig. 8 The structure shows out-of-plane auxetic material behavior in xz- and yz- planes. The deformation is magnified and the red lines denote the profile of the undeformed unit cell for a better visualization

4 Summary

In conclusion, studying micromechanical properties of biological as well as engineering materials through advanced modeling techniques promises to unlock new avenues in the design of high-performance, damage-tolerant materials. By bridging the gap between natural and synthetic materials, we develop innovative solutions that meet the demanding requirements of modern engineering materials. Our work significantly contributes to the development of high-performance materials by advancing the understanding and design of complex material structures.

We engage in close collaboration with a diverse team of experts, including material scientists, safety engineers, engineers, zoologists, chemists, physicists, and mathematicians from around the globe (for example, Australia, Austria, China, Great Britain, Italy, Japan, Korea, Malaysia, Netherlands, Romania, Singapore, South Africa, Sweden, Turkiye, Ukraine, USA). This interdisciplinary approach ensures that our research integrates the latest advancements from various fields, leading to innovative solutions and enhancements in material performance and safety.

Acknowledgements Figures 6 and 8 were prepared by Xiang-Long Peng, Ph.D., a former member of the Chair of Solid Mechanics at University of Wuppertal, Germany, now at TU Darmstadt. SB gratefully acknowledges funding by the Deutsche Forschungsgemeinschaft (DFG, German Research Foundation) – Projektnummer: 529221905.

References

B. Alheit, S. Bargmann, D. Reddy *Computationally modelling the mechanical behaviour of turtle shell sutures – a natural interlocking structure*, Journal of the Mechanical Behavior of Biomedical Materials 100, 103973, 2020

B. Alheit, S. Bargmann, D. Reddy *Dynamic mechanical behaviour of suture interfaces as inspiration for architectured hierarchical interlocking composites,* Journal of the Mechanics and Physics of Solids 157, 104620, 2021

B. Alheit, S. Bargmann, D. Reddy *How suture networks improve the protective function of natural structures: A multiscale investigation,* Acta Biomaterialia 145, 283–296, 2022

S. Bargmann, I. Scheider, T. Xiao, E. Yilmaz, G. Schneider, N. Huber *Towards bio-inspired engineering materials: Modeling and simulation of the mechanical behavior of hierarchical bovine dental structure*, Computational Materials Science 79, 390–401, 2013

S. Bargmann, B. Klusemann, J. Markmann, J. Schnabel, K. Schneider, C. Soyarslan, J. Wilmers *Generation of 3D representative volume elements for heterogeneous materials: A review,* Progress in Materials Science, 96, 322–384, 2018

F. Barthelat, H. Tang, P.D. Zavattieri, C.M. Li, H.D. Espinosa *On the mechanics of mother-of-pearl: a key feature in the material hierarchical structure*, J. Mech. Phys. Solids 55 (2) (2007) 306–337

P. Fratzl and R. Weinkamer *Nature's hierarchical materials*, Progress in Materials Science, 52(8), 1263–1334, 2007

I. Gitman, H. Askes, L. Sluys *Representative volume: existence and size determination*, Eng Fract Mech, 74(16), 2518–34, 2007

E. Griffiths, J. Wilmers, S. Bargmann, D. Reddy *Nanoporous metal based composites: giving polymers strength and making metals move*, J Mech Phys Solids 137, 103848, 2020

R. Hill *Elastic properties of reinforced solids: some theoretical principles* J Mech Phys Solids11(5), 357–72, 1963

K.S. Katti, B. Mohanty, D.R. Katti *Nanomechanical properties of nacre*, J. Mater. Res. 21 (5), 1237–1242, 2006

P. Kowol, S. Bargmann, P. Görrn, J. Wilmers *Strain relief by controlled cracking in highly stretchable multi-layer composites,* Extreme Mechanics Letters 54, 101724, 2022

J. Lamsfuss, S. Bargmann *Skeletal muscle: Modeling the mechanical behavior by taking the hierarchical microstructure into account,* Journal of the Mechanical Behavior of Biomedical Materials 122, 104670, 2021

J. Lamsfuss, S. Bargmann *Computational modeling of damage in the hierarchical microstructure of skeletal muscles*, Journal of the Mechanical Behavior of Biomedical Materials 134, 105386, 2022

J. Lamsfuss, S. Bargmann *Mechanisms of thermal treatments in trigger points of the skeletal muscle: computational microstructural modeling*, European Journal of Mechanics/A Solids 99, 104906, 2023

C. Leuter, S. Bargmann, J. Wilmers *Generation of 3D representative volume elements (RVEs) of nacre*, software 17, 100534, 2023

S. Ma, I. Scheider, S. Bargmann *Anisotropic constitutive model incorporating multiple damage mechanisms for multiscale simulation of dental enamel*, Journal of the Mechanical Behavior of Biomedical Materials 62, 515–533, 2016

S. Ma, I. Scheider, S. Bargmann *Continuum damage modeling and simulation of hierarchical dental enamel*, Modelling and Simulation in Materials Science and Eng. 24(4), 045014, 2016b

A. McBride, S. Bargmann, D. Pond, G. Limbert *Thermoelastic modelling of skin at finite deformations*, Journal of Thermal Biology 62, 201–209, 2016

R. Menig, M.H. Meyers, M.A. Meyers, K.S. Vecchio *Quasi-static and dynamic mechanical response of Haliotis rufescens (abalone) shells*, Acta Mater. 48, 2383–2398, 2000

S. H. Oh, J.-K. Kim, Y. Liu, M. Wurmshuber, X.-L. Peng, J. Seo, J. Jeong, Z. Wang, J. Wilmers, C. Soyarslan, J. Kim, B. Kittiwirayanon, J. Jeong, H.-J. Kim, Y. H. Huh, D. Kiener, S. Bargmann, H. Gao *Limpet teeth microstructure unites auxeticity with extreme strength and high stiffness*, Science Advances 8 (48), eadd4644, 2022

I. Okulov, J. Wilmers, S.-H. Joo, S. Bargmann, H. Kim, H. Kato *Anomalous compliance of interpenetrating-phase composite of Ti and Mg synthesized by liquid metal dealloying*, Scripta Materialia 194, 113660, 2021

M. Ostoja-Starzewski *Material spatial randomness: from statistical to representative volume element*, Probab Eng Mech 21(2), 112–32, 2006.

X.-L. Peng, S. Bargmann *A novel hybrid-honeycomb structure: enhanced stiffness, tunable auxeticity and negative thermal expansion,* International Journal of Mechanical Sciences 160, 106021, 2021

X.-L. Peng, S. Lee, J. Wilmers, S. Oh, S. Bargmann *Orientation-dependent micromechanical behavior of nacre: in situ TEM experiments and finite element simulations,* Acta Biomaterialia 147, 120–128, 2022

X.-L. Peng, S. Bargmann *Nacre-inspired auxetic interlocking brick-and-mortar composites*, Composites Communications 48, 101892, 2024

J. W. Pro and F. Barthelat *Discrete element models of tooth enamel, a complex three-dimensional biological composite*, Acta Biomaterialia 94: 536–552, 2019

S. H. Raza, C. Soyarslan, S. Bargmann, B. Klusemann *Computational modeling of amorphous polymers: A Lagrangian logarithmic strain space formulation of a glass-rubber constitutive model*, Computer Methods in Applied Mechanics and Engineering 344, 887–909, 2019

J. M. Rensberger and H. U. Pfretzschner, *Enamel structure in astrapotheres and its functional implications*, Scanning Microscopy 6(2):477–495, 1992

C. Renteria, J. Fernández-Arteaga, J. Grimm, E. Ossa, D. Arola *Mammalian enamel: A universal tissue and diverse source of inspiration*, Acta Biomaterialia 136, 402–411, 2021

I. Scheider, T. Xiao, E. Yilmaz, G. Schneider, N. Huber, S. Bargmann *Damage modeling of small scale experiments on dental enamel with hierarchical microstructure*, Acta Biomaterialia 15, 244–253, 2015

K. Schneider, B. Klusemann, S. Bargmann *Automatic three-dimensional geometry and mesh generation of periodic representative volume elements for matrix-inclusion composites*, Advances in Engineering Software 99, 177–188, 2016

F. Song, A.K. Soh, Y.L. Bai, *Structural and mechanical properties of the organic matrix layers of nacre*, Biomaterials 24 (20), 3623–3631, 2003

D. Sopu, C. Soyarslan, B. Sarac, S. Bargmann, M. Stoica, J. Eckert *Structure-property relationships in nanoporous metallic glasses*, Acta Materialia 106, 199–207, 2016

U. G. K. Wegst and M. F. Ashby. *The mechanical efficiency of natural materials*, Philosophical Magazine, 84(21), 2167–2186, 2004

Z. Weng, Z. Liu, R. Ritchie, D. Jiao, D. Li, H. Wu, L. Deng, and Z. Zhang, *Giant panda's tooth enamel: Structure, mechanical behavior and toughening mechanisms under indentation*, Journal of the Mechanical Behavior of Biomedical Materials 64:125–138, 2016

J. Wilmers, S. Bargmann, *Functionalisation of metal-polymer-nanocomposites: Chemoelectromechanical coupling and charge carrier transport*, Extreme Mechanics Letters 21, 57–64, 2018

J. Wilmers, S. Bargmann *Nature's design solutions in dental enamel: Uniting high strength and extreme damage resistance*, Acta Biomaterialia 107, 1–24, 2020

J. Wilmers, M. Waldron, S. Bargmann, *Hierarchical Microstructure of Tooth Enameloid in Two Lamniform Shark Species, Carcharias Taurus and Isurus oxyrinchus*, Nanomaterials 11, 969, 2021

M. Wurmshuber, J. Wilmers, J. Kim, S.H. Oh, S. Bargmann, D. Kiener *Lower hardness than strength: the auxetic composite microstructure of limpet tooth,* Acta Biomaterialia 166, 447–453, 2023

J. Wilmers, M. Wurmshuber, Ch. Gescher, C. Graupp, D. Kiener, S. Bargmann, *Unraveling the orientation dependent mechanics of dental enamel in the red-necked wallaby,* Acta Biomaterialia 185, 254–265, 2024

H.-P. Yu, Y.-J. Zhu, *Guidelines derived from biomineralized tissues for de-sign and construction of high-performance biomimetic materials: from weak to strong*, Chemical Society Reviews 53(9):4490–4606, 2024

C. Ziscovici, P. W. Lucas, P. J. Constantino, T. G. Bromage, and A. van Casteren, *Sea otter dental enamel is highly resistant to chipping due to microstructure*, Biology Letters 10(10):20140484, 2014

Safe Production and Processing of Reactive Powder Materials in Additive Manufacturing

Silja-Katharina Rittinghaus and Bilal Gökce

Contents

Abstract

Additive manufacturing (AM) has established itself as a revolutionary technology that makes it possible to create complex components directly from digital models. For industrial high-tech applications, reactive powder materials such as titanium and aluminum are commonly used, though their fine particle form can pose risks of fire and explosion. The key considerations and conditions for the production (atomization) and metallurgical processing of such powder materials using AM techniques are presented, taking into account both safe working environments and the preservation of high material quality. Additionally,

S.-K. Rittinghaus · B. Gökce (✉)
Chair of Materials Science and Additive Manufacturing,
School of Mechanical Engineering and Safety Engineering,
University of Wuppertal, Wuppertal, Germany
e-mail: goekce@uni-wuppertal.de

S. Bracke and R. Pieper (eds.), *Safety Engineering*,
https://doi.org/10.1007/978-3-658-47473-7_7

aspects of sustainability are discussed, which are of great importance for the development of ecological material cycles. Best practice examples are presented as the part of work flows recommended for both research labs and production environments. Critical work steps are highlighted and rules to be taken into account are explained. In addition, current challenges and strategies with regard to sustainability are discussed.

Keywords

Additive Manufacturing · Reactive powders · Safety practices

1 Introduction

Additive Manufacturing (AM), often referred to as 3D printing, has transformed the manufacturing landscape by enabling the creation of complex and customized parts with unprecedented precision. Especially in the production and processing of reactive powder materials, such as metal or ceramic powders, it offers immense flexibility and enables the realization of innovative designs (Fig. 1).

Reactive powder materials, which are particularly sensitive to temperature, pressure, or moisture due to their chemical properties, are attractive candidate fuels for various propulsion and energy-conversion applications [1]. However, when used as feedstock in AM, they pose significant safety challenges. Safe handling, storage, and processing of these materials are crucial to prevent risks such as explosions or unintended chemical reactions and thus to prevent accidents and ensure the well-being of workers and the integrity of manufacturing facilities.

This involves a comprehensive understanding of their properties, including pyrophoricity, explosibility, and reactivity with environmental factors.

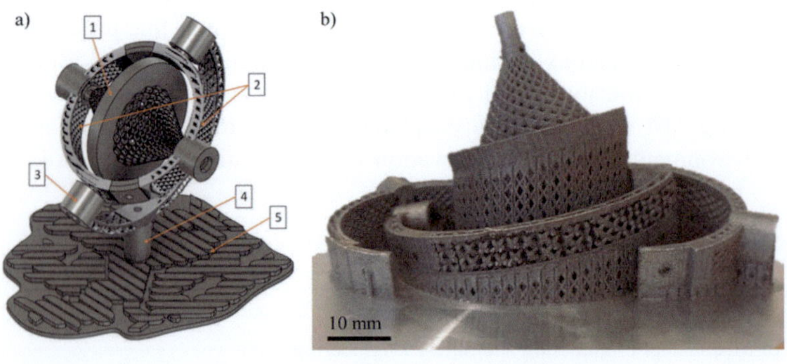

Fig. 1 Complex parts designed by the students J. N. Petenati and J. Vehrs (**a**) and printed (**b**) within the lecture "Additive Manufacturing" at the University of Wuppertal. The task was to design a 3D-printable gyroscope under consideration of design rules and both functional and economic criteria

Implementing stringent safety protocols, such as proper storage conditions, controlled environments, and the use of personal protective equipment (PPE), is essential. Additionally, advanced techniques like inert gas environments and automated handling systems can mitigate risks associated with human error.

This paper aims to explore the critical aspects of safety in the production and processing of reactive powder materials in AM, highlighting best practices, technological advancements, and an overview of necessary measures that ensure a safe and efficient manufacturing process.

2 Reactive Powder Materials

Reactive materials undergo significant changes in their physical or chemical properties when exposed to specific environmental conditions, such as temperature, pressure, or chemical stimuli. These materials are integral to various advanced technological applications due to their unique and dynamic properties [1]. However, this high reactivity also brings some drawbacks and requires special precautions when handling micro- to nanosized powder materials (Fig. 2).

Reactive powders like titanium and aluminum can spontaneously ignite when exposed to air due to their high surface area and reactivity (pyrophoricity). Pyrophoricity thus is a major safety issue during synthesis, handling, and storage, and necessitates handling these materials in inert atmospheres, such as argon or nitrogen, to prevent accidental ignition.

Fig. 2 SEM image of powder particles typically used in AM: a powder used for PBF-LB/M, ranging from about 15–63 μm diameter in size

Furthermore, reactive powders in fine particles sizes, as e.g., remaining in filters of AM machines, increases the risk of dust explosions. The minimum ignition energy (MIE) and the lower explosive limit (LEL) are critical parameters that need to be understood and controlled. For instance, aluminum powder has a low MIE, making it highly susceptible to ignition from static electricity or mechanical sparks.

It has also to be considered that reactive powders can react with moisture or oxygen, leading to oxidation or even violent reactions. For example, titanium powder can form titanium dioxide when exposed to air, which can affect the material properties and safety.

Typical reactive materials being crucial for various high-performance applications and thus attractive for AM are mainly titanium and aluminum (Fig. 3), but also refractory-based alloys, which is reflected by analyzed research trends [2]. They are therefore considered in more detail below. It should be mentioned, however, that advanced materials based on, for example, zirconium can also be processed by AM [3]. However, most of these are even further away from industrial application.

2.1 Titanium and Alloys

Titanium is renowned for its high strength-to-weight ratio, good biocompatibility, excellent corrosion and wear resistance, and ability to withstand high temperatures. These properties make it ideal for aerospace [5], biomedical applications [6], automobile [7] and chemical processing industries [8]. It is one of the most

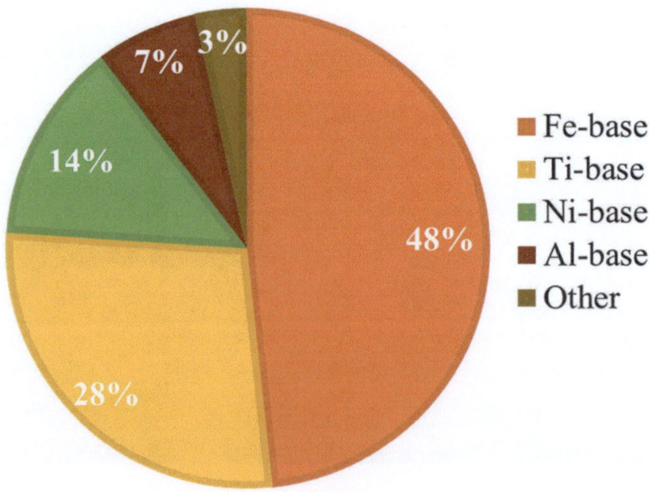

Fig. 3 Metal powder processed in AM, data from: [4]

used materials in AM, despite challenges in handling caused by both high-quality demands in the applying sectors and its inherent reactivity.

Titanium powder is classified as a hazardous material due to its susceptibility to oxidation, combustion, and explosion. It is considered a flammable and explosive dust, which, in the presence of an open flame, heat sources, or during chemical reactions, can ignite or even lead to an explosion.

Titanium and its alloys exhibit a strong affinity for oxygen, forming a native, passivating TiO_2 layer of about 2–6 nm thick, almost immediately when the metal surface is exposed to air at room temperature. Investigations show that the oxide layer thickness of Ti and Ti6Al4V decreases only slightly during longer storage, preventing further reactions of the underlying material [9]. Upon heating, the material might reach the point of thermal runaway or combustion, either locally or uniformly [10]. During the processing and handling of titanium powder, dust clouds are often generated. If these clouds accumulate under certain conditions and come into contact with an ignition source, they can cause combustion or explosions by rapid oxidation as reaction with ambient air, posing significant risks of severe safety incidents in production environments. The MIE for titanium powder can range from 3 to 30 mJ, while generally temperatures between 300–700 °C can trigger the ignition of titanium powder [11].

2.2 Aluminum and Alloys

Aluminum is another widely used reactive material, valued for its lightweight, high conductivity, and resistance to oxidation. Its applications range from aerospace and automotive industries to packaging and construction, with Al–Si alloys being the most investigated alloy systems for additive manufacturing [12].

The reactivity of aluminum with oxygen is high, and dependent not only on site but also morphology and microstructure of the powder [13]. Studies indicate that the critical size for aluminum particles to be pyrophoric is 32 nm and even smaller when passivated [14], and thus way below the typical particle size fraction used in AM of 10 µm and larger. However, as aluminum also shows a high tendency to evaporate during AM, fine nanosized dust particles are likely to occur and deposit on machine surfaces and in filters, causing risks. Also, during atomization much smaller particles sizes are produced, even if not being part of the used powder product. The MIT of aluminum dust was investigated to be around 330 °C, dependent on cloud density [15].

2.3 Refractory Metal Alloys

Refractory metals alloys attractive for consideration for AM include base materials like tungsten [16], molybdenum [17, 18], and tantalum [19], which are known for their ability to retain strength and stability at very high temperatures. These alloys are essential in industries that require materials to perform under extreme heat, such as in jet engines, nuclear reactors, and space exploration. Challenges are not only the high melting point, posing the necessity to process these materials at high temperatures but the fact that most refractory metals exhibit poor oxidation resistance. The result is the temperature dependent rapid formation of non-protective or volatile oxides, leading to the rapid degradation of the material [20]. The specific characteristics of the respective elements and their alloys differ [21], for AM applications, especially protection from oxidation is mandatory not only during handling but also storage, as these materials degrade so rapidly.

2.4 Magnesium and Alloys

AM of magnesium alloys is of growing interest in both research and industry due to its tremendous potential for development of both biodegradable implants [22] and lightweight structural parts for automotive and aerospace [23].

Magnesium poses a significant challenge for additive manufacturing due to its highly reactive properties. In its pure form, magnesium oxidizes rapidly and uncontrollably, requiring it to be stored in conditions that prevent exposure to oxygen. Due to the challenges associated with the handling of magnesium powder, currently, wire-based wire arc additive manufacturing (WAAM) is in the focus of R&D activities [24], demonstrating good success in the production of dense and homogeneous components [25].

3 Production of Reactive Powders in Micron Size

3.1 Atomization

The typical method for the production of reactive powder materials for AM is inert gas atomization. This process produces high-quality metal powders with controlled particle size and shape, reducing the risk of both oxidation and contamination. The powders are produced in an inert gas environment, ensuring their stability and safety. Particle size, shape, chemistry and even microstructure are decisive criteria for AM powders [26]. Moreover, also ecological and economical concerns are to be considered, which means that the highest possible throughput and low shielding gas consumption are also important objectives.

In general, two variants for the production of reactive AM powders have been established, vacuum inert gas atomization (VIGA) and electrode induction gas atomization (EIGA). In VIGA, the liquid metal is contained in a crucible with an outlet at the bottom, allowing for a controlled flow through a nozzle, where the liquid is atomized by inert gas streams. The so formed small droplets cool down rapidly, and reach a collecting container in solid condition. EIGA, especially required for the use of materials that would react with the crucible material, is based on the continuous melting of the tip of a previously produced electrode. Here, too, the resulting melt flow is atomized via a nozzle.

Especially in research, ultrasonic atomization (UA) techniques are being adopted, as they provide the possibility to produce small batches of powder and are very flexible on terms of composition. In UA, melt is dropped onto a platform excited with ultrasound. Standing waves are formed, which result in the ejection of micron sized droplets, which upon rapid cooling end up as powders suitable for AM [27] (Fig. 4).

High melting temperatures combined with high reactivity and sensitivity to interstitial elements can pose significant challenges when producing powder from reactive materials, specifically refractory metal alloy powders. A recent study compared the production of Mo-20Si-52.8Ti (at.%) with an UA and an EIGA process [28], where the EIGA electrode was PM manufactured from elemental powders to provide compositional flexibility. Both processes were demonstrated to yield powders fulfilling the requirements for AM.

Subsequent to atomization, further steps are necessary: removal of the powders from the collection container, sifting to separate fine dusts, and sieving to classify them into the required particle size fraction. All this is done with reactive powders under the strictest safety precautions and often in the absence of oxygen again under inert gas atmosphere—right down to the packaging. In order to achieve high powder qualities, careful and therefore time-consuming cleaning of the complex systems is of the utmost importance.

Fig. 4 EIGA process for the production of AM powder from a Ti-based electrode, image provided by GfE Metalle und Materialien GmbH, Nürnberg

3.2 Passivation

To ensure the safe handling of powder materials, passivation can be applied. This involves the controlled exposure of spontaneously oxidizing materials to an oxygen-containing atmosphere. For example, titanium and aluminum quickly form a passivating oxide layer on their surface, which protects them from further reactions, making processes like sieving or feeding into AM machines safer.

Variants applied for highly reactive metal powders include heat treatment in an oxygen-rich atmosphere, where the fresh powder is heated in an oxygen-containing atmosphere (such as air or oxygen gas) at a specific temperature. For instance, titanium or aluminum powders are often exposed to air at moderate temperatures, which leads to the formation of the desired thin, stable oxide layer on the surface of the powder particles. Metals can also be passivated in an inert gas atmosphere, such as argon or nitrogen, which prevents unwanted reactions while still allowing for the formation of a protective oxide layer under controlled conditions.

Less common is the application of a thin coating or layer of a passivating material, such as a polymer or a thin metal film to protect the powder particles from oxidation [29]. However, this may have unwanted effects of the final part properties.

3.3 Transport and Storage

Reactive powder materials are classified as hazardous substances, which places strict requirements on storage, but also on documentation and transportation companies. These materials must be stored and transported in airtight containers to prevent exposure to air and moisture, avoiding oxidation or spontaneous ignition. Storing powders under inert gases like argon or nitrogen can further reduce the risk of ignition and powder degradation. The selection of suitable containers however may be difficult, as certain aspects like shipping or on-site handling favor small quantities, while compositional batch control and traceability as well as the possible need to fill multiple containers in one machine, one of the riskier process steps, could indicate advantages in using larger containers.

Larger storage areas should be equipped with explosion-proof ventilation systems and fire suppression systems. Cupboards in accordance with DIN EN 14470-1 are usually considered suitable for the storage of any type of AM powders. In the immediate vicinity of heat and spark sources, such as AM systems, only the daily consumption quantities actually required should be stored. During transportation, whether by external carriers or within facilities, powders should be kept in sealed containers and handled carefully to prevent spills and static discharges. Also, only keeping small amounts per container diminishes the risk of critical self-containing reactions.

Any amounts of reactive powder materials transported and stored should be rigorously documented, to enable appropriate safety measures, such as the use of the correct fire extinguishers, in the event of a fire or traffic accident.

4 Additive Manufacturing

4.1 Powder Handling

Typical steps of powder handling at local facilities include transfer to smaller containers, tumble mixing to ensure a homogenous particle size distribution, sieving—especially when powder is being re-used or a special particle size fraction needed—and feeding the respective AM machine. These steps are preferably to be performed in controlled environments, e.g., using glove boxes or other enclosed chambers with inert gas atmospheres to maintain a stable and sage atmosphere for handling reactive powders. By this, the risk of ignition and explosion can be minimized. Ventilation systems, including explosion-proof exhausts, help to manage any accidental releases of powder and prevent the accumulation of flammable dust.

Automated handling systems can minimize exposure of employees and surfaces to reactive powders. These are nowadays capable of handling transfer, mixing and loading of powders, thus excluding incidents caused by human errors. However, these are most suitable for industrial environments, where large quantities are handled in a repeatable manner. Even if there are systems available that allow machine feeding without any air contact, in most cases, this is performed by pouring or scooping the powder in the respective machine internal storage before closing the machine and activating an inert atmosphere. Using anti-static materials and coatings for tools and surfaces are comparatively simple measures to further reduce the risk of static discharge, while also gentle and patient pouring can contribute to avoiding dust explosions. In general, all equipment and containers should be properly grounded to prevent static electricity buildup, which can ignite reactive powders. Using explosion-proof equipment and electrical systems in areas where reactive powders are handled can prevent ignition sources. Additionally, fire suppression systems, such as inert gas flooding or dry chemical extinguishers, can quickly control any fires that do occur.

Naturally, employees should be equipped with appropriate personal protective equipment (PPE) to provides barriers against potential fires and explosions and prevent direct contact with powder material.

4.2 Processing

Nowadays, numerous AM techniques for the processing of metals powders are established at least in research. Most prominent are powder-based techniques using high-energy electron beams (electron beam melting or electron beam powder bed fusion, PBF-EB) or laser beams (selective laser melting or laser beam powder bed fusion, PBF-LB) to selectively fuse powder particles from a so called "powder bed". Nozzle based welding techniques, such as direct energy deposition (DED) are also suitable to build 3D-structures from reactive metal powders. These processes are usually conducted in controlled environments, reducing the risk of ignition and explosion and ensuring contamination free processing. This typically means either global shielding of the whole process environment with an inert gas (PBF-LB) or vacuum (PBF-EB) atmosphere, or local shielding of the immediate processing zone provided by an inert gas flow (DED). While e.g., also PBF-LB in vacuum is being researched [30], or DED can be performed in a glove box [18], in industrial environments typically the aforementioned setups can be found, as these are also the most frequent ones offered by AM machine manufacturers. All these AM processes have in common, that digital models are, via respective data processing, transferred into the layer-wise build of 3D parts on a respective build plate (substrate). Thus, powder-bed-based processing, as illustrated in Fig. 5, essentially involves three steps: powder spreading, selective remelting, and platform lowering. These steps are repeated until the desired build height is achieved.

Typical powder particles sizes are about 10–63 µm for PBF-LB, 20–90 µm in DED up to 250 µm in PBF-EB. As the whole process is performed in enclosed chambers and fully automated, risks with regard to reactive powder materials are very low, and the processing itself can be considered of the safest steps in the whole manufacturing chain. However, system failure like interruption in gas supply or power outage can happen, potentially leaving a heated-up powder bed unprotected. To prevent incidents, advanced sensors and monitoring systems can detect changes in environmental conditions, such as oxygen levels and humidity, in real-time. These systems can trigger alarms and safety protocols to prevent accidents. Usually, at least oxygen levels and temperature are controlled by

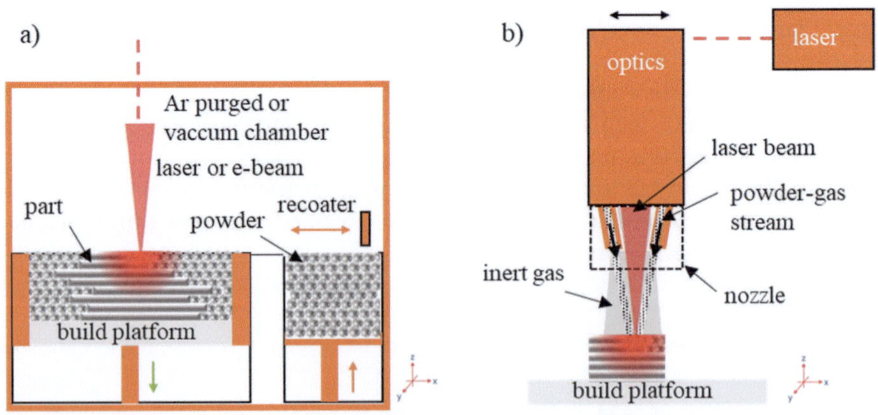

Fig. 5 AM processes, schematically: **a** PBF-LB/EB, **b** DED

default, and cause the machine to shut down the beam-source if critical levels are exceeded.

During the AM melting steps, very high local energies are applied, typically causing parts of the powder to be evaporated [31]. In alloys this often occurs on a larger scale, e.g., when titanium alloys also contain aluminum, which has a significantly lower evaporation temperature compared to titanium. This type of evaporation can easily reach up to 1–2 at.%. Evaporated material solidifies as nano-sized dust particles and is deposited on all internal surfaces and in filters, and thus causes significantly greater inconvenience and safety risks than the micro sized AM powder.

4.3 Post-Processing

The post-processing related to AM includes all steps from opening the machine, to removing and de-powdering part, removing the part from the substrate plate and all surface finishing steps. With regard to the aforementioned powder related risks, part removal and de-powdering are the most critical ones. These can be but are rarely automated, as most AM machines are used to produce rather individual than serial parts, which means that investing in automatic systems is rarely economically worthwhile. Dependent on the requirements, post-processing can be very demanding. If e.g., the remaining powder is supposed to be reused, it has to removed from the build area without contaminating it. For specific powders, this might even require transporting it in a sealed container without any air contact. When removing the parts on the build plate, it has to be ensured that it has sufficiently cooled down. If no automatic system is installed, gentle and slow handling can prevent dust formation, while suitable equipment (cf. Sect. 5) protects the employee. Remaining powders in the vicinity or cavities of the parts should be removed completely and under inert conditions before further processing the parts, generally, the rules for powder handling apply. The methods for this may vary, as they are highly dependent on the process used, part geometry and size, and desired surface properties. Typical methods include e.g., sandblasting for outer surfaces, or the use of compressive air and subsequent bathing in ethanol to clean cavities and channels.

4.4 Cleaning, Filter Exchange and Waste Disposal

Usually, antistatic vacuum cleaners, for reactive powders, explosion-protected vacuum cleaners that do not generate sparks are used to remove remaining powder particles after removal of the parts from the AM machine. They need to be equipped with a HEPA filter or another specialized filtration system to efficiently capture fine particles and prevent them from being released into the air. Naturally, the filters should be regularly checked and replaced to maintain optimal dust filtration. It has also to be taken into account, that using the same vacuum cleaner for multiple materials can cause additional risks and is to be avoided. E.g., mixing titanium powder and aluminum powder can lead to dangerous chemical reactions that are similar to the thermite reaction. The exothermic reaction can result in hazardous fires, explosions, or dust explosions.

For certain materials as e.g., titanium, wet vacuum cleaners can be used to diminish the risk of dust dispersion or static buildup. It has to be made sure that the liquid used is non-reactive with the respective powder. Some wet vacuum systems also have explosion-proof components, which are essential when handling hazardous materials like titanium powder. Additionally, the wet vacuum should be cleaned thoroughly after use to avoid cross-contamination or corrosion.

As previously described, the inner surfaces of the AM machine are likely to be contaminated with dust, which thus has to be removed which great care. Wiping cloths soaked with ethanol are suitable to clean accessible surfaces, while tubes and the recoaters might have to be replaced between build if dissimilar materials are processed on one machine.

Safe filter exchange is a particularly critical aspect of maintaining a safe working environment in AM, especially when dealing with reactive powder materials. The particles contained in filters are extremely fine (nano-sized), have a high surface area-to-volume and ratio and thus a way higher reactivity compared to the micron sized particles serving as input in AM. Additionally, small particles are less affected by gravity, making it easier for them to stay airborne, which increases the risk of dust clouds forming.

Best practice procedures typically start with a good preparation and thus isolation of the system. Although it may sound trivial, it is important to ensure that the machine is completely switched off and that there are no sources of heat or sparks in the vicinity when the filters are to be removed. If possible, it is desirable to perform the filter exchange in an inert atmosphere to minimize the risk of ignition. In best case, the filter exchange is performed in a controlled environment, such as a glove box or an enclosed chamber with proper ventilation. However, this is not feasible in many cases, especially with large filters. One variant that has been used successfully in practice is therefore to allow the filters to react (burn off) in a controlled manner. For this, the filter has to removed and transported to a suitable area first, which can be a large free outdoor space with fireproof ground. Using gentle handling and anti-static tools help to minimize dust generation and thus risk of explosion. In any case, the used filters are to be disposed in accordance with hazardous waste regulations, ensuring they are sealed and labeled correctly. The whole area used for filter exchange has to be cleaned up thoroughly in order to remove any residual particles. New filters should be inspected before putting them in place, and machinery should be tested to ensure correct and installation and full functionality. During all steps, the use of suitable Personal Protective Equipment (PPE) (cf. Sect. 5) is mandatory, with additional flame-resistant full-body clothing and face shields to protect against potential fires and explosions (Fig. 6).

Fig. 6 Employees in full PPE, carefully preparing the exchange of a PBF-LB filter

All waste produced during cleaning and maintenance of the powder AM machines, including wipes, gloves, filters etc. is usually considered hazardous and has to be handled as such, stored in fire-proof containers, separated from regular waste and properly labelled and disposed by respective companies.

4.5 Incidents in Additive Manufacturing

Incidents in AM have recently been studied in order to derive suitable measures for respective prevention in future [32]. Several risks have been identified that can be linked to reactive powder materials in particular and highlight the importance of stringent safety measures. One reported incident involved a dust explosion in a facility handling aluminum and titanium powders. The explosion was triggered by static electricity, emphasizing the need for proper grounding and bonding of equipment [33]. Another incident occurred when a dust collector containing aluminum dust was ignited and revealed a significant lack of proper training also for fire inspectors, and unsuitable machinery equipment [34]. AM processes being digital by nature also imposes less apparent risk regarding cyber security. A comprehensive threat analysis revealed that additive manufacturing systems are vulnerable to cyber-attacks, which can not only lead to operational but physical disruptions [32]. This underscores the need also for robust cybersecurity measures in AM facilities.

By adhering to best practices and learning from past incidents, the risks associated with AM can be significantly mitigated, ensuring a safer working environment for all involved.

5 Health Risks and Personal Protective Equipment

The handling of AM powders and, more specifically, reactive materials in powders form poses well-known risks for persons working in the field [35]. While the flammable character of reactive powders and the associated risk for explosions and fires can be minimized by process and organizational measures, certain steps of AM processes still bring the staff in direct contact with powders, mostly during pre- or post-processing steps or cleaning activities. This imposes certain risks specifically with regard to the human respiratory system and skin.

A study focusing on titanium powders found that AM powders investigated were inhalable and contained thoracic and respirable sized particles with the potential to deposit beyond the larynx and alveoli if inhaled [36].

Even more crucial, nano-sized particles may be emitted during the manufacturing process and subsequent cleaning of machine parts. These particles can be also inhaled into the respiratory system or potentially penetrate the body through the skin or eyes. While one-time exposure might be harmless, the human body has difficulty metabolizing most of powders, and their accumulation can rapidly reach toxic levels [37].

Possible health effects include fume fever, a condition induced by breathing metal fumes that produces symptoms such as fever, headache, chills, and fatigue. Inflammation and breathing difficulties can arise (asthma), and even chronic lung disorders, especially when long-term exposure is given [35].

Consequently, a very important personal protective measure that every employee can take is to wear PPE consistently. Every employer is obliged to provide these, while the employee is responsible for wearing them and ensuring that they function properly. Typically, the PPE consists of the components flame-resistant clothing, which could also be supported by wearing cotton materials instead of synthetic fabric in general, suitable gloves, respiratory protection (masks) to prevent inhalation of fine powder particles, and anti-static footwear. Depending on the work area and task this can and should be complemented by face shields (e.g., during filter exchange), laser safety eyewear and further equipment not specifically related to reactive powders but working in AM lab environments. A clear separation of work and private clothing is necessary in order to avoid contamination of the latter as far as possible. From an organizational point of view, the installation of a separate changing room as an interlock has proven its worth. Also, regular cleaning of the clothing worn during work with reactive powders has to be organized; both of which can be a challenge, especially for smaller organizations.

The best protection for employees, however, is provided by proper training and sensitization to the risks and correct behavior in the event of an incident. As a bonus point, this is almost free of charge.

6 Sustainability and Recycling

The AM process chain and specifically powder production is very energy consuming. Thus, the best method to increase sustainability is to increase material efficiency, which a challenging task with regard to reactive powders. Unused AM powders from the build can often be sieved and reconditioned for reuse in subsequent builds. This practice not only reduces waste but also lowers material costs. Unfortunately, many reactive powder materials are single-use, while e.g., steel powders can be reused multiple times without part quality issues.

One of the most apparent reasons why reuse of reactive powders in AM presents several unique challenges is powder degradation. Reactive powders like titanium and aluminum are prone to oxidation when exposed to air during the AM process. When using or even storing reactive powders, they tend to degrade whenever in contact with in most cases oxygen. Thus, every handling step not only increases risks but also potential material losses, as oxidation can alter the materials' chemical composition and reduce their effectiveness in subsequent uses [38].

Specifications for e.g., Ti-6Al-4 V, allow for a max. of 0.2 wt% (Grade 5) or even below 0.13 wt% (Grade 23) oxygen [39], which means a very narrow tolerance. On the other hand, several studies have already demonstrated that standard Ti alloy powder can be reused multiple times in AM. However, they also highlight the limitations, particularly concerning the increasing oxygen content in Ti alloy powder with each reuse [40–42]. This elevated oxygen level can result in surface oxidation of the particles, leading to issues such as lack of fusion during PBF, ultimately decreasing the mechanical performance of the produced parts.

A further issue is the cyclic reheating being native to AM. Repeated exposure to high temperatures can cause changes in the microstructure of the powders, affecting their flowability and packing density [38]. Also, contamination with foreign particles may occur during the reconditioning process, e.g., through dust or debris from the build chamber or sieving device, which can affect the quality of the final product. In general, using the same equipment for different materials without proper cleaning can lead to cross-contamination, which can compromise the integrity of the reused powder [43].

Processing the same powders multiple times also raises issues with regard to the particle sizes being inconsistent. The AM process can lead to changes in particle size distributions, as larger particles may break down, and smaller particles may agglomerate. In addition, not only the size but also the shape of the powder particles can change due to mechanical stresses during the AM process, affecting their flowability and packing behavior [38]. In one study [44], the reuse of Ti6Al4V in PBF-EB was investigated over 30 cycles. The researchers observed an increase in surface hardness with more cycles and raised concerns about the powder's flowability and spreadability due to morphological changes in the particles.

Costs may limit the reusability of powders as well. Ensuring the quality of recycled powders requires rigorous monitoring and testing. This includes assessing particle size distribution, chemical composition, and mechanical properties to ensure they meet the required standards [38]. Additionally, the cost of reconditioning reactive powders can be high due to the need for specialized equipment and processes to handle these materials safely [43]. Even then, current reconditioning techniques, such as sieving and blending, may not fully restore the original properties of the powders.

Production-wise counter-strategies to increase sustainability in AM thus include nesting strategies to fill the build space the most efficient way and the use of machines with build spaces suitable for the respective production needs. Also, in combination with modern monitoring methods that provide inline detection of failed parts or processes, the direct material efficiency can be increased.

If the reduction in the amount of consumed powder and the increase in the amount of reused powder cannot be further optimized, recycling remains an important factor.

Proper sorting of powders plays a crucial role hereby. By reducing raw material extraction through the recycling of reactive powders, significant environmental benefits can be achieved. This includes lowering the carbon footprint and reducing the impact on natural resources. Additionally, not only powder but also scrap metal from AM can be recycled into new powder, e.g., from removed support structures. However, sorting powders by type is difficult, categorizing them as hazardous substances requires special disposal methods, and clear separation is difficult in, for example, research facilities that process many different but similar materials. Due to contamination, many powders are also no longer suitable as input material for the production of the same qualities, and downcycling takes place.

Material science strategies to improve sustainability of AM processes focus the development of more sustainable alloys. Researchers are developing alloys from up to 100% recycled feedstock E.g., Continuum Powders has created nickel superalloys from purely secondary materials [45]. Another rising research trend not limited to AM is to develop alloys consisting of smaller numbers and more environmentally friendly elements, so called lean alloys, or alloys with a higher tolerance against contamination [46].

By integrating sustainable practices and recycling into the handling of reactive powder materials in AM, the industry can prospectively reduce the environmental footprint of AM technologies while maintaining high standards of production quality and efficiency.

7 Regulatory Frameworks

Numerous regulations related to work place safety, AM and specific materials have been released over the years. However, some of the most important guidelines for the safe handling of reactive powders are briefly described below.

Important fundamentals are summed up in the international standard ISO/ASTM 52900:2021 [47]. It defines key terminology related to AM and was developed jointly by the International Organization for Standardization (ISO) and the American Society for Testing and Materials (ASTM) to provide a common language and framework for AM technologies. Besides other aspects, the standard also addresses materials used in AM, providing guidance on their characteristics and how they are classified, which helps in ensuring consistency across material data and usage.

In ISO/ASTM 52920:2023 [48], the requirements for industrial additive manufacturing processes and production sites. It includes guidelines for ensuring the safety and quality specifically of AM processes, particularly when dealing with reactive powder materials.

Furthermore, the ATEX (Atmosphères Explosibles) [49] directives of the European Union are of utmost importance. The main goal of these directives is to ensure the safety of equipment used in environments where explosive gases, vapors, or dusts might be present. They define the requirements for equipment design, installation, and maintenance to minimize the risk of explosions and to protect both people and property.

In ATEX 99/92/EC the minimum requirements for improving the safety and health of workers potentially at risk from explosive atmospheres are specified, while in ATEX 2014/34/EU guidelines for equipment and protective systems intended for use in explosive atmospheres, covering aspects such as design, production, and testing, are provided. The REACH (Registration, Evaluation, Authorisation and Restriction of Chemicals) [50] regulation of the European Union is supposed to ensure the safe use of chemicals by requiring companies to register and assess the risks associated with their substances. This regulation is essential for managing the risks associated with reactive powder materials in AM.

In the United States of America, the National Fire Protection Association (NFPA) 484: Standard for Combustible Metals [51] provides guidelines for the safe handling and processing of combustible metals, including reactive powders like titanium and aluminum. It includes requirements for fire prevention, explosion protection, and safe storage. In addition, the Occupational Safety and Health Administration (OSHA) [52] provides regulations and guidelines for handling hazardous materials, including reactive powders. OSHA standards are set to ensure that workplaces are safe and humans protected, and give information about e.g., required PPE.

In Germany, the regulatory framework for the safe production and processing of reactive powder materials in AM is governed by additional regulations. One of the most important ones is the Occupational Safety and Health Act (Arbeitsschutzgesetz—ArbSchG) [53], which provides the general framework for ensuring the safety and health of employees in all industries, including those involved in additive manufacturing. Also, the Hazardous Substances Ordinance (Gefahrstoffverordnung—GefStoffV) [54] is meaningful when reactive powder materials are involved, as it regulates the handling of hazardous substances. It requires employers to conduct risk assessments, implement control measures to minimize exposure, and ensure proper labeling and storage of hazardous materials. Furthermore, the Technical Rules for Hazardous Substances (Technische Regeln für Gefahrstoffe—TRGS) [55] provide detailed guidelines on the safe handling of hazardous substances. For reactive powders, TRGS 727 (Prevention of Dust Explosions) and TRGS 900 (Occupational Exposure Limits) are particularly relevant. These rules specify measures to prevent dust explosions and limit exposure to hazardous substances.

By adhering to these international and national regulations and standards, institutions can ensure the safe production and processing of reactive powder materials in additive manufacturing, protecting both people and the environment. This list is of course not exhaustive, but can serve as an initial guide to designing safe processes and working environments for AM of reactive powders.

8 Conclusions

In conclusion, the utilization of reactive powder materials in AM presents both significant opportunities and inherent challenges. From materials like titanium, aluminum, and magnesium alloys to refractory metals, the properties and behavior of these powders demand careful consideration throughout the entire manufacturing process. The production of reactive powders, particularly through atomization, requires specialized techniques to ensure powder quality, while passivation and proper storage are essential to mitigate risks associated with their highly reactive nature.

In AM, proper powder handling, processing, and post-processing are key to achieving high-quality results while maintaining safety and operational efficiency. However, the unique characteristics of reactive powders such as their susceptibility to ignition underscore the importance of implementing robust safety protocols, including regular cleaning, filter exchange, and waste management practices. Furthermore, incidents in additive manufacturing involving reactive powders highlight the critical need for enhanced training and precautionary measures to prevent accidents.

Health risks associated with exposure to reactive powders in AM processes necessitate the use of appropriate PPE to safeguard workers. These risks also emphasize the need for ongoing research and improvements in safety standards. In parallel, sustainability and recycling efforts are becoming increasingly important, as the responsible management of powder materials not only reduces environmental impacts but also enhances the overall efficiency of manufacturing processes. Research plays a crucial role in addressing these challenges. Improved manufacturing techniques and better recycling methods can significantly reduce the risks associated with reactive powders. Additionally, materials research can lead to the discovery of more sustainable alternatives that are both cost-effective and environmentally friendly, thus contributing to a circular economy in the industry.

Finally, the regulatory frameworks governing the use of reactive powders in AM are crucial to ensuring that safety and environmental standards are consistently met. As these frameworks evolve, institutions must remain adaptable and proactive in complying with both current and future regulations. In sum, while reactive powders offer immense potential for advanced manufacturing applications, their safe and sustainable use requires a comprehensive approach that integrates technical, safety, health, and environmental considerations.

References

1. Sundaram, D., Yang, V., Yetter, R.A.: Metal-based nanoenergetic materials: Synthesis, properties, and applications. Progress in Energy and Combustion Science, 61, 293–365 (2017). https://doi.org/10.1016/j.pecs.2017.02.002.
2. Kusoglu, I.M., Gökce, B., Barcikowski, S.: Research trends in laser powder bed fusion of Al alloys within the last decade. Additive Manufacturing, 36, 101489 (2020). https://doi.org/10.1016/j.addma.2020.101489.
3. Goetz, I.K., Kaplan, M., Hans, M., Ström, P., Jansson, U., Hjörvarsson, B., Schneider, J.M.: Reactive metal additive manufacturing: Surface near ZrN–metallic glass composite formation and mechanical properties. Additive Manufacturing, 66, 103457 (2023). https://doi.org/10.1016/j.addma.2023.103457.
4. Sculpteo: 3D Printing Materials & Industry Report 2020. Sculpteo, 2020. https://www.sculpteo.com/en/3d-printing-industry-report/. Accessed: 01. Mar. 2021.
5. Gomez-Gallegos, A., Mandal, P., Gonzalez, D., Zuelli, N., Blackwell, P.: Studies on titanium alloys for aerospace application. Defect and Diffusion Forum, 419–423, Trans Tech Publ (2018).
6. Kaur, M., Singh, K.: Review on titanium and titanium-based alloys as biomaterials for orthopedic applications. Materials Science and Engineering C, 102, 844–862 (2019).
7. You, S.H., Lee, J.H., Oh, S.H.: A study on cutting characteristics in turning operations of titanium alloy used in automobile. International Journal of Precision Engineering and Manufacturing, 20, 209–216 (2019).
8. Gurrappa, I.: Characterization of titanium alloy Ti-6Al-4V for chemical, marine and industrial applications. Materials Characterization, 51(2–3), 131–139 (2003). https://doi.org/10.1016/j.matchar.2003.10.006.
9. Hryha, E., Shvab, R., Bram, M., Bitzer, M., Nyborg, L.: Surface chemical state of Ti powders and its alloys: Effect of storage conditions and alloy composition. Applied Surface Science, 388(Part A), 294–303 (2016). https://doi.org/10.1016/j.apsusc.2016.01.046.
10. Erickson, K.L. Jr., Rogers, J.W., Ward, S.J.: Titanium oxidation kinetics and the mechanism for thermal ignition of titanium-based pyrotechnics. In: Internal Pyrotechnics Seminar, pp. 679–697 (1986).
11. Schulz, O., Eisenreich, N., Kelzenberg, S., Schuppler, H., Neutz, J., Kondratenko, E.: Non-isothermal and isothermal kinetics of high temperature oxidation of micrometer-sized titanium particles in air. Thermochimica Acta, 517, 98–104 (2011).
12. Zhu, Z., Hu, Z., Seet, H.L., Liu, T., Liao, W., Ramamurty, U., Nai, S.M.L.: Recent progress on the additive manufacturing of aluminum alloys and aluminum matrix composites: Microstructure, properties, and applications. International Journal of Machine Tools and Manufacture, 190, 104047 (2023). https://doi.org/10.1016/j.ijmachtools.2023.104047.
13. Esposito, P.-H., Leroux, C., Heresanu, V., Neisius, T., Madigou, V., Denoyel, R., Coulet, M.-V.: Influence of texture and microstructure on the reactivity of aluminum powders. Materialia, 14, 100880 (2020). https://doi.org/10.1016/j.mtla.2020.100880.
14. Sundaram, D.S., Puri, P., Yang, V.: Pyrophoricity of nascent and passivated aluminum particles at nano-scales. Combustion and Flame, 160(9), 1870–1875 (2013). https://doi.org/10.1016/j.combustflame.2013.03.031.

15. Deng, J., Qu, J., Wang, Q., Zhai, X., Xiao, Y., Cheng, Y., Shu, C.-M.: Minimum ignition temperature of aluminium dust clouds via the Godbert–Greenwald furnace. Process Safety and Environmental Protection, 129, 176–183 (2019). https://doi.org/10.1016/j.psep.2019.07.006.
16. Talignani, I., Seede, R., Whitt, A., Zheng, S., Ye, J., Karaman, I., Kirka, M.M., Katoh, Y., Wang, Y.M.: A review on additive manufacturing of refractory tungsten and tungsten alloys. Additive Manufacturing, 58, 103009 (2022). https://doi.org/10.1016/j.addma.2022.103009.
17. Rebesan, P., Ballan, M., Bonesso, M., Campagnolo, A., Corradetti, S., Dima, R., Gennari, C., Longo, G.A., Mancin, S., Manzolaro, M., Meneghetti, G., Pepato, A., Visconti, E., Vedani, M.: Pure molybdenum manufactured by Laser Powder Bed Fusion: Thermal and mechanical characterization at room and high temperature. Additive Manufacturing, 47, 102277 (2021). https://doi.org/10.1016/j.addma.2021.102277.
18. Schmelzer, J., Rittinghaus, S.-K., Wilms, M.B., Michael, O., Krüger, M.: Strength-ening of additively manufactured Me-Si-B (Me = Mo, V) by Y_2O_3 particles. Inter-national Journal of Refractory Metals and Hard Materials, 101, 105623 (2021). https://doi.org/10.1016/j.ijrmhm.2021.105623.
19. Aliyu, A.A., Poungsiri, K., Shinjo, J., Panwisawas, C., Reed, R.C., Puncreobutr, C., Tumkanon, K., Kuimalee, S., Lohwongwatana, B.: Additive manufacturing of tantalum scaffolds: Processing, microstructure and process-induced defects. International Journal of Refractory Metals and Hard Materials, 112, 106132 (2023). https://doi.org/10.1016/j.ijrmhm.2023.106132.
20. DiStefano, J.R., Pint, B.A., DeVan, J.H.: Oxidation of refractory metals in air and low pressure oxygen gas. International Journal of Refractory Metals and Hard Materials, 18, 237–243 (2000). https://doi.org/10.1016/S0263-4368(00)00026-3.
21. Won, C.W., Nersisyan, H.H., Won, H.I., Lee, J.H.: Refractory metal nanopowders: Synthesis and characterization. Current Opinion in Solid State and Materials Science, 14(3–4), 53–68 (2010). https://doi.org/10.1016/j.cossms.2009.10.001.
22. Chen, Y., Xu, Z., Smith, C., Sankar, J.: Recent advances on the development of magnesium alloys for biodegradable implants. Acta Biomaterialia, 10, 4561–4573 (2014). https://doi.org/10.1016/j.actbio.2014.07.005.
23. Bai, J., Yang, Y., Wen, C., Chen, J., Zhou, G., Jiang, B., Peng, X., Pan, F.: Applications of magnesium alloys for aerospace: A review. Journal of Magnesium and Alloys, 11(10), 3609–3619 (2023). https://doi.org/10.1016/j.jma.2023.09.015.
24. Chen, J., Chen, B.: Progress in additive manufacturing of magnesium alloys: A review. Materials (Basel), 17(15), 3851 (2024). https://doi.org/10.3390/ma17153851.
25. Kim, J.D., Jun, H.U., Cheon, J., Kim, J.H., Ji, C., Kim, Y.D.: Effect of additional side shielding on the wire arc additive manufacturing of AZ31 magnesium alloy. Journal of Materials Research and Technology, 25, 6567–6577 (2023). https://doi.org/10.1016/j.jmrt.2023.07.147.
26. Gruber, K., Smolina, I., Stopyra, W.: Assessing metal powder quality for additive manufacturing using diffuse light spectroscopy. Powder Technology, 434, 119366 (2024). https://doi.org/10.1016/j.powtec.2024.119366.
27. Dobkowska, A., Żrodowski, Ł., Chlewicka, M., Koralnik, M., Adamczyk-Cieślak, B., Ciftci, J., Morończyk, B., Kruszewski, M., Jaroszewicz, J., Kuc, D., Święszkowski, W., Mizera, J.: A comparison of the microstructure-dependent corrosion of dual-structured Mg-Li alloys fabricated by powder consolidation methods: Laser powder bed fusion vs pulse plasma sintering. Journal of Magnesium and Alloys, 10(12), 3553–3564 (2022). https://doi.org/10.1016/j.jma.2022.06.003.
28. Hinrichs, F., Kauffmann, A., Schliephake, D., Seils, S., Obert, S., Ratschbacher, K., Heilmaier, M.: Flexible powder production for additive manufacturing of refractory metal-based alloys. Metals, 11(11), 1723 (2021). https://doi.org/10.3390/met11111723.
29. Feng, Z., Yang, F., Li, Y., Shao, Y., He, X., Chen, C., Guo, Z.: Passivation coating on titanium powder surface via $SnCl_4$-Ti gas–solid fluidization reaction. Advanced Powder Technology, 34(10), 104167 (2023). https://doi.org/10.1016/j.apt.2023.104167.

30. Kaserer, L., Bergmueller, S., Braun, J., et al.: Vacuum laser powder bed fusion—track consolidation, powder denudation, and future potential. International Journal of Advanced Manufacturing Technology, 110, 3339–3346 (2020). https://doi.org/10.1007/s00170-020-06071-6.
31. Wang, L., Guo, Z., Peng, G., Wu, S., Zhang, Y., Yan, W.: Evaporation-induced composition evolution in metal additive manufacturing. Advanced Functional Materials, 2412071 (2024). https://doi.org/10.1002/adfm.202412071.
32. Kumar, M., Epiphaniou, G., Maple, C.: Comprehensive threat analysis in additive manufacturing supply chain: A hybrid qualitative and quantitative risk assessment framework. Production Engineering Research & Development (2024). https://doi.org/10.1007/s11740-024-01283-1.
33. OSHA: News Release: Region 1 (2014). https://www.osha.gov/news/newsreleases/region1/05202014. Accessed: 11/10/2024.
34. CSB: CSB Determines Fatal 2003 Incident at Hayes Lemmerz Plant in Indiana Most Likely Caused by Explosion in Dust Collection System (2024). https://www.csb.gov/csb-determines-fatal-2003-incident-at-hayes-lemmerz-plant-in-indiana-most-likely-caused-by-explosion-in-dust-collection-system-company-did-not-identify-or-control-hazards-of-aluminum-dust/. Accessed: 11/10/2024.
35. Modupeola, D., Popoola, P.: Safety practices and occupational hazards of the additive manufacturing of high entropy alloys. Safety and Extreme Environments, 6, 139–146 (2024). https://doi.org/10.1007/s42797-023-00097-1.
36. du Preez, S., de Beer, D.J., du Plessis, J.L.: Titanium powders used in powder bed fusion: Their relevance to respiratory health. The South African Journal of Industrial Engineering, 29(4), 94–102 (2018). https://doi.org/10.7166/29-4-1975.
37. Chen, R., Yin, H., Cole, I.S., Shen, S., Zhou, X., Wang, Y., Tang, S.: Exposure, assessment and health hazards of particulate matter in metal additive manufacturing: A review. Chemosphere, 259, 127452 (2020). https://doi.org/10.1016/j.chemosphere.2020.127452.
38. Soundarapandiyan, G., Johnston, C., Khan, R., Chen, B., Fitzpatrick, M.E.: A technical review of the challenges of powder recycling in the laser powder bed fusion additive manufacturing process. Journal of Engineering, 2021, 97–103 (2021). https://doi.org/10.1049/tje2.12013.
39. ASTM F2924-14: Standard Specification for Additive Manufacturing Titanium-6 Aluminum-4 Vanadium with Powder Bed Fusion. ASTM International (2014), 1–9. https://doi.org/10.1520/F2924-14.2.
40. Wei, C., Ma, X., Yang, X., Zhou, M., Wang, C., Zheng, Y., Zhang, W., Li, Z.: Microstructural and property evolution of Ti6Al4V powders with the number of usage in additive manufacturing by electron beam melting. Materials Letters, 221, 111–114 (2018). https://doi.org/10.1016/j.matlet.2018.03.124.
41. Ghods, S., Schultz, E., Wisdom, C., Schur, R., Pahuja, R., Montelione, A., Arola, D., Ramulu, M.: Electron beam additive manufacturing of Ti6Al4V: Evolution of powder morphology and part microstructure with powder reuse. Materialia, 9, 100631 (2020). https://doi.org/10.1016/j.mtla.2020.100631.
42. Meier, B., Skalon, M., Warchomicka, F., Belei, C., Görtler, M., Kaindl, R., Sommitsch, C.: Effect of the reuse of powder on material properties of Ti6Al4V processed by SLM. AIP Conference Proceedings, 2113(1), 150006 (2019). https://doi.org/10.1063/1.5112682.
43. Koushik, T., Shen, H., Kan, W.H., Gao, M., Yi, J., Ma, C., Lim, S.C.V., Chiu, L.N.S., Huang, A.: Effective Ti-6Al-4V powder recycling in LPBF additive manufacturing considering powder history. Sustainability, 15, 15582 (2023). https://doi.org/10.3390/su152115582.
44. Montelione, A., Ghods, S., Schur, R., Wisdom, C., Arola, D., Ramulu, M.: Powder reuse in electron beam melting additive manufacturing of Ti6Al4V: Particle microstructure, oxygen content and mechanical properties. Additive Manufacturing, 35, 101216 (2020). https://doi.org/10.1016/j.addma.2020.101216.
45. Continuum Powders. https://www.continuumpowders.com/. Accessed: 10 Oct 2024.

46. Li, X., Lu, K.: The high performance of alloy materials can be maintained with compositional "plainification". Science, 364(6442), 733–734 (2019). https://doi.org/10.1126/science.aaw9905.
47. ISO/ASTM 52900:2021: Additive manufacturing—General principles—Terminology. International Organization for Standardization (ISO) and American Society for Testing and Materials (ASTM), 2021. https://www.iso.org/standard/73739.html.
48. ISO/ASTM 52920:2023: Additive manufacturing—Material characteristics—Requirements for materials used in additive manufacturing. International Organization for Standardization (ISO) and American Society for Testing and Materials (ASTM), 2023. https://www.iso.org/standard/73739.html.
49. ATEX (Atmosphères Explosibles) directives: Directive 2014/34/EU of the European Parliament and of the Council of 26 February 2014 on equipment and protective systems intended for use in potentially explosive atmospheres. Official Journal of the European Union, 2014.
50. REACH Regulation (EC) No 1907/2006: Regulation (EC) No 1907/2006 of the European Parliament and of the Council of 18 December 2006 concerning the Registration, Evaluation, Authorization and Restriction of Chemicals (REACH), establishing a European Chemicals Agency. Official Journal of the European Union, 2006.
51. NFPA 484: Standard for Combustible Metals. National Fire Protection Association (NFPA), 2019.
52. U.S. Department of Labor, Occupational Safety and Health Administration (OSHA): OSHA Standards for General Industry. 29 CFR Part 1910, U.S. Government Printing Office, Washington, D.C. (2021).
53. Occupational Safety and Health Act (ArbSchG): Act on the Protection of Employees at Work, Federal Law Gazette (BGBl.) I, p. 123, last amended by Art. 3 of the Act of December 20, 2022, Federal Law Gazette (BGBl.) I, p. 2653.
54. Hazardous Substances Ordinance (GefStoffV): Ordinance on the Protection Against Hazardous Substances, Federal Law Gazette (BGBl.) I, p. 2118, last amended by Art. 3 of the Act of December 20, 2022, Federal Law Gazette (BGBl.) I, p. 2653.
55. Technische Regeln für Gefahrstoffe (TRGS): Technical Rules for Hazardous Substances, published by the Federal Ministry of Labour and Social Affairs, last amended by the 2022 regulation, available at: https://www.baua.de.

How Can the Understanding of Droplet Movement Contribute to the Safety in Mechanical Engineering?

Sebastian Burgmann, Martin Rohde, Beawer Barwari, Veronika Brox and Uwe Janoske

Contents

Abstract

Component malfunctions caused by corrosion or the penetration of fluids in electronic components, are responsible for a huge economic loss. This damage is often caused by minimal amounts of liquid that can enter components due to various effects, e.g. flow or vibration. Particularly due to the increasing complexity of components and the increased use of electronic components, the damage will increase significantly in the coming years if no targeted measures are taken. Despite the clear effects of water penetration, it is often unclear how this occurs. Due to complex flow conditions combined with superimposed vibrations and movements of the parts, it is difficult to predict the movement of droplets. In this paper, the basics of the movement of droplets under the influence of various forces are examined and the criteria for the onset of droplet movement are described in detail. Both numerical and experimental methods are presented.

S. Burgmann · M. Rohde · B. Barwari · V. Brox · U. Janoske (✉)
Faculty of Mechanical and Safety Engineering,
University Wuppertal, Wuppertal, Germany
e-mail: uwe.janoske@uni-wuppertal.de

S. Bracke and R. Pieper (eds.), *Safety Engineering*,
https://doi.org/10.1007/978-3-658-47473-7_8

Keywords

Droplet movement · CFD · Vibration · Contact angle

1 Introduction

Corrosion is a process which leads typically to a deterioration of metals based on chemical reactions. The impact of corrosion on all aspects of daily life are tremendous. The economic impact is estimated to be around 2.5 trillion dollars in each year summarizing costs for maintenance as well as the costs caused by production losses and safety hazards [1]. Therefore, a deep knowledge of corrosion and especially of measures to prevent corrosion is required. As corrosion is an electrochemical process, beside the anode and cathode an electrolyte as well as oxygen is needed for the corrosion processes to be inhibited. The electrolyte is responsible for establishing the electrical connection between the anode and cathode, resp. the different kinds of metal and is one of the key factors for corrosion. The conductivity of the liquid can be enhanced by additional salts, acids, etc.

The electrolyte often appears as small droplets that adhere to the surfaces of machinery, housings, and similar structures. These droplets frequently encounter gas flows that pass over the surfaces where they reside. The additional force from this airflow can be utilized to selectively remove droplets, as seen in applications like water management in fuel cells [2, 3] or cleaning processes [4]. Generally, a droplet will move when the external force exceeds the adhesive force holding it in place. Various external factors, such as surface vibration [5–7], gravity [8, 9], magnetic and electric fields [10, 11], and shear flow [12–16], have been studied for their impact on droplet behaviour. In the case of shear flow, research by Dussan [13] and Durbin [12] indicates that a droplet begins to move once a critical air flow velocity is reached. Before detachment, the droplet undergoes contour deformation, with larger droplets displaying more visible deformation compared to smaller ones [17–20]. Additionally, droplet oscillation with characteristic frequencies has been observed [21–24], although the exact mechanisms and conditions that initiate droplet motion are not fully understood.

The following provides a brief overview of droplet characteristics on surfaces and their detachment mechanisms. For a liquid droplet on a horizontal substrate, three interfaces form due to adhesion and cohesion forces: 1) between the fluid and the surrounding gas, 2) between the fluid and the substrate, and 3) between the gas and the substrate. The angle between the substrate surface and the tangent along the droplet surface is defined as the static contact angle (θ_S). Three wetting behaviours are observed depending on the contact angle: complete wetting ($\theta_S = 0°$), partial wetting ($0 < \theta_S < 90°$), and non-wetting ($90° < \theta_S < 180°$). These behaviours can be influenced by the liquid's surface tension (σ_{lg}), the substrate's surface energy (σ_{sg}), and the interfacial tension between the fluid and the substrate (σ_{sl}). Research by Milne and Amirfazli [25], and Moghtadernejad et al. [26], used surfaces with varying wettabilities to analyse how the droplet contact angle affects the detachment mechanism. They found that droplets on hydrophilic surfaces exhibit higher adhesion than on hydrophobic surfaces under the same conditions, requiring less force for detachment on hydrophobic surfaces. Bixler and Bhushan [27] demonstrated that the lotus effect, induced by nanostructured coatings, reduces drag, increases the contact angle, and enhances self-cleaning efficiency. Experiments by Shastry et al. [28] and Li et al. [29] showed that droplet transport can be controlled by introducing a wetting gradient on the surface, allowing adhered droplets to move easily towards areas of higher wettability under external forces. Hao et al. [30] conducted material analysis on an inclined plate using rough hydrophobic substrates and found that water droplet sliding under gravity begins with the detachment of the rear contact line. Hao et al. [31] confirmed this dynamic mechanism by studying droplet behaviour in wall-parallel airflow on superhydrophobic horizontal surfaces. Hu et al. [32] experimentally analysed surface topography using micro-grooved surfaces with different orientations and dimensions, determining that the critical velocity for droplet movement is higher for vertically oriented micro-grooves (relative to the airflow direction) compared to parallel grooves. Several experimental studies have focused on porous materials or gas diffusion layers to better understand water management in fuel cells. Research by Kumbur et al. [33], Cai et al. [34], and Theodorakakos et al. [35] indicated that the channel flow rate and surface properties of the gas diffusion layer affect droplet detachment, with more hydrophobic surfaces enhancing droplet removal. Ha et al. [36] observed that droplets on dry gas diffusion surfaces can be removed with lower airflow compared to wet surfaces. Das et al. [37] studied the aging behaviour of gas diffusion layers and concluded that aged materials make droplet removal more difficult than on unused surfaces.

The movement of droplets also depends on fluid properties, as shown in various studies. Thoreau et al. [38] used oil droplets in laminar water flow to investigate the influence of surfactants on droplet detachment, concluding that surfactants reduce the critical velocity by generating surface tension gradients along the droplet surface. Gupta and Basu [39] studied a water–oil disperse system and found that oil droplets with lower interfacial tension exhibit more visible deformation compared to those with higher interfacial tension. Seevaratnam et al. [40] and Madani and Amirfazli [41] explored the instability of oil droplets of different viscosities in laminar water flow, identifying three regimes—sliding, crawling, and detachment by lift-off—depending on droplet size, flow rate, and viscosity ratio. They noted that lift-off detachment becomes more pronounced as droplet viscosity increases. Fan et al. [42] investigated droplets of pure water, pure glycerine, and a water-glycerine solution in turbulent channel gas flow, finding that the critical velocity is strongly dependent on contact angle hysteresis and droplet volume, but weakly on droplet viscosity. Fu et al. [43] confirmed these findings in their study of water-glycerine droplets in a flat plate boundary layer flow, noting that the critical velocity depends strongly on contact angle hysteresis and droplet size.

In addition to experimental studies, several numerical studies have been conducted to accurately model droplet behaviour. Various methods have been proposed to simulate the pinning effect of the contact line. For example,

Theodorakakos et al. [35] used a sinusoidal law and an iterative procedure to predict the evolution of the contact angle within the hysteresis range, achieving local pinning of the contact line. Fang et al. [44] implemented contact angle hysteresis by calculating the current contact angle as a function of the previous time step, requiring the evaluation of numerous nonlinear equations. Dupont and Legendre [45] eliminated contact line velocity during the hysteresis phase by iteratively calculating the necessary contact angle value from the local momentum balance within the Volume of Fluid method (VoF). Similar methods were applied by Spelt [46] and Ding and Spelt [47] for the level set method. Annapragada et al. [48] used an analytical equation for contact angle variation derived by ElSherbini and Jacobi [49] to model droplet shape changes before movement, calculating the contact angle as a function of the maximum (advancing) and minimum (receding) contact angles. Linder et al. [50] proposed switching the boundary condition for the liquid phase fraction to pin the contact line, using Dirichlet boundary conditions with a fixed phase fraction value for contact angles within the hysteresis range, while allowing the contact line to propagate with Neumann boundary conditions if the contact angles are outside this range. He and Yao [51] numerically modelled contact angle hysteresis using pseudo-line tensions to describe surface heterogeneities. In this study, the feedback deceleration technique (FDT), developed by Park and Kang [52] for the level set method and later implemented by Krämer et al. [53] for the volume of fluid method, was used to simulate droplet adhesion.

This paper presents results on droplet motion influenced by aerodynamic and vibration using both experimental and numerical methods.

2 Methods and Materials

In the following section, the experimental setup as well as the numerical method are presented.

2.1 Experimental Setup

The flow channel is a rectangular Plexiglas structure with an internal cross-section of 22×22 mm and a length of 1000 mm, which is sufficient to ensure a fully developed flow profile. Two electromagnetic vibration exciters (shakers) are used to produce two-dimensional harmonic surface vibrations. Shaker 1 (LDS V406) operates in the horizontal plane, while Shaker 2 (LDS V455) operates vertically. Both shakers have a total harmonic distortion of 0.2% at nominal power. Shaker 1 can be positioned in two orientations relative to the flow direction: in Case 1, it is parallel to the flow direction (as shown in Fig. 1), while in Case 2, it is perpendicular to the flow direction (indicated by an orange dashed arrow).

Two accelerometers (BKSV-4513B, Brüel & Kjaer) are used to control the shakers, and a triaxial sensor (PCB-356A32, PCB Piezotronics) monitors lateral acceleration. The substrate movement is induced by the shakers' harmonic

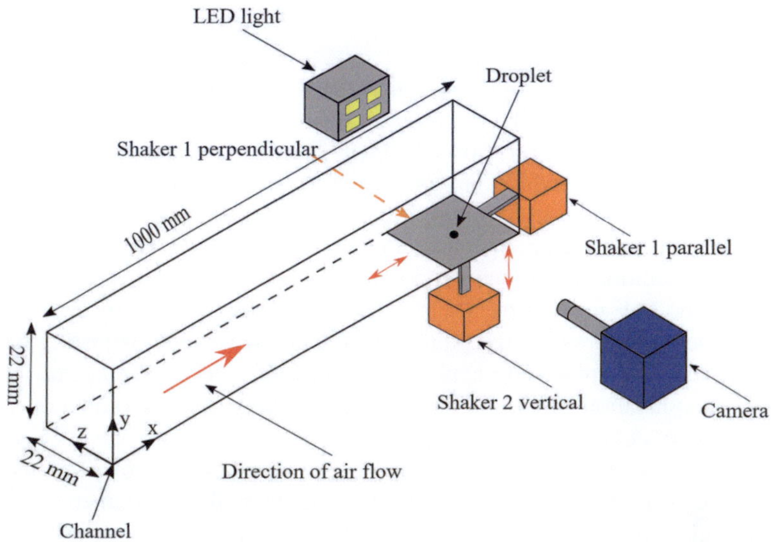

Fig. 1 Experimental setup of the flow channel with the arrangement of the two electro-magnetic shakers and the high-speed camera

sinusoidal excitation, with a constant amplitude and frequency throughout the experiment. Further details can be found in Rohde et al. [54, 55].

All tests are conducted on acrylic glass (PMMA), which has hydrophilic properties and forms a static contact angle of approximately 74° with demineralized water droplets. The advancing and receding contact angles are measured at 82° and 55°, respectively, resulting in a contact angle hysteresis of 27°. Three droplet volumes are tested: 7.8 µl, 15.6 µl, and 23.4 µl. The blockage factor (Ap/Ac), determined from the projected area of the static drop (Ap) relative to the channel's cross-sectional area (Ac), is 0.007 for the 7.8 µl droplet, 0.011 for the 15.6 µl droplet, and 0.015 for the 23.4 µl droplet.

A mass flow controller (Bürkert 8626) regulates the volume flow, with the flow velocity increasing at a rate of 0.258 m/s per second. The experiments are recorded with a high-speed camera (EoSens Cube 7, Mikrotron) at up to 1000 fps. The captured grayscale images are analysed using custom software. The critical inflow velocity, v_{crit}, is defined as the air velocity required to initiate droplet movement. This is determined using a dimensionless approach, normalizing the droplet velocity by the airflow velocity and the horizontal displacement of the droplet's centroid by the current droplet width. The critical inflow velocity, indicated by a characteristic change in the curve slope, signifies the point at which droplet detachment occurs. A detailed method for determining this critical velocity is provided in Barwari et al. [56].

2.2 Simulation Model

The numerical simulation described in this study was conducted using a customized solver called hysteresisInterFoam, developed within the open-source computational fluid dynamics (CFD) software OpenFOAM®. This solver is an adaptation of the interFoam solver, which is typically used for simulating multiphase flows of incompressible, isothermal immiscible fluids using the Volume of Fluid (VoF) method. The customization includes implementing droplet adhesion using the feedback deceleration technique proposed by Park and Kang [52], which was adapted for the VoF method by Maurer [57] and further refined by Krämer et al. [53]. This method allows for the pinning of the contact line by adjusting the contact angle, and it is universally applicable across different fluid systems. A comprehensive explanation of the numerical methodology and its validation can be found in Krämer et al. [53].

The solver was validated using various droplet sizes, liquid properties, and wetting characteristics. It was compared against experimental measurements involving static droplets, droplets in shear flow, and droplets on inclined surfaces. A brief overview of the numerical method is provided, followed by a description of the numerical model, including the geometry, boundary conditions, and numerical schemes used in this study.

A notable feature of the hysteresisInterFoam solver is the implemented contact angle hysteresis, which allows for the pinning of the contact line for Reynolds number values below a critical threshold, as discussed in Barwari et al. [58]. This pinning effect is crucial for studying adhering droplets in shear flow.

The feedback deceleration technique from Park and Kang numerically implements the adhesion effect by adjusting the contact angle for the next time step θ^{n+1} based on the current contact angle θ^n and the current contact line velocity $\boldsymbol{u}_{cl}$ The following equations describe the contact angle adjustment:

$$\Theta^{n+1} = \begin{cases} \Theta^n + \delta\Theta & \boldsymbol{u}_{cl} > 0 \\ \Theta^n & \boldsymbol{u}_{cl} = 0 \\ \Theta^n - \delta\Theta & \boldsymbol{u}_{cl} > 0 \end{cases}$$

These equations indicate that when the contact line moves $\boldsymbol{u}_{cl} \neq 0$, a correction increment $\delta\theta$ is applied, inducing a counteracting velocity due to changes in interface curvature, thereby pinning the contact line. The contact line cells are defined by a water volume fraction α within the range $0.000001 < \alpha < 0.999999$. It is crucial to correct all contact line cells within this range to achieve the desired contact angle value, minimizing unnecessary corrections.

The contact angle increment $\delta\theta$ is a key factor and was extensively analyzed by Krämer et al. [53]. It depends on the change in contact angle hysteresis $\theta_A - \theta_R$ where θ_A and θ_R are the advancing and receding contact angles, respectively and the changes in tangential surface tension forces within one time step. The formula for $\delta\theta$ is:

$$\delta\theta = \frac{\theta_A - \theta_R}{|cos\theta_A - cos\theta_R|} \frac{\Delta x}{\Delta y} \frac{\mu_{liquid}}{\sigma} |\boldsymbol{u}_{cl}| \tag{1}$$

where μ_{liquid} is the viscosity of the liquid, σ is the surface tension, and $\mathbf{u}_{cl}$ is the contact line velocity. The maximum aspect ratio for the contact line cells ($\Delta x/\Delta y = 10$) was chosen to ensure numerical stability and prevent droplet shape distortion. The receding and advancing contact angles are measured experimentally for each material system. For this study, values from Maurer [57] were used: $\theta_R = 33.8°$ and $\theta_R = 83.3°$. The contact line velocity $\mathbf{u}_{cl}$ is computed for each grid cell and time step, with the liquid properties chosen to match the experimental setup ($\mu_{water} = 0.89$ mPa·s, $\sigma = 71.96$ mN/m).

Once the droplet overcomes the contact angle hysteresis and begins to move, a simple dynamic contact angle model is applied, using θ_R and θ_A for $\mathbf{u}_{cl} < 0$ and $\mathbf{u}_{cl} > 0$, respectively, as described by Fujimoto et al. [59] and Li et al. [60]. Thus, the advancing or receding contact angles are applied depending on the direction of the contact line velocity.

The simulation model features a rectangular channel cross-section. No-slip boundary conditions are applied at the channel walls, with a total pressure boundary condition at the outlet. The wall on which the droplet is placed uses a contact angle boundary condition that permits numerical slip and utilizes the modified feedback deceleration technique to ensure droplet adhesion until the critical velocity is reached. The droplet shape is initialized with a static contact angle of 74.4° (PMMA), as observed experimentally. The inlet velocity profile is assumed based on measurements and fitted to a parabolic function in the inlet cross section, accounting for both laminar and turbulent flow conditions. Sensitivity studies indicate minimal influence from using two-equation turbulence models on velocity profile changes and critical droplet velocities. Thus, no additional turbulence model was employed in the simulations. A block-structured grid was used for the numerical mesh. The droplet diameter for a volume of 20 μl is represented by approximately 16 cells in the computational mesh, with a similar number of cells used to represent the droplet's height. To accurately capture the velocity profile near the droplet, the mesh was refined using cell expansion near the bottom surface and the inlet. A grid independence study conducted by Krämer et al. [53] confirmed that reliable reproduction of droplet dynamics can be achieved with around 770,000 cells, ensuring adequate resolution with 16 cells across the droplet diameter. The current study employs the same numerical schemes as described by Krämer et al. Specifically, the Gauss theorem is used for interpolating values from cell centres to face centres. Gradient terms are interpolated linearly, while the advection term is discretized using the limited linear scheme. For Laplace operators, the linear corrected scheme is applied. Time derivatives are discretized using the implicit Euler method. Additionally, the average Courant number is kept below 0.5 to control the time step size and maintain simulation stability.

3 Results and Discussion

3.1 Critical Droplet Detachment Velocity

Based on the definition and analysis presented above, for all investigated cases the critical velocity is determined which is the local air velocity at the level of the droplet height when the movement of the droplet starts. Figure 2 shows the individual critical velocities as a function of the droplet volume and the mass fraction for water-glycerine droplets on PMMA.

The error bars represent the spreading of the results as a consequence of the repetition of each experiment. As can be seen, the statistical deviation is low. There are some findings that can be directly deduced from these graphs. In all cases, the critical air velocity required to initiate a droplet movement decreases as the droplet volume increases and this result agrees with literature [25]. All investigated fluid-material systems show a similar trend. The 39.9 μl droplet needs a 15–25% lower air flow velocity to start moving as compared to the 7.8 μl droplet. This is because a larger droplet is subjected to a higher local velocity, as the highest droplet contour surface is further away from substrate compared to the smaller droplet.

Figure 3 shows the critical droplet Reynolds number which is defined with the material properties of the liquid (density ρ_d, dynamic viscosity μ_d, the droplet

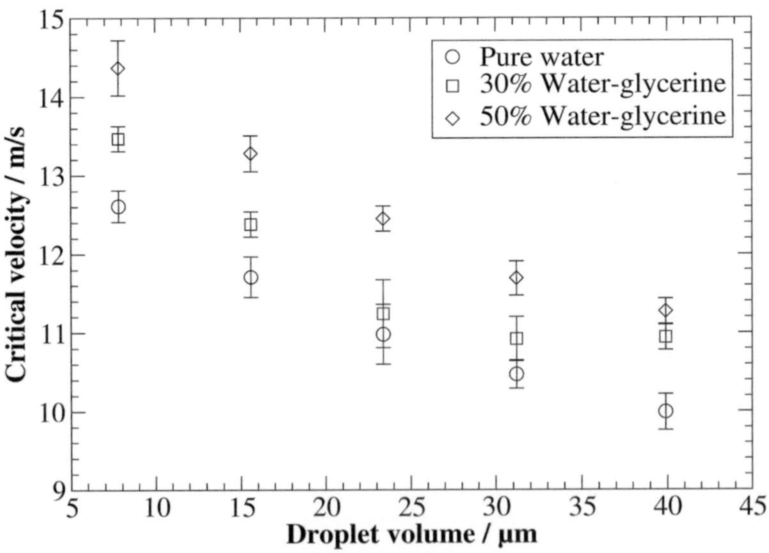

Fig. 2 Critical velocity required to initiate droplet movement as a function of the droplet volume (left) for water-glycerine mixtures on PMMA (top)

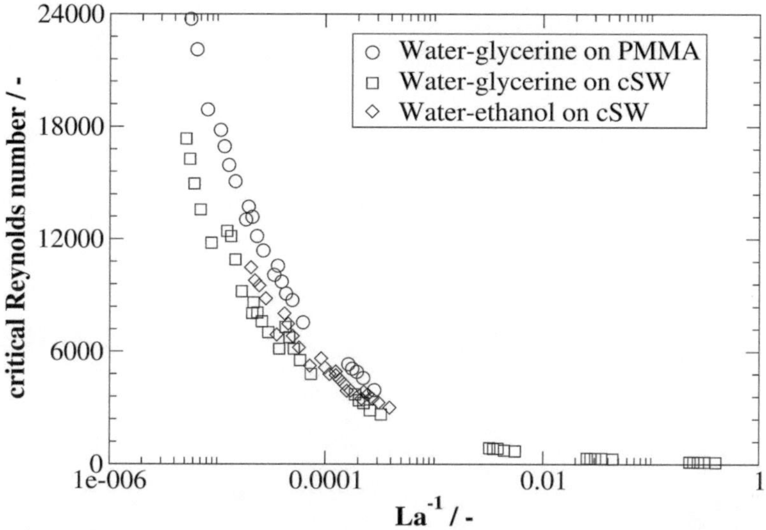

Fig. 3 The critical droplet Reynolds number for the droplet detachment as a function of the reciprocal value of the Laplace number for all investigated droplets

height d_h and the critical velocity of the gas v_{crit}, which is the velocity at the level of droplet height as a function of the inverse of the Laplace number. The Laplace number is defined with the length of the droplet d_w and the surface tension σ_{sl}.

$$Re_{crit} = \frac{\rho_d d_h v_{crit}}{\mu_d} \tag{2}$$

$$La = \frac{\rho_d \sigma_{sl} d_w}{\mu_d^2} \tag{3}$$

The surface tension σ_{sl} can be derived by the surface energy of the material σ_{sg} and the surface tension σ_{lg} and the static contact angle H_s (index: l liquid, g gas, s solid). The critical Reynolds number can be fitted to an exponential function

$$Re_{crit} = 62.713\left(\frac{1}{La}\right)^{-0.483} \tag{4}$$

3.2 Superposition of Shear Flow and Oscillation Excitation

Three droplet volumes (7.8, 23.4, 39.9 µl) are analysed. In Fig. 4, the critical velocity ratio Ψ is plotted over the dimensionless frequency factor f/f_0, where Ψ is given by $\Psi = v_{crit}(SP)/v_{crit}(SF)$. The critical velocity $v_{crit}(SP)$ determined from the superposition experiment (surface vibration and shear flow) is normalized to

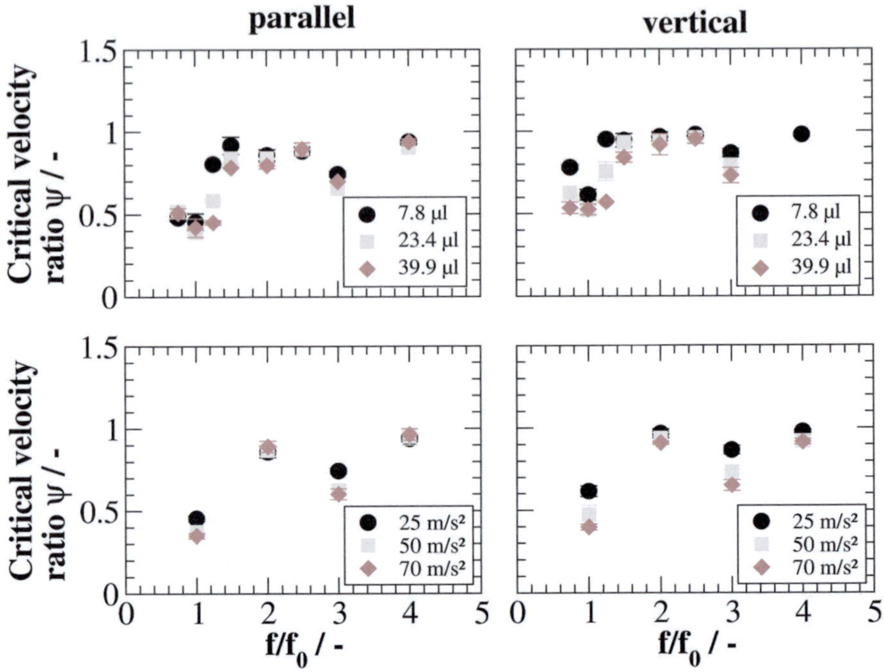

Fig. 4 Frequency dependence of the critical velocity ratio for different droplet volumes with a=25 m/s^2 (top) and for different accelerations for a constant droplet volume (7.8 µl) (bottom) for a vibration in parallel (E||) or perpendicular (E⊥) direction

the critical velocity $v_{crit}(SF)$ determined from experiments with an shear flow. In the experiments, a frequency range from $0.75<f/f_0<4$ is covered. The droplets are excited with an acceleration of 25 m/s^2, 50 m/s^2 and 70 m/s^2. The experimental uncertainty is given by the standard deviation. All of the critical velocity ratio values shown in Fig. 4 are less than or equal to one, indicating that in all cases the superposition of oscillations results in droplet detachment at lower velocities than for a shear flow without vibration. The excitation direction parallel ($E\|$) to the channel is shown on the left side of Fig. 4 (top), while the excitation direction perpendicular ($E\perp$) to the channel is shown on the right. Interestingly, the droplets show similar behaviour over the entire frequency spectrum, independent of the excitation direction. An influence of the superimposed vibration can be observed in the range $f/f_0=1$ and $f/f_0=3$, with the strongest decrease of the critical velocity ratio at $f/f_0=1$. In the range $1.5<f/f_0<2.5$ and $f/f_0=4$, Ψ is close to unity, which means that in this frequency range superimposed vibration excitation has no significant effect on Ψ. Slight differences in the magnitude of Ψ are apparent for the different droplet volumes. Comparing both excitation directions $E\|$ and

$E\perp$, the parallel excitation direction always reveals the lowest values for Ψ. This can be shown by the example of the 7.8 µl droplet. While Ψ for $E\|$ assumes a value of 0.45 for $f/f_0=1$, it is 0.61 for $E\perp$. The same applies für $f/f_0=3$. Here Ψ for $E\|$ is 0.61 while it is 0.86 for $E\perp$. The effect of increasing the excitation acceleration is shown in Fig. 4 (bottom) for a 7.8 µl droplet. For $f/f_0=1$ and $f/f_0=3$ a further decrease of the critical velocity ratio with increasing acceleration can be seen, while for $f/f_0=2$ and $f/f_0=4$ no significant difference can be observed. The increased acceleration to 75 m/s^2 leads to a further decrease for Ψ from 0.45 to 0.35 ($E\|$) at $f/f_0=1$, while a decrease from 0.61 to 0.4 is observed for $E\perp$. At $f/f_0=3$ the critical velocity ratio Ψ drops from 0.74 to 0.6 for $E\|$ and from 0.86 to 0.63 for $E\perp$. An explanation of the observed droplet behaviour is given in the following paragraphs. The fact that the critical velocity ratio Ψ is lowered for $E\|$ and $E\perp$ in the same frequency range can be explained by the droplet deformation (see Rohde et al. [54]).

Figure 5 shows the comparison between the critical velocity ratio with and without vibration in vertical direction for experimental and numerical data for two different droplet sizes. The curves show for both droplet volumes the same characteristic reduction of the velocity ratio for the second eigenfrequency. For the first and third eigenfrequency the vertical vibration has a negligible influence on the critical velocity. The numerical and experimental data show a good agreement for different droplet volumes and frequency ratios.

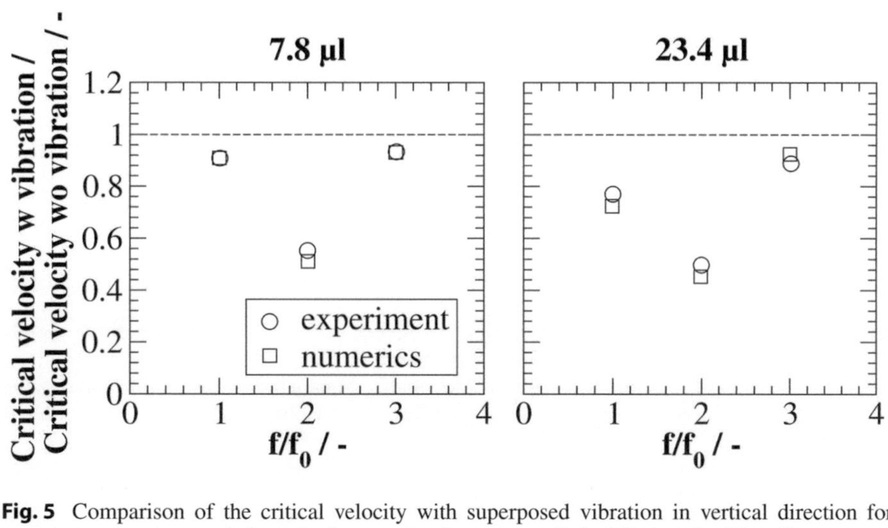

Fig. 5 Comparison of the critical velocity with superposed vibration in vertical direction for experimental and numerical results for two droplet volumes (left: 7.8 µl, right: 23.4 µl)

4 Conclusion and Perspective

The droplet movement as one of the key factors to transport electrolyte which may lead to corrosion processes is a complex phenomenon dependent on various factors. In this paper, the influence of aerodynamic and vibrational forces on the droplet shape as well as on the critical velocity has been demonstrated. By a superposition of vibrational forces, the critical velocity of the airflow can be reduced up to 50% to initiate a droplet movement. The influence of the direction of the vibration (horizontal or vertical) has been demonstrated for different droplet volumina.

In further studies the influence of superposed vibrations in different directions as well as the influence of obstacles placed in the channel is studied.

References

1. G. Koch, J. Varney, N. Thompson, O. Moghissi, M. Gould, J. Payer, NACE international impact report: International measures of prevention, application, and economics of corrosion technologies study, NACE Int., Houston, TX, USA, Tech. Rep (2016).
2. F.Y. Zhang, X.G. Yang, C.Y. Wang, Liquid water removal from a polymer electrolyte fuel cell, Journal of The Electrochemical Society 153 (2005) A225.
3. S. Burgmann, M. Blank, O. Panchenko, J. Wartmann, μPIV measurements of two-phase flows of an operated direct methanol fuel cell, Exp Fluids 54 (2013).
4. C. Mohan Kumar, P. Elsen, B. Berger, U. Janoske, Numerical modeling and analysis of viscous media removal from grooved surfaces with rotating impinging jets, Int. J. CMEM 4 (2016) 573–582.
5. X. Noblin, R. Kofman, F. Brochard-Wyart, Vibrations of sessile drops, European Physical Journal Special Topics 166 (2009) 7–10.
6. X. Noblin, R. Kofman, F. Celestini, Ratchetlike motion of a shaken drop, Physical review letters 102 (2009) 194504.
7. S. Daniel, M.K. Chaudhury, Rectified motion of liquid drops on gradient surfaces induced by vibration, Langmuir: the ACS journal of surfaces and colloids 18 (2002) 3404–3407.
8. C.W. Extrand, Y. Kumagai, Liquid drops on an inclined plane: the relation between contact angles, drop shape, and retentive force, Journal of colloid and interface science 170 (1995) 515–521.
9. D. Quéré, M.-J. Azzopardi, L. Delattre, Drops at Rest on a Tilted Plane, Langmuir 14 (1998) 2213–2216.
10. C.G. Cooney, C.-Y. Chen, M.R. Emerling, A. Nadim, J.D. Sterling, Electrowetting droplet microfluidics on a single planar surface, Microfluid Nanofluid 2 (2006) 435–446.
11. J.R. Dorvee, A.M. Derfus, S.N. Bhatia, M.J. Sailor, Manipulation of liquid droplets using amphiphilic, magnetic one-dimensional photonic crystal chaperones, Nature materials 3 (2004) 896–899.
12. P.A. Durbin, Free-streamline analysis of deformation and dislodging by wind force of drops on a surface, The Physics of Fluids 31 (1988) 43–48.
13. E.B. Dussan V., On the ability of drops to stick to surfaces of solids. Part 3. The influences of the motion of the surrounding fluid on dislodging drops, J. Fluid Mech. 174 (1987) 381–397.
14. B. Barwari, M. Rohde, O. Wladarz, S. Burgmann, U. Janoske, Motion of adhering droplets induced by overlapping of gravitational and periodical acceleration, International Journal of Multiphase Flow 135 (2021) 103537.

15. B. Barwari, Fluiddynamische Instabilitäten haftender Tropfen unter Gravitation, mechanischer Schwingungsanregung und aerodynamischer Krafteinwirkung, Shaker, 2021.
16. B. Barwari, S. Burgmann, M. Rohde, U. Janoske, Effects of fluid rheology and wetting properties on the droplet motion by a shear flow, PAMM 19 (2019) e201900068.
17. B. Barwari, S. Burgmann, U. Janoske, Deformation and movement of adhering droplets in shear flow, in: 5th Int. Conf. on Experimental Fluid Mechanics, 2018.
18. B. Barwari, S. Burgmann, U. Janoske, Strömungsinstabilitäten eines haftenden Tropfens unter Anströmung, in: GALA 18, 2018.
19. S. Burgmann, B. Barwari, U. Janoske, Inner flow structure of an adhering oscillating droplet in shear flow, in: 19th International Symposium on the Application of Laser and Imaging Techniques to Fluid Mechanics, 2018.
20. Z. Lin, X. Peng, Inner flow behavior in an oscillating liquid droplet, Heat Trans. Asian Res. 38 (2009) 485–491.
21. S. Burgmann, B. Barwari, U. Janoske, Oscillation of adhering droplets in shear flow, in: 5th Int. Conf. on Experimental Fluid Mechanics, 2018.
22. S. Burgmann, V. Krämer, M. Dues, J. Steinbock, L. Büttner, J. Czarske, U. Janoske, Flow-measurements in the wake of an oscillating sessile droplet using laser-Doppler velocity profile sensor, tm—Technisches Messen 89 (2022) 178–188.
23. S. Burgmann, V. Krämer, M. Rohde, M. Dues, U. Janoske, Inner and outer flow of an adhering droplet in shear flow, International Journal of Multiphase Flow 153 (2022) 104140.
24. S. Burgmann, B. Barwari, T. Maurer, U. Janoske, Hydrodynamische Instabilitäten eines Tropfens auf einer Platte unter Anströmung und Vibration (2017).
25. A. Milne, A. Amirfazli, Drop shedding by shear flow for hydrophilic to superhydrophobic surfaces, Langmuir: the ACS journal of surfaces and colloids 25 (2009) 14155–14164.
26. S. Moghtadernejad, M. Mohammadi, M. Jadidi, M. Tembely, A. Dolatabadi, Shear driven droplet shedding on surfaces with various wettabilities, SAE International Journal of Aerospace 6 (2013) 459.
27. G.D. Bixler, B. Bhushan, Bioinspired rice leaf and butterfly wing surface structures combining shark skin and lotus effects, Soft Matter 8 (2012) 11271.
28. A. Shastry, M.J. Case, K.F. Böhringer, Directing droplets using microstructured surfaces, Langmuir: the ACS journal of surfaces and colloids 22 (2006) 6161–6167.
29. Y. Li, L. He, X. Zhang, N. Zhang, D. Tian, External-Field-Induced Gradient Wetting for Controllable Liquid Transport: From Movement on the Surface to Penetration into the Surface, Advanced materials (Deerfield Beach, Fla.) 29 (2017).
30. P. Hao, C. Lv, Z. Yao, F. He, Sliding behavior of water droplet on superhydrophobic surface, EPL 90 (2010) 66003.
31. P. Hao, C. Lv, Z. Yao, Droplet detachment by air flow for microstructured superhydrophobic surfaces, Langmuir: the ACS journal of surfaces and colloids 29 (2013) 5160–5166.
32. H. Hu, S. Huang, L. Chen, Displacement of liquid droplets on micro-grooved surfaces with air flow, Experimental Thermal and Fluid Science 49 (2013) 86–93.
33. E.C. Kumbur, K.V. Sharp, M.M. Mench, Liquid droplet behavior and instability in a polymer electrolyte fuel cell flow channel, Journal of Power Sources 161 (2006) 333–345.
34. Y.H. Cai, J. Hu, H.P. Ma, B.L. Yi, H.M. Zhang, Effects of hydrophilic/hydrophobic properties on the water behavior in the micro-channels of a proton exchange membrane fuel cell, Journal of Power Sources 161 (2006) 843–848.
35. A. Theodorakakos, T. Ous, M. Gavaises, J.M. Nouri, N. Nikolopoulos, H. Yanagihara, Dynamics of water droplets detached from porous surfaces of relevance to PEM fuel cells, Journal of colloid and interface science 300 (2006) 673–687.
36. T. Ha, B. Kim, H.-S. Kim, K. Min, Investigation on the liquid water droplet instability in a simulated flow channel of PEM fuel cell, Journal of Mechanical Science and Technology 22 (2008) 1030–1036.

37. P.K. Das, A. Grippin, A. Kwong, A.Z. Weber, Liquid-water-droplet adhesion-force measurements on fresh and aged fuel-cell gas-diffusion layers, Journal of The Electrochemical Society 159 (2012) B489.
38. V. Thoreau, B. Malki, G. Berthome, L. Boulange-Petermann, J.C. Joud, Physico-chemical and dynamic study of oil-drop removal from bare and coated stainless-steel surfaces, Journal of Adhesion Science and Technology 20 (2006) 1819–1831.
39. A.K. Gupta, S. Basu, Deformation of an oil droplet on a solid substrate in simple shear flow, Chemical Engineering Science 63 (2008) 5496–5502.
40. G.K. Seevaratnam, H. Ding, O. Michel, J.Y.Y. Heng, O.K. Matar, Laminar flow deformation of a droplet adhering to a wall in a channel, Chemical Engineering Science 65 (2010) 4523–4534.
41. S. Madani, A. Amirfazli, Oil drop shedding from solid substrates by a shearing liquid, Colloids and Surfaces A: Physicochemical and Engineering Aspects 441 (2014) 796–806.
42. J. Fan, M.C. Wilson, N. Kapur, Displacement of liquid droplets on a surface by a shearing air flow, Journal of colloid and interface science 356 (2011) 286–292.
43. S.C. Fu, W.T. Leung, C.Y.H. Chao, Detachment of droplets in a fully developed turbulent channel flow, Aerosol Science and Technology 48 (2014) 916–923.
44. Chen Fang, Carlos Hidrovo, Fu-min Wang, John Eaton, Kenneth Goodson, 3-D numerical simulation of contact angle hysteresis for microscale two phase flow, International Journal of Multiphase Flow 34 (2008) 690–705.
45. Jean-Baptiste Dupont, Dominique Legendre, Numerical simulation of static and sliding drop with contact angle hysteresis, Journal of computational physics 229 (2010) 2453–2478.
46. Peter D.M. Spelt, A level-set approach for simulations of flows with multiple moving contact lines with hysteresis, Journal of computational physics 207 (2005) 389–404.
47. H. Ding, P.D.M. Spelt, Onset of motion of a three-dimensional droplet on a wall in shear flow at moderate Reynolds numbers, J. Fluid Mech. 599 (2008) 341–362.
48. S.R. Annapragada, J.Y. Murthyand, S.V. Garimella, International Journal of Heat and Mass Transfer 5-6 (2012) 1457–1465.
49. A.I. ElSherbini, A.M. Jacobi, A Model for Condensate Retention on Plain-Fin Heat Exchangers, Journal of Heat Transfer 128 (2005) 427–433.
50. N. Linder, A. Criscione, I.V. Roisman, H. Marschall, C. Tropea, 3D computation of an incipient motion of a sessile drop on a rigid surface with contact angle hysteresis, Theoretical and Computational Fluid Dynamics 29 (2015) 373–390.
51. P. He, C.-W. Yao, Simulating contact angle hysteresis using pseudo-line tensions, MRS Communications 9 (2019) 1060–1066.
52. J.K. Park, K.H. Kang, Numerical analysis of moving contact line with contact angle hysteresis using feedback deceleration technique, Physics of Fluids 24 (2012).
53. V. Krämer, B. Barwari, S. Burgmann, M. Rohde, S. Rentschler, C. Holzknecht, C. Gmelin, U. Janoske, Numerical analysis of an adhering droplet applying an adapted feedback deceleration technique, International Journal of Multiphase Flow 145 (2021) 103808.
54. M. Rohde, B. Barwari, S. Burgmann, U. Janoske, Droplet motion induced by superposition of shear flow and horizontal surface vibration, International Journal of Multiphase Flow 155 (2022) 104163.
55. M. Rohde, S. Burgmann, U. Janoske, The impact of a two-dimensional vibration excitation on the critical incident flow velocity of a sessile droplet, International Journal of Multiphase Flow 171 (2024) 104663.
56. B. Barwari, S. Burgmann, A. Bechtold, M. Rohde, U. Janoske, Experimental study of the onset of downstream motion of adhering droplets in turbulent shear flows, Experimental Thermal and Fluid Science 109 (2019) 109843.
57. Thomas Maurer, Experimentelle und numerische Untersuchung der Tropfenbewegung unter Einfluss von äußeren Kräften, Shaker Verlag, Aachen, 2017.
58. B. Barwari, S. Burgmann, U. Janoske, Hydrodynamic Instabilities of Adhering Droplets Due to a Shear Flow in a Rectangular Channel, Chemie Ingenieur Technik 91 (2019) 991–1000.

59. H. Fujimoto, Y. Shiotani, A.Y. Tong, T. Hama, H. Takuda, Three-dimensional numerical analysis of the deformation behavior of droplets impinging onto a solid substrate, International Journal of Multiphase Flow 33 (2007) 317–332.
60. Y. Li, J. Zhou, M. Hu, D. Jing, Whole Contact Line Pinning for Droplets Impacting on a Hydrophobic Surface Due to Hydrophilic TiO(2) Nanoparticle Addition, Langmuir: the ACS journal of surfaces and colloids 37 (2021) 6673–6680.

Contribution of Materials Technology to Achieving Sustainability Goals

Arne Röttger and Michael Blüm

Contents

Abstract

The challenge of current climate change requires strategies for sustainable consumption, prioritizing products produced with minimal resource use and CO_2-emissions. Shifting from a linear to a circular economy is essential, utilizing strategies such as repurposing, repairing, remanufacturing, and recycling. Successful implementation requires thoughtful product design and a comprehensive understanding of materials and manufacturing processes.

With the focus on the material, this article uses selected examples to show different strategies to manufacture products with low use of resources and CO_2-emission. The examples are questions from successfully implemented research projects. For example, by adapting the alloy design, the material properties could be improved, and the product usage phase could be extended. Other examples focus on reducing critical and cost-intensive elements to minimize costs and delivery

A. Röttger (✉) · M. Blüm
Chair of New Manufacturing Technologies and Materials,
Bergische Universität Wuppertal, Solingen, Germany
e-mail: roettger.fuw@uni-wuppertal.de

S. Bracke and R. Pieper (eds.), *Safety Engineering*,
https://doi.org/10.1007/978-3-658-47473-7_9

dependencies. Further examples deal with the repurpose approach or the joining of steel-aluminum connections to produce a hybrid multi-material mix to manufacture high-strength and lightweight structures for the automotive sector.

Keywords

Materials technology · Sustainability · Tool steel · Repurpose · Recycle · Circular economy

1 Climate Change, Sustainability, and Circular Economy

Throughout Earth's history, the climate has been subject to strong temperature fluctuations due to various causes (luminosity of the sun, albedo effect, Earth's orbit, volcanic eruptions). Since the beginning of the industrial age, the temperature has risen more rapidly in a short period without the influence of an extreme event (e.g., volcanic eruption) than ever before in geological history. The increase in temperature and the associated climate change since the beginning of the industrial era are primarily due to anthropogenic activities (burning fossil fuels, agriculture, deforestation, etc.), which are accompanied by the release of climate-damaging gases such as carbon dioxide, nitrous oxide, methane and fluorinated hydrocarbons (F-gases). The change in the concentration of these gas components in our atmosphere causes a change in the Earth's energy balance (shift in the tropopause, less infrared radiation is emitted into space), so there is currently an energy surplus of 0.6 Ws/m^2, which causes global warming. This global warming, in turn, is associated with weather extremes, such as more prolonged periods of heat/drought or heavy rain events [1].

While greenhouse gases like methane, nitrous oxide, and SF_6 have a more significant impact on the greenhouse effect than CO_2, the comparative impact of all greenhouse gases is considered in terms of their CO_2 equivalent, known as the greenhouse gas potential. CO_2 emissions contribute the largest share (75%) compared to F-gases (2%), N_2O (4%), and methane (18%). Notably, CO_2 is released during the combustion of fossil fuels for energy generation (34%) and in industrial processes (24%), indicating considerable potential for CO_2 savings in the transport (15%), building (6%), and agricultural (22%) sectors [2, 3]. To achieve an expansion of renewable energy sources and reduce CO_2 emissions in the energy industry, various saving options exist in industrial production [4]. Import dependency on critical raw materials in the industrial sector, especially rare earth elements like LREE and HREE, is increasing. Elements such as tungsten, cobalt, vanadium, and niobium, essential for product production, are also vital but receive less attention in public discussions [5, 6]. Despite the assessment of raw material criticality, it is undeniable that raw material availability is limited and finite. Hence, raw materials must be used efficiently to ensure the prosperity of future generations, leading to the concept of sustainability. To minimize raw material usage in product manufacturing and reduce CO_2 emissions, the linear economy needs to transition to a circular one. This transition, which emphasizes the reuse and recycling of raw materials within the economic cycle, is a crucial step towards sustainability (Fig. 1).

In the context mentioned, Potting et al. have provided recommendations for action to improve "circularity". These recommendations are divided into three areas: "R0–R2 = Smarter Product use and Manufacture", "R3–R7 = Extended Lifetime", and "R8–R9 = Sensible Use of Materials" [7]. The effectiveness of the sustainability aspects decreases from R0 (Circular Economy) to R9 (Linear Economy). The first area (R0–R2) focuses on product development, aiming to design products that require minimal use of raw materials and manufacturing steps. The R3–R7 area aims to extend product durability through re-use, repair, refurbishment, remanufacturing, and repurposing. Finally, the terms "Recycle" and "Recover" are listed in R8 and R9. When developing technical manufacturing systems to ensure a failure-prone operation, measures such as a simple design, good access options, options for easy component integration through a platform design, or the simple repair or replacement of components are often considered [8]. These concepts are currently only partially implemented in products for private purposes. Motivated by the goal of high corporate profits, the durability of consumer products is kept artificially low, so that consumers are motivated to buy new products with improved functionality, which is referred to by the term obsolescence (artificial or natural product aging) [9]. Particularly when it comes to electronic products, such as smartphones, Ylä-Mella et al. state that artificial product aging transforms these products into electronic scrap before the actual product lifespan, as they no longer satisfy the users' needs in terms of performance, status, or due to changed usage behavior [10]. Measures such as refurbishment or repair can enable an extended product usage phase, especially for customers with lower demands so that obsolescence can be effectively countered. In particular, repairing or installing new components requires an appropriate design. Delnaey et al. say that design has an impact of around 70 to 80% on the sustainability of a product [11]. The design, material selection, and manufacturing technologies are directly linked when developing technical products. The product's intended use determines

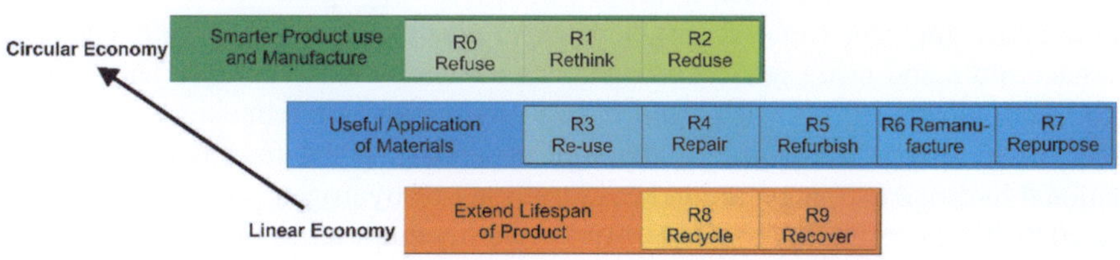

Fig. 1 Schematic representation of possibilities for increasing the circular economy based on [7]

the design and the material to be used. In turn, the design and the material define the manufacturing technologies to be used, whereby significant economic constraints in product production must be considered.

2 Basic Measures in the Field of Materials Technology for Climate-Friendly and Sustainable Production

2.1 Material Production, Recycling and Material Development

About 40% of all greenhouse gases can be linked to metal production [12]. Metal production is both resource- and energy-intensive, with fossil fuels often used to generate energy. Approximately 10% of the primary energy generated worldwide is needed for processing and producing metallic products, with the global resource requirement for metallic ores totaling 3248 million t in 2019. Iron ore, accounting for 94% of all mined minerals at 3040 million t, is the most important, followed by aluminum at approximately 62.9 million t [13]. Considering this, it is important to focus on these materials and implement measures to reduce greenhouse gases and energy usage in metal production and processing.

The blast furnace process is considered the primary process worldwide (approx. 70% of all steel) in steel production. The iron ores are reduced with coke, resulting in CO_2 as the reaction product. Based on Germany as a business location with a crude steel production of 40 million t, 55 million t of CO_2 was emitted in 2019 (corresponding to 28% of total industrial CO_2 emissions in the Federal Republic of Germany) [14]. The CO_2 emissions resulting from crude steel production account for approximately 65 to 70% of the total CO_2 emissions from the production and processing of steel products, so strategies will have to be developed in the future to reduce CO_2 emissions from crude steel production. The savings in CO_2 emissions in the steel industry over the last 7 decades are enormous. From 1960 to 2008, the energy required for steel production was reduced by 40%. In particular, by reducing the consumption of reducing agents and improving the quality of coke, CO_2 emissions were reduced by 23% between 1990 and 2007 [12]. From the previous considerations, it is clear that the decarbonization of Germany as an industrial location is strongly linked to the decarbonization of steel production. Solutions include using green hydrogen as a reducing agent in the blast furnace route and using renewable energy sources in electric steel production. The blast furnace route and using green hydrogen can save around 2 t of CO_2 per t of steel produced. However, to switch from coke to hydrogen as a reducing agent, 2 million t of hydrogen are required, although according to the federal government's national hydrogen strategy, around 0.42 million t of hydrogen should be produced by 2030. To cover the hydrogen demand in the medium term, in addition to green hydrogen production through water electrolysis, gray hydrogen production via steam reforming from methane will be used, which, however, results in the release of CO_2 emissions. The release of CO_2 emissions is avoided when producing blue hydrogen through CCS strategies [15].

In the short and medium term, focusing on using renewable energy sources in steel production, processing, and recycling seems to make more sense. The production of 1 t of steel requires an energy input of approx. 20–25 GJ [16]. In comparison, aluminum production using the Bayer (2.8–3.4 GJ/t) and Hall-Héroult processes (47–54 GJ) requires significantly higher energy requirements. Based on the current energy mix in the Federal Republic of Germany, aluminum production emits 6.17–6.68 t CO_2/t Al. In comparison, the recycling of aluminum scrap only requires an energy requirement of only 5% of the primary aluminum production, so that when recycling 1 t of aluminum, CO_2 emissions can be reduced to 0.3 t CO_2/t Al. If the material steel is considered, approximately 1.67 t of CO_2 can be saved by recycling steel scrap, and even 4.3 t of CO_2 can be saved by recycling stainless steel [17].

From the previous consideration, it is clear that appropriate technologies have been developed to reduce CO_2 in material production and processing. Leverage to reduce CO_2 emissions in material production is the cost-effective production of sufficient quantities of green hydrogen and the use of green electricity from renewable energy sources. It is also clear that the recycling of scrap metal strongly influences a resource-efficient circular economy, as energy for the extraction and processing of metal products can be significantly reduced. However, it should be noted that due to dissipation and the rapidly growing demand (growing global wealth, increasing world population) for metal products, the need cannot be met through recycling and the resulting circular economy alone. Currently, around 1/3 of the material requirements are covered by circular economy concepts, with 2/3 of all metal products being produced from primary raw materials to satisfy demand. By 2060, forecasts suggest that this ratio will reverse, underscoring the need for a circular economy but simultaneously demonstrating that metal ores will continue to be mined and processed into metals via the primary routes described previously [18].

Successful recycling requires scrap metal to be sorted according to the alloy composition. This is necessary to avoid contamination and adverse effects on the properties of the recycled metal (down-cycling). Because separation by alloy type is often not possible, unwanted alloying elements are unintentionally introduced when scrap is remelted, which can have a negative impact on the material properties. In this context, reference is made to the corrosion-resistant stainless steels like AISI304 and AISI316L, in which the content of Cu and Zn has increased in recent decades through successive recycling. The element Zn, in particular, can form brittle phases even with small alloy additions which reduces the materials ductility or decrease corrosion resistance by acting as a local element. Attempts are being made to develop measures (more precise measurement of the chemical composition of steel scrap, separation according to alloy type, and avoidance of cross-contamination) that will enable better separation of these undesirable elements from scrap. The collection of steel scrap is carried out by the metal processing industry, the customer, or the recycling centers as the products reach their end of service life. Problems arise when the different substances are not recorded, separated, or contaminated during processing. Separating processes such as machining with a geometrically determined (drilling, milling) or geometrically undetermined cutting edge (grinding) are often used in product manufacturing. For economic reasons, metal cutting is carried out with the help of a cooling lubricant, which has to be removed before the chip materials are recycled. This is sometimes complex and not economically worthwhile, so the grinding sludge is landfilled. Separating cooling lubricants is technologically possible but usually associated with high costs. In addition to simple recycling strategies, it is essential to develop solutions to make this waste usable as recyclable materials. Not only contamination from cooling lubricants but also the mixing of the metal waste products to be recycled or the introduction of foreign substances counteract the recyclability. During grinding with SiC, cBN, and Al_2O_3, the chip waste is contaminated by abrasive particles that cannot be removed in a subsequent processing step.

The abrasive particles introduced can be viewed not only as foreign substances but also as valuable alloying elements. However, this requires knowledge of the exact chemical composition of the chip waste so that new alloys with corresponding properties can be derived, considering the foreign substances introduced (which are now considered alloying elements). The situation is similar to cross-contamination with other chip waste due to a lack of pure sorting. By mixing chip waste, the elements bound in the different scrap waste are also mixed. Remelting the chip materials into new semi-finished products would result in a completely new alloy with different material properties. To avoid the problem of cross-contamination, the use of so-called lean alloys is recommended [19]. These are alloys whose properties can be adjusted by the processing technology or an appropriate heat treatment over a wide range using a minimum number of different alloying elements. Concerning the user, it is essential to examine to what extent the number of different steels can be reduced, for example, to improve scrap sorting directly during production or later at the recycling centers. However, using lean alloys can result in lower sustainability for components subject to severe wear and/or corrosion during operation. The addition of alloying elements is necessary to ensure sufficiently high corrosion (passive layer formation, e.g., due to Cr) or wear resistance (hard phases due to alloying with Cr, V, Mo, W, etc.). Otherwise, severe corrosion or wear will occur, resulting in rapid material removal and reduced product durability. The raw materials are lost in this case, which must be compensated through an energy-intensive primary process. Another strategy is to develop dirty alloys, which are less susceptible to the influence of foreign substances introduced from the recycling process.

2.2 Materials Processing

Saving energy and raw materials in manufacturing is closely linked to the product design and processing materials [20]. As previously stated, the highest CO_2 emissions in steel production are due to the high energy use and the reduction route.

However, the statements made for steel production can be transferred to other manufacturing processes. Maximum CO_2 savings can be expected if the manufacturing processes are designed to be energy efficient on the one hand and green electricity is used on the other [21]. For example, there is a high potential for energy savings when heating materials for hot forming or heat treatment. In addition to choosing the right furnace dimensions, preference should be given to heating sources in which the heat is not generated via fuel combustion and then via heat conduction and heat convection but rather directly in the material. In this context, heating via induction with up to 90% efficiency should be mentioned [22]. To further reduce the energy requirement for heat treatment, you can use design measures (more delicate components for faster heating) or materials that can be hardened from lower austenitization temperatures. The latter leads again to the question of the right choice of material. Another energy-saving potential lies in bundling production steps. For example, the component can be hardened directly from the forging heat, or a structure with the appropriate properties can be set using suitable temperature–time control from the hot forming process, considering the material-specific time–temperature transformation behavior.

In addition to save energy and process heat during heat treatment or hot forming, additive and powder metallurgical manufacturing processes enable the production of near-net-shape products with less material use [23]. Once powder metallurgy processes have become established in producing complex-shaped mass-produced parts, the slow build-up rates of additive manufacturing processes (here, powder bed processes like Powder Bed Fusion-Laser Beam/Metal PBF-LB/M) are counteracting widespread industrial use. In the PBF-LB/M process, the powder can also oxidize by reacting with the residual oxygen content in the process gas atmosphere. Particle oxidation changes the laser coupling, so the process parameters must be linked to the oxidation state of the powder. To avoid this problem, recycling the powders for the PBF process, for example, by annealing in a reducing atmosphere, is currently being considered [24]. In order to avoid the low build-up rates of the PBF-LB/M process, faster processes can be used to enable higher build-up rates (e.g., metal extrusion, directed energy deposition laser/arc). However, the higher build rates are accompanied by increased surface roughness of the additively manufactured components, so the amount of machining required increases. Current hybrid manufacturing systems allow combinatorial additive and subtractive manufacturing, so that setup times, machine setup times, etc. are reduced. With the implementation of hybrid manufacturing systems using additive and subtractive manufacturing as an example, high-end products can be manufactured economically even in comparison to low-wage countries [25, 26].

3 Selected Examples—Contribution of Materials Technology to Resource and Climate-Friendly Product Production

In the past, the authors have dealt with aspects of resource efficiency in the field of materials technology both at the Chair of New Manufacturing Technologies and Materials at the University of Wuppertal and at the Chair of Materials Technology at the Ruhr University Bochum. The focus of the work was usually on the development of new materials, considering the relevant manufacturing technologies. The goal was to reduce the use of raw materials, optimize manufacturing processes, or derive new alloys that possess a lower content of critical elements and conflict materials. Selected results will be presented below to demonstrate the importance of materials technology for sustainable business and the achievement of climate protection goals.

3.1 Extending Product Durability by Developing New Materials with Better Properties

Tools made from tool steel are used to manufacture and process products. If the material is processed at temperatures up to approx. 200 °C, cold work steels (CWS) are used, which are divided into three groups: CWS1 (tough), CWS2 (hard), and CWS3 (carbide-rich). The carbide-rich materials in the CWS3 group are the ledeburitic cold-work steels. In the heat-treated state, their microstructure consists of a tempered metal matrix in which carbides have different sizes and morphologies, depending on the original forming process (casting + formed = broken-up linear carbide networks; HIP = spherical carbides), are present. The carbides are mostly Cr- and V-rich carbides. The Cr-rich carbides of type M_7C_3 have a hardness of 10.55 to 21.9 GPa and a low fracture toughness of 1.99 to 3.99 MPa$\sqrt{m}$ [27, 28]. Even if the properties are sufficient for most purposes, the service life of tools can be effectively increased by producing hard phases that have higher hardness and are also cost-effective. In this case, Fe-rich borides of type M_2B should be mentioned, which have a hardness of about 1550 HV1 [29] (Fig. 2).

In addition to the higher hardness, these borides also have the advantage compared to Cr-rich carbides, that the hard phase formation is decoupled from the martensitic hardness through the element carbon. This decoupling makes it possible to develop new materials with a greater variation in the hard phase volume content independently of the carbon-martensitic hardening of the metal matrix.

In the last few decades, several works have dealt with the description of the ternary system Fe-C-B, the knowledge of which is essential to derive new alloying concepts. However, these phase diagrams, usually determined based on experimental methods, are insufficient to develop material based on computer-aided methods using the CalPhad method (Calculation of Phase Diagram). Lentz

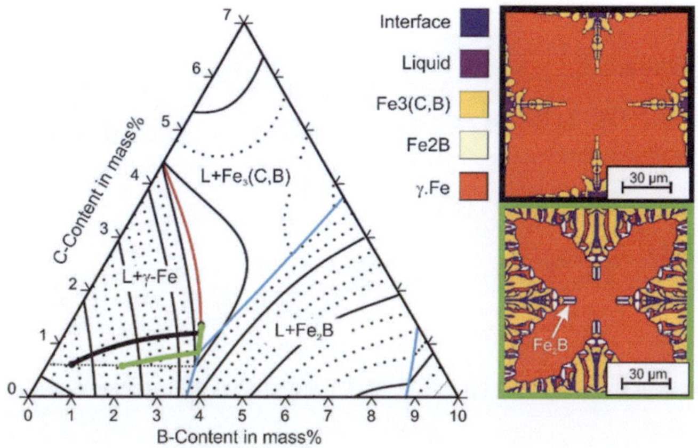

Fig. 2 Description of the solidification behavior in the hypoeutectoid region of Fe-C-B alloys: black curve = solidification sequence of Fe0.6C0.6B; green curve = solidification sequence of the alloy Fe0.6C1.8B

et al. investigated the hypoeutectoid system Fe-C-B in. They could describe the solidification behavior of Fe-C-B alloys using the phase field method in [30]. It was found that solidification occurs primarily through the formation of austenitic cells. Due to the limited solubility of C and B in the austenite, both elements are enriched in the residual melt. Depending on the local B/(C + B) ratio, the eutectic borides Fe_2B and/or Fe_3CB are formed. The so-called B cementite (Fe_3CB) with a B solubility of up to 20 at.-% represents the low-melting phase. With the help of thermodynamic calculations and in-situ XRD measurements, the phase formation kinetics in the solid state were investigated in [31]. The formation of the $Fe_{23}CB_6$ phase from Fe_3CB was demonstrated, which, however, partially binds the element carbon and thereby counteracts the desired decoupling of hard phase formation from martensitic hardening (Fig. 3).

In [32], the mechanical properties of the three hard phases of type Fe_2B, Fe_3CB, and $Fe_{23}CB_6$ were measured compared to the Cr-rich carbide of type M_7C_3 using nanoindentation. It was found that the hardness of the Fe_2B type boride (19.04 GPa) is higher than that of the M_7C_3 type carbide (16.43 GPa), the carboborides $Fe_{23}CB_6$ (10.39 GPa) and Fe_3CB (11.18–12.24 GPa, depending on the B/C + B ratio) but have a lower hardness. The aim was, therefore, to suppress the formation of the softer carboborides by specifically adding alloying elements and to promote stability and, thereby, the formation of the harder boride of the Fe_2B type. In [33, 34], it was proven that the tetragonal Fe_2B changes into the orthorhombic structure by adding Cr and exceeding 14.7 at.%. In addition, both hardness (21.5 GPa → 27.0 GPa) increased in the direction of a higher solute Cr content without much loss in fracture toughness (3.5 → 3.26 MPa$\sqrt{m}$), so that the Cr-rich M_2B with orthorhombic structure has a similar hardness to monocarbides

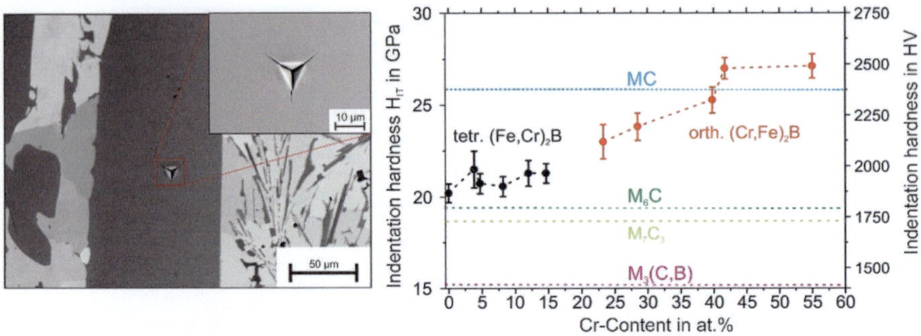

Fig. 3 Hardness of different hard phases in steels, determined using nanoindentation

as V_8C_7 for example. With the knowledge of the effect of corresponding alloying elements and the microstructure formation process during casting solidification, new FeCrCB and FeCrCBMo alloys could be derived in [35], which have 70% lower alloy costs compared to steel D2 (X155CrMoV12-1), and better wear behavior in the cast/HIP and heat-treated condition.

3.2 Development of New Materials with Improved Recycling Properties and Lower Content of Critical Raw Materials

Co-based materials are often used if components are exposed to tribo-mechanical or chemical loads at higher temperatures. These materials come from the CoCr system and can be adjusted in terms of their corrosion and wear properties depending on the addition of the elements W, Mo, and C. Carbon additions cause the formation of hard Cr-rich carbides of the M_7C_3 type, with additions of the element W additionally forming W-rich carbides of the M_6C type [36, 37]. If higher corrosion resistance is required, lower-C Co-based alloys are used so that in combination with a sufficiently high Cr content, it remains dissolved and not set in carbides in the metal matrix. The addition of the element Mo promotes the formation of the passive protective layer and promotes corrosion resistance. Even if Co base materials have excellent properties, their use is questionable due to the high content of the essential element cobalt, which is considered critical, especially since cobalt is sometimes obtained using child labor and represents a conflict material [38]. Therefore, there is a high motivation to replace Co-based materials with those having lower content of critical or conflict elements. The disadvantage, however, is that frequently used Fe or Ni-hard alloys experience a substantial drop in hardness and strength above 600 °C, which is because corresponding strength-increasing mechanisms (drop in dislocation density, Ostwald ripening, and coarsening of precipitates and large-angle grain boundaries) diminish in their effectiveness.

If hard alloys are loaded beyond the yield point, the softened metal matrix begins to flow plastically, whereby the hard phases cannot follow the elongation amounts specified by the matrix, which will result in a brittle fracture of the hard phases. The result is the formation of a characteristic tribolayer, which, depending on the interaction with the counterbody in the tribosystem, is referred as a mechanically mixed layer (MML) [39, 40, 41]. In order to counteract the strong hard phase fragmentation, the matrix strength must be increased. Thus, the plastic deformation of the metal matrix during wear loading can be counteracted. According to Fig. 4, Co-based hard alloys not only have a slight drop in hardness over temperature, but the metal matrix can transform from the fcc structure (alpha-Co) to the hdp structure (beta-Co) [37]. This transformation is accompanied by a reduction in the number of active sliding systems from 12 to 3 so that the plastic flow of the metal matrix and any hard phase fragmentation can be effectively countered. In order to substitute Co-based alloys for high-temperature applications, the key is to develop a hard alloy with a higher high-temperature strength so that plastic flow and the resulting hard phase fragmentation are counteracted. In addition, the requirements regarding strength, oxidation, and wear resistance must be met. Brackmann et al. and Krell provided a first approach in [41, 42] by successively substituting the Co content with Fe in the Stellite 6 alloy. It was found that with increasing Co-substitution by Fe, the hot hardness decreased, and with a Fe content of 45 mass%, a metal matrix with bcc-structure was present. The Co-substitution by Fe also causes an increase in the stacking fault energy so that the fcc-hdp transformation becomes more difficult or does not occur as the

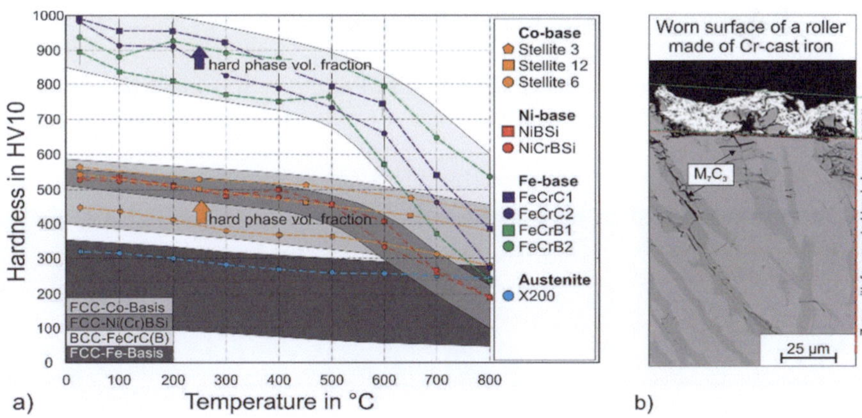

Fig. 4 **a** Hot hardness of different Fe-, Ni- and Co-based hard alloys, **b** worn roller surface with a pronounced mechanically mixed layer (MML) and highly fragmented hard phases below the tribo-loaded surface

Fe content increases (see EBSD measurements below a surface subject to wear in Fig. 5). As a result, despite a sufficiently high hot hardness of 220–260 HV10 at 800 °C, there was insufficient resistance to hot wear.

Based on the concept of superalloys, a further alloying approach considered the development of a precipitation-hardenable Fe-Ni base alloy [43]. In order to ensure a sufficiently high corrosion and oxidation resistance, Cr contents of 25 mass% were considered. In addition, Cr with the other alloying elements Nb (1 mass%), W (2 mass%), V (4 mass%) causes the formation of hard carbides of the type M_7C_3 ($M_{23}C_6$), MC and M_6C. In order to promote a sufficiently high supporting effect from the matrix on the hard phases, in addition to 30 mass% of Ni, the element Al was also regarded in this alloying concept in amounts of up to 3 mass% to produce a high-volume content of the NiAl precipitate through an adapted heat treatment. However, this precipitate is only formed above a Ni content of approx. 20 mass% [44]. In addition, high Ni contents must be used to counteract the stabilizing effect of the elements Cr, Al, Nb, C, and W on the ferrite phase since recovery and recrystallization would result in a strong drop in hardness and strength at temperatures above 650 °C. Through solution annealing, at 1100 °C followed by aging at 650, 700, and 750 °C, the hot hardness of the newly developed alloy was higher than that of the reference alloy Stellite 6 in the range of 500 to 800 °C due to the precipitation of the NiAl phase. Wear tests confirm that the developed alloy, aged at 650 °C, has a similar wear rate compared to the reference alloy Stellite 6. However, if the alloy was aged at higher temperatures (700/750 °C), Ostwald ripening of the NiAl precipitates occurred, which resulted in a drop in high-temperature strength and an increase in the wear rate [42] (Fig. 6).

3.3 Recycling of Metal Matrix Composites

Even if the recycling of products according to Potting et al. should be one of the last measures to be considered in the context of the circular economy, important

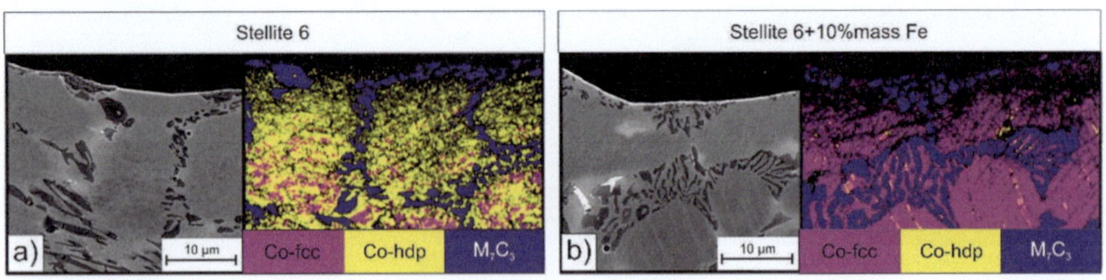

Fig. 5 Microstructure with local phase representation to depict the fcc-hpd transformation of the Co matrix, **a** Stellite 6 alloy, **b** Stellite 6 with 10 mass% Fe where Fe increases the stacking fault energy and counteracts the fcc-hdp transformation

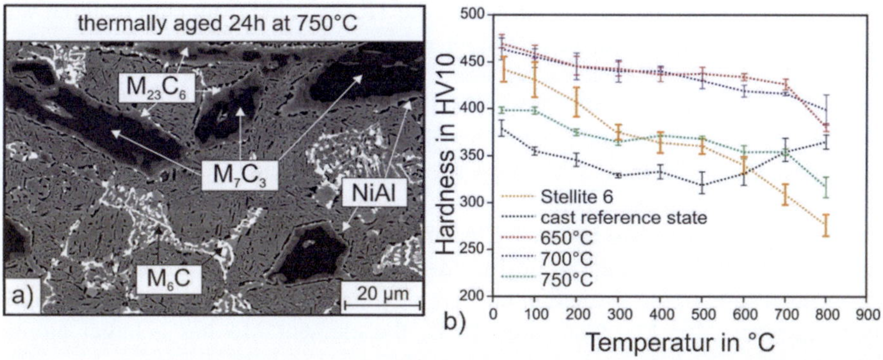

Fig. 6 **a** Microstructure of the developed Fe-based alloy as a replacement for Stellite 6, with a microstructure consisting of the hard phases $M_{23}C_6$, M_7C_3, M_6C, which are present in a precipitation-hardened (NiAl) γ-Fe matrix, **b** Hot hardness of the developed alloy depending on the aging temperature to adjust the amount and size of NiAl-precipitates

and sometimes critical raw materials can be kept in the recycling cycle. In this context, reference should be made to the material "cemented carbide", which preferably consists of a Co-based matrix with a high-volume content of tungsten carbide (WC). Since around 69% of the element tungsten for the production of cemented carbide is obtained in the Republic of China, there is a high delivery dependency. Thus, recovering this raw material through recycling is essential to decoupling supply dependencies [45]. Through the consistent recycling of cemented carbide, 42% of the tungsten can be covered by recycling in the European Union. Not only tungsten but also other materials are limited available and expensive to produce, so recycling them is also worthwhile. In this context, the metal matrix composite "Ferrotitanit" should be mentioned. Ferrotitanit consists of a metallic matrix in which approx. 22–34 mass% of cost-intensive titanium carbides (TiC) are added as a reinforcing phase to increase wear resistance against abrasion. Ferrotitanit can be machined in the soft-annealed state and, in addition to good wear properties, also has good corrosion resistance if sufficiently alloyed with the elements Cr and Mo. For this reason, this material is ideal for the production of complex-shaped tools for plastics processing. When developing a sensible recycling strategy, not only the chip material but also the TiC present in worn tools at the end of their service life must be considered.

In the project "RecyTiC" funded by the BMBF, FOSTA developed a recycling strategy that was scaled up to an industrial scale and tested at company DEW [46]. The metal matrix is removed using an acid, thus releasing the TiC, which can be made usable again after washing and drying. The material considered was the Fe-based material Nikro128, which contains 30% TiC, 13.5% Cr, 9% Cr, 4% Ni, and 6% Mo (in mass%). Comparing the TiC removed from the chips or worn tools with the TiC used during initial production highlights a change in the morphology and size of the recycled TiC. In particular, the TiC recycled from chip material was much finer because it fractured to smaller fragments during machining. The TiC recycled from the bulk material possess a coralline-like morphology and a bigger size due to sintered TiC. OES (optical emission spectrometry) measurements also revealed a change in the chemical composition of the TiC before use (composition 80 mass% Ti and 20 mass% C). In particular, the recycled TiC has an increased Mo content of 6.22 mass% (Fig. 7).

The enrichment of Mo in the TiC occurs during the primary processing of the material Nikro128 and can be attributed to a pronounced Mo diffusion into the TiC. This results in the formation of thermodynamically more stable TiMo mixed mono-carbides [46, 47]. Compared to Ti, the element Mo has a larger atomic diameter and a higher number of valence electrons, so with increasing Mo diffusion, the degree of metallicity increases in the MC, and the hardness decreases because the Ti-C and Mo-C bond lengths also increase. In addition to the drop-in hardness, reusing the TiMoC recyclates was ineffective, as they could only be incorporated into the metal matrix in a force-fitting manner. Thus, the previously described Mo diffusion from the metal matrix into the TiC is essential to promote metallurgical integration. In order to solve this issue, the Nikro128 alloy was improved by alloying with elements with a stronger tendency to form monocarbides with titanium. This serves two purposes: integrating metallurgical titanium carbide into the metal matrix and retaining molybdenum in the metal matrix to enhance corrosion resistance. The element Nb was considered in the further investigations because, as a homologous element like Ti, it forms monocarbides and has a high solubility in TiC [47]. When dissolved in TiC, Mo causes a reduction in the bond length, which is accompanied by an increase in the covalent bond strength. The result is a higher shear hardness and fracture toughness, as proven with DFT

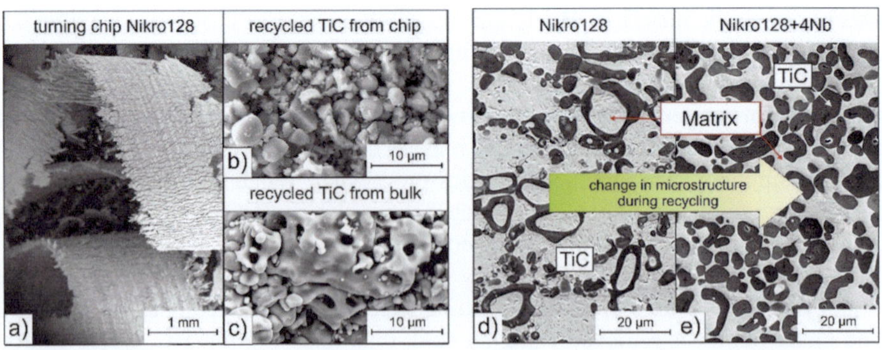

Fig. 7 Recycling of TiC and microstructure of Nikro 128, **a** chip before recycling from soft annealed Ferrotitanit Nikro128, **b** TiC particles recycled from chip material, **c** TiC particles recycled from worn tools (bulk material), **d** reference microstructure of Nikro128 produced for the first time, **e** microstructure of Ferrotitanit with recycled TiC and 4 mass% Nb additions

calculations (see Fig. 8). The hardness increased up to 25 at.% dissolved Nb in the TiNbC. However, the highest fracture toughness of approx. 2.05 $\text{MPa}\sqrt{\text{m}}$ was measured at a Nb content of 10 at.% and decreases towards higher Nb contents. Building on these more theoretical considerations, Nikro128 samples with varying Nb-contents (0–4 mass%) were produced. The measurement of the chemical composition of the TiC showed that both the elements Mo and Nb accumulate in the TiC hard particles during the material processing. At the same time, the element Mo (up to 2 mass%) remains increased in the matrix and promotes the corrosion resistance. Adding the element Nb results in an overall increase in the hard phase volume fraction (Table 1).

The hardness of TiC increases with minor Nb additions from 3114 ± 93 HV0.05 (reference state TiC) to 3468 ± 85 HV0.05 (TiNbC in sample Nikro128 + 2 mass% Nb) [47]. However, if higher Nb content is added (3 and 4 mass%), the hardness drops to 3200 HV0.05, which contradicts the previous DFT calculation. An explanation can be found in the fact that the element carbon was not added similarly to form stoichiometric MC. So, only Nb/Mo was added, which led to the formation of sub-stoichiometric MC, which is associated with a lower covalent bonding character and hardness. In addition to adjusting the hardness, adding Nb also

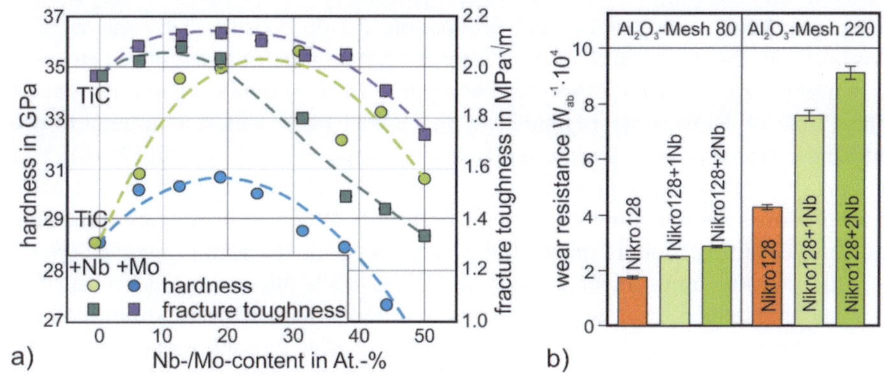

Fig. 8 Properties of TiC and bulk Ferrotitanit; **a** Influence of the elements Mo and Nb on the hardness and fracture toughness of TiC, **b** Wear properties of Ferrotitanit in the initial state (Nikro128 without recycled TiC) and with recycled TiC and adapted alloy design (additions of Nb) against the abrasive Al_2O_3 with different sizes

Table 1 Chemical composition of TiC in mass%, measured in the Ferrotitanit as a function of the added Nb content

Material/element	Ti	C	Mo	Nb
Nikro128	74.09 ± 0.61	18.44 ± 0.20	6.22 ± 0.37	–
Nikro128 + 2Nb	71.33 ± 2.12	17.88 ± 0.41	5.16 ± 1.19	5.57 ± 2.80
Nikro128 + 4Nb	67.08 ± 1.61	17.18 ± 0.42	4.15 ± 0.36	11.59 ± 1.51

changed the morphology and size of the MC carbides. If the material Nikro128 contains MC with a corona-like morphology, these form into a more spherical morphology with a larger diameter through Nb additions. Mohr discussed the change in the TiC morphologies and their size caused by the formation of coherence stresses if the element Nb is added [47]. This coherency stresses in the TiC surface promote carbide dissolution and re-precipitation, resulting in coarser carbides. These coarser carbides provide greater wear resistance against coarser abrasives. Mohr was also able to prove that with an adapted heat treatment, the corrosion properties of the material Nikro128 with recycled TiC have a higher corrosion resistance, which is attributed to the higher Mo content in the metal matrix. This material development impressively shows that it is possible with the derivation of a fundamental understanding of the material changes during recycling and a correspondingly adapted material design that better material properties of the recycled materials can result, representing up-cycling.

3.4 Repurpose and Remanufacturing

The steel industry is responsible for around ¼ of all CO_2 emissions. Measures focusing on reducing CO_2 emissions can preferably be attributed to recycling or using gray/blue hydrogen in steel production. As demonstrated by the work of Potting et al. described in *Sect.* 1, measures such as repairing, avoiding, etc., lead to a higher degree of circularity in the context of the circular economy concept. The Circle of Tools project (funded by the BMBF) considered a strategy to produce new products from worn circular knives made of steel D2 (X153CrMoV12) as part of a circular economy. Conventionally, the circular knives are sharpened multiple times for maintenance (remanufacturing). When an insufficient low diameter of the cutting tools is reached, they are recycled. Recycling involves remelting to produce semi-finished products. These semi-finished products must be separated, reshaped, heat-treated, and ground again, for example, to produce new circular knives. In addition to preserving the material through recycling, there is considerable potential for resource and CO_2 savings, as remelting and the complex production of further products can be avoided. As an example of the production of new knives or tools, worn circular knives were considered as semi-finished products, from which tool blanks such as chisels, roofing felt cutting knives, or knife blanks were removed using different cutting processes (water jet cutting, laser cutting). Figure 9 shows the process of the repurposing strategy under consideration by the manufacturing of a chisel for example. Although the repurposing strategy under consideration requires energy to be used to remove the tool blanks from the worn circular knives and the tools also have to be processed by grinding, the costs, energy, and thus CO_2 emissions can still be reduced by avoiding further remelting, forming, cutting and hereby the necessary heat treatment to adjust the material properties can be saved [49]. Chisels are conventionally made from 80CrV2 steel, as this steel offers a good compromise between material properties and costs. Due to the higher alloy costs of the ledeburitic cold work steel X153CrMoV12, an

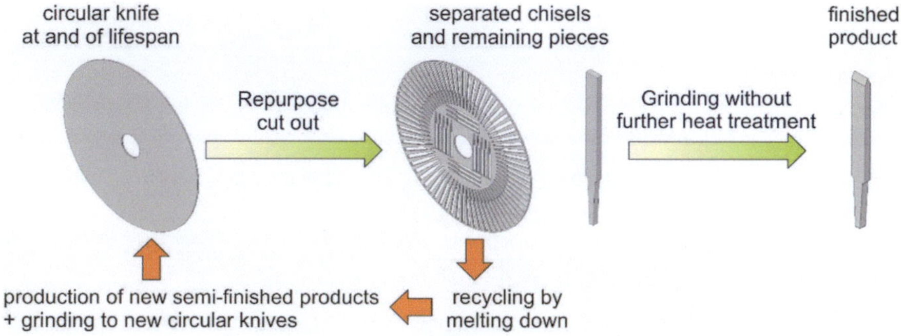

Fig. 9 Schematic description of the repurpose and recycling approach for circular knives [48]

increase in production costs would be expected, which the customer would not be willing to pay, considering the costs of competing products.

Taking the repurpose strategy into account, in turn, requires the necessary manufacturing steps to produce the product (elimination of forging processes, heat treatment) so that costs can be saved using the repurpose approach. Even if the production is carried out using the repurposing strategy with similar production costs compared to conventional production using 80CrV2 steel, there are secondary advantages in product use. Kronenberg et al. in [48] prove that in practical chisel tests against a hardwood material, the cutting edge of the chisel made of D2 wears less than the standard material 80CrV2. Based on the geometric measurement of the edge rounding and the associated material removal of the chisels, the D2 steel could be assigned a service life that is 3 to approx. 6 times longer than the frequently used 80CrV2 steel. The high variation in the service life extension was attributed to the large influence of the carbide orientation [48]. The alignment of the carbides in the worn circular knives can be traced back to the actual circular knife production. For example, blanks are first removed from a larger semi-finished product and then cross-rolled to the required diameter or thickness. The forming process produces a characteristic rolling texture associated with a carbide orientation. The microstructure texture is adopted depending on the direction in which the repurpose blanks are removed. It is, therefore, essential that the microstructure of the starting material is known precisely so that the repurposed semi-finished products can be removed so that the microstructure is optimal for the later direction of loading.

While only the worn circular knife and the possibilities of repurposing have been discussed, grinding sludge has already been produced during the actual circular knife production due to the machining. The chisels produced by repurpose also receive their cutting edge (see Fig. 10) by grinding with abrasives such as Al_2O_3, SiC, or cBN. In addition, a cooling lubricant is used during grinding so that the resulting grinding slurries consist of grinding slurries, abrasives, and cooling

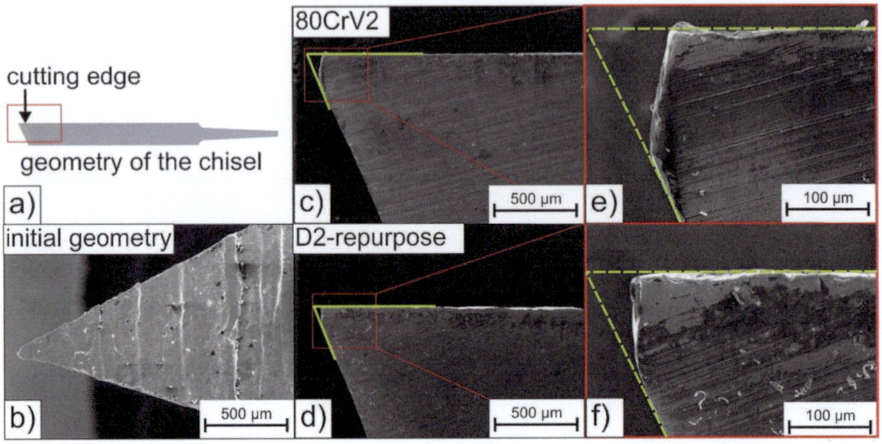

Fig. 10 Description of the wear behavior of the chisels during technological chipping tests against hardwood; **a** schematic representation of the chisel and the loaded cutting edges, **b** cutting edges in the ground and unused condition, **c** and **e** cutting edge wear of material 80CrV2; **d** and **f** Cutting edge wear when chiseling made from steel D2 by repurpose

lubricants [50]. Recycling requires a complex separation of the individual components, which is often not worthwhile due to the complex system technology and the associated costs. Therefore, current work attempts to process the grinding sludge and make it usable again as starting materials for the production of new semi-finished products. As part of the BMWK "GENESIS" project, high-speed steels and magnetic materials are being examined to determine whether new tools or magnetic materials can be produced with similar or better material properties using advanced sintering methods (FAST, SPC, HIP) after appropriate processing of the recyclates [51]. A research project funded by the DBU examines whether processed grinding sludge can be used as starting materials for the production of cored wires, which can subsequently be processed using deposition welding or additive manufacturing methods. The melting of the recycled grinding sludge and producing new semi-finished products via the smelting metallurgical route should also be considered. The analysis of the chemical composition of the recycled grinding sludge showed that, in particular, the Si and C content increased from 1.53 mass% of C and 0.35 mass% of Si to 2.18–2.23 mass% of C and 1.22–1.34 mass% of Si. Subsequently, in the computer-aided material development, the Si and C contents introduced were not viewed as disruptive variables but rather as raw materials and, therefore, important alloying elements for up-cycling.

If the purely recycled microstructure has a hardness of 56.2 HRC in the secondary hardened state (reference D2 = 61.2 HRC), the hardness could be increased to 58.5 HRC by adapting the alloy design. Although the procedure described illustrates in principle the possibility of recycling grinding sludge and reprocessing it into semi-finished products using melt metallurgy, this requires precise knowledge of the chemical composition of the grinding slurry and is, therefore, less practical for technological applications. Future developments aim to develop materials less susceptible to introduced contaminants (so called dirty alloys) and can, therefore, be recycled more easily without alloying adjustments (Table 2).

The intelligent choice of the grinding abrasives used in the grinding process and their influence on the subsequent recycling behavior can also be more closely examined. It was proven that when Al_2O_3 is used as a grinding material, and the inert behavior of Al_2O_3 in Fe melts or their lower density, they rise in the melt pool as a result of the buoyancy forces and are deposited on the surface as a slag layer. However, this requires a sufficiently long process during remelting to allow Al_2O_3 particles to float completely. Otherwise, these oxides remain in the matrix and significantly reduce the mechanical and cyclic-mechanical material properties (Fig. 11).

3.5 Multi-Material Design and Lightweight Construction

In the field of avionics and automotive, there are great efforts to reduce vehicle weight in order to minimize the energy used when accelerating the mass. In addition to the primary effects (saving fuel and reducing emissions), reducing the mass of vehicles generally also has secondary saving effects on the drive unit, brakes, drive train, etc.. In addition to structural lightweight construction (e.g., topology-optimized design), low-density materials (aluminum, magnesium) have often been considered in structural design. However, suppose high stiffness or energy absorption is required in the event of a crash. In that case, structures made of aluminum and magnesium must be designed accordingly massively due to the low strength and stiffness of these materials, accompanied by increased use of materials. Most steels, on the other hand, have a higher stiffness and strength compared to the previously mentioned materials. In order to combine the advantages of different materials, high-strength steels are used in the automotive sector for structures that must ensure high strength and crash safety. In contrast, structures that require lower strength can be made from light metals. In order to be able to manufacture structures from extremely high-strength steels, the technology "press hardening" is used. Previously cut and, if necessary, pre-formed steel sheets made of 22MnB5 (ferritic-pearlitic microstructure), are austenitized at 920 °C for 6 min., transferred into a water-cooled tool and are formed and hardened in one processing step. The forming of the sheets takes place in an austenitic and, therefore, easily deformable state before the martensitic transformation occurs, if the sheets are cooled down in the cooled hot forming tool below the martensitic start temperature. This gives press-hardened steels a complex shape with tensile strengths of approx. 1800 MPa (Fig. 12).

The joining technologies represent a crucial technology for the production of the multi-material design. However, due to the high strength of the press-hardened sheets, only press or fusion welding can be used. Even if the steel 22MnB5

Table 2 Chemical composition of steel X153CrMoV12 (base material) in the states of nominal composition, after recycling and sintering and after recycling and melt metallurgical processing (in mass%)

Condition	C	Si	Mn	Cr	Mo	V	Fe
Norm	1.53	0.35	0.40	12.00	0.85	0.85	Bal.
Sintered	2.23 ± 0.39	1.22 ± 0.44	0.07 ± 0.03	11.09 ± 1.34	0.82 ± 0.20	0.54 ± 0.02	Bal.
Melted	2.18 ± 0.09	1.34 ± 0.01	0.31 ± 0.00	10.00 ± 0.03	0.77 ± 0.01	0.67 ± 0.01	Bal.

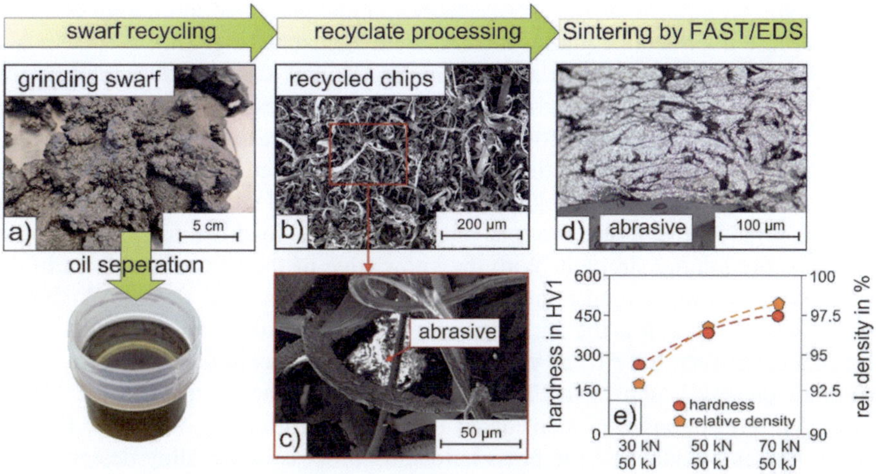

Fig. 11 Schematic representation of the recycling process of grinding sludge with SiC and cooling lubricants from machining production: **a** grinding sludge, **b** and **c** reclined grinding swarf, **d** microstructure compacted using electro-discharge sintering (EDS) using recyclates, **e** density and hardness of the EDS-compacted recyclates depending on the process parameters used

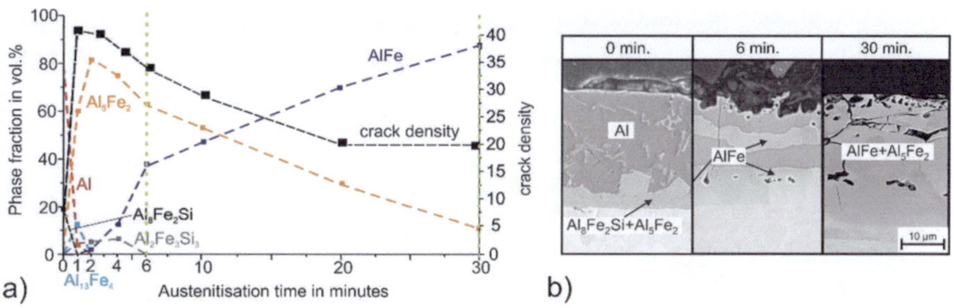

Fig. 12 Phase formation kinetics in the AlSi10Fe3 layer on 22MnB5 steel applied by hot-dipping during press hardening; **a** Quantitative description of the phases that form and their local assignment in image **b** [52]

can be welded despite a martensitic microstructure and high hardening in the heat-affected zone, the combination with the Al base material and the AlSiFe layer present on the press-hardening steels is still a challenge. The AlSiFe layer is applied to the press-hardening sheets by hot-dipping (layer thickness approx. 15–30 µm) to protect the sheets against oxidation during austenitization before the actual press hardening. They also promote paintability in subsequent process steps, as a primer is not required due to their porous morphology and pronounced surface topography. Regardless of the AlSiFe layer, joining steel and aluminum is ambitious because intermetallic phases form in the fusion zone (Fig. 13).

These intermetallic phases (IMP) are brittle and cause failure during the cooling phase from the joining heat, as the internal stresses that form cannot be reduced by plastic material deformation [53]. For this reason, joining is only possible if the layer of the IMP that forms does not exceed a thickness of 10 µm. There should also be a low crack density within the AlSiFe layer, as present cracks cause internal stress concentrations and reduce the load-bearing cross-section and, thus, the joining strength. With the knowledge of the phases that form in the AlSiFe layer during the austenitization of the press-hardened sheets, Windmann et al. discovered that it is the Al_5Fe_2 phase, in particular, that causes cracking. The correlation between the Al_5Fe_2 volume content that forms in the AlSiFe layer and the crack density is shown in Fig. 12. To reduce the crack density in the AlSiFe layer, [54, 55] options were considered to avoid the formation of the $Al_{13}Fe_4$ and Al_5Fe_2 phases by adapting the press hardening process or the alloy design of the AlSiFe layer (influence Si) and promote AlFe-formation. It was proven that thinner AlSiFe layers, a higher temperature and time during austenitization, as well as an adapted alloy concept, promote the diffusion of the elements Al and Fe and the formation of the AlFe phase according to the reaction $Al_{13}Fe_4 \rightarrow Al_5Fe_2 \rightarrow (Al_2Fe) \rightarrow AlFe$ [54].

In addition to explicitly influencing the phase formation present in the AlSiFe layer and the associated crack density, the joining of the Al or Mg workpiece to the AlSiFe layer must also be considered. During welding, the workpieces are reheated, whereby the renewed application of heat can cause the formation of undesirable brittle IMP at the steel (AlSiFe layer)-Al/Mg joining zone. In order

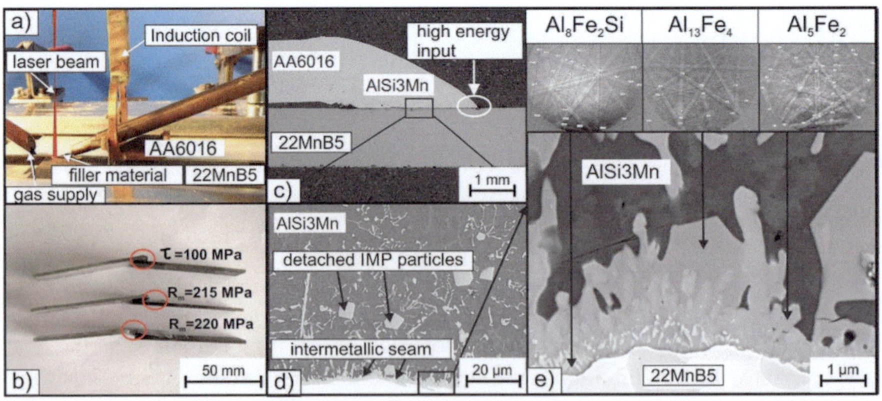

Fig. 13 Joining steel and aluminum using a combined soldering/welding process: **a** Structure of the welding process, **b** AA6016-22MnB5 joining connections and their strength, tested in heavy testing, **c–e** Formation of the microstructure at the joining interface [58]

to avoid this problem, several options are presented in the literature to enable low heat input during joining. In addition to the CMT (Cold Metal Transfer) process, hybrid laser welding with inductor-assisted heat input was considered in [56, 57]. Using different energy sources and locally targeted heat input, only enough heat could be introduced so that the Al/Mg filler material melts, but the steel substrate or the AlSiFe layer remains in the solid state. In combination with a sufficiently long length of the weld seam, high connection strengths could be achieved in the tensile shear test for the joint connections AlSi3Mn-22MnB5 ($\sigma = 215$–220 MPa) and AZ31-22MnB5 ($\sigma = 220$ MPa), with failure occurring in the heat-affected zone of the light metal.

4 Conclusion

Materials technology plays an important and essential role in collective efforts to achieve sustainability and climate protection. The development of lean alloys (with reduced usage of alloying elements) or dirty alloys (more resistant to contamination) is a key strategy for enhancing the recyclability of materials without compromising their properties. Improved material properties can significantly extend the service life of components susceptible to wear or corrosion during operation. It's important to determine if adding specific elements to improve resistance against corrosion (such as Cr and Mo) and wear (such as Cr, Mo, V, Nb, W, etc.) makes sense for each application, considering factors like criticality since these elements are lost when materials are removed. For instance, there's potential in developing new material concepts containing hard phases, leading to superior properties than traditional alloy concepts while using fewer raw materials, as demonstrated by B-alloyed tool steels. Similarly, suitable Stellite replacement alloys with a low content of critical elements (such as Co and W) could also be developed. However, understanding the material behavior in use and the required properties is crucial for material development.

Repurposing products at the end of their lifespan can lead to energy and production steps savings, as seen with circular knives. By repurposing, there's no need for complex product recycling, such as re-melting. Ideally, product design should be done across companies so that at the end of its lifespan, the product becomes an ideal semi-finished product for subsequent production by repurposing. However, questions about product return also need to be addressed.

The joining of steel and light metals (Al and Mg) is particularly relevant for automotive applications, as it allows the manufacturing of lightweight structures using high-strength steels and aluminum/magnesium alloys with low density. This joining of both materials is a key technology, and adjustments to both the material and manufacturing technology are necessary to achieve a Fe-Al/Mg connection.

Acknowledgements The results presented were gathered in various research projects carried out by the German Research Society (Project number 232707733), the Federal Ministry for Economic Affairs and Climate Action of Germany (BMWK; project no. 03X3567F), the Federal

Ministry of Education and Research (BMBF; project no. 03X3032, project no. 033R230A, project no. 03XP0025) and the European Regional Development Fund (EFRE; project no. EU-2-1-0005B). We thank for the funding granted.

References

1. Murphy, D. M., et al.: (2009), An observationally based energy balance for the Earth since 1950, *J. Geophys. Res.*, 114, D17107
2. Shah, I. H., et al.: Comprehensive review: Effects of climate change and greenhouse gases emission relevance to environmental stress on horticultural crops and management, J. of Environ. Man., Vol. 351, 2024, 119978
3. William F. Lamb, *et al.:* 2021 *Environ. Res. Lett.* 16 073005
4. van Vuuren, D.P., et al.: Stabilizing greenhouse gas concentrations at low levels: an assessment of reduction strategies and costs. Climatic Change 81, 119–159 (2007)
5. Mathieux, F., et al.: Critical Raw Materials and the Circular Economy—Background report. JRC Science-for-policy report, EUR 28832 EN, Publications Office of the European Union, Luxembourg, 2017
6. European Commission: Directorate-General for Internal Market, Industry, Entrepreneurship and SMEs, Grohol, M. and Veeh, C., *Study on the critical raw materials for the EU 2023—Final report*, Publications Office of the European Union, 2023, https://data.europa.eu/doi/10.2873/725585
7. Potting, J., et al.: (2017): Circular Economy: Measuring innovation in the product chain. Hg. v. PBL Netherlands Environmental Assessment Agency. PBL Netherlands Environmental Assessment Agency. The Hague, last checked on 23.01.2021
8. Van den Berg, M. R., and C. A. Bakker. "A product design framework for a circular economy." *Product lifetimes and the environment* (2015): 365–379
9. Barros, M., et al.: From Planned Obsolence to the Circular Economy in the Smartphone Industry; An Evolution of Strategies Emboided in Product Features, Proc. Of Des. Society, 2021, 1, 1607-1616
10. Ylä-Mella, J., et al.: 2022. "End-of-Use vs. End-of-Life: When Do Consumer Electronics Become Waste?" *Resources* 11, no. 2: 18
11. Delaney, E., et al.: The Investigation of Environmental Sustainability within Product Design, a critical review, Design Science, 2022, 8, e15
12. Dierk Raabe: *Chemical Reviews* **2023** *123* (5), 2436–2608
13. British geological survey: All the Metals We Mined in One Visualization (visualcapitalist.com)
14. Treibhausgasemissionen (THG) der deutschen Industrie 2017. Quelle: Eigene Darstellung gemäß Emissionsdaten des Umweltbundesamts (UBA), des Wuppertal Instituts sowie der Verbände VDZ und WV Stahl 2017
15. Merten, F. & Scholz, A. (2023). Metaanalyse zu Wasserstoffkosten und -bedarfen für die CO_2-neutrale Transformation. Wuppertal Institut. Meta-Analysis of the Costs of and Demand for Hydrogen in the Transformation to a Carbon-Neutral Economy (wupperinst.org)
16. L.D. Danny Harvey, A bottom-up assessment of recent (2016–20) energy use by the global iron and steel industry constrained to match a top-down (International Energy Agency) assessment, Energy, Vol. 293, 2024, 130675
17. Sáez-Guinoa, J., *et al.:* The effects of energy consumption of alumina production in the environmental impacts using LCA. *Int J Life Cycle Assess* **29**, 380–393 (2024)
18. Oda, J., et al.: Long-Term Global Availability of Steel Scrap. *Resour. Conserv. Recycl.* 2013, *81*, 81– 91
19. Oro R., et al.: Lean alloys in PM: From design to sintering performance. *Powder Metallurgy.* 2012; 55(4):294–301

20. Sun, X., et al.: Sources of greenhouse gas emission reductions in OECD countries: Composition or technique effects, Ecological Economics, Vol. 193, 2022, 107288
21. Awasthi, A., et al.: Chapter eleven—Sustainability and survivability in manufacturing sector, Editor(s): Kaushik Kumar, J. Paulo Davim, In Woodhead Publishing Reviews: Mechanical Engineering Series, Modern Manufacturing Processes, Woodhead Publishing, 2020, pp 205–219
22. Frogner, K., et al.: (2011). Industrial heating using energy efficient induction technology. In K. Frogner (Ed.), Proceedings of the 44th CIRP International Conference on Manufacturing Systems
23. Hegab, H., et al.: Design for sustainable additive manufacturing: A review, Sustainable Materials and Technologies, Vol. 35, 2023, e00576
24. Powell, D., et al.: Understanding powder degradation in metal AM to allow the upcycling of recycled powders, J. of Cleaner Production, Vol. 268, 2020, 122077
25. Wang, Z., et al.: Surface roughness evolution of 1Cr12Ni3MoVN alloy prepared by additive/subtractive hybrid manufacturing, Optics & Laser Tech., Vol. 181, 2025, 111743
26. Baqershahi, M. H., et al.: Design optimisation for hybrid metal additive manufacturing for sustainable construction, Engineering Structures, Vol. 301, 2024, 117355
27. Zum Gahr KH (1980) Fracture analysis of white cast iron. IJMR 71(2):103–109
28. Berns H (1995) Microstructural properties of wear-resistant alloys. Wear 181:271–279
29. Zhifu, H., et a.: Improving fracture toughness and hardness of Fe_2B in high boron white cast iron by chromium addition, Mat. & Design, Vol. 31, Issue 6,2 010, pp 3084–3089
30. Lentz, J., et al.: Solidification and phase formation of alloys in the hypoeutectic region of the Fe-C-B system, Acta Mat., Vol. 99, 2015, pp. 119–129
31. Lentz, J., et al.: Mechanism of the $Fe_3(B,C)$ and $Fe_{23}(C,B)_6$ solid-state transformation in the hypoeutectic region of the Fe-C-B system, Acta Mat., Vol. 119, 2016, pp. 80–91
32. Röttger, A., et al.: Boron-alloyed Fe-Cr-C-B tool steels—Thermodynamic calculations and experimental validation, Materials & Design, Vol. 88, 2015, pp. 420-429
33. Lentz, J., et al.: Hardness and modulus of Fe_2B, $Fe_3(C,B)$, and $Fe_{23}(C,B)_6$ borides and carbo-borides in the Fe-C-B system, Mat. Charact., Vol. 135, 2018, pp. 192–202
34. Lentz, J., et al.: Enhancement of hardness, modulus and fracture toughness of the tetragonal $(Fe,Cr)_2B$ and orthorhombic $(Cr,Fe)_2B$ phases with addition of Cr, Materials & Design, Volume 156, 2018, Pages 113–124
35. Lentz, J., et al.: (2020), Microstructures, Heat Treatment, and Properties of Boron-Alloyed Tool Steels. steel research int., 91: 1900416
36. Ahn, DG.: Hardfacing technologies for improvement of wear characteristics of hot working tools: A Review. *Int. J. Precis. Eng. Manuf.* **14**, 1271–1283 (2013)
37. Krell, J., et al.: Comprehensive investigation of the microstructure-property relationship of differently manufactured Co–Cr–C alloys at room and elevated temperature, Wear, Volumes 444–445, 2020, 203138
38. Raphael D., et al.: Conflict minerals and battery materials supply chains: A mapping review of responsible sourcing initiatives, The Extractive Industries and Society, Vol. 8, Issue 4,2021,100935
39. Winkelmann, H., *et al.:* Wear Mechanisms at High Temperatures. Part 3: Changes of the Wear Mechanism in the Continuous Impact Abrasion Test with Increasing Testing Temperature. *Tribol Lett* **37**, 419–429 (2010)
40. Varga, M.: High temperature abrasive wear of metallic materials, Wear, Volumes 376–377, Part A, 2017, Pages 443–451
41. Brackmann, L., *et al.:* Influence of a Partial Substitution of Co by Fe on the Phase Stability and Fatigue Behavior of a CoCrWC Hard Alloy at Room Temperature. *Metal Mater Trans A* **53**, 2708–2723 (2022)
42. Krell, J.: Entwicklung warmverschleißbeständiger Legierungskonzepte zur Substitution von Co-Basislegierungen. PhD diss., Bochum, Ruhr-Universität Bochum, 2020

43. Krell, J., et al.: Influence of precipitation hardening on the high-temperature sliding wear resistance of an aluminium alloyed iron-nickel base alloy. *Tribology International* 148 (2020): 106342
44. Letzig, D., et al.: Screening of NiAl-base Ni-Fe-Al alloys for structural high temperature applications and development of a new Ni-30Fe-10Al-Cr alloy. *Int.l J. of Mat. Research* 90, no. 9 (1999): 712–721
45. Tkaczyk, A. H., et al.: Sustainability evaluation of essential critical raw materials: cobalt, niobium, tungsten and rare earth elements. *J. of Physics D: Applied Physics* 51, no. 20 (2018): 203001
46. Mohr, A., et al.: Characterization of recycled TiC and its influence on the microstructural, tribological, and corrosion properties of a TiC-reinforced metal matrix composites. *J. of Com. Mat.* 51, no. 26 (2017): 3611–3621
47. Mohr, A., et al.: Adaptation of TiC Hard Particles Properties and Morphology in Metal Matrix Composites by Refractory Elements. *Key Eng. Mat.* 742 (2017): 99–105
48. Kronenberg, P., et al.: Repurpose–How to upgrade tools and save resources. *Cleaner Waste Systems* 6 (2023): 100114
49. Hagedorn, W., et al.: More than recycling–The potential of the circular economy shown by a case study of the metal working industry. *J. of Cleaner Production* 377 (2022): 134439
50. Großwendt, F., et al.: A novel powder-metallurgical eco-friendly recycling process for tool steel grinding sludge. *J. of Cleaner Production* 392 (2023): 136329
51. Jäger, S., et al.: Potential of the recycling of grinding sludge by various powder metallurgical processes. *Procedia CIRP* 104 (2021): 893–899
52. Windmann, M., et al.: Phase formation at the interface between a boron alloyed steel substrate and an Al-rich coating. *Surf. and Coat. Techn.* 226 (2013): 130–139
53. Windmann, M., et al.: "Mechanical properties of Al_XFe_Y intermetallics in Al-base coatings on steel 22MnB5 and resulting wear mechanisms at press-hardening tool steel surfaces." *Surf. and Coat. Tech.* 321 (2017): 321–327
54. Windmann, M., et al.: "Formation of intermetallic phases in Al-coated hot-stamped 22MnB5 sheets in terms of coating thickness and Si content." *Surf. and Coat. Tech.* 246 (2014): 17–25
55. Windmann, M., et al.: "Removal of oxides and brittle coating constituents at the surface of coated hot-forming 22MnB5 steel for a laser welding process with aluminum alloys." *Surf. and Coat. Tech.* 285 (2016): 153–160
56. Windmann, M., et al.: "Laser beam welding of magnesium to coated high-strength steel 22MnB5." *Int. J. of Adv. Manu. Tech.* 87 (2016): 3149–3156
57. Windmann, M., et al.: "Laser beam welding of aluminum to Al-base coated high-strength steel 22MnB5." *J. of Mat. Proc. Tech.* 217 (2015): 88–95
58. Windmann, M., et al.: "Microstructure and mechanical properties of the heat-affected zone in laser-welded/brazed steel 22MnB5–AA6016 aluminum/AZ31 magnesium alloy." *J. of Mat. Proc. Tech.* 247 (2017): 11–18

Dustiness of Powders—Characterization, Causes, Change

Eberhard Schmidt

Contents

Abstract

The property of a powder to release fine particles from the bulk under certain stresses can be called dustiness. The release of dust into the surrounding air is usually undesirable for several reasons. On the one hand, it is associated with risks to the environment, fire protection and the health of employees in production or application. On the other hand, it can be associated with a loss of valuable materials. The dustiness can be measured qualitatively and often also quantitatively using various methods described below. These differ mostly in the way the powders are stressed. The dustiness is always determined by both the powder properties and the type of stress; it is not an intrinsic property. This is also clear from the treatment of the subject in standards and directives. And it is reflected in the treatment of dustiness in science and research. Several examples of completed and ongoing work are presented.

Keywords

Dustiness · Particulate Matter · Environment Health and Safety Risks

E. Schmidt (✉)
Institute of Particle Technology, BUW, Wuppertal, Germany
e-mail: eberhard@uni-wuppertal.de

S. Bracke and R. Pieper (eds.), *Safety Engineering*,
https://doi.org/10.1007/978-3-658-47473-7_10

1 Dustiness in General

The dusting tendency of a disperse solid can be understood as the property (in the sense of a sum parameter) of releasing a particle fraction of a specific quantity and size distribution in a gaseous environment—at least briefly—in a gaseous state when handled in a certain way. This release of dust is generally undesirable, as it is associated with a loss of material (valuable resources or products) and often with exposure or danger to the people involved and pollution of the working environment or the environment; see for example Fig. 1.

The dusting tendency is therefore an important, but previously neglected, product property in terms of application technology, which can change along the process route. Comminution, agglomeration, classification or mixing of the solids can for example be involved. Initial property functions that describe this dusting tendency—which is often also variable over time and thus integrated into the dynamics of the processes—as a function of the distributions of particle size, particle shape and particle interaction of the disperse systems during certain handling are being further developed as part of completed [2…12] and ongoing research projects.

2 Characterization of Dustiness

2.1 Free Fall and Impact in Still Air—Dusting in Bulk

A powder sample of 30 g falls in a tube approx. 0.5 m into a cubic chamber with an edge length of about 20 cm and hits the floor generating a transient dust cloud (see modified "DustView" in Fig. 2). The course of the changing light intensity

Fig. 1 Dust release during construction work [1]

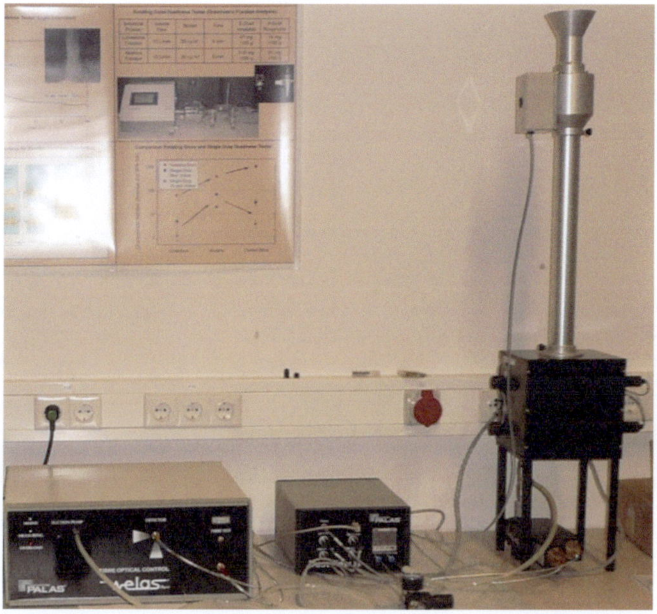

Fig. 2 Apparatus for analyzing the dusting tendency during drop and impact, equipped with additional optical measurement technology for on-line particle size analysis [1, 13]

of a laser beam recorded over 40 s is the primary signal. A maximum value and a 30-s-value are obtained as characteristic values. Conclusions about particle size distribution of liberated dust would be possible, however, are actually not used, due to measurement inaccuracy. Using two extinction measurement systems, the temporal development of the resulting dust cloud in the largely sealed chamber can be tracked. In addition, samples can be taken at different points to analyze the particle size.

2.2 Sliding and Falling Movement in a Rotating Drum—Dusting During Mixing

The "Dustmeter" consists of a control unit, a metal drum of volume 2 L, a glass bottle of volume 1 L, and selected dust collection units (see Fig. 3). The sample—usually 100 g—is filled into the air flow-through drum. The rotating drum generates dust. The released dust is transferred to the glass bottle and collected in the unit behind. The glass bottle acts as a pre-separator to collect large particles. The smaller ones are collected on the glass fiber filter inside the separation unit. Airflow, rotational frequency, and stressing time are freely adjustable or can be chosen according to technical standards. The ratio of the amount of dust measured by a separation unit (here: cyclone plus filter) and the sample mass itself, is the dimensionless dust value.

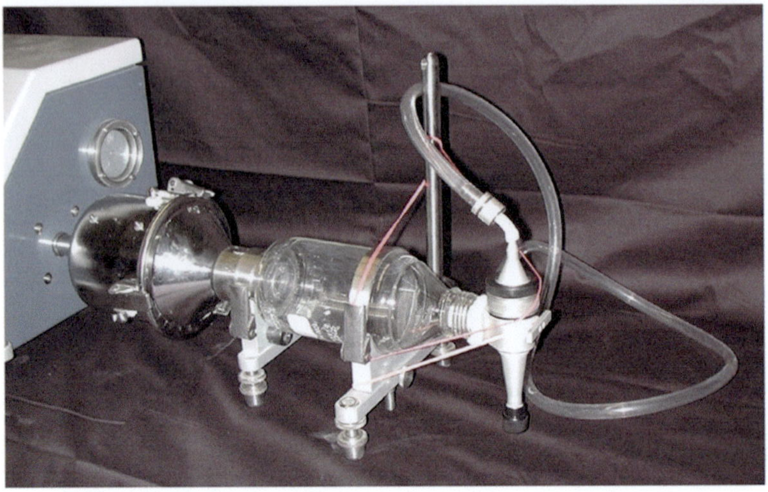

Fig. 3 Metal, slowly rotating, air-flowing drum, here with glass pre-separator and downstream flow-fractionated main separation in a cyclone with protective filter [1, 13, 14]

The conversion of the drum for the technical realization of special investigations has been successfully carried out several times. To characterize nanoparticles, for example, the aerosol is fed to a mobility analyzer via a modified buffer (cf. Fig. 6). At the same time, raw data from old investigations are analyzed and compared from the perspective of the property function.

2.3 Dispersion by Compressed Air Blast—Dusting During Dispersion

Less than 10 mg of the powder sample is sucked in by means of a propellant jet and dosed into a largely sealed chamber in dispersed form (see Fig. 4). Dust is separated from this chamber in size fractions using special sampling probes and thus made accessible for gravimetric analysis. An extension of the apparatus for the connection of on-line particle measurement technology (light scattering and mobility analyzers) is planned.

2.4 Whirling up in the Air Flow—Dusting with Whirling up

A fan directs air of a defined volume flow from the inner tube onto the dusty surface, where it is deflected and transported upwards between the inner and outer tubes together with the whirled up dust. By means of sampling, a partial flow is fed to the scattered light spectrometer for analysis. The main flow is deflected back into the inner tube, creating a loop flow (see Fig. 5).

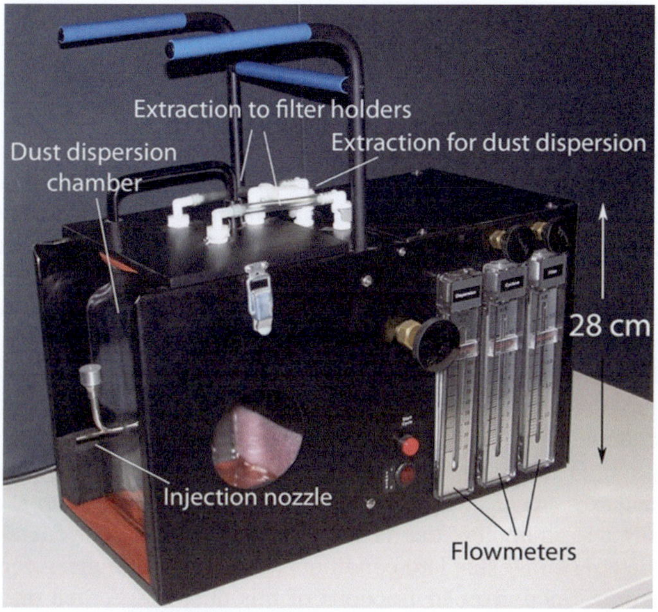

Fig. 4 "UNC Dustiness Tester", here for the determination of inhalable and respirable fine dust fractions of smallest powder quantities [1, 13, 15]

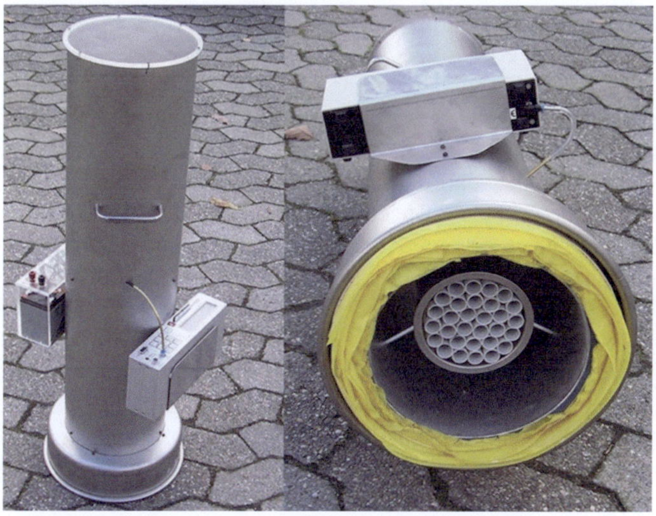

Fig. 5 Dust swirl double tube for determining the dusting potential of contaminated surfaces, here with aerosol spectrometer for on-line analysis [1, 13, 16]

3 Dustiness in Standardization

DIN EN 15051 [17] specifies test procedures that make a classification of the dusting behaviour of solid bulk materials in health-relevant fractions. This classification serves to inform users (for example producers, manufacturers, industrial hygiene professionals, and employees) about the potential of dust emissions that occur when handling or preparing bulk material in the workplace. The classification provides the manufacturer of the bulk material with information that can be useful to improve the products. Also, persons handling bulk material can use the classification to estimate the influence of prior treatments and to select products that generate less dust, to the extent that they are available.

Part 1 of the series of standards specifies the environmental conditions, the handling of samples and the analysis procedure as well as the procedure to calculate and present the results. A justification is provided for why more than one procedure is required. Help is provided for the selection of the procedure. Part 2 of the series of standards specifies the test procedure with a rotating drum as well as the corresponding test equipment that is used for the reproducible generation of dust from bulk material under standard conditions; in addition, the measurement, of the inhalable, thoracic and alveolar fractions of this dust is specified or, as the case may be, reference is made to existing European standards. Part 3 of the series of standards analogously specifies the test procedure with a continuous drop as well as the corresponding test equipment.

The standard was prepared in the CEN/TC 137 "Assessment of workplace exposure to chemical and biological agents" (Secretariat: DIN). The responsible German standardisation committee is the research group NA 095-03-01-01 AK "Dust" of the working committee "Measurement strategies and requirements for test methods" of the DIN standards committee safety design principles (NASG).

DIN 33897 [18] specifies the foundations for routine procedures that can be applied in daily lab practice alternatively to the reference test procedure specified in DIN EN 15051 for the reproducible determination of the dusting procedure of bulk material for the purpose of occupational health and safety. The part 1 establishes the framework for the content-related arrangement of individual procedures that are specified in additional parts of this series of standards and respectively represents concrete examples for the implementation of the specifications of this part of DIN 33897.

Part 3 specifies a routine procedure for the practical simulation of dust generation processes in the bakery trade. However, the procedure can also be applied to determine the dusting behaviour of other powdery substances. For this, a powder disperser generates a dust-air mixture and an air-bulk source current is defined that corresponds to the technical conditions of a bakery in a reproducible way. The determination of the dusting procedure is based on the measurement of the corresponding dust fractions in the dust-air mixture with a constant volume flow of 4 m^3/h. The dusting of the test material takes place in a dust channel with specified dimensions. The masses on the filters of the collection heads that were gravimetrically determined after the end of the measuring duration in relation to the dusted mass give the dusting parameters for the inhalable and respirable (and possibly also thoracic) dust fraction.

The routine procedures described in part 4 of the standard uses movement in a rotating drum lasting multiple minutes to simulate dust generation processes that include both continuous fall processes as well as abrasion processes as they occur during handling (for example emptying of sacks, manual filling, or transferring or conveying/transport) of powdery or granulated bulk material and during which dust is released during falling, through whirling up on impact, and/or through mechanical abrasion of the particles. The procedure also provides parameters for the dusting behaviour of bulk material for the inhalable and respirable dust fraction. In routine operation, it can be applied to determine the dusting behaviour of all powdery or granulated substances with the exception of such powdery substances that tend to granule or felt formation through continuous movement in a rotating drum without the addition of auxiliary agents.

The procedures described in DIN 55992 [19] serve to determine the dust portion generated under process or handling conditions of solids, in particular of powdery or granulated pigments and fillers. The measured values for dust generation obtained by testing using this standard do not provide an indication about the total dust contained in the tested products. Rather, the determined measured value for the dust generation represents a parameter for the assessment of the dust behaviour of pigments and fillers.

The rotation procedure described in part 1 was tried in a ring trial in which six test facilities took part and in which seven different pigments were tested. During this trial it was found that the determined measured value for dust generation depends on the dominant humidity and temperature at the test location. The variation coefficients of individual measuring laboratories differ depending on the properties of the tested substance in a range from 1.65% to 22.75% (with a mean of approx. 10%). The variation coefficients of the results of different measuring points range from 9.1% to 38.4%. The values determined according to this standard are therefore to be seen preferentially as the relative values of a test facility.

The VDI 2263 guideline, part 9 [20] on the subject of "Dust fires and dust explosions—Hazards—assessment—protective measures—Determination of dustiness of bulk materials" describes a procedure to determine the dusting behaviour of flammable dusts, with which practical dust generation processes (conveying, delivering, filling, loading, unloading, etc.) are simulated. The described procedure thus provides a parameter value for the dusting behaviour of dusts that can be consulted primarily for the assessment of dust explosion hazards.

Fugitive dust emissions occur to a significant extent during the storage, handling and transportation of bulk materials. The guideline VDI 3790, part 3 [21] refers to such fugitive dust emissions. The aim of the guideline is to estimate emission factors for this. The guideline does not allow the definition of emission limits.

The list of standards and guidelines could certainly be expanded, especially if other organizations and other industries are taken into account. In any case, one must be aware that the measured values for dust must always be considered together with the appliance, or more precisely with the type of stress in the appliance. Further information on this can be found in [22…30].

4 Dustiness in Science and Research

4.1 Release of Nanoparticles

The work of Hamelmann [2] is the first to provide a summarized overview of the influence of dust emissions in the areas of occupational safety and environmental protection, as well as fire and explosion protection and health protection. It summarizes the technical methods used to determine dust emissions from the past to the present and analyses them in terms of their sample exposure. Previous work has always been limited to the mere presentation of the methods used. The further development of a proven method for dust measurement in order to be able to detect workplace-relevant dust fractions in a practical manner represents significant scientific progress in this work.

The last part of the thesis describes a new model for mapping the dust behavior of nanodisperse materials, which was developed as part of this work and not only takes into account the requirements for the representation of ultrafine particles, but also presents a new size-dependent representation of dust behavior. The scientific progress resulting from the combination of the method of investigation and the chosen representation is to be able to prove that significant dust release occurs during the handling of nanodisperse material systems. In previous studies, no attention was paid to the release of dust during the handling of nanodisperse material systems. This was always justified by the relative increase in the van der Waals adhesion force with decreasing particle diameters.

The two dustiness testers "DustView" and "Dustmeter" were used to characterize a special fraction of fumed silica, see Fig. 6.

The buffer container has a volume of 35 L, the bag inside of 20 L. For each test a new bag was used; to get 15 L of aerosol the tests last 180 s for the low gas flow rate of 5 L per minute and 45 s for the high gas flow rate of 20 L per minute. Each bag allows 5 SMPS-measurements after being filled.

Figure 7 shows that alongside the differences in the total number concentration, also significant differences in the size distribution of the released particles can occur, depending on how the powder is subjected to a stress. According to the used particle measurement techniques, for small particles the mobility equivalent diameter of the SMPS system and for large particles the light scattering equivalent diameter of the OPC system is depicted. This intrinsic particle measurement problem has no effect on the general conclusions drawn. Total measurement range here: 30 nm to 45 μm.

When the powder is dropped, the rise of the cumulative number distribution with particle size is uniform; however, the rotation produces a bimodal distribution. This is even more evident in the smoothed distribution density by number given by Eq. 1 and shown in Fig. 8.

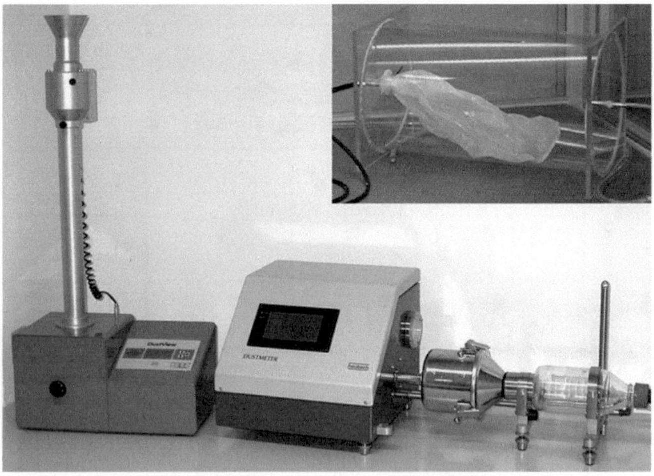

Fig. 6 Single drop apparatus ("DustView", left), rotational slip and drop measuring device ("Dustmeter", bottom right), and buffer container (self-made, top right) for the quantitative characterization of the ability of a powder (fumed silica) to produce fine airborne dust [31, 32]

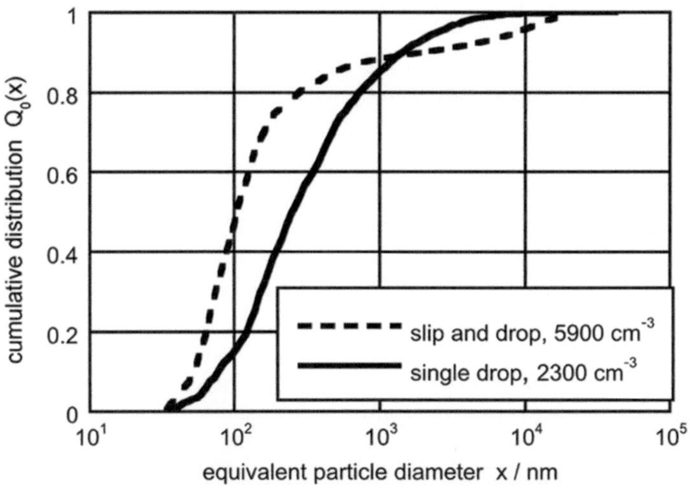

Fig. 7 Cumulative size distribution by number of released particles when fumed silica powder subjected to a rotational movement in a drum or when dropped in a box [31, 32]

$$q_0 * (x) = q_0(x)x = dQ_0(x)/d(\ln x) \tag{1}$$

There is a particularly large amount of fine (in the range of around 100 nm) though also comparatively large (in the region of around 10 μm) particles found in the dust released. The bimodal distribution differs significantly compared to the

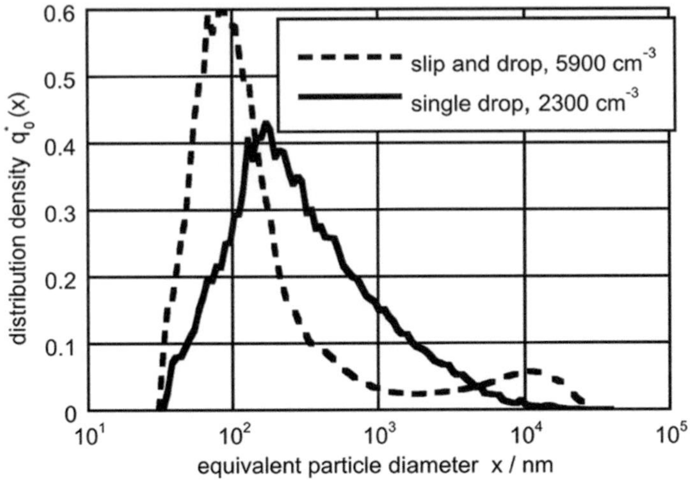

Fig. 8 Distribution density by number of released particles when fumed silica powder subjected to a rotational movement in a drum or when dropped in a box (compare Fig. 7) [31, 32]

monomodal distribution found in the case of single drop. This measurement result also makes clear again that the characterization of the ability of powders to produce fine airborne dust depends on the stress type. Knowledge of the size of the primary particles and their state in respect of aggregation and agglomerate alone is insufficient here. Together with time-varying stress functions (given by equipment properties) and strength functions (given by material properties), the transient processes of dustiness can be predicted quantitatively.

It was shown that it is possible to reliably characterize powder in respect of its ability to release (nano-scaled) particles when handled. The large number of size classes makes it possible, if required, to focus on parts of the entire particle size range from 5 nm to 45 μm. However, the results are only relevant to occupational health and safety if the type of stress to which the powder is subjected in the test device corresponds to the stress when handled in the real workplace. Recognized methods for the adapted selection of device still have to be found.

4.2 Comparison of Measuring Devices

The work of Bach [5] shows the limits to the applicability of specific methods for measuring the dustiness of powders and bulk solids. It thus contributes to a better understanding of the physical principles of dustiness and demonstrates the applicability of the available techniques. By systematically varying the parameters drop height, mass of the charged material and particle size distribution of the charged material, the relationship between these parameters and the extinction result was quantified.

The work also shows a new approach to derive information about the inhalable particle size fraction from the results of a single-drop method. For the first time, it offers the possibility of using a simple measurement technique to determine indicative but reproducible results for the proportion of inhalable dust in the total dust released.

It must also be noted that, based on the input data on which this work is based (22 powders), it is not possible to predict the dusting tendency with the aid of material-specific properties such as solid density, moisture, bulk density or the particle size distribution of the starting material. The multiple linear regression method from the field of multivariate data analysis was selected for processing the input parameters.

Table 1 shows examples of data for twelve different materials whose dusting tendency was quantified using different methods. Based on this, the materials were divided into dusting classes, which now only allow qualitative statements to be made.

To extend the usability of a rotation method for measuring the dusting tendency, a concrete proposal is made for the derivation of dusting classes from the dust dimensions when measuring very coarse-grained bulk materials. This allows a weighting factor required in VDI Guideline 3790, Sheet 3 for the calculation of emission factors for handling and transportation processes with coarse-grained bulk materials to be determined in a reproducible and uniform manner [27].

Table 1 Classifications for the inhalable dust fraction according to three methods [30]

Test material	Reference test method B, inhalable fraction	Heubach Dustmeter, 12 test materials, inhalable fraction	Palas Dustview, 11 test materials, inhalable fraction
Aluminium hydroxide	Moderate	Low	Low
Aluminium oxide	Low	Moderate	Moderate
Barium sulphate A	Low	Low	Low
Barium sulphate B	Very low	Very low	Low
Bentonite A	High	Moderate	
Bentonite B	High	High	Moderate
Silica dust	Moderate	Moderate	Moderate
Talc A	Moderate	Moderate	Moderate
Talc B	Moderate	Moderate	Moderate
Calcium carbonate A	Low	Moderate	Moderate
Calcium carbonate B	Low	Moderate	Low
Wheat flour	Moderate	Low	Low

4.3 Dustiness of Powder Mixtures

The work of Parey [7] presents, applies and discusses a model for evaluating the dusting tendency of powders and binary powder mixtures based on their material and disperse properties and uses model calculations to determine the dusting tendency of different powders. For binary mixtures of two materially different components, the next step is to predict the composition of the resulting dust.

For the model calculations and comparative experimental investigations, powders are used whose properties enable a wide range of dustable fractions as well as adhesive force situations between materially and dispersively dissimilar particle fractions. The model is thus tested for a particle size range from a few micrometers to a few hundred micrometers and the real dusting behavior of the powders and powder mixtures is shown in practice.

The shape and the general appearance of the utilized powders are shown by SEM pictures given in Fig. 9. Thereby the difference in shape is obvious: while Eskal shows relatively round geometrics with rugged surface, Korund is predominantly shaped angular with nibs and partial smooth surfaces. Table 2 gives data for eight fractions analyzed. Finally, the dustiness measurement results for the binary inter-modal mixtures containing Eskal and Korund each are depicted in Fig. 10.

Clearly, this investigation requires the differentiated consideration of both the overall dustiness and the released dust's composition. Considering the four different sized powders of each material, a 16 elements matrix occurs in which the diagonal again is the representation of the intramodal mixture special case. Figure 10 presents the related investigation results, where the heights of the columns represent the overall dustiness and the different colored columns' sections correspond to the percentage of respective component in the collected dust (see Fig. 9).

Table 2 Raw materials (Eskal E and Korund K fractions) designation and characteristics [7]

Indicator	E 10	K 10	E 20	K 20	E 60	K 60	E 80	K 80
Mass median value / µm	12.3	12.4	23.2	23.7	64.0	67.2	87.3	90.4
$x_{10,3}$ / µm	7.3	7.6	16.3	15.6	44.8	45.5	63.7	60.5
$x_{90,3}$ / µm	19.4	18.9	32.3	34.6	91.0	97.0	125.6	139.7
Standard deviation of mass PSD/ µm	11.8	4.5	6.4	7.8	19.7	21.8	27.0	34.1
Density / $g \cdot cm^{-3}$	2.61	3.94	2.61	3.94	2.61	3.94	2.61	3.94
Hardness (Mohs)	3	9	3	9	3	9	3	9

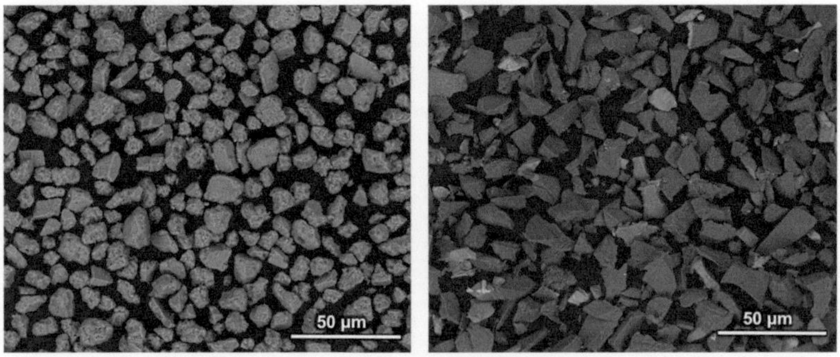

Fig. 9 SEM picture of Eskal $CaCO_3$ (left) and Korund Al_2O_3 (right) [7]

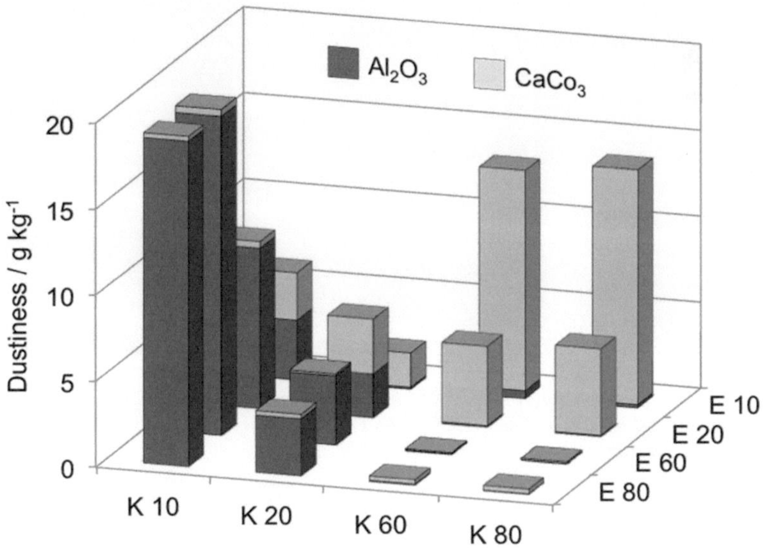

Fig. 10 Dustiness and dust composition from Al_2O_3 and $CaCO_3$ mixtures [7, 33]

4.4 Forecast Functions

In the work of Londershausen [8, 34, 35], the dusting tendency of nine different materials as depicted in Fig. 11 was investigated using three different test methods (drop apparatus, rotation apparatus, UNC Dustiness Tester). Care was taken in the selection process to ensure, that the selected test materials will cover a wide range of material- and distribution-specific properties.

Fig. 11 Scanning electron micrographs of nine test materials: Arizona test dust A1 to A4, Construction site test dust mix (Bau), ECE R-45 (P1), KSL 11,047 (P2), Pural NF (F1), Ulmer Weiss XMF (F2) [34, 35]

For each material, sorption isotherms for water were derived from experiments with a climatic test chamber. By varying environmental parameters such as relative humidity, air pressure and moisture content, their influence on the dusting tendency was investigated. Based on the results obtained and taking into account the respective measurement methodology and the geometry of the equipment, characteristic limiting velocities for the movement behavior of the particles were determined. Based on these limit velocities, limit particle diameters were derived for the relevant environmental parameters.

Based on these results, characteristic values for the dusting tendency were determined, taking into account distribution and material-specific parameters. With the help of these characteristic values, it was possible to establish empirical prediction functions for the test methods used and the environmental parameters investigated.

The dustiness of a disperse solid can be understood as a property, which when handled in a gaseous environment, behaves similar to an aerosol, releasing the respective particle fraction of given quantity and size distribution. The dustiness is therefore a product property, which might change along the process path, for example through comminution, agglomeration, classification or mixing of solids involved.

Property functions which describe time variable dustiness integrated in dynamic processes as a function of the distribution of particle size, particle shape, and particle interaction during a certain handling, were determined as part of this project. For this purpose, experiments with laboratory equipment such as "free fall in still air", "moving in a rotating drum", "dispersion, pressure surges method", or "airflow dispersion" were performed at well defined boundary conditions and physically based models were established. The prediction functions were successfully implemented in the flow sheet simulation DYSSOL [9]. Together with the description of time-dependent changes of the related strain-functions (apparatus properties) and rigidity-functions (material properties) this approach will help to better predict transient processes of dustiness in future.

For each testing apparatus several forecast functions were developed, taking into account relative humidity, particle distribution parameters as mass median $x_{50,3}$ and Sauter diameter d_{32}. For example, in case of the single drop method prognosis functions the two attenuations of light Emax and E30 are predicted by equations using the apparatus specific calculation factor KSD as shown in Eq. (2), where e is bulk porosity; d_{32} is Sauter diameter; $x_{i,2}$ is a percentile of the particle size distribution by area.

$$K_{SD} = (2/3) \cdot (d_{32}/\mu m) \cdot (\varepsilon/(1-\varepsilon)) \cdot \left(x_{50,2} - x_{10,2}\right) / x_{50,2} \quad (2)$$

Releasing particles does not only represent loss of material and thus a cost factor but also a potential threat to human health as well as to the environment. Because of this it may be relevant to know the dust formation tendency of a simulated substance for classification issues according to the European Chemicals Regulation REACH (Registration, Evaluation, Authorization and Restrictions of Chemicals).

In DYSSOL dynamic flow sheet [9], dust formation tendency prognosis based on single drop, rotating drum and UNC dustiness tester are part of simulation results. The dustiness forecast module is an integral part of the simulation core (cf. Fig. 12), connected to relevant simulation results obtained from the product stream leaving the simulated units. The prediction functions offer development potential beyond their use in DYSSOL, for instance in the development of a more efficient particle reservoir model using Exner equation based sediment mass balance approach. Such a reservoir model may improve the performance of the published fractionate release model [36] and other geoscientific dust release models.

Another application of such a reservoir model may be in further development of the adhesion-based assessment of the dustiness of powders and binary powder mixtures [33]. Irrespective of such development options further experimental research is needed in case of fractionate release of particles from strained reservoirs as for instance powders grinded with grinding aids.

Another approach to the development of forecasting functions is proposed by Woschny et al. [11, 12]. In the end, the investigations will lead to a so named Reference Test Bulk Material (RTBM). Figure 13 shows the desirable structure of the RTBM with its main components and their designations [37].

The production process of the reference material blends is based on the single step using a 3D shaker mixer bringing the fractions in contact. Here, only the duration of mixing can be varied. Consequently, the duration is the only factor influencing the structure of the powder layer on the surface of the metallic spheres. The appropriate mixing time is determined using the images from the reflected light microscope. By means of the comparison of the powder layer using a blend of the metallic spheres and a limestone fraction (Esk15) with a powder mass fraction of 1% powder, the appropriate time was identified. The shortest duration of mixing was 1 min, the longest 10 min and the third duration was 5 min.

In Fig. 14, for each duration a single bulk particle is shown representatively. The bulk particles were taken from a blend of 100 g in total. The three images reveal that a certain time of mixing is necessary to cover the sphere with a powder

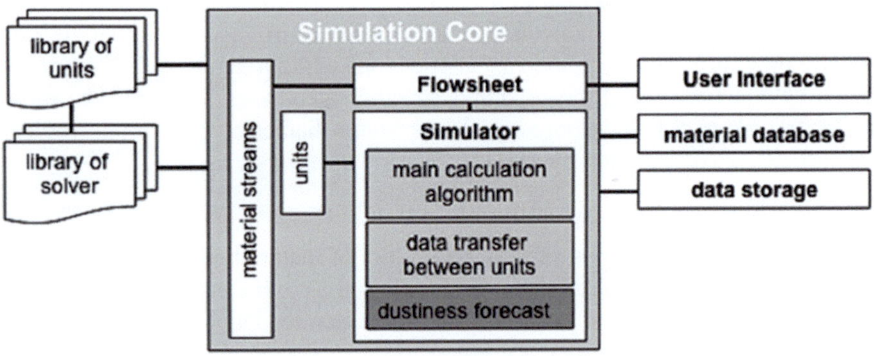

Fig. 12 DYSSOL flowsheet simulation schematics [9]

Fig. 13 Composition of the Reference Test Bulk Material (RTBM) [37]

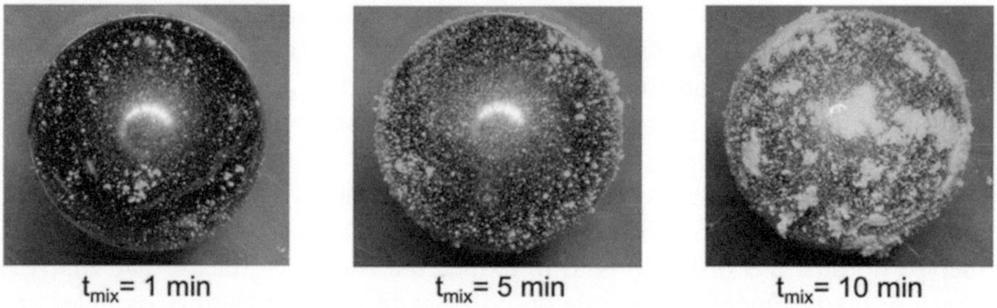

Fig. 14 Metal sphere attached with limestone particles shown after three various mixing times t_{mix} of 1, 5 and 10 min [12]

layer. After a mixing time of one minute, the sphere surface is not completely covered. Further, the increase of the duration of mixing reveals the effect of the formation of agglomerated layers on the surface. Consequently, the higher the mixing time the higher the rate of agglomeration. The duration of five minutes is a compromise of both effects.

4.5 Fractional Release Rate

The subject of the analysis are two model systems, which are presented in [6, 36]. In system A, a powder mass $m_{p,tot}$ (t) depending on time t with a given particle size distribution $q_{3,p}(x,t)$ is put under a stress in such a way that there is a particle mass $m_{ae,tot}$ (t) (or, as the case may be $dm_{ae,tot}$ (x,t) given an incremental analysis based on particle size x) depending on time within the closed system boundary of the gas volume $V(t)$; this results in the particle mass volume concentration $c_{Rm,tot}$ (t). In system B, the volume occupied by the powder is subjected to a flow of gas, so that a change of the particle mass transport concentration $c_{Tm,tot}$ (t) in the gas volume flow $V^{\cdot}(t)$ between inlet and outlet becomes possible. The entire particle mass flow in the aerosol is described by $m^{\cdot}_{ae,tot}$ (t), the incremental particle mass flow is described by $dm^{\cdot}_{ae}(x,t)$. A possible dependency of time t and the particle size x must be taken into account. The pressure, the temperature, the relative gas humidity and the absolute product humidity can initially be assumed to be unchangeable.

The concentrations can be indicated with regard to the particle mass or the particle number, both for the total quantity as well as incrementally for the individual particle sizes (classes). In general, the following sizes serve to describe the particles in the aerosol: Total and incremental particle masses—or particle number concentrations $c_{Rm,tot}$ (t), $dc_{Rm}(x,t)$, $c_{Rn,tot}$ (t), $dc_{Rn}(x,t)$ with regard to the space (system A) and the transport (system B) $c_{Tm,tot}$ (t), $dc_{Tm}(x,t)$, $c_{Tn,tot}(t)$, $dc_{Tn}(x,t)$.

The linking of the quantities (number/mass) with the concentrations (volume / transport) takes place via the fractional release rate $R(x,t)$ which is dependent on particle size and time and the volume or the volume flow, as the case may be. An integration weighted with the particle size distributions delivers the total release rates, which are only dependent on time, with regard to particle number $R_{0,tot}(t)$ or, as the case may be, the dust mass $R_{3,tot}(t)$. The time-dependent particle size distributions are presented as follows for the aerosol or powder quantity and in relation to mass: $q_{0,ae}(x,t)$, $q_{3,ae}(x,t)$, $q_{0,p}(x,t)$, $q_{3,p}(x,t)$.

The respective equation systems are set up for the two variations A and B to describe the indicated concentrations as a function of the particle size and the time. To do this, the first step is to introduce the fractional release rate $R(x,t)$ as a given property function. This rate describes with what probability particles of a given size can pass from the powder into the gas. Together with the actually present powder mass and their particle size distribution, the actual quantity passing from powder to gas can be calculated as a function of time, depending on the particle size. With this selective process, the particle size distribution of the original will change; this dynamic reaction must be analysed according to the formulation of the time-dependency.

For the case of model system A and the analysis of the particle mass, the following Eqs. (3) and (4) can be used, for example:

$$m_{ae,tot}(t) = \int_{x=0}^{\infty} dm_{ae}(x,t) \tag{3}$$

$$c_{Rm,tot}(t) = \frac{m_{ae,tot}(t)}{V(t)} \tag{4}$$

It should be noted that the integrand is the added particle mass of size x released over time, which means it is to be determined using an integration itself. Analogously, the equations are to be set up for the particle number and the model system B, which we omit here for reasons of brevity.

The fractional release rate R(x,t) indicates the probability that particles of a specific size x can be released in the time dt. The dust quantity per particle size actually released in dt is obtained by multiplying it with the quantity present in the powder. For example, in relation to the mass, this results in the ratio presented in Eq. (5) for the momentarily completely released share of the mass:

$$R_{3,tot}(t) = \int_{x=0}^{\infty} R(x,t) \cdot q_{3,p}(x,t) \cdot dx \tag{5}$$

The fractional release rate can be calculated from functions for the description of the binding and separating forces F_h or F_t on particles of a given size x at the time t in a powder according to Eq. (6) that can initially be assumed as known.

$$R(x,t) = \mathrm{f}(F_t(x,t), F_h(x,t)) \tag{6}$$

Since these functions for the description of particle interactions have to be assumed to be temporally flexible and must have been able to be distributed with regard to individual particle sizes, a calculation of $R(x,t)$ is certainly not trivial. Special types of distribution functions are to be developed which on the one hand must be able to depict real processes as a model with sufficient accuracy, but on the other hand must also make a stable solution of the equation systems described above possible.

For three variations the temporal progression of the mass-dependent particle volume concentration is presented in Fig. 15 as an example. Since neither sink terms (e.g. due to sedimentation) nor interactive terms (e.g. due to agglomeration) are implemented in this simulation, concentration and particle size distribution remain constant for given times. If the binding force is increased by a factor of 2 while at the same time the time-dependency is reduced (circular symbols), initially a little less dust is released; after this the concentration will not be constant until times $t > 20$ s have been reached. An exclusive increase in the adhesive force on the other hand leads to a lower plateau level being reached in a shorter time (diamond symbols).

The presented introduction and application of a fractional release rate $R(x,t)$ is to be developed further. To achieve this, dust release trials should be carried out in lab facilities under meticulously defined conditions to ensure that all relevant values can be measured and recorded accurately. Using special conversions, different model systems can be analysed. The pressure, the temperature, the relative gas humidity and the absolute product moisture must be recorded as changeable values and must be analysed regarding their effect on the property function or the fractional release rates.

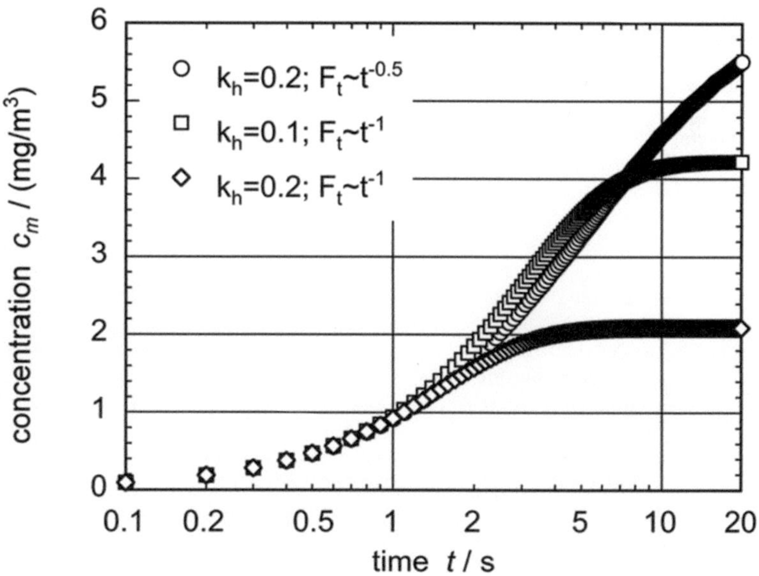

Fig. 15 Mass-dependent particle volume concentration in the aerosol for different dependencies of the binding and separating forces according to Eq. (6) [36]

References

1. Schmidt, E.: Staubfreisetzung bei der Handhabung von Pulvern und Schüttgütern. Sicher ist Sicher 68 (11), 492–497 (2017).
2. Hamelmann, F.: Untersuchung zum Staubungsverhalten bei der Handhabung disperser Stoffsysteme und die Weiterentwicklung etablierter Messverfahren zur Bestimmung der Staubfreisetzung. Shaker, Aachen (2005).
3. Reznik, G., Klenk, U., Schmidt, E.: Untersuchungen zur Staubungsneigung von Braunkohle unterschiedlicher Feuchte. Chem.-Ing.-Tech. 78 (12), 1885–1889 (2006).
4. Reznik, G., Klenk, U., Schmidt, E.: Messung der Staubungsneigung von Braunkohle als Funktion der Feuchte. GRdL 66 (7/8), 309–310 (2006).
5. Bach, S.: Beurteilung des Staubungsverhaltens von Pulvern und Schüttgütern. Shaker, Aachen (2013).
6. Schmidt, E.: Fraktionsfreisetzungsgrad – Ein neuer Ansatz zur Quantifizierung der Staubungsneigung von Pulvern. Chem.-Ing.-Tech. 87 (5), 638–643 (2015).
7. Parey, M.: Ein Modell zur haftkraftbasierten Bewertung der Staubungsneigung von Pulvern und binären Pulvermischungen. Shaker, Aachen (2017).
8. Londershausen, T.: Entwicklung von Prognosefunktionen zur Abschätzung der Staubungsneigung von trockenen und feuchten Schüttgütern. Shaker, Aachen (2018).
9. Vaupel, K., Londershausen, T., Schmidt, E.: Property function to compute the dustiness of powders. In: Dynamic flowsheet simulation of solids processes, pp. 413–432. Springer Nature, Cham Switzerland (2020).
10. Schwindt, N., Schmidt, E.: Auf dem Weg zur Vorhersage der Staubfreisetzung – Ein Ansatz zur Entwicklung einer Prognosemethode. Schüttgut&Prozess 1, 17–21 (2020).
11. Woschny, N., Schulz, D., Kruggel-Emden, H., Schmidt, E.: Ein experimentell-numerischer Ansatz zur Entwicklung von Staubfreisetzungsfunktionen. Chem.-Ing.-Tech. 93 (8), 1278–1286 (2021).
12. Woschny, N., Schulz, D., Kruggel-Emden, H., Schmidt, E.: Predicting dust emissions—Experimental study compared to coupled DEM/CFD simulations using a reference test bulk material. Advanced Powder Technology 32 (9), 3189–3206 (2021).
13. Schmidt, E.: Charakterisierung der Staubungsneigng von Pulvern. Silo World 2, 30–39 (2016).
14. Hamelmann, F., Schmidt, E.: Vorhersage der arbeitsschutzrelevanten Staubfraktionen von Pulvern durch Adaptierung bewährter Probennahmesysteme an ein Rotationsmessgerät. GRdL 64 (1/2), 35–40 (2004).
15. Bach, S., Eickmann, U., Schmidt, E.: Comparison of established systems for measuring the dustiness of powders with the UNC dustiness tester developed especially for pharmaceutical substances. Ann. Occup. Hyg. 57 (8), 1078–1086 (2013).
16. El-Waraki, S., Schmidt, E.: Gerät zur Charakterisierung des Feinstaubpotenzials von Straßenoberflächen. Chem.-Ing.-Tech. 79 (8), 1227–1228 (2007).
17. DIN EN 15051 (Entwurf): Exposition am Arbeitsplatz – Messung des Staubungsverhaltens von Schüttgütern – Teil 1: Anforderungen und Auswahl der Prüfverfahren – Teil 2: Verfahren mit rotierender Trommel – Teil 3: Verfahren mit kontinuierlichem Fall. DIN Media (2024).
18. DIN 33897: Arbeitsplatzatmosphäre – Routineverfahren zur Bestimmung des Staubungsverhaltens von Schüttgütern – Teil 1: Grundlagen – Teil 3: Verstaubung in ruhender Luft – Teil 4: Wiederholter Fall im Querstrom. DIN Media (2016).
19. DIN 55992: Bestimmung einer Maßzahl für die Staubentwicklung von Pigmenten und Füllstoffen – Teil 1: Rotationsverfahren. DIN Media (2006).
20. VDI 2263 Blatt 9: Staubbrände und Staubexplosionen – Gefahren – Beurteilungen – Schutzmaßnahmen – Bestimmungen des Staubungsverhaltens von Schüttgütern. DIN Media (2008).
21. VDI 3790 Blatt 3: Umweltmeteorologie – Emissionen von Gasen, Gerüchen und Stäuben aus diffusen Quellen – Lagerung, Umschlag und Transport von Schüttgütern. DIN Media (2010).

22. Hamelmann, F., Schmidt, E.: Messverfahren zur Bestimmung des Staubungsvermögens von dispersen Pulvern. Chem.-Ing.-Tech. 74 (12), 1666–1676 (2002).
23. Hamelmann, F., Schmidt, E.: Methoden zur Charakterisierung des Staubungsverhaltens von Pulvern. Chem.-Ing.-Tech. 75 (6), 696–700 (2003).
24. Hamelmann, F., Schmidt, E.: Methods for characterizing the dustiness estimation of powders. Chem. Eng. Technol. 27 (8), 844–847 (2004).
25. Hamelmann, F., Schmidt, E.: Methods for dustiness estimation of industrial powders. China Particuology 3 (1-2), 90–93 (2005).
26. Hamelmann, F., Schmidt, E.: Methods of estimating the dustiness of industrial powders—A review. KONA Powder and Particle 21, 7–18 (2003).
27. Bach, S., Schmidt, E., Bach, B., Volkhausen, W.: Bestimmung von Staubmaßzahlen zur Ermittlung von konkreten Gewichtungsfaktoren für diffuse Staubemissionen aus Umschlagvorgängen in einem Hüttenwerk. GRdL 72 (3), 114–119 (2012).
28. Bach, S., Schmidt, E.: Ein Vergleich zweier Sedimentationsverfahren zur Bestimmung des Staubungsverhaltens von Pulvern. Chem.-Ing.-Tech. 80 (3), 343–349 (2008).
29. Bach, S., Schmidt, E.: Bestimmung des Staubungsverhaltens von Pulvern – Vergleichsmessungen mit einem Prüfverfahren nach DIN EN 15051. GRdL 68 (1/2), 25–31 (2008).
30. Bach, S., Schmidt, E.: Determining the dustiness of powders—A comparison of three measuring devices. Ann. Occup. Hyg. 52 (8), 717–725 (2008).
31. Hamelmann, F., Schmidt, E.: Freisetzung von Nanopartikeln bei der Handhabung pyrogener Kieselsäuren. GRdL 66 (11/12), 475–476 (2006).
32. Schmidt, E.: Release of nanoparticles when handling fumed silica. Chem. Eng. Technol. 38 (11), 2096–2098 (2015).
33. Parey, M., Schmidt, E.: A model to evaluate the dustiness of powders and binary powder mixtures. Chem. Eng. Technol. 41(01), 35–43 (2018).
34. Londershausen, T., Schmidt, E.: Development of forecast formulas for assessing the dust formation tendency of moist bulk materials using the UNC dustiness tester. Chem. Eng. Technol. 41 (11), 2259–2265 (2018).
35. Londershausen, T., Schmidt, E.: Forecast formulas for bulk material dustiness under low vacuum conditions in a single-drop device. Chem. Eng. Technol. 44 (01), 58–64 (2021).
36. Londershausen, T., Schmidt, E.: Dustiness quantification of powders by the fractional release rate. Chem. Eng. Technol. 40 (01), 116–121 (2017).
37. Woschny, N.: Ein Beitrag zur verbesserten Beschreibung der Schüttgutstaubfreisetzung unter Verwendung eines Referenzmaterials. Shaker, Düren (2025, in press).

Optimization Algorithms for Supporting Passive Safety Engineering of Mobility Systems

A. Schumacher, K. Weider, P. Clemens and C. Gutgar

Contents

Abstract

Mathematical and semi-mathematical optimization methods are presented, which improve the lightweight structure of crash-relevant components of mobility systems. This is about the fulfillment of all requirements from laws or consumer protection tests. The article is a compilation of relevant methods for optimization that can be used to support the development process of complex mechanical systems. To evaluate the designs, crash simulations based on the finite elements are used in all presented methods. Topology optimization is mainly shown. So it's

A. Schumacher (✉) · K. Weider · P. Clemens · C. Gutgar
Chair for Optimization of Mechanical Structures, University of Wuppertal, Wuppertal, Germany
e-mail: schumacher@uni-wuppertal.de

S. Bracke and R. Pieper (eds.), *Safety Engineering*,
https://doi.org/10.1007/978-3-658-47473-7_11

about the optimization of the location and arrangement of structural elements and material. Methods for the efficient determination of necessary sensitivities are also presented. The article is intended to show the variety of different options for improving passive safety. The quality and performance of the used finite element simulation models are considered, especially for the presented methods using model reduction procedures and automatic generated sub-models.

Keywords

Lightweight crash structures · Crash simulation · Structural Optimization

1 Introduction

In the case of an unavoidable traffic accident, passive safety restraint systems help to mitigate the consequences of the accident. These systems are used more or less in all mobile systems (primarily vehicles and aircraft, but also ships). The automotive industry was and is a pioneer in the development of restraint systems for relevant accident scenarios. For this reason, this article focuses on the development of vehicles as an example. In addition to seatbelt and airbag systems, the mechanical car body lightweight structure plays a decisive role (Fig. 1).

The mechanical structure converts the kinetic energy of the accident into deformation energy. Ideally, the deceleration values of the persons involved should not exceed certain values. The deformation paths should be as short as possible in order to be able to build a compact structure. A good development process ensures a uniform reduction in speed. The structural behavior is highly non-linear, so the development of crash elements is very complex. The finite element based crash simulation used is correspondingly complex.

For several years, research has been carried out into methods that support this complex development of mathematical optimization methods [1]. The methods differ in the type of design variables **x** (e.g. wall thickness, shape or topology), the

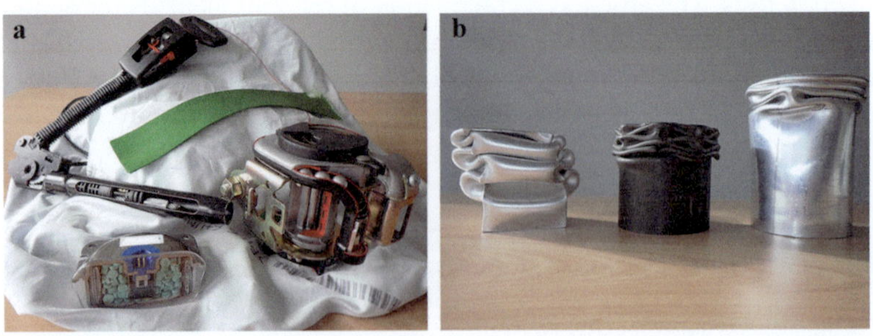

Fig. 1 **a**) Airbag, belt, belt tensioner in a retractor, belt tensioner in the seat belt buckle, **b**) Crashelements

type of objective functions f(**x**), the inequality constraints **g**(**x**) and equality **h**(**x**) as well as the type of optimization method.

An optimization task is always formulated in the following form:

$$\min f(\mathbf{x}) \quad \text{(objective function)}, \tag{1}$$

while considering the following constraints:

$$g_j(\mathbf{x}) \leq 0 \; j=1,m_g \text{ (inequality constraints)}$$
$$h_k(\mathbf{x}) = 0 \; k=1,m_h \text{ (equality constraints)}$$
$$x_i^l \leq x_i \leq x_i^u \; i=1,n \text{ (side constraints, upper and lower bounds)}$$

The restriction to "min f(**x**)" in the formulation does not pose a problem in practice, because the problem can easily be reformulated for maximization tasks with

$$\max f(\mathbf{x}) = -\min -f(\mathbf{x}).$$

In this book contribution, we present some optimization methods that were developed at the Chair of Optimization of Mechanical Structures (OMS) at the Faculty of Mechanical Engineering and Safety Engineering at the University of Wuppertal or in which the Chair played a major role. The book contribution begins with a schematic compilation of the requirements on the passive safety behavior in Chap. 2. Chapter 3 then discusses the current analysing software tools. Chapters 4 to 8 describe a selection of optimization methods. The article concludes with an outlook on research activities in the coming years.

2 Requirements on Passive Safety Behavior of Automobiles

There are many different laws and consumer protection tests around the world that must be taken into account in global vehicle development (Fig. 2).

The test structures and evaluation schemes are described in detail in all regulations [2]. The specified legal requirements must either be fulfilled or the specified criteria in the consumer protection tests are incorporated into an evaluation scheme with which the ratings (stars) are determined.

Regarding to the crash-loaded structures with highly non-linear behavior, there are a lot of sub-objectives and sub-constraints:

- Consideration of special acceleration values like the head injury criterion (e.g. the HIC-value),
- Realize energy absorption possibilities,
- Fulfill special force levels in defined position of the automotive,
- Realize smooth force-displacement curve and smooth acceleration-time curve,
- Enable special force paths for special load cases,
- Make high stiffness of special parts possible, e.g. parts in a main force paths in the passenger area,

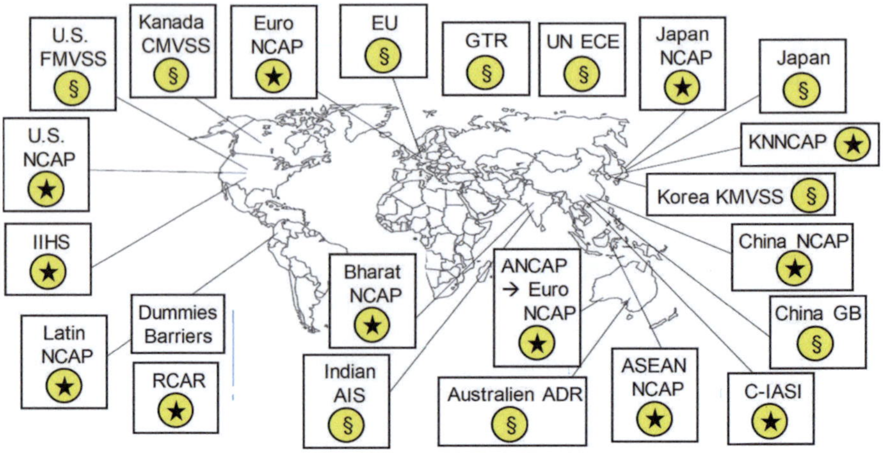

Fig. 2 Overview of the laws to be taken into account (§) and consumer protection tests (*) [2]

- Make low stiffness of special parts possible, e.g. at positions of the head contact of a pedestrian,
- Special safety criteria, e.g. intrusion of structural components in the battery system of an automotive.

3 Software Tools for Analysing the Crash Behavior of Mechanical Structures

Currently, finite element models with a finite element count in the range of 10 million finite elements are used for entire vehicles. Commercial solvers such as LS-DYNA®, PAM-Crash® or Radioss® are used. Good material definitions and good algorithms for describing the contacts during the crash are important. With the help of explicit time integration, the calculation is carried out with small time steps through the entire crash phase. Calculation times of one day for a simulation of a vehicle design are not uncommon. For this purpose, the mainframe computer Pleiades [3] with almost 20,000 CPUs is available at the University of Wuppertal.

It may make sense to perform the calculation with submodels (SM) rather than full models (FM). Routines for the automatic generation of submodels have been developed for this purpose [4, 5]. These routines automatically select the domains of the submodel from the overall vehicle. They then characterize the boundary conditions using different methods (Fig. 3).

In addition to the structure of the crash models and the calculation, efficient evaluation is of fundamental importance. All necessary results can be evaluated with post-processors such as Animator®.

If crash simulations have been carried out for different designs, information on clustering and outliers can also be obtained [6, 7]. Therefore, algorithms have been developed that can automatically transfer the results to another evaluation space.

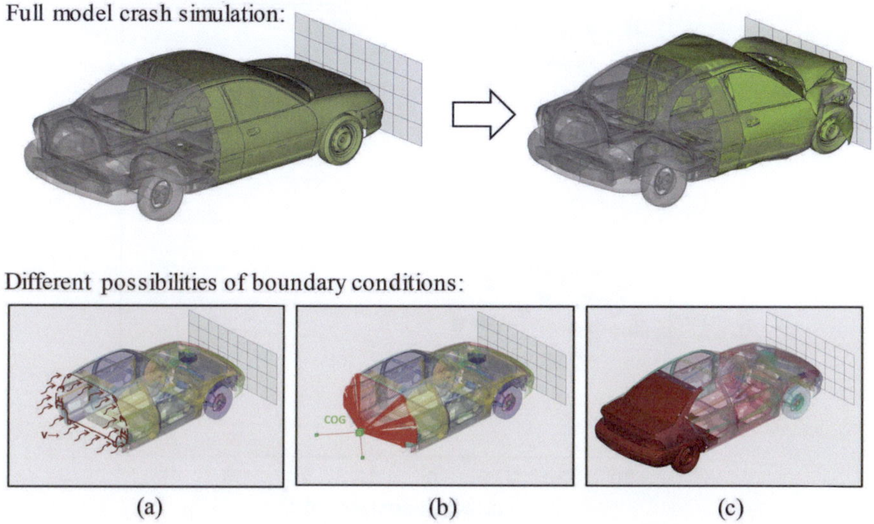

Fig. 3 **a**) Automatic submodel generation—Undeformed and deformed submodel with different boundary conditions: Displacements model approach (**a**), Kinematic mass approach (**b**), Rigid model approach (**c**)

Principle Component Analysis (PCA) is now a widely used method for model reduction. Figure 4 shows an example of a result plot. This makes it easy to recognize which designs are inliers and which designs are outliers.

4 Automatic Dimensioning of Crash Structures in a Multidisciplinary Development Framework

Figure 5 shows the wall thicknesses of a vehicle body that can still be varied at a late stage of development [8, 9]. In a multidisciplinary optimization (MDO), these wall thicknesses should be optimized so that the constraints are met for all load cases. Even in a late development phase, the wall thicknesses of a vehicle body can still vary. The objective function is usually the weight of the car body.

The following requirement list is typical [8]:

<u>Front crash</u> (rigid wall, full overlap, 0°, 56 km/h—FMVSS 208, USA)

1. longitudinally ascending force absorption of the longitudinal member, left and right side
2. maximal intrusions of the tank-room at several points
3. maximal intrusion of the bulkhead at several points
4. maximal relative displacement of the A- and B-pillar
5. real occupant load criterion

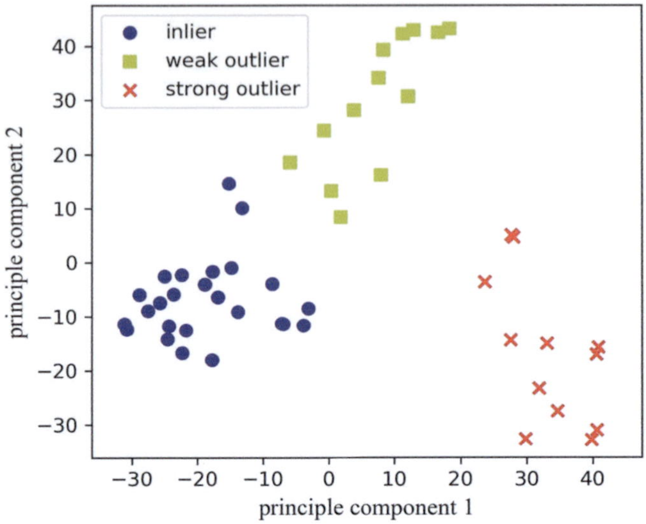

Fig. 4 Exemplary results of a Principle Component Analysis (PCA) [7]

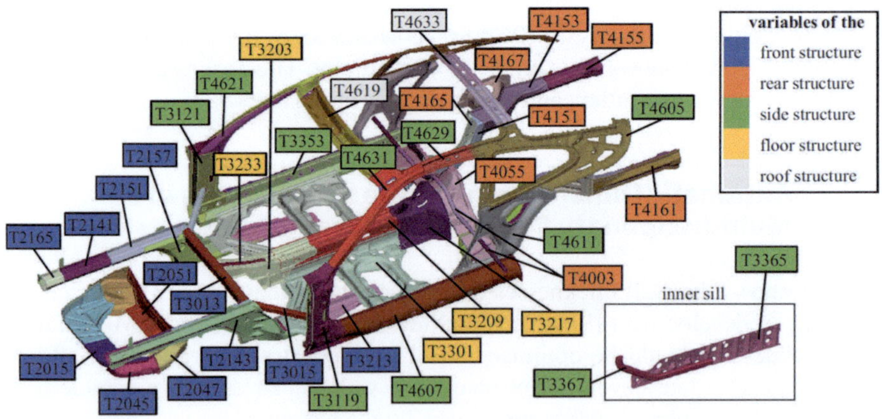

Fig. 5 Definition of the wall thickness design variables for a multidisciplinary optimization [8]

<u>Front ODB—Offset-Deformable Barrier</u> (40% overlap, 0°, 56 km/h—ECE-R 94/03, Europe)

1. longitudinally ascending force absorption of the longitudinal member, left side
2. maximal intrusions of the tank-room at several points

3. maximal intrusion of the bulkhead at several points
4. maximal relative displacement of the A- and B-pillar

Side pole (rigid pole, 90°, 29 km/h—EuroNCAP)

1. maximal angle between two seat cross members
2. maximal relative displacement between tunnel and sill at specific point

Rear ODB (deformable and moveable barrier, 70% overlap, 0°, 80 km/h (barrier)—FMVSS 301, USA)

1. maximal relative displacement at several points of the second row of seats, left side
2. maximal relative displacement of the A- and B-pillar
3. maximal relativ displacement of the car power unit

Roof crush (rigid and moveable plate—FMVSS 216a, USA)

1. maximal force of the structure at the contact point to the plate at 127 mm intrusion

Frequency

1. two bending modes
2. two torsion modes

As can be seen from this list, only very few constraints are in the required range in the initial design. The MDO is also used to find a design that is within the permissible range. The MDO usually proceeds according to the following scheme:

Step 1: Design of Experiments
Step 2: Finite Element analyses for all load cases
Step 3: Extraction of the design criteria (constraints)
Step 4: Generation of metamodels
Step 5: Optimization
Step 6: Verification

This complex process was significantly accelerated with the following extensions:

- Regarding step 1: The selection of sensitive design variables and relevant load cases is done with the so called global sensitivity matrix [9]. Therefore, the usage of robustness campaigns or the creation of a new database with usage of existing company knowledge have to be involved. The first result is a scheme for extracting relevant outputs out of simulations. The second result is a scheme for fitting metamodels and getting sensitivity information by using global sensitivity analysis (sobol indices).

- Regarding step 2: Integration of the content of Chap. 3 for generating of submodels [4, 5]
- Regarding step 4: The optimization success depends on approximations, bad metamodels leads to bad optimization results. Creating high quality structural approximations with even small sample sizes is needed. Different metamodel type have to stand in a competition for ever simple output. This metamodel types have to be fundamentally different: e.g. linear regression, gaussian process, radial basic function, neuronal network, decission tree. Each metamodel type requires an internal hyperparameter optimization.

These extensions create a very efficient MDO process that is equipped with many automatic correction loops and automatic adjustments (Fig. 6).

5 Finding Topological Derivatives for Crash Structures

The topological derivative indicates how a functional changes when a cut-out is introduced. The topological derivatives for shell structures in dynamic loadcases, which are often used in lightweight designs, are developed at the chair of optimization of mechanical structures. These sensitivities consider the highly nonlinear

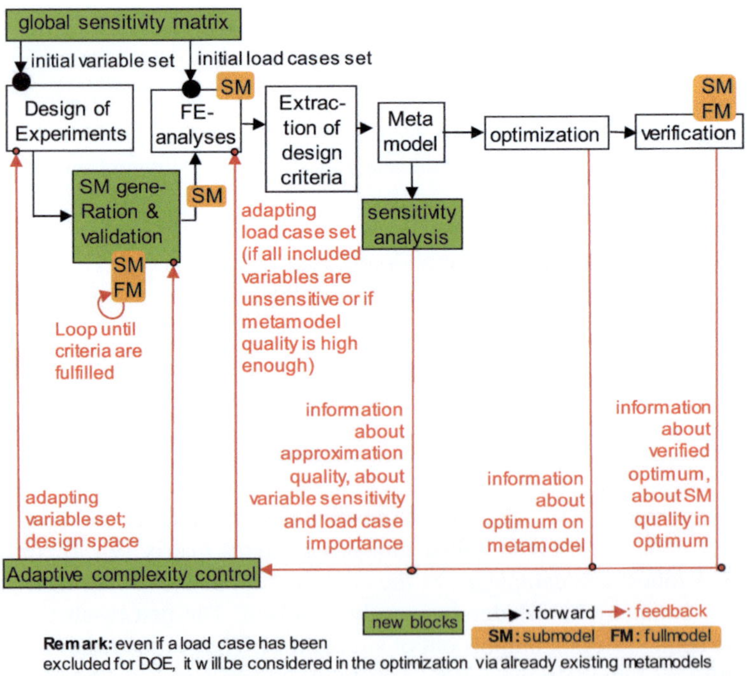

Fig. 6 New efficient MDO process

material behavior and large deformations of the structure. The aim of the investigation is the support of an efficient topology optimization for crash-loaded structures. The considered functionals can in general be described by displacements, velocities and accelerations. In particular, the semi-analytical sensitivity calculations for the internal deformation energy and for single point displacements in arbitrary directions in the shell structure are presented. These functionals cover the most important functions for the development of crash structures. Using the material derivative in combination with the adjoint method, a terminal value problem has to be solved. Depending on whether first differentiation and then discretization in the time domain is performed or first discretization and then differentiation, a separate solution scheme for the adjoint is derived. For the numerical evaluation of the topological derivatives, the basic equations for large deformed shell elements are included as well as a phenomenological material interpolation, which represents the tangential material behavior resulting from plastic strain and isotropic hardening. The elaborated equations and the procedure for the calculation of the topological derivatives are written in [10, 11]. Exemplary, Fig. 7 shows the geometry and the loading conditions of the cantilever. In Fig. 8 the topological derivative for the single point displacements are shown. For a targeted softening of the cantilever, the topological derivative reveals the most sensitive region in the center of the beam directly before the force application. This region contributes to the support of the single point at the center of the beam, benefiting from the bending stiffness at the mounting. The topological derivative of the displacement functional of the point at the leading edge shows a different characteristic. The point is supported by a circumferential frame. Between mounting and force application, directly before the force application, the cantilever beam has the lowest sensitivity.

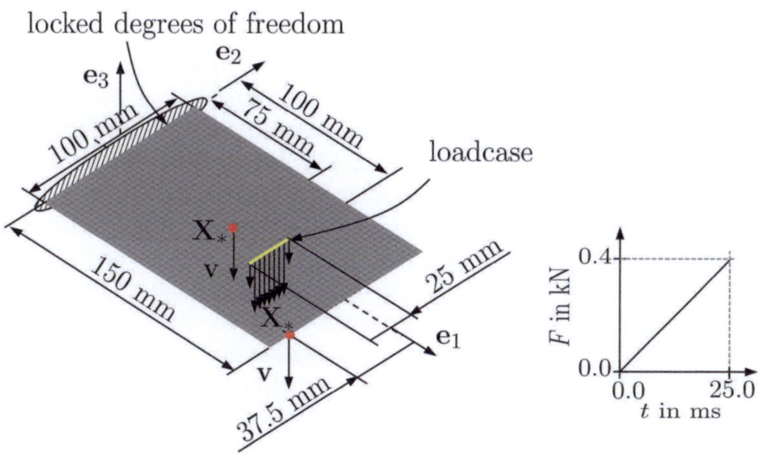

Fig. 7 Geometry and loading conditions of the cantilever [10]

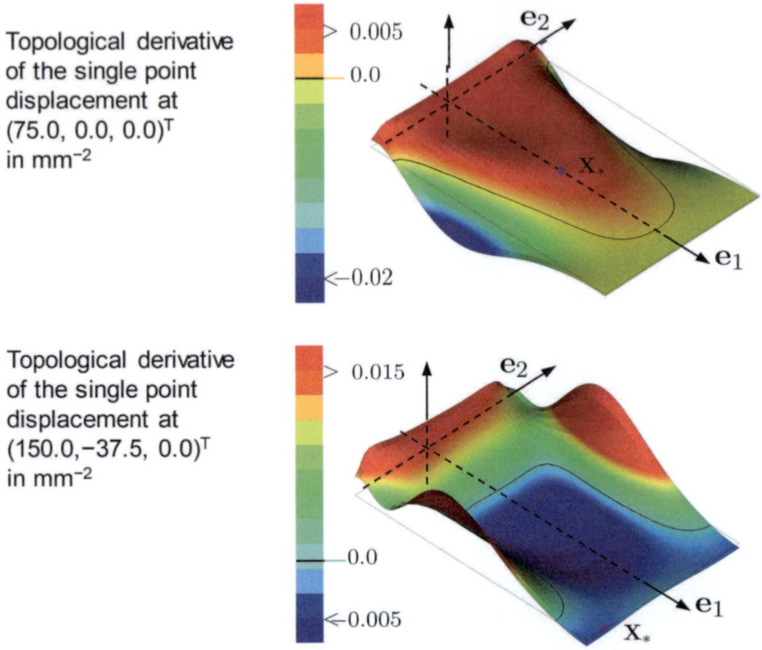

Fig. 8 Topology derivatives for different requirements [10]

6 Topology Optimization of Crash-Loaded Sheet Metals

The most commonly used crash-loaded structures are sheet-metals. For this reason, it is necessary to invest in an optimization process that makes it possible to optimize the topology and shape of individual components in the entire crash structure. At the chair a new method for the optimization of deep-drawn shell components in crash by using substitute load cases in established sensitivity based static optimization has been developed [12, 13]. The optimization process is split into an inner and an outer loop. In the outer loop the contact forces during crash are determined. These contact forces are used as substitute load cases in a static topology optimization, which builds the inner loop of the method. The method was inspired by the Equivalent Static Loads Method [14], but differs primarily in two aspects. These are the use of the contact forces as alternative to equivalent static loads in the linear static topology optimization and the extraction of a shell model that represents the voxel result obtained from topology optimization. A manufacturing constraint is applied in the topology optimization so that the results are shell components. This manufacturing constraint is enabling the mentioned extraction of a shell model [15, 16].

In the following, the optimization of a cantilever sheet structure shown. In Fig. 9 the mechanical model of the example is shown. The objective of the optimization is the minimization of the intrusion, so that manufacturing constraints are fulfilled:

- constant wall thickness
- no undercuts in deep-drawing direction
- no ribs

Figure 10 shows the optimization history and the final design.

The method has shown to be numerical stable and suitable for the shown examples when the intrusion is the objective of the optimization. The intrusion was reduced significantly. The key novelty of the method is the ability to perform a static topology optimization of a voxel design space for a non-linear dynamic optimization model with the result being a shell model that can be analyzed for crash.

7 Graph and Heuristic Based Topology Optimization of Crash-Loaded Profiles

In practical tasks, objectives and constraint functions must be treated, which are currently not solvable for topology optimization with purely mathematical methods. For this reason, the Graph and Heuristic Based Topology Optimization of Crash-Loaded Profiles (GHT) was developed [17–20]. The geometry structure is controlled with mathematical graphs. Heuristics help the mathematical

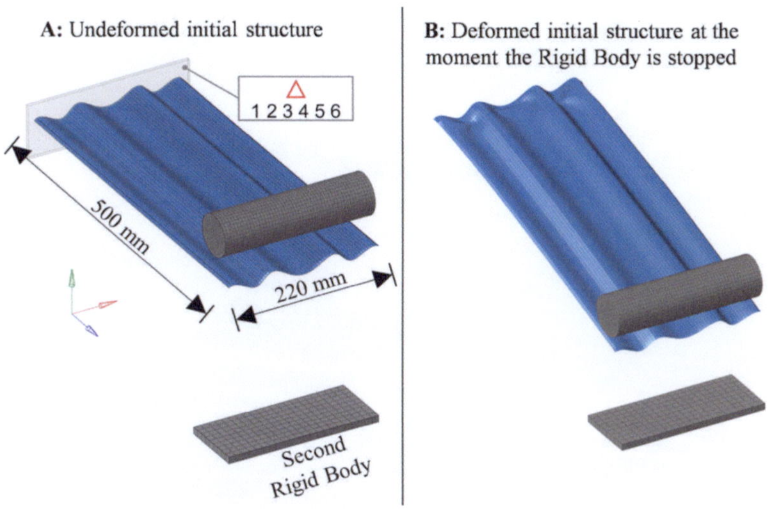

Fig. 9 Mechanical model of the sheet metal optimization example [12]

optimization algorithms. These heuristics were derived from expert knowledge. In the following, 6 of these heuristics are presented (Fig. 11), which, unless otherwise stated, compete with each other:

1. Heuristic "Delete Needless Components": The simulation data used are the total internal energy densities of the individual walls. The wall with the lowest energy density is removed.
2. Heuristic "Support Buckling Components": It is about the homogenization of the buckling or deformation behavior of the structure. The heuristic is used for a) walls or profiles that offer little resistance to intrusion by other bodies or b) unstable walls or profiles. The simulation data used are the velocity vectors of the FE nodes, which can identify instabilities very well [19].
3. Heuristic "Balance Energy Density": By connecting loaded with unloaded components, the distribution of the internal energy density should be homogenized. The simulation data used are the total internal energy density of the individual walls or profiles.
4. Heuristic "Remove Small Chambers": The heuristic reduces the complexity of the structure.
5. Heuristic "Smooth Structure": Simplification of the structure by smoothing kinks inside the structure. This heuristic does not compete with other heuristics because no topology change is made. It is activated after each topology change of the structure.

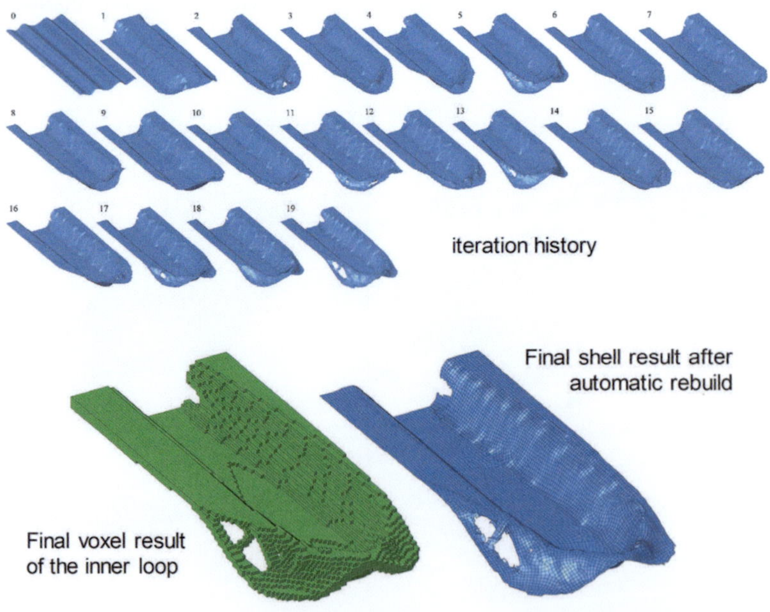

Fig. 10 Geometry and loading conditions of the cantilever [12]

6. Heuristic "Scaling Wall Thicknesses": Scaling the wall thicknesses of the structure to achieve a given mass. This heuristic also does not compete with other heuristics since no topology change is made. It is also activated after each topology change of the structure, so that the mass of the structure does not change due to the topology change (constant mass development).

The evaluation of the simulation data used can be very difficult and requires some investigations to ensure a safe process of automatic structure optimization. As an example, for the heuristic "Support Buckling Components" the correlation between the very high velocity of the structural nodes and the buckling is shown in Fig. 12.

These 6 heuristics are used as basic heuristics in all applications. However, they have their origin in the topology optimization of laterally loaded profiles in a crash. Additional heuristics have been developed for special applications, which are described below.

The process of the GHT is based on the Bubble method [21], which changes the topology of the structure in the outer loop and optimizes the shape in the inner loop (Fig. 13). In the mechanically linear range, an analytical hole positioning criterion was developed for the bubble method, which evaluates the local stresses in the component. This hole positioning criterion has been replaced by heuristics in the GHT.

So the inner loop is executed with mathematical optimization algorithms while the outer loop uses the heuristics. The GHT process has a modular structure. For example, shape optimizations in individual iterations can be dispensed with and/or several heuristics can be pursued in parallel.

The basis for changing the geometry through the optimization software and for the automatic creation of input decks is a flexible description of the geometry

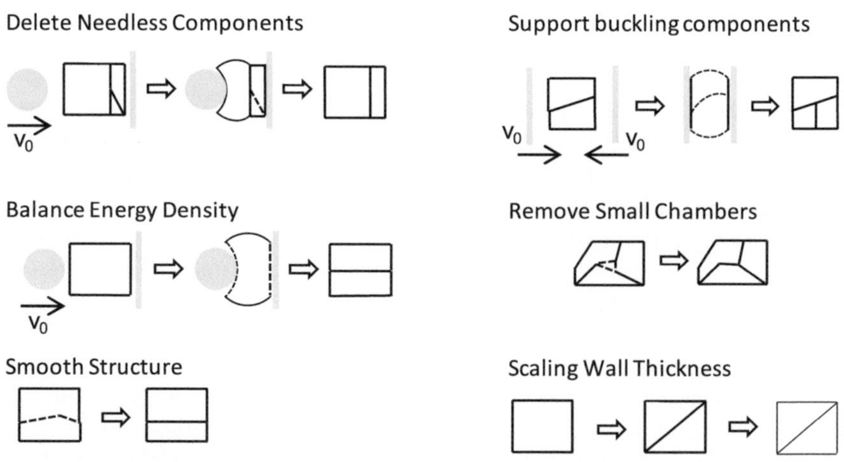

Fig. 11 Heuristics using the example of the design of profile structures in crash load cases [20]

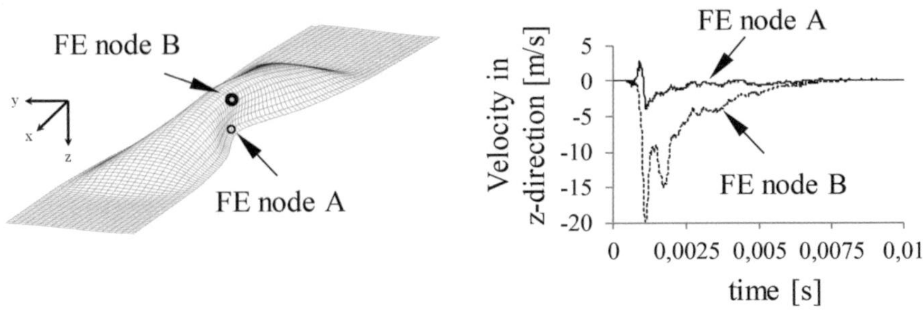

Fig. 12 Velocity of two nodes during the buckling of a wall [18]

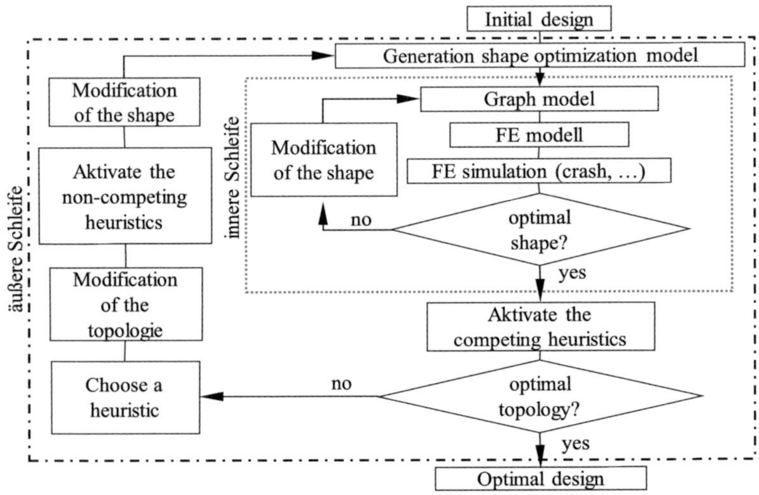

Fig. 13 Scheme of the GHT [18]

using the mentioned mathematical graphs. The first scope of application for the GHT is the optimization of profile cross-sections, which can be described using a two-dimensional graph. However, the structure itself and all finite element simulations carried out are three-dimensional.

The first application is the optimization of the cross-section of a vehicle rocker [18]. The manufacturing process is extrusion and the material is aluminum. Figure 14 shows a sub-model of the vehicle rocker hitting a pile.

The evaluation of the strains on the initial design in LS-DYNA®, i.e. the empty rocker profile, can be seen in Fig. 15.

The optimization task is to find the optimal topology and shape of the cross section of the rocker profile by solving the following optimization task:

Minimize the maximal force at a moved rigid wall (velocity v0), so that functional constraints

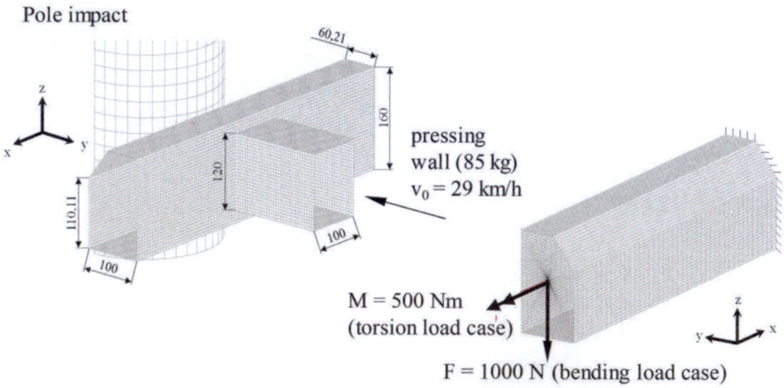

Fig. 14 Load cases of the rocker example [18]

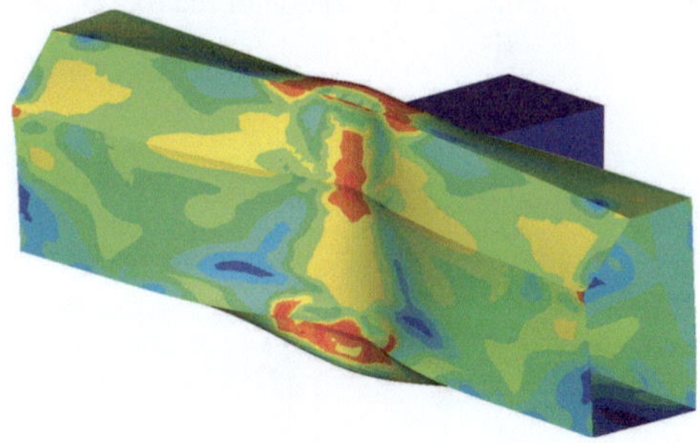

Fig. 15 Evaluation of the strains in the initial design (blue: no strain, red high strain) [18]

- mass ≤2801 kg
- intrusion of the pole ≤70 mm
- stiffness (bending and torsion) ≥50% stiffness initial design

and the manufacturing constraints

- 1.6 mm ≤ wall thickness ≤3.5 mm
- distance of walls ≥10 mm
- connection angle of walls ≥15°
- maximum chamber size ration 1: 20

are fulfilled.

Figure 16 shows the optimization history. In each iteration, the five best designs were followed up and shape optimization was only carried out in the 3rd iteration.

Figure 17 shows the force-time curve of the starting design, the optimal design and the theoretical optimum (constant level during the crash time). The force-time curve of the optimal design is very close to the theoretical optimum, where the acceleration rises very quickly to a constant level and quickly drops back to zero.

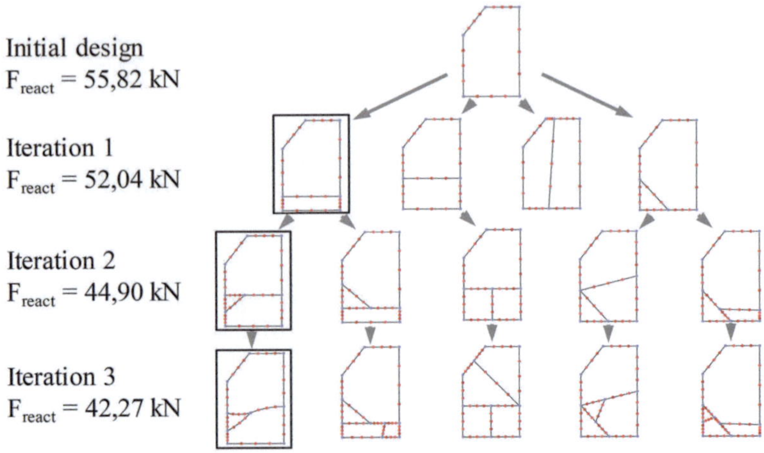

Fig. 16 Topology optimization of the rocker: Optimization history [18]

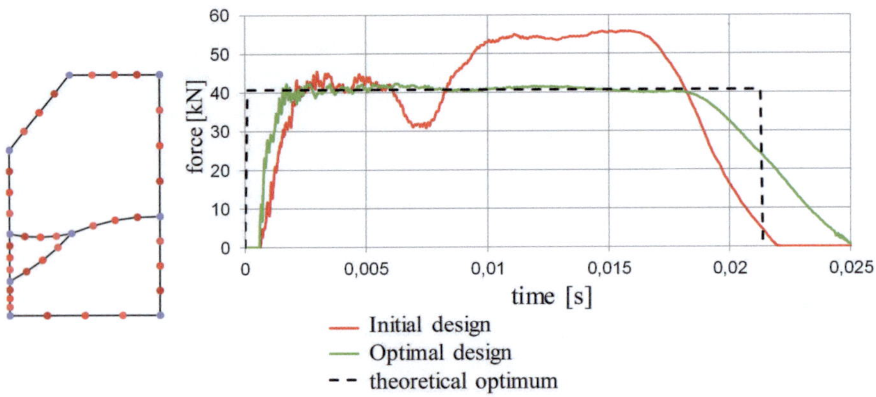

Fig. 17 Topology optimization of the rocker: Optimization result [18]

Several extensions of the GHT are developed:

- Topology optimization for axial-loaded profiles [22],
- Topology Optimization for profiles build with shell metals [23],
- Topology Optimization for for composite materials [24],
- Finding new heuristics with artificial intelligence procedures [25, 26].

8 Graph and Heuristic Based Topology Optimization for Finding Good Car Body Layouts for Crash

The idea of the GHT for profiles was expanded to find good car body layouts. For this purpose, the geometry description was expanded and the heuristics were adapted to the 3D space [27, 28]. The application example presented here is about the optimization of a three-dimensional frame structure (Fig. 18). Profiles can be arranged in the shown design space. These profiles should all have the same cross-section. In Fig. 19 the deformation of the initial design is shown. The optimization task reads as follows:

Minimize the maximum occurring contact force, so that functional constraints

- Intrusion of the pole ≤ 150 mm
- Mass ≤ 0.5 kg

and the manufacturing constraints

- 0.8 mm ≤ wall thickness ≤ 3.5 mm
- Length of the profiles ≥ 35 mm
- distance of walls ≥ 35 mm
- connection angle of walls ≥ 15°
- Symmetry to y-axis

are fulfilled.

The optimization result can be seen in Fig. 20. In the two diagrams, contact force and displacement are plotted over the iterations. At the bottom of the picture the final structure is shown. The outer frame shape has been tapered in the shape optimization in the front upper area. The framework can be divided into two areas.

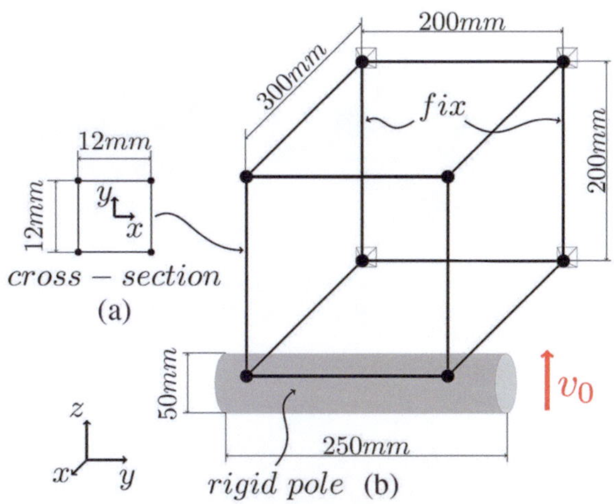

Fig. 18 Initial frame structure with quadratic cross-sections (12 × 12 mm) (**a**) impacted by a rigid pole (**b**) [27]

Fig. 19 Initial frame impacted by rigid pole: Deformed state at t=0.020 s [27]

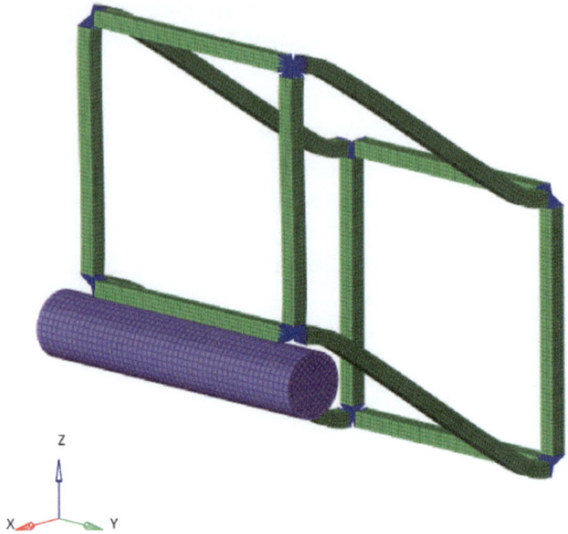

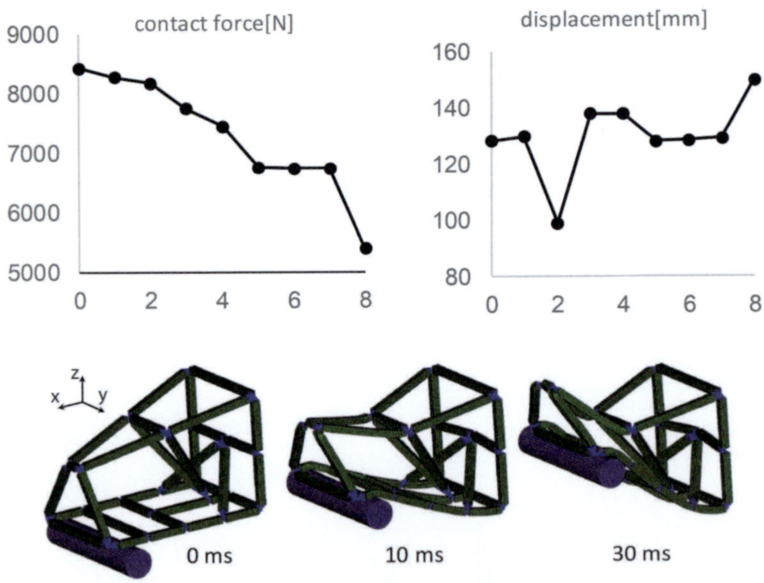

Fig. 20 Optimum of rigid pole model: Minimize the contact force. Initial state at t_0=0.000 s (a). Deformed state at t=0.010 s (b). Deformed state at t=0.030 s (c) [27]

The front area near the impact is more flexible, while additional profiles have developed in the rear area near the clamping. As a result, the rear of the frame provides the rigidity needed to stop the cylinder before 150 mm of displacement. Figure 21 shows the diagram of the contact force over the displacement.

9 Next years Activities

For all optimization methods presented here, there are research projects at the chair for Optimization of Mechanical Structures (OMS) with which the optimization of crash -based structures are further pursued. It is often about the integration of certain manufacturing processes, such as roll-forming of sheet metals or additive manufacturing.

Acknowledgements The authors would like to thank Christopher Ortmann, David Kracker, Florian Beyer, Jana Büttner, Jens Trilling, Johannes Sperber, Robert Dienemann, Simon Link and Sven Wielens. With the methods they developed in their doctoral theses, they have contributed to making this book contribution to such a comprehensive collection of tools for the engineering of lightweight structures considering crash loadcases. The research was supported by the German Research Foundation - DFG (Schu 915_3-1, Schu 915_4-1), the Federal Ministry of Education and Research – BMBF (01IS13025A-F), Volkswagen AG, Dr. Ing. h.c. F. Porsche AG and Altair Engineering Inc..

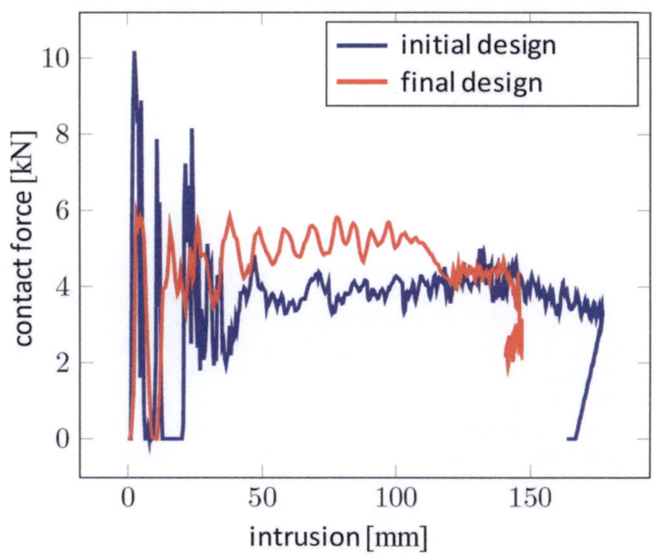

Fig. 21 Force-displacement behavior of the initial design and the final design [27]

References

1. Schumacher, A.: Optimierung mechanischer Strukturen - Grundlagen und industrielle Anwendungen, 3. Auflage, Springer-Verlag, Berlin, Heidelberg, ISBN 978-3-662-60327-7, https://rd.springer.com/book/10.1007/978-3-662-60328-4 (2020)
2. Hoffmann, R. et. al.: Safety Companion, CARHS, Alzenau, https://www.carhs.de/en/safety-companion-overview.html (2024)
3. N.N.: Scientific Computer Center PLEIADES, https://pleiades.uni-wuppertal.de/en/ (2024)
4. Wielens, S., Benito Cia, L., Schumacher, A.: Using automatically generated submodels in vehicle body crash optimization, International Journal of Crashworthiness, https://doi.org/10.1080/13588265.2022.2109873 (2022)
5. Wielens, S.: Automatische Erstellung von Submodellen für die Crashoptimierung von Fahrzeugkarosserien, Dissertation University of Wuppertal, Shaker Verlag, ISBN: 978-3-8440-8717-8 (2022)
6. Kracker, D., Dhanasekaran, R.K., Schumacher, A., Garcke, J.: "Method for automated detection of outlier in crash simulation, Intl. Journal of Crashworthiness, Volume 28(1), 96–107, https://doi.org/10.1080/13588265.2022.2074634 (2023)
7. Kracker, D.: Automatisierte Auswertung von Crashsimulationen unterschiedlicher Fahrzeug-Entwicklungsständen mit Methoden des maschinellen Lernens. Dissertation University of Wuppertal, Shaker Verlag, ISBN: 978-3-8440-9424-4 (2024)
8. Büttner, J., Schumacher, A., Bäck, T., Schwarz, S., Krause, P.: Making multidisciplinary optimization fit for practical usage in car body development, Structural and Multidisciplinary Optimization 66:62 https://doi.org/10.1007/s00158-023-03505-z (2023)
9. Büttner, J.: Effiziente Lösungsansätze zur Reduktion des numerischen Ressourcenbedarfs für den operativen Einsatz der Multidisziplinären Optimierung von Fahrzeugstrukturen, Dissertation University of Wuppertal, Shaker Verlag, ISBN: 978-3-8440-8560-0 (2022)
10. Weider, K.: Topologische Ableitung zur Optimierung crashbelasteter Strukturen, Dissertation University of Wuppertal, Shaker Verlag, ISBN: 978-3-8440-8248-7 (2021)
11. Weider, K., Schumacher, A.: "Topological derivatives for shell structure optimization considering crash loadcases with material nonlinearities and large deformations", Structural and Multidisciplinary Optimization (2024) 67:114, https://doi.org/10.1007/s00158-023-03711-9 (2024)
12. Clemens, P., Schumacher, A.: Nested loop approach for topology and shape optimization of crash-loaded deep-drawn components using contact forces for the inner loops, Structures, Volume 55, September 2023, Pages 2013-2022, https://doi.org/10.1016/j.istruc.2023.06.015 (2023)
13. Clemens, P.: Topologie- und Formoptimierung crashbeanspruchter Tiefziehblech-Bauteile, Dissertation University of Wuppertal, to appear in Shaker Verlag (2025)
14. Choi, W.S; Park, G.J.: Structural optimization using equivalent static loads at all time intervals. In: Computer Methods in Applied Mechanics and Engineering 191; Jg. 2002 (19), S. 2105–2122. https://doi.org/10.1016/S0045-7825(01)00373-5 (2002)
15. Dienemann, R., Schumacher, A., Fiebig, S.: "Using topology optimization for finding shell structures manufactured by deep-drawing", Journal of Structural and Multidisciplinary Optimization 56:473-485 (2017)
16. Dienemann, R.: Entwicklung einer Optimierungsmethodik für die Form- und Topologieoptimierung von tiefziehbaren Blechstrukturen, Dissertation University of Wuppertal, Shaker Verlag, ISBN: 978-3-8440-6196-3 (2018)
17. Schumacher, A.: "How Mechanically Inspired Design Rules Help in the Topology Optimization of Structures with Highly Nonlinear Behavior", in: H. Altenbach, J. Hohe, C. Mittelstedt: Progress in Structural Mechanics, Springer Nature Switzerland AG, ISBN 978-3-031-45553-7 (2024)

18. Ortmann, C.; Schumacher, A.: "Graph and heuristic based topology optimization of crash loaded structures", Journal of Structural and Multidisciplinary Optimization 47:839-854 (2013)
19. Ortmann, C.: Entwicklung eines graphen- und heuristikbasierten Verfahrens zur Topologieoptimierung von Profilquerschnitten für Crashlastfälle, Dissertation University of Wuppertal, Shaker Verlag, ISBN: 978-3-8440-3746-3 (2015)
20. Ortmann, C., Sperber, J., Schneider, D., Link, S., Schumacher, A.: Crashworthiness design of cross-sections with the Graph and Heuristic based Topology Optimization incorporating competing designs, Structural and Multidisciplinary Optimization 64:1063–1077 (2021)
21. Eschenauer, H.A., Kobelev, V.V., Schumacher, A.: "Bubble method for topology and shape optimization of structures". Journal of Structural Optimization 8, 42–51. 2021, 202 (1994)
22. Sperber, J.: Graphen- und Heuristikbasierte Topologieoptimierung axial belasteter Crashstrukturen, Dissertation University of Wuppertal, Shaker Verlag, ISBN: 978-3-8440-8634-8 (2022)
23. Link, S.: Graphen- und heuristikbasierte Topologieoptimierung von crashbelasteten Profilstrukturen aus verbundenen BlechenDissertation University of Wuppertal, to appear in Shaker Verlag (2025)
24. Schneider, D.: Graphen- und heuristikbasierte Topologieoptimierung von Profilstrukturen aus Faser-Kunststoff-Verbunden in Crashanwendungen. Dissertation University of Wuppertal, Shaker Verlag, ISBN: 978-3-8440-8884-7 (2023)
25. Trilling, J., Schumacher, A., Zhou, M.: "Reinforcement learning based agents for improving layouts of automotive crash structures", Applied Intelligence https://doi.org/10.1007/s10489-024-05276-6 (2024)
26. Trilling, J.: Unterstützung der Graphen- und Heuristikbasierten Topologieoptimierung crashbelasteter Strukturen durch Reinforcement Learning. Dissertation University of Wuppertal, Shaker Verlag, ISBN: 978-3-8440-9554-8 (2024)
27. Beyer, F., Schneider, D., Schumacher, A.: Finding three-dimensional layouts for crashworthiness load cases using the graph and heuristic based topology optimization, Journal of Structural and Multidisciplinary Optimization, 63:59–73, https://doi.org/10.1007/s00158-020-02768-0 (2021)
28. Beyer, F.: Entwicklung eines Optimierungsverfahrens für crashbelastete 3D-Rahmenstrukturen mit der Graphen- und Heuristikbasierten Topologieoptimierung. Dissertation University of Wuppertal, Shaker Verlag, ISBN: 978-3-8440-9475-6 (2024)

One-Dimensional View of the Structural Changes in the Filtering Layer During the Clogging-Phase of Filter Media—A New Approach to a Holistic View of the Filtration Process

Qian Zhang

Contents

Abstract

Due to the discrepancy between the previous theoretical description of the filtration process and the actual structural changes in the filter layer, a new holistic approach to describing the filtration process is introduced, which is based on an analytical description of the structural changes of the filter packing. This article presents some new aspects for the description of filtration processes that result from this new approach. Here, a new specific one-dimensional spatial distribution function for newly deposited particles in the filter packing during the filtration process is introduced to describe the deposition profile of newly deposited particles of any filtration process in dimensionless form. While the initial deposition profile is determined by the filtration conditions and the filter medium together, the final deposition profile is independent of the given filter medium and a function determined by the given filtration conditions. During

Q. Zhang (✉)
Institute of Particle Technology, University of Wuppertal, Wuppertal, Germany
e-mail: zhang@uni-wuppertal.de

S. Bracke and R. Pieper (eds.), *Safety Engineering*,
https://doi.org/10.1007/978-3-658-47473-7_12

the clogging phase, the one-dimensional deposition profile gradually takes on its final form. The systematic change in the one-dimensional deposition profile during a continuous filtration process can be evaluated using a correlation analysis to describe the kinetics of this dynamic process.

Keywords

Clogging Phase · Dust Separation · Filtration Kinetics

1 Introduction

1.1 Background

Separating solid particles from gases is one of the most frequent tasks in the process technology as well as in the field of environmental protection and occupational health and safety. Filtering separators are often used for this purpose. Filtering separators can be divided into two main groups: depth filters and surface filters, whereby materials made of fiber layers are by far the most commonly used filter media for the two groups. Depth filters are designed for dust separation inside the filter medium; the corresponding filtration process is called depth filtration. In the case of surface filters, a particle layer which is also known as dust cake or filter cake, forms on the surface of the filter medium; the corresponding filtration process is called surface filtration or cake-forming filtration.

A new, clean filter medium with a porous structure for surface filtration always starts with an initial phase in which particle deposition can be described according to the principle of depth filtration in the given medium. In the literature, this initial filtration phase is referred to as the clogging phase. A filter medium developed for surface filters can act as a depth filter for a long time at significantly reduced dust concentration. Conversely, a filter medium developed for depth filters can quickly enter the cake-forming filtration phase at significantly increased dust concentration. Although neither of the two examples represents a meaningful real-world application, both demonstrate the fact: From a purely theoretical point of view, the boundary between depth and surface filtration is fluid. This means that a continuous filtration process using a clean filter medium always begins with depth filtration, goes through a clogging phase and finally transitions into the surface filtration phase. A unified theoretical approach for a holistic view of the filtration process, which would allow a generally valid description of a smooth transition from depth filtration to surface filtration, is not available using the classical theory of depth filtration and surface filtration. *Zhang* recently introduced a new approach for a holistic view of the filtration process by systematically analyzing and describing the structural changes of the filter layer during continuous dust deposition by a one-dimensional consideration in the gas flow direction [1, 2]. This article presents some new aspects for the description of filtration processes that result from this new approach.

1.2 Classical Theory of Depth Filtration and Surface Filtration

Depth Filtration and Filtration Kinetics. The theoretical basis for the evaluation of particle separation is formed in all filtering separators by the so-called single-collector separation theory for depth filtration. With the model of depth filtration, the entire inner collector surface—total projection area facing the gas flow—of a given filter layer is used for the calculation of the total separation performance of the filter layer. If the individual separation mechanisms and the geometric structure of the filter layer are described appropriately, a satisfactory agreement can be achieved between the results of the model calculations using analytical models and the results of the experimental investigations.

This only applies to the initial state of particle separation, as long as the resulting separation efficiency of the original single-collectors is not changed to an extent that can no longer be neglected by the resulting separation efficiency of the particles that are deposited in (and on) the filter medium. The term filtration kinetics refers to the behavior over time of the separation performance of a filter layer, which includes both the initial clean filter medium and the particles deposited in and on the filter medium.

Model Approaches to Describe the Filtration Kinetics. Analytical models for particle separation are used exclusively to calculate the separation efficiency or the separated particle quantity. The calculated particle quantity is assigned to the given filter layer. All previous model approaches to describe the filtration kinetics of a filter layer—usually a fiber packing—therefore have in common that the given initial packing is always the focus of the analysis. The structure of the initial packing determines the size of the filter layer, which no longer changes with the deposited particle quantity; both the so-called dendrite growth model and the fiber growth model for modifying or expanding the original single-collectors describe only structural changes within the given packing. The problem with this is that all calculations of these model approaches do not lead to a phase transition of the filtration process without a previously defined, model-independent termination criterion. Specifically, this means that both the end of the filtration kinetics of depth filtration and the beginning of surface filtration cannot be defined by this type of model consideration itself.

Surface Filtration. When a dust cake forms on the surface of the filter medium, it takes over the function of the active filtering layer with its significantly increased and steadily rising separation efficiency compared to the original filter medium, while the underlying filter medium is converted into a supporting layer. Therefore, surface filtration is always represented in the classical model description by a macroscopic growth of the particle layer and not described with the basic separation theory for filtering separation of particles.

Model Approaches for the Transition into Surface Filtration. If a conventional modeling of the clogging phase of a given filter medium is to be continued into the surface filtration phase in order to calculate the further development of the temporal change of the pressure loss of the filter layer in this phase, this transition can so far only be achieved by an additional, independent add-on model, which in turn requires empirically determined parameters to describe the dust cake. As described above, the formation of the dust cake is not modeled here by the separation of particles on a microscopic scale, but only by the generation of macroscopic growth of the particle layer using structural data of the dust cake obtained from other investigations.

For special cases, the conventional model for the filtration kinetics of a given filter medium and the macroscopic model for the cake-forming filtration can be combined in such a way that the add-on model for surface filtration is superimposed on the model for the clogging phase of the initial packing at the right point in time for the progress of the overall process, and thus a modeling result of the pressure loss over time that conforms to the measuring data is generated. However, the two independent models for the different filtration phases do not answer the important questions in a physically correct way, when and how a dust cake actually formed from the initial porous packing.

1.3 Description of the Actual Structural Changes of the Filter Packing During the Clogging Phase as a Key to a Holistic View of the Filtration Process

Initial Evaluation of the Results from 3D Simulation of Cake Formation. Using high-quality CFD simulation and calculation of particle trajectories, the actual structural changes of a given filter packing can be examined during continuous dust deposition. Here, simulation calculations that build up a dust cake by simulating the deposition process of the particles on the basis of a high-resolution flow calculation on a micro scale enable a corresponding quantitative investigation.

On the basis of the obtained 3D particle structures during the process of building up the dust cake, one-dimensional packing profiles of all deposited particles throughout the entire filter layer (see Fig. 1) and deposition profiles formed by newly deposited particles (see Fig. 2) in different phases during a filtration process can be evaluated [1]. Here, the particle structure formed by all the deposited particles up to a certain filtration time (without filter medium) is regarded as a granular bed with a varying one-dimensional packing profile in the gas flow direction.

Figure 1 shows exemplarily the evaluated one-dimensional packing profile of all deposited particles throughout the entire filter layer at various times during a continuous filtration process that begins with a new, clean filter medium and continues until a dust cake is built up [1, 2]. It can be seen clearly in Fig. 1 that the dust cake generated in the filtration process has a significantly more porous packing section on its upstream side surface, and that this highly porous packing section has a characteristic, stable s-shaped packing profile as a function of the spatial coordinate in the gas flow direction.

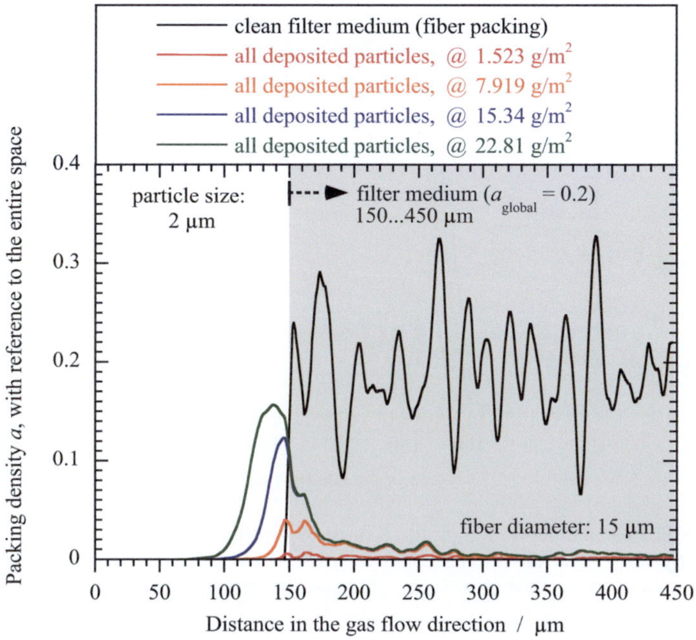

Fig. 1 One-dimensional packing profile of all deposited particles throughout the entire filter layer by variation of the dust area load on the filter medium, represented as the local packing density in relation to the spatial coordinate in the gas flow direction. Evaluation results from a 3D dust cake build-up generated on a model filter (fiber packing) using CFD simulation and calculation of particle trajectories [1]

Figure 2 shows the corresponding deposition profiles formed by newly deposited particles (see Fig. 2) in different phases during the same filtration process as in Fig. 1, which are evaluated by calculating the absolute changes in the packing profile in a short period during the filtration process compared to an earlier filtration time.

This targeted evaluation, shown in Fig. 2, now clearly shows that surface filtration is always a special, highly effective depth filtration in the highly porous packing section on the upstream side of the dust cake. In the first deposit profile in the early phase of the filtration process (change of the dust loading from 0 to 1.523 g/m^2), it can be clearly seen that a measurable proportion of the deposited particles are deposited outside the packing section of the filter medium. In the phase of surface filtration, the highly porous packed section on the upstream side of the dust cake acts as the actual active deposition zone of the entire filter layer for the subsequent dust particles in the inlet gas, since hardly any particles penetrate into the underlying section of the dust cake—the packing density in the core section of the dust cake no longer changes.

Conclusion for a New Holistic View of the Filtration Process. The evaluation of the actual structural changes of the filter packing makes clear that during

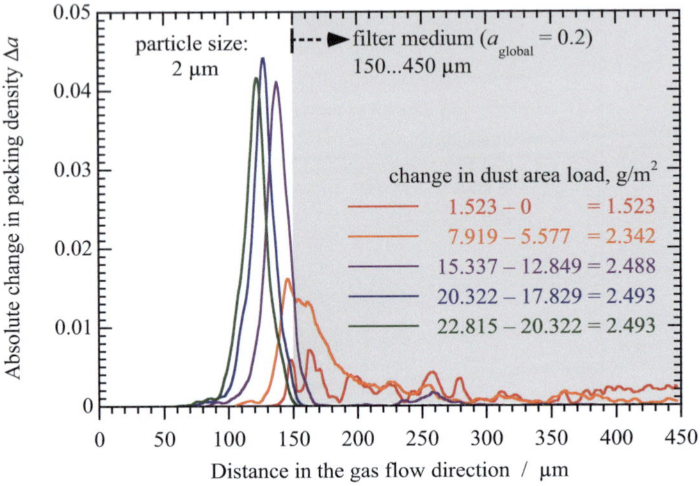

Fig. 2 One-dimensional deposition profile formed by newly deposited particles in the active filtering layer by variation of the filtration time, represented as the absolute change in local packing density in relation to the spatial coordinate in the gas flow direction (cf. Fig. 1.) [2]

the process of dust cake formation from a given porous system, there is no clear phase boundary for particle deposition on the microscale; and there is a continuous change of the filter packing, which includes continuous changes inside the filter packing as well as a continuous change of the packing size—a continuous expansion of the filter packing against the gas flow direction—. From the perspective of the structural changes in the filter packing, the entire filtration process can be described as follows: For the filtering system, the filtration process is a dynamic process of the structural changes in the filter packing. While the initial deposition profile in the depth filtration phase describes the initial state of this dynamic process, the stable deposition profile in the surface filtration phase represents the final state of this process. The systematic description of these structural changes creates a new opportunity to view the entire filtration process holistically.

2 Theoretical Description of the Structural Changes of the Filter Packing During the Filtration Process

2.1 A Specific One-Dimensional Spatial Distribution Function for Newly Deposited Particles as the Deposition Profile

A new specific one-dimensional spatial distribution function for newly deposited particles in the filter packing during the filtration process is introduced to describe the deposition profile of newly deposited particles of any filtration process in dimensionless form [1, 2]. In this context, "dimensionless" means that both the spatial coordinate indicating the location of the particle deposition and the quantity of deposited particles are dimensionless.

Based on the description of particle size distribution, this specific one-dimensional spatial distribution function for newly deposited particles (monodisperse particles of size x) in the filter packing during the filtration process is defined as follows (Eq. (1)):

$$Q_r\left(\frac{h}{x}\right) = \frac{\text{Partial quantity of newly deposited particles in the range of } \frac{h_{\text{min}}}{x} \ldots \frac{h}{x}}{\text{Total quantity of newly deposited particles in the range of } \frac{h_{\text{min}}}{x} \ldots \frac{h_{\text{max}}}{x}} \quad (1)$$

Here, the dimensionless spatial coordinate in the direction of gas flow is described by the relative distance with respect to the particle size h/x. h_{min} is the spatial coordinate of the first deposited particle in the direction of gas flow, and h_{max} the spatial coordinate of the last deposited particle in the direction of gas flow.

Accordingly, the following definition of the density function of this distribution (Eq. (2)) applies. And the normalization condition applies to this distribution (Eq. (3)).

$$q_r\left(\frac{h}{x}\right) = \frac{dQ_r\left(\frac{h}{x}\right)}{d\frac{h}{x}} \quad (2)$$

$$\int_{\frac{h_{\text{min}}}{x}}^{\frac{h_{\text{max}}}{x}} q_r\left(\frac{h}{x}\right) d\frac{h}{x} = Q_r\left(\frac{h_{\text{max}}}{x}\right) = 1 \quad (3)$$

The spatial coordinate in the direction of gas flow can be defined by flexibly shifting the respective zero point ($h \rightarrow h_{\text{sliding}}$) to allow conclusive comparison between different distributions, e.g., during an ongoing filtration process, or when the filter medium or the particle size changes (see Fig. 3 and 4).

2.2 Initial One-Dimensional Deposition Profile in the Depth Filtration Phase of a Clean, Homogenous Filter Medium

For a clear presentation of spatial data, a new notation form for the well-known grade efficiency—also called fractional separation efficiency—$T(x)$ ($T(x)$: the ratio of the quantity of particles of the size x that is separated by a filter layer to the quantity of particles of the same size x in the inlet gas) is used:

$$T_x\left(\frac{h}{x}\right) = \text{resulting grade efficiency for the particle size } x \text{ to a depth of } \frac{h}{x} \quad (4)$$

For a clean, homogenous filter medium made from fibers with the porosity ε, the single fiber collection efficiency $\phi(x)$ and the volume specific projection area of

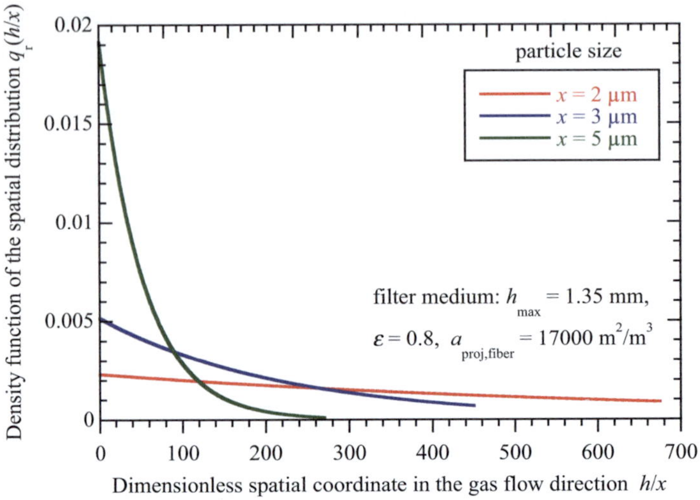

Fig. 3 Initial one-dimensional deposition profile in the depth filtration phase of a clean, homogenous filter medium from fibers by variation of the particle size, calculated by Eq. (7)

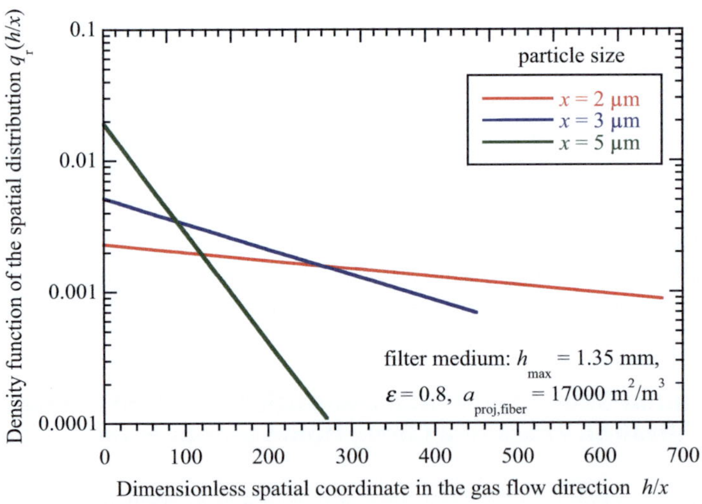

Fig. 4 Initial one-dimensional deposition profile in the depth filtration phase of a clean, homogenous filter medium from fibers by variation of the particle size, calculated by Eq. (7). Logarithmic scale is used for the y-axis (cf. Fig. 3.)

fibers facing the inlet gas flow $a_{proj,\ fiber}$, the resulting grade efficiency can be calculated by Eq. (5):

$$T_x\left(\frac{h}{x}\right) = 1 - e^{-\frac{a_{proj,fiber} \cdot \varphi(x) \cdot h}{\varepsilon}} \tag{5}$$

The cumulative spatial distribution of newly deposited particles of the size x in any filtering layer with thickness h_{max} can be described by definition according to Eq. (1) as follows:

$$Q_r\left(\frac{h}{x}\right) = \frac{T_x\left(\frac{h}{x}\right)}{T_x\left(\frac{h_{\max}}{x}\right)}, \text{ with } 0 \leq h \leq h_{\max} \tag{6}$$

For a clean, homogenous filter medium made from fibers, the density spatial distribution of newly deposited particles of the size x at the beginning of the filtration process, i.e., the initial deposition profile, is then calculated by Eq. (7):

$$q_r\left(\frac{h}{x}\right) = \frac{\mathrm{d}T_x\left(\frac{h}{x}\right)}{T_x\left(\frac{h_{\max}}{x}\right) \cdot \mathrm{d}\frac{h}{x}} = \frac{1}{1 - e^{-\frac{a_{\text{proj,fiber}} \cdot \varphi(x) \cdot h_{\max}}{\varepsilon}}} \cdot \frac{a_{\text{proj,fiber}} \cdot \varphi(x) \cdot x}{\varepsilon} \cdot e^{-\frac{a_{\text{proj,fiber}} \cdot \varphi(x) \cdot h}{\varepsilon}} \tag{7}$$

Figure 3 shows exemplarily the initial deposition profile for the filtration of three different particle sizes ($x = 2$ µm, 3 µm, 5 µm) with the same filter medium and under the same filtration conditions. Here, the particle size x and the corresponding single fiber collection efficiency $\phi(x)$ are the only two variables for calculations with Eq. (7). The higher the value of the single fiber collection efficiency $\phi(x)$, the closer and more concentrated the deposited particles are to the upstream side of the filter medium.

Figure 4 shows the same quantitative relationships with a logarithmically scaled y-axis. When using logarithmic scale for the y-axis, the exponentially falling curve of the density function of the spatial distribution in the initial depth filtration phase (cf. Eq. (7)) can be clearly seen.

2.3 One-Dimensional Deposition Profile in the Surface Filtration Phase

In the surface filtration phase, the highly porous packed section on the upstream side of the dust cake with its stable, characteristic packing profile acts as the actual active deposition zone of the entire filter layer for the subsequent dust particles in the inlet gas. As a result, the corresponding one-dimensional deposition profile of newly deposited particles—defined by the previously introduced spatial distribution function—also has a stable, characteristic curve shape in this phase, which can be described by a logistic function as follows [1, 2]:

$$Q_r\left(\frac{h_{\text{sliding}}}{x}\right) = \frac{1}{1 - e^{-c_1 \cdot \frac{h_{\text{sliding}}}{x}}} \tag{8}$$

Here, the zero point of the x-axis is shifted to the mode of the distribution function to make a conclusive comparison. The parameter c_1, also called logistic growth rate, is the only parameter needed to describe the shape of the curve. c_1 is only depended on the filtration conditions and not on the original filter medium. Given the filtration conditions, c_1 can theoretically be calculated with the corresponding single particle collection efficiency using classical separation theory (see [2]). For dust cakes created by high-quality CFD simulation and calculation of particle trajectories, the parameter c_1 can be evaluated by a corresponding curve fitting (see [1, 2]). Figure 5 and 6 show exemplarily the results of the evaluation from high-quality CFD simulation and calculation of particle trajectories with the same filtration conditions as in the cases shown in Fig. 3, in the form of cumulative spatial distribution and its density function.

3 Correlation Analysis of the Change in the One-Dimensional Deposition Profile During the Entire Filtration Process

3.1 Modeling Results of the Continuous Change in the One-Dimensional Deposition Profile During the Entire Filtration Process Using a Multi-Layer Model with Offset for Dust Cake Growth

Zhang recently introduced a new multi-layer model with offset for dust cake growth, which enables the description of particle separation during the entire filtration process—from the initial depth filtration with a given filter medium to the cake-forming filtration phase—without a separate add-on model for switching between depth and surface filtration [2, 3]. The modeling results of this model can be used to calculate the introduced spatial distribution function for newly

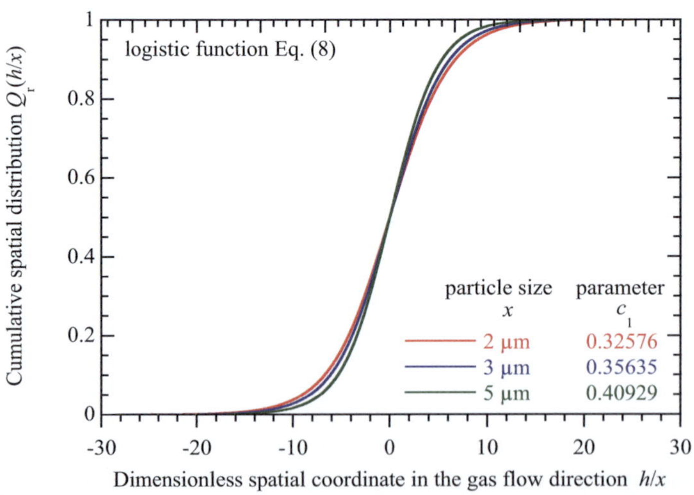

Fig. 5 One-dimensional deposition profile in the form of cumulative spatial distribution in the surface filtration phase by variation of the particle size. Evaluation of the results from high-quality CFD simulation and calculation of particle trajectories with the same filtration conditions as the cases shown in Fig. 3 [1, 2]

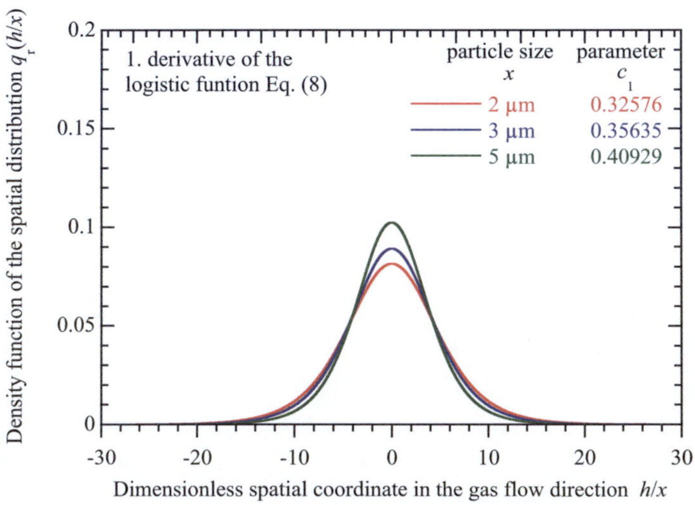

Fig. 6 One-dimensional deposition profile in the form of density function of the spatial distribution in the surface filtration phase by variation of the particle size. Evaluation of the results from high-quality CFD simulation and calculation of particle trajectories with the same filtration conditions as the cases shown in Fig. 3 [1, 2]

deposited particles during the entire filtration process. This allows the continuous change of the one-dimensional deposition profile to be evaluated and analyzed during the entire filtration process.

Figure 7 shows an example of the result of a model calculation with this model for a filtration process with the same filtration conditions as in the case for the particle size $x = 3$ µm shown in Fig. 3 and 6. Here it can be seen that the curve of the one-dimensional deposition profile approaches its final shape with increasing dust area load. After a critical value of the dust area load (in this case 15.98 g/m^2), the curve of the one-dimensional deposition profile takes on a shape that hardly differs from its final shape. A quantitative analysis of this dynamic change process can be carried out using a correlation analysis, as presented below.

3.2 Using Correlation Analysis to Describe the Dynamic Process of the Structural Change During the Entire Filtration Process

With the introduction of the specific one-dimensional spatial distribution function for newly deposited particles in the filter packing, a new feature is developed to characterize the filtration process. A quantitative evaluation of the change in the one-dimensional deposition profile during a continuous filtration process can be carried out using the methods of correlation analysis, which is widely used

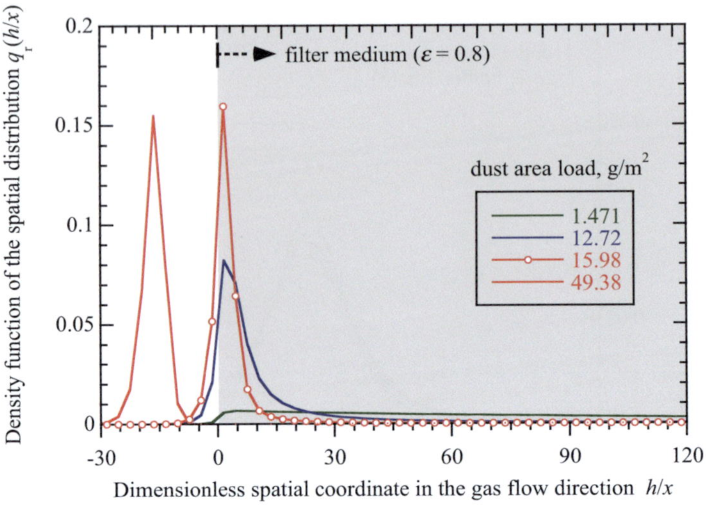

Fig. 7 Selected one-dimensional deposition profiles in the form of density function of the spatial distribution in correlation with dust area load during a continuous filtration process. Evaluation of the results from a model calculation with a multi-lay model with offset for dust cake growth introduced by Zhang [2]

in digital signal processing and image processing. Details of this analysis were described by *Zhang* (see [2]).

Here, the so-called normalized cross-correlation function of the one-dimensional deposition profile, using the density function, can be evaluated with regard to the final deposition profile to give the similarity between the curve progressions: In the surface filtration phase, the maximum similarity between the curve progressions of the deposition profile is reached; the normalized cross-correlation function takes on the value 1. At the beginning of a filtration process with a clean filter medium, the similarity between the initial deposition profile and the final deposition profile is low: The lower the similarity, the lower the corresponding value of the normalized cross-correlation function.

In the case of filtration processes with a clean, homogenous filter medium, the evaluated normalized cross-correlation function of the one-dimensional deposition profile can be described with a linear approximation function as follows:

$$r_{q_{r,\tau\ \max};q_{r,\tau}}(\tau) = \begin{cases} r_{q_{r,\tau\ \max};q_{r,\tau}}(\tau = 0) + c_2 \cdot \tau & \text{for } 0 \leq \tau \leq \tau_{\text{critical}} \\ 1 & \text{for } \tau_{\text{critical}} < \tau \end{cases} \tag{9}$$

Both the resulting dust area load and the so-called filter parameter, which indicates the ratio of the total projection area of deposited particles to the filter face area, can be used as variables τ, which indicates the progress of the dynamic process.

Figure 8 shows the results of this evaluation for three filtration processes by variation of the dust particle size and using the resulting dust area load to indicates the progress of the dynamic process. With the exception of the different dust particle size, the same filtration conditions apply to all filtration processes here. As can be seen here, the smaller the dust particles are, the less the resulting mass of deposited particles to reach the cake-forming filtration phase.

Figure 9 shows the results of this evaluation for the same filtration processes, but using the resulting filter parameter of deposited particles to indicates the progress of the dynamic process. Compared to the illustration in Fig. 8, it can be clearly seen here that the corresponding filter parameter is the more meaningful variable to indicate the progress of the dynamic process instead of the particle mass. With different particle sizes, the gradients of the three linear curves only differ slightly. The differences of varying degrees between the respective initial deposition profile and the final deposition profile in the case of different particle sizes are overcome at a similar rate. This can be explained conclusively using appropriate filtration theory.

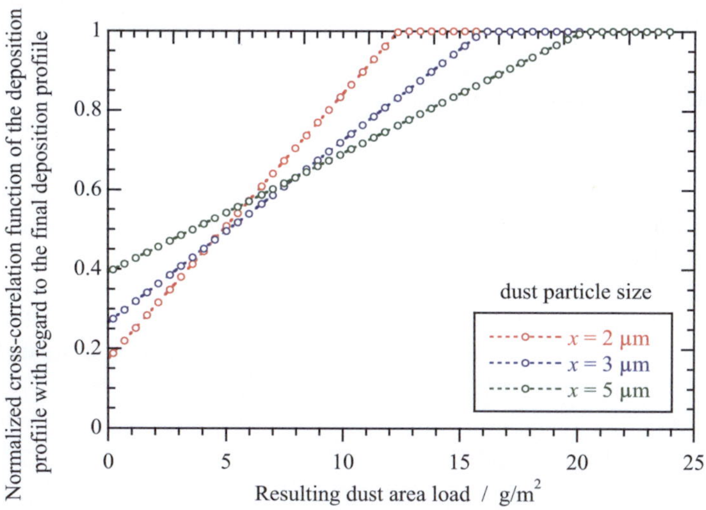

Fig. 8 Evaluation of the normalized cross-correlation function of the one-dimensional deposition profile with regard to the final deposition profile as a function of resulting dust area load for filtration processes by variation of the dust particle size

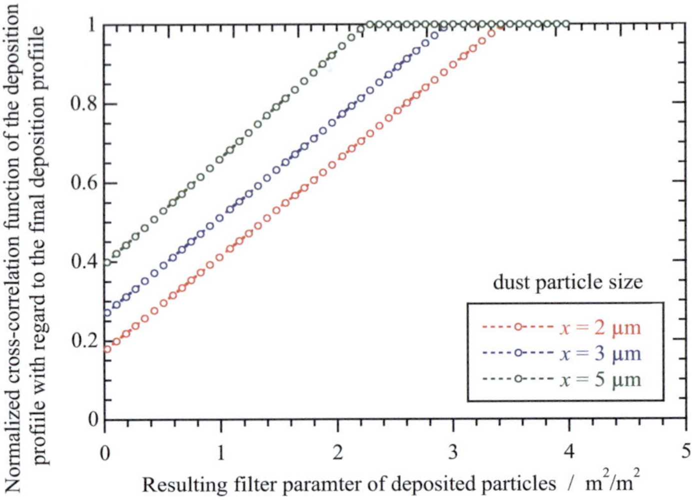

Fig. 9 Evaluation of the normalized cross-correlation function of the one-dimensional deposition profile with regard to the final deposition profile as a function of resulting filter parameter of deposited particles for filtration processes by variation of the dust particle size

4 Summary

Due to the discrepancy between the previous theoretical description of the filtration process and the actual structural changes in the filter layer, a new holistic approach to describing the filtration process is introduced, which is based on an analytical description of the structural changes of the filter packing in the direction of gas flow. A new specific one-dimensional spatial distribution function for newly deposited particles in the filter packing during the filtration process is introduced to describe the deposition profile of newly deposited particles of any filtration process in dimensionless form. The systematic change in the one-dimensional deposition profile during a continuous filtration process can be evaluated using a correlation analysis to describe the kinetics of the dynamic process.

References

1. Zhang, Q.: One-dimensional view of the structural changes in the filtering layer during continuous dust deposition – A new approach to a holistic view of the filtration process, Chem. Ing. Tech., 96 (8), 1131–1137 (2024).
2. Zhang, Q.: Strukturänderung der filternden Schicht und Änderung der Filtrationskinetik während der Clogging-Phase eines Filtermediums bei der Staubabscheidung – Eine holistische Betrachtung des Filtrationsprozesses, Habilitationsschrift, Bergische Universität Wuppertal, Shaker Verlag, Düren (2024).
3. Zhang, Q.: Modeling the entire filtration process from depth filtration to surface filtration with a multi-layer construction and an offset for dust cake growth, Chem. Ing. Tech., 97 (4), 334–343 (2025).

Fire Prevention

Challenges in Fire Spread Modeling—From Bench Scale up to Wildfire Scale

Fabian Brännström, Hamza Jamil, Mehrdad Nouroozyan and Johannes Sailer

Contents

Abstract

Safety Engineering is a broad field with significant societal impact. Within this discipline, fire safety engineering plays a critical role, frequently utilizing fire simulations across diverse scenarios and scales—from small laboratory settings to large-scale wildfire modeling. This chapter addresses key challenges in fire spread modeling through selected simulation examples, including pyrolysis, radiative heat transfer, and a burning tree, each illustrating specific challenges that substantially affect the accuracy of fire spread simulations. Additionally, the chapter introduces a potential approach for managing uncertainties in *Computational Fluid Dynamics* models using *Polynomial Chaos Expansion*, applied here to a compartment fire scenario. This chapter is not intended as an exhaustive review of the topic; rather, it aims to underscore the need for further research in the field of fire spread modeling.

Keywords

Fire spread · Wildfire · Pyrolysis · Radiative heat transfer · Simulation · Uncertainties

F. Brännström (✉) · H. Jamil · M. Nouroozyan · J. Sailer
Chair of Fire Dynamics, University of Wuppertal, Wuppertal, Germany
e-mail: braennstroem@uni-wuppertal.de

S. Bracke and R. Pieper (eds.), *Safety Engineering*,
https://doi.org/10.1007/978-3-658-47473-7_13

1 Introduction

Fire safety engineering is a critical discipline within the broader field of *Safety Engineering*, addressing frequent and potentially destructive risks in buildings, industrial facilities, and public spaces. As an integral component of safety engineering, fire safety emphasizes the prevention, control, and mitigation of fire-related hazards, aiming to protect lives, property, and the environment. And in fire safety engineering fire simulation has become a valuable tool, particularly for predicting smoke and temperature distribution in various scenarios. These simulations rely on a predefined heat release rate time curve, often termed a *design fire*, to model the behavior of a fire. However, the complexity of fire phenomena, involving complex processes of combustion, heat transfer, and turbulence, makes accurate simulation a challenging task. Common applications of fire simulations include assessing the performance of smoke exhaust systems in buildings like malls and sports arenas [1]. While these simulations can provide valuable insights, their accuracy is heavily dependent on accurate boundary conditions, initial conditions, and a realistic representation of the building geometry and its materials. A significant challenge lies in defining relevant fire scenarios, especially for complex structures like multi-story hospitals or large-scale industrial facilities. The scenario definition itself involves a substantial portion of the boundary and initial conditions, often necessitating assumptions due to the inherent uncertainty of real-world fire events.

A greater challenge arises when simulating fire spread itself. In these simulations, only the initial ignition source is predefined, while the heat release rate (which defines the design fire curve) is calculated based on local conditions. Nearby materials may ignite or pyrolyze due to heat transfer from the ignition source. Such simulations are commonly used in the railway industry for decision-making during the design phase, such as choosing appropriate materials for seats and linings [2]. These simulations require many assumptions, which can significantly influence the results. Despite the wide adoption of fire safety codes and standards, there is currently no common standard for fire spread simulations.

The concept of flammability, which is directly related to fire spread, further illustrates the complexity of fire simulations. A general definition of flammability includes a material's ease of ignition, the intensity of its burning, heat release, fire spread propensity, and the rate at which it generates smoke and toxic products during combustion [3]. Determining ignition itself is often challenging. A more precise definition of ignition is based on the critical mass flow of volatiles (pyrolyzate) needed to sustain a flame capable of transferring heat to the surface without being extinguished. However, this mass flux is difficult to determine, leading to the common use of the *ignition temperature*, although this value varies across environments and lacks.

Given the complexities involved, one might question whether such detailed fire simulations are truly necessary outside of fundamental research. In the railway industry, for instance, there is a need for design curves during the vehicle design process. Instead of simulations, large-scale testing of passenger compartments could be done toward the end of the design phase. However, a single test would not provide enough data to fully address the complex scenarios, and the current practice favors simulations as complementary tools to existing fire safety concepts.

Another important application of fire spread simulations is in the context of wildfires, which are increasingly linked to climate change. Even a slight rise in temperatures can increase the frequency and intensity of wildfires, which in turn contribute to greenhouse gas emissions and impact both local and global climates. Wildfires occur even in moderate climate zones; for example, Arctic smoldering fires in 2015 released an estimated 143 Mt of carbon [4,5], while the 2023 wildfires in Canada emitted an estimated 647 Mt of carbon which is about four times its total fossil fuel emissions [6]. Fire spread predictions are not only used for fire safety planning but also as operational tools during active wildfires. In addition to local knowledge and firefighter experience, simulations help inform evacuation plans and firefighting strategies. Fire spread models are also used for preparing prescribed burns (e.g., for forest management) and conducting forensic analysis after wildfire events. These simulations do not need to provide real-time results, so a variety of modeling approaches can be employed.

The following sections will present an examination of fire spread simulation techniques, with a particular focus on the complexities involved in modeling such intricate phenomena. Section 2 provides an overview of the various models that are commonly utilized in fire simulation, and it offers a more detailed introduction to fire spread modeling, including an examination of the most prevalent methods for determining material parameters from bench scale experiments. The section concludes with an examination of wildfire modeling, with a particular focus on fire spread processes.

This section also discusses the distinct challenges encountered in large-scale scenarios, such as uncertain environmental conditions, variations in fuel characteristics, and dynamic changes throughout the progression of fires, particularly in the case of wildfires. Following this basic discussion of fire spread methods, Sect. 3 examines three specific applications that underscore key simulation challenges. These applications include comprehensive pyrolysis modeling in Sect. 3.1, the role of numerical techniques in radiative heat transfer in Sect. 3.2, and fuel distribution and its impact on heat release rates of burning trees in Sect. 3.3. Finally, before the conclusion, Sect. 4 presents a potential approach to address some of the challenges highlighted by these applications. This section introduces a method to propagate input uncertainties in complex CFD simulations, exemplified through a case study that, while distinct from previous applications, demonstrates how to enhance risk understanding through uncertainty analysis.

2 Fire Spread Modeling

Fire spread is a complex, multi-scale phenomenon that typically begins at small scales, such as ignition from a cigarette or an electric spark, and can evolve to larger scales, encompassing rooms, buildings, or even entire landscapes. This process involves intricate interactions between solid materials and the surrounding gas, all heavily influenced by uncertainties. The solid-phase process of pyrolysis is particularly complex, poorly understood, yet critical for driving flame and fire spread. Pyrolysis involves the thermal decomposition of materials, producing intermediate substances like char that alter heat transfer conditions throughout the process. This process depends on multiple factors, including diffusion, convective heat transfer, liquid movement in solid pores, heterogeneous reactions, and in-depth radiation, all of which are impacted by temperature-dependent properties. Additional challenges such as structural changes from fractures or cracking are frequently overlooked in models.

The processes occurring in the gas phase are similarly complex and difficult to model. Gas-phase dynamics often include buoyancy-driven, reactive, non-premixed, sooty flames, which vary between laminar, transitional, or turbulent states depending on the scale. Convective and radiative heat transfer create a complex coupling between the gas and solid phases. Despite, or fortunately, depending on the view point, an extensive body of literature exists on fire spread simulations, which not only aids in estimating fire spread in specific cases but also significantly enhances general understanding [7–9]. These simulations support research by refining experiment accuracy, comparing results with CFD models, and prompting more precise measurements. The *Measurement and Computational Fire Phenomena* (MaCFP) group, established in 2017, exemplifies this effort. Initially focused on providing high-resolution gas-phase fire experiments for simulation validation, MaCFP has since expanded to include subgroups dedicated to solid-phase modeling and radiative heat transfer phenomena.

In fire modeling, *Large Eddy Simulation* is commonly employed due to the generally moderate *Reynolds* numbers involved. For practical applications, a *Very Large Eddy Simulation* approach is also frequently utilized, with grid cell sizes on the order of 10 cm to 20 cm for modeling smoke and temperature distribution and extending to 10 m to 100 m for large-scale wildfire simulations.

In fire spread modeling, gas-phase reactions are often simplified to a few global reaction steps, as precise details on local fuel composition and ambient conditions are typically unavailable. This approach relies on mixture-based combustion methods, primarily derived from the eddy break-up concept and its subsequent adaptations. However, accurately predicting soot and toxic gas formation under these simplifications remains challenging and is strongly simplified.

Heat transfer—radiative, convective, and conductive modes—drives all these processes, making accurate heat transfer prediction critical for fire spread modeling. However, obtaining precise, unbiased measurements and implementing detailed heat transfer models is complicated by the coarse mesh used in simulations, unknown species concentrations in both gas and solid phases, the impact of pyrolysis, and the porous, heterogeneous structure of solid materials. As a result, convective heat transfer is often estimated using *Nusselt* number correlations. For radiative heat transfer, most models rely on *Finite Angle Methods* with gray radiative properties, which can introduce numerical issues such as ray effects and false scattering [10, 11] may lead to inaccuracies in radiative heat flux predictions, particularly when the spectral characteristics of radiation are ignored, as discussed in studies by [12] and [13]).

Methods for simulating flame and fire spread range from simplified threshold-temperature approaches, which use a temperature threshold with prescribed local heat release rates, for detailed pyrolysis calculations that incorporate approximately 4 to 10 reactions for the solid phase and employ eddy dissipation-based models for gas-phase reactions. This modeling approach remains largely consistent across different scales relevant to building fires, as similar models are applied to both small- and large-scale simulations, including rooms and entire buildings.

Detailed pyrolysis modeling typically relies on an Arrhenius-based first-order reaction model, where global reaction processes are chosen according to the specific needs of the simulation (e.g., serial, parallel, or competing reactions). Input parameters for kinetic and thermal properties are primarily obtained through inverse modeling techniques based on small-scale analyses, such as *Thermal Gravimetric Analysis* (TGA), *Differential Scanning Calorimetry*, *Simultaneous Thermal Analysis*, *Micro Combustion Calorimetry*, *Controlled Atmosphere Pyrolysis*, and *Cone Calorimetry* under various heating and heat flux conditions [9,14,15]. Often, the heating rates applied in TGA do not match those observed in real fire scenarios, which may vary widely from 10 K/min to 1000 K/min depending on material depth and heat source intensity. This mismatch in heating rates and strong simplifications in occuring pyrolysis reactions can lead to non-physical parameters in inverse modeling [16]. There is ongoing debate within the field regarding the best approach to address this challenge.

Several simulation tools are available for flame and fire spread modeling, including Fire Dynamics Simulator [17], gpyro [18], and fireFoam [19]; all are open-source platforms. Each of these tools offers unique features and simplifications: gpyro, for example, is particularly comprehensive for pyrolysis modeling but lacks gas-phase calculation capabilities. For pyrolysis simulations in this document, PATO [20] is employed; developed by NASA and based on OpenFOAM, PATO was originally designed for space re-entry applications. It is coupled with fireFoam, allowing it to be used in flame and fire spread simulations.

Wildfire modeling encompasses a significantly broader range of temporal and spatial scales, extending from millimeter-scale phenomena to landscape scales where interactions with the atmosphere can influence local wind patterns and even regional weather conditions (Fig. 1).

Wildfire spread simulation approaches are typically divided into three primary modeling classes: empirical, semi-empirical, and physics-based models. Additionally, wildfire modeling is categorized into operational and research-focused applications. Operational modeling predominantly uses empirical or semi-empirical models, while physics-based models are mainly applied in research contexts, though efforts are underway to adapt these more sophisticated models for operational use in tasks such as prescribed burns.

Physics-based models themselves vary widely in their level of detail and methodology. These range from comprehensive, first-principles models that incorporate as many physical and chemical processes as feasible to hybrid models that combine empirical and physics-based components to different extents. Many quasi-physical

Fig. 1 Time and length scales occuring in wildfires

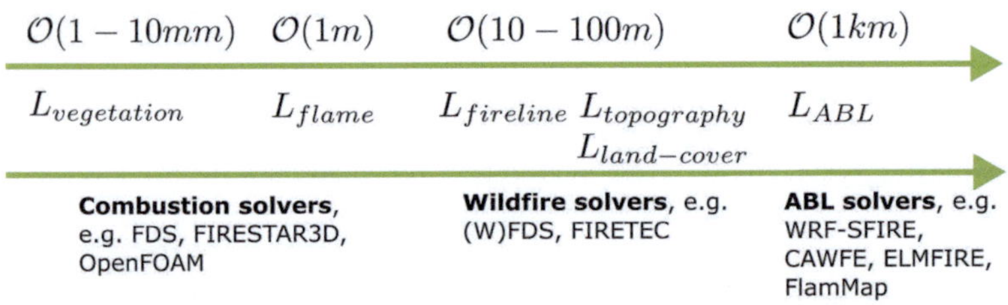

Fig. 2 Multi-scale problem in wildfire simulation (based on [23])

modeling approaches also exist; these incorporate empirical insights to simulate specific wildfire behaviors with varying degrees of approximation. [21]

When focusing specifically on CFD solvers, modeling methods can be categorized into several distinct classes [22]. *Combustion solvers* are designed to capture fire dynamics at the scales of vegetation and flame, while *Wildfire solvers* address dynamics at the fireline and landscape levels. At the largest, meteorological scales, *Atmospheric Boundary Layer solvers* (ABL) are developed to simulate interactions on a regional scale. Figure 2 presents examples of these solvers alongside their corresponding scales.

These varying scales are also associated with distinct modeling approaches and tools in wildfire simulation. Figure 2 provides a brief overview of the different methods and their respective scales.

Physical modeling methods often utilize *Lagrangian* particles, which streamline the simulation process by representing complex tree structures, including trunks and leaves, as a distribution of particles that mimic the effective behavior of actual vegetation. These particles can be thermally thin or thick and can assume specific aerodynamic properties, allowing for the application of drag coefficients and *Nusselt* number correlations. However, this approach is not suitable for modeling thermally thick tree trunks, as *Lagrangian* particles do not meet the continuum requirements necessary for accurate trunk representation. Consequently, trunks in realistic tree models are often omitted, and tree distributions are simplified for practical purposes

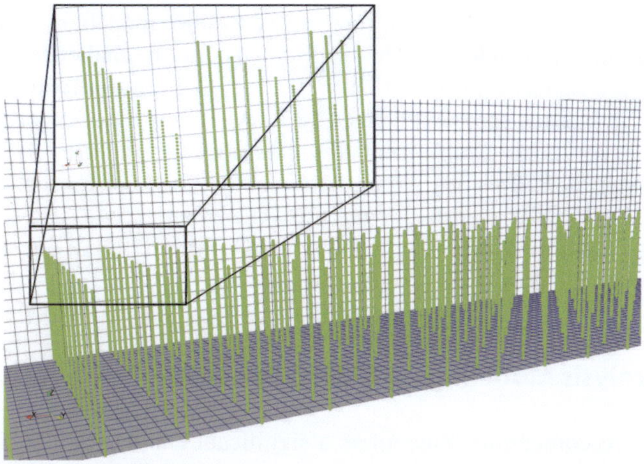

Fig. 3 Example particle distribution of grass for calculation with particle method; see also the experiments by Finney et al. [26] and Ahmed [22]

[24]. This particle-based approach can be effective for simulating thermally thin fuels, such as grassland [25]. Figure 3 illustrates a sample particle configuration that could be applied in grassland fire spread simulations. Similar controlled fuel arrangements have been experimentally investigated by Finney et al. [26] using cellulosic materials in a series of experiments on flame spread dynamics.

Simplified shapes are frequently used to model detailed trees in simulations [24,27,28], primarily due to the limited availability of precise geometric data (both locally and globally) regarding actual trees or plants. Consequently, trees are generally represented by simplified particle distributions in these models. While this approach can influence the fire spread dynamics, it may still yield acceptable results when the particle distribution is calibrated alongside adjustments to their thermal and kinetic properties. In such simulations, ignition sources are typically positioned beneath or at the edge of the tree. However, when examining fire spread across multiple trees, the placement of ignition sources and the modes of heat transfer vary from tree to tree, challenging the efficacy of using simplified geometry to represent complex tree structures. Currently, no definitive solution exists, and further research is required to address these complexities.

3 Challenges—Selected Examples

A limited set of fire scenarios is examined to illustrate key challenges in fire spread simulation. The first case, discussed in Sect. 3.1, addresses the influence of model-dependent parameters in comprehensive pyrolysis models, using the *MaCFP Gasification* case (see [29]) as an example. The second case, in Sect. 3.2, demonstrates how numerical methods can significantly affect heat flow calculations, evaluated here through radiative heat transfer from a small burner flame to surrounding walls.

The third case, covered in Sect. 3.3, focuses on wildfire modeling, specifically the numerical burning of a single tree. This case underscores the challenges of accurately representing tree fuel and its impact on fire spread. Section 4 introduces a potential approach for addressing these challenges by showing how input uncertainties propagate through outputs in a simple room fire scenario, exemplified by the well-known *Steckler* compartment fire [30]. While this method does not resolve all issues, it offers a valuable tool for uncertainty analysis. The emergence of *Scientific Machine Learning* also presents additional methods for propagating input uncertainties and predicting field behaviors in fire scenarios.

3.1 Pyrolysis Modeling—Model Dependent Parameters

The fire science community encounters a significant challenge in that kinetic and thermal parameters of materials are not readily available. Typically, these properties are derived through inverse modeling. Moreover, due to the limited understanding of the complex reactions and thermal behavior of materials, these parameters are often regarded as effective estimates rather than as true intrinsic properties. Consequently, these effective parameters—particularly when derived through an inverse modeling process—are typically tailored to the specific simulation tool used during that process. To illustrate the challenges associated with model-dependent parameters in comprehensive pyrolysis modeling, the *MaCFP Gasification* setup (see Fig. 4a) is simulated using the PATO framework. This setup simplifies the scenario by suppressing flaming combustion through the use of nitrogen as the ambient gas, thereby eliminating additional uncertainties that would otherwise be introduced by flame radiation following the ignition of pyrolysis gases. Figure 4a depicts the simulation domain, featuring a wedge-shaped mesh with corresponding boundary conditions tailored to this geometry.

Since flaming combustion is not a factor in this case, the simulation domain can be simplified by modeling only the solid component, with a predefined heat flux applied to the top surface of the PMMA sample. For this scenario, a uniform heat

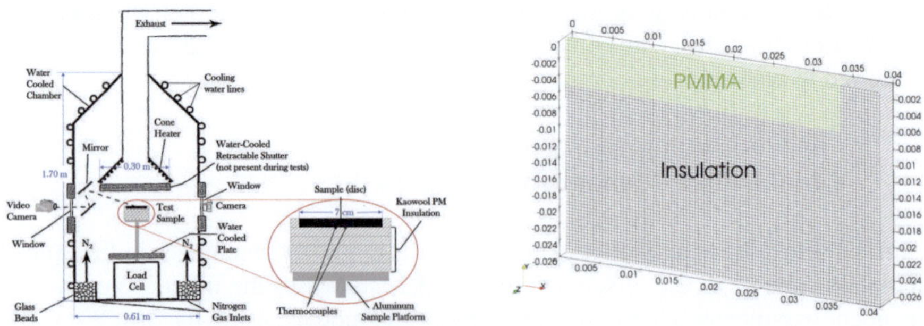

(a) Experimental setup ([31]) (b) Mesh setup

Fig. 4 Description of *MaCFP Gasification* setup

flux of 25 $\frac{\mathrm{kW}}{\mathrm{m}^2}$ is applied, without accounting for any specific spatial distribution of incident heat flux, while all other boundaries are maintained at ambient conditions.

This section provides a selective overview of the system of equations; further details are available in [20] and [32]. The pyrolysis reactions for N_p materials (or phases) and P_i sub-materials (sub-phases) are modeled using Arrhenius laws. The progress χ_{ij} of the pyrolysis reaction for sub-material j within the main material i is expressed as:

$$\partial_t \chi_{ij} = A_{ij} \exp\left(\frac{-E_{ij}}{RT}\right)(1 - \chi_{ij}), \quad \forall i \in [1, N_p], \quad \forall j \in [1, P_i] \tag{1}$$

with the pre-exponential factor A_{ij} and activation energy E_{ij}. The overall pyrolysis-gas production rate is evaluated as:

$$\pi_{\mathrm{tot}} = \sum_{i}^{N_p} \sum_{j}^{P_i} \rho_i \epsilon_i F_{ij} \partial_t \chi_{ij} \tag{2}$$

where ρ_i represents the intrinsic density, ϵ_i denotes the volume fraction and F_{ij} is mass fraction of reaction j within phase i. Here, we have PMMA as the phase 1 and not sub-phase. With local thermal assumption the energy conservation is given as: In this case, PMMA is considered as phase 1 and no sub-phase is introduced. Under the assumption of local thermal equilibrium, the energy conservation equation is expressed as:

$$\rho c_p \partial_t T + \partial_x \cdot (\mathbf{k_s} \cdot \partial_x T) + -h\pi_{tot} = 0 \tag{3}$$

where, $\mathbf{k_s}$ is a second-order tensor accounting for conduction in the solid.

The limitations of using readily available parameters derived from other models become apparent when attempting to apply parameters from various MaCFP participants who utilized different inverse modeling algorithms, parameter ranges, and simulation tools [31]. For instance, the *mass loss rate* (MLR) curve predicted by PATO—a tool not included in the original inverse modeling—does not align with the experimental MLR curves (see Fig. 5). Contrarily, when the parameters are derived directly within PATO, the results differ significantly. In this study, only a single setup with a specified heat flux was considered for the inverse derivation process.

The resulting MLR curve, shown in Fig. 6, demonstrates strong agreement between the MLR calculated by PATO for the *MaCFP Gasification* setup and the experimental results. Thus, these outcomes are not predictions in the strict sense but are instead based on an inverse modeling process.

In general, increasing model complexity often amplifies sensitivity to these effective parameters. For several years, the use of effective parameters has been debated within the fire science community (see, e.g., [16]), as more complex reaction models do not necessarily yield better predictive accuracy across the diverse scales relevant for flame and fire spread. Currently, determining the required number of reactions and the extent of temperature-dependent thermal properties remains case-specific. Even setting thermal conductivity accurately can be challenging, as factors such as

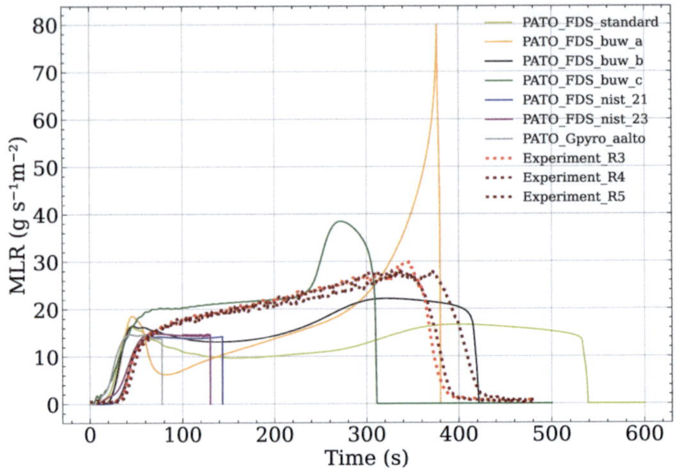

Fig. 5 Mass loss rate calculated with PATO based on input from MaCFP participants

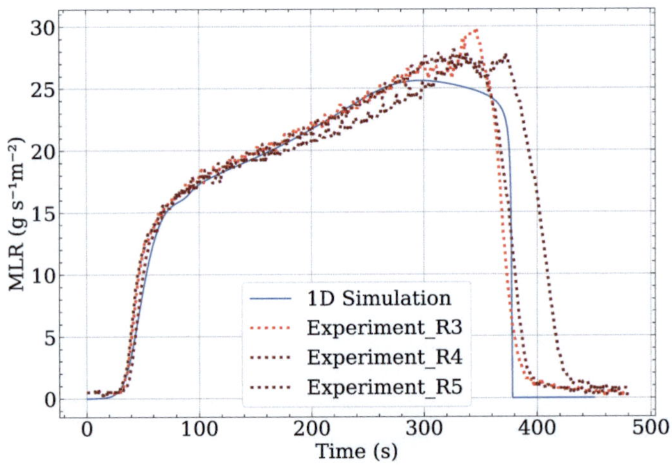

Fig. 6 Mass loss rate calculated with PATO after inverse modeling approach for relevant parameters

overlapping local radiation and conduction in porous materials, anisotropic behavior, material heterogeneity, unknown contact resistance between solids, and decomposition across a wide temperature range affect the material's behavior.

The most effective current approach is to conduct inverse modeling across multiple scales using varied heating rates or heat fluxes, with the same model later used in production, to establish a robust set of effective parameters.

This section underscores several challenges in fire spread modeling with detailed pyrolysis models. The unknown characteristics of solid fuels and their volatile products after decomposition necessitate inverse modeling, leading to the use of effective parameters derived from this approach.

3.2 Modeling of Radiative Heat Transfer—Numerical Effects

It is well known that numerical diffusion, dispersion, and dissipation significantly affect computed fields in simulations of turbulent and transitional laminar-turbulent flows. Each of these factors can impact simulations to varying degrees, depending on the specific scenario and simulation setup. For thermal radiative heat transfer, additional numerical challenges arise, including false scattering (essentially a form of numerical diffusion) and the ray effect. The ray effect occurs in *Finite Angle Methods*, where intensity is calculated only for a finite number of predefined directions. While increasing the number of angles can reduce the ray effect, it also requires solving additional intensity equations across the entire domain. Each ray direction introduces a new transport equation, often resulting in 400 to 800 additional equations that must be solved.

The following selected example illustrates the challenges that arise particularly in fire modeling, where the domain may contain both optically thin and thick regions, complicating accurate simulation.

Here the so called *FM Burner* test case from the MaCFP Radiative Heat Transfer Phenomena sub-group [33] is used for highlighting numerical challenges for the calculation of radiative heat transfer. The following Fig. 7 displays the actual simulation domain with the burner which is placed in the middle; it is an ethylene burner set up to a heat release rate of about 15 kW for standard oxygen concentration in the environment. The radiation absorption model of the gas is set to a *Planck Mean* value for the purpose of demonstrating the impact of numerical methods on radiative heat transfer. Further details about the simulation are available in Paul et al. [12].

In this study, the *FM Burner* test case from the *MaCFP Radiative Heat Transfer Phenomena* sub-group [33] is used to highlight numerical challenges in calculating radiative heat transfer. Figure 7 illustrates the simulation domain, where the burner is centrally positioned. This setup involves an ethylene burner operating at a heat release rate of approximately 15 kW under standard atmospheric oxygen conditions. For this demonstration, the gas radiation absorption model is set to a Planck mean value to examine mainly the influence of the numerical method on radiative heat transfer. Further details regarding the simulation setup are provided by Paul et al. [12]. Figure 8 presents a comparison between the *Finite Angle Method* (FAM) calculation and the solution from a detailed *Photon-Monte-Carlo* (PMC) solver, which serves as accurate validation data. For this comparison, FAM was configured with only 80 ray directions. The ray effect is clearly visible in the FAM setup, showing a significant impact on the incident radiative flux on the surrounding walls when compared to the *Photon-Monte-Carlo* results. These local peaks and troughs lead to variations in radiative heat transfer ranging from 500 W/m^2 to 3000 W/m^2 in this setup, which could strongly influence temperature development and, consequently, fire spread in these regions. Achieving a comparable heat flux distribution would require a much more computationally expensive approach, using approximately 1200 rays, though this requirement is highly case-dependent.

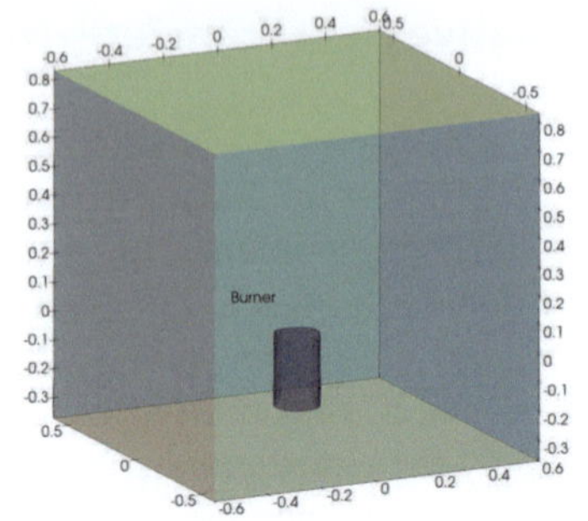

Fig. 7 *FM Burner* experimental setup

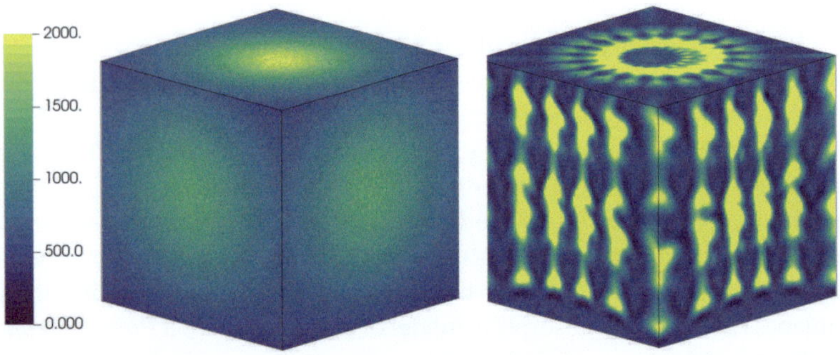

Fig. 8 Demonstration of ray effect for *FM Burner* test case by visualizing the calculated heat flux (W/m^2) on the surrounding walls of the domain; left: PMC benchmark [12], right: FAM with 80 directional angles

3.3 Wildfire Modeling—Distribution of Particles

Thus far, we have examined challenges in pyrolysis modeling and the numerical impacts of the radiation solver, though it is worth noting that both can significantly affect fire spread simulations, influencing predictions of heat release rates and emissions.

In contrast, wildfire modeling presents an entirely different level of uncertainty. While we may apply the most accurate pyrolysis and combustion models with known boundary conditions, challenges arise with fuel distribution. In most controlled scenarios of building fires, a CAD model specifies fuel locations; however, in wildfire modeling, we often have only a rough understanding of spatial fuel distribution. The importance of accurate fuel distribution depends heavily on the specific scenario and the objectives of the simulation. To explore this, we simulate a burning tree using three different configurations, each with the same fuel load, total mass, and moisture content but differing in overall shape.

The three configurations are as follows: (1) a standard conical shape, taken from the FDS Validation Guide [24], (2) a cylindrical shape with matched volume and fuel mass, and (3) a detailed tree shape with similar volume and mass. One might

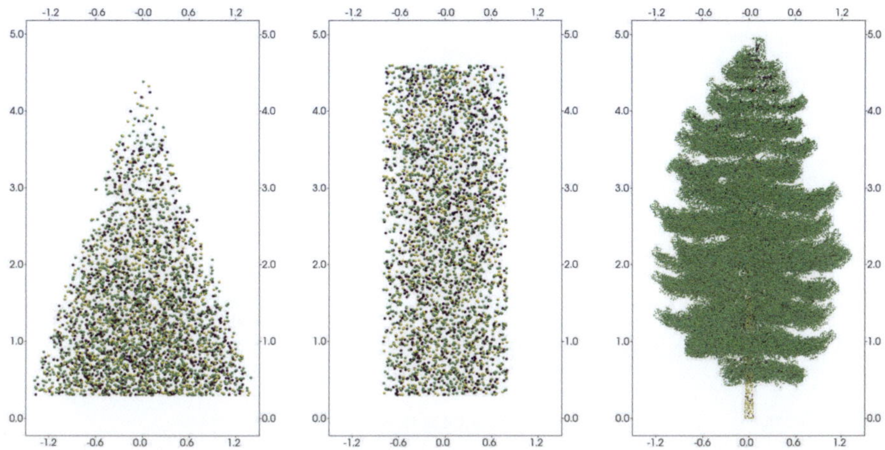

(a) Cone shape (b) Cylinder shape (c) Detailed shape

Fig. 9 Selected tree shapes and distribution of fuel; numbers in the images indicate positions in meters

initially assume that the detailed tree configuration would yield the most accurate results, but as shown in the results section, this may not be the case, deserving further discussion. In wildfire modeling, fitting fuel load based on mass alone may not be optimal; instead, specific parameters, such as fuel load per surface area, may be more relevant depending on the scenario.

The cone shape, taken directly from the FDS Validation Guide, remains unaltered and all trees exclude trunks, as particle-based models generally do not accurately represent large solid objects, adding another layer of simplification across the setups. Figure 9 illustrates the particle distributions for the three shapes: conical, cylindrical, and realistic tree configurations. While each setup maintains a nearly identical total fuel mass, their spatial distributions vary, which is known to influence fire spread dynamics, particularly in wildfire scenarios. Determining the correct distribution remains a challenge, as various factors depend on the simulation's application. For detailed modeling, a particle-based approach is typically used, requiring that simplified particles represent each leaf or needle's pyrolysis behavior and interact aerodynamically with the tree's surrounding environment. This must be achieved on a relatively coarse mesh, which may not fully resolve essential processes like heat transfer and gas or solid-phase combustion. Additionally, the *Lagrangian* particle approach does not accommodate large trunks, which would require a continuum approach. This limitation is sometimes bypassed by assuming that larger trunks do not directly participate in fire spread. However, this assumption neglects important flow effects, such as vortex shedding around tree trunks, which can lead to phenomena like fire whirls or tornadoes.

For all three cases, the simulation domain (in all direction a length of 9.6 m) and same mesh setup ($96 \times 96 \times 96$ cells) was chosen. The ignition was achieved by an ring-shape ignitor with a radiaus of 0.6m and a heat release rate of $103\,\mathrm{kW/m^2}$; the ignitor was placed below the trees at a height of $z = 50$ cm above ground. Moisture

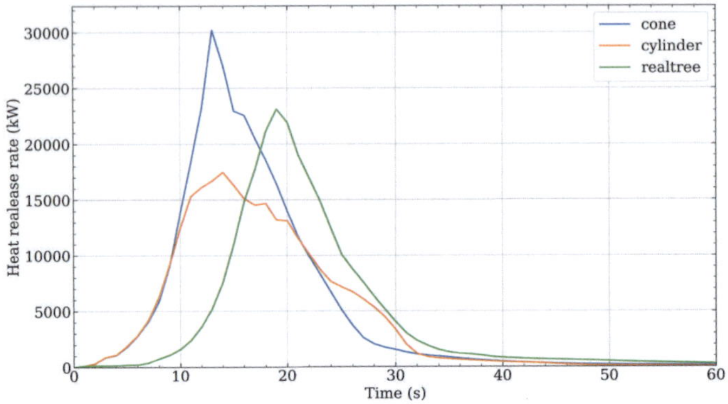

Fig. 10 Comparison of calculated heat release rates (kW) for different shapes

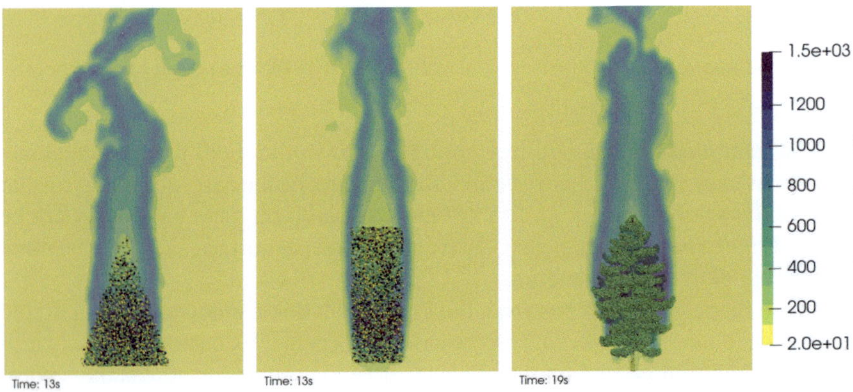

Fig. 11 Snapshots of temperature field for different tree shapes at selected times; left: cone shape, middle: cylinder shape, right: detailed tree

content, thermal and kinetics properties were used based on NIST Douglas Firs FDS validation setup [24].

Figure 10 presents a comparison of the heat release rate (HRR) curves for the three different shapes.

As expected, the fuel distribution significantly affects both the timing and peak values of the calculated HRR for each shape. In this setup, the maximum HRR ranges from approximately 17 MW for the cylindrical shape to around 23 MW for the detailed tree shape, and reaches about 30 MW for the conical shape. Additionally, there is a time shift of about 6 s in the peak HRR values between the generic shapes and the detailed tree shape, which is notable given the overall burning duration of roughly 40 s. These differences are expected to influence fire spread calculations in larger-scale scenarios involving numerous trees with varied shapes and non-uniform spatial distributions. Figure 11 provides a snapshot comparison of the temperature distribution for each tree shape at times close to their respective peak HRR values.

The overall temperature distributions in these selected snapshots appear quite similar across the full domain. However, closer inspection reveals significant local temperature differences. While this comparison provides limited insight for detailed analysis—since these snapshots are not directly comparable—it highlights the complexities inherent in accurately setting up tree models for this study. This analysis only compares the heat release rate for a single tree. An open question remains: how does this variability affect fire spread among multiple trees, each with slightly different fuel distributions? Additionally, fluid dynamic interactions occur between trees, which are strongly influenced by fuel distribution. The chosen distribution shapes flow patterns that, in turn, influence heat transfer to surrounding leaves, trunks, and soil.

Another critical factor is the specific location of ignition, as ignition points are relative to the fuel distribution and not precisely the same in each of the three scenarios here. This variation also applies to real-world experiments, as discussed in [27], which examines the role of ignition in fire spread among trees. The field of wildfire modeling requires considerably more research. Current efforts are focused on transferring detailed forest fuel data from LIDAR measurements to semi-detailed, voxel-based tree models for particle-based simulations (see [28]).

4 Handling of Uncertainties in Fire Simulations

How, then, can we address the diverse challenges across various scales, modeling as well as numerical approaches, and initial and boundary conditions? Ultimately, we must contend with uncertainties for the different challenges. Despite these challenges, fire simulations continue to play a crucial role in supporting decisions in both project planning and field applications. There is no singular solution to address these complexities, underscoring the need for high-quality experiments and rigorous validation studies. The *MaCFP* initiative [34] marks an important step in this direction, and expanding access to laser diagnostics could further enhance our understanding of species distribution, temperature gradients, and boundary conditions in smaller-scale experiments. However, this still leaves the fundamental challenge of scaling—from small ignition sources to land-scale wildfires. Research must continue at large scales, despite the increased uncertainties in boundary and initial conditions and the variability in fuel distribution and structure. This need extends across other scales as well. In addition, there is a pressing need for methods and tools that can analyze global sensitivities and propagate input uncertainties in computationally demanding CFD simulations (see, e.g., [35]).

In the next two chapters, we explore an approach for uncertainty analysis based on *Polynomial Chaos Expansion* (PCE). PCE offers a mathematical framework to manage costly CFD simulations by approximating the ensemble flow field with a surrogate model. Following a brief theoretical overview in Sect. 4.1, we demonstrate how PCE could be applied to routine project work. For illustration, we use a compartment fire scenario, here it is the well-known *Steckler* case (see Sect. 4.2). While this example is not directly related to the cases discussed earlier, it demonstrates an effective approach for handling uncertainties. In this case, we assume uncertainties in the burner's heat release rate and the surrounding wall density. A comprehensive global sensitivity analysis is not included here, as the focus is on demonstrating a potential method for managing uncertainties in fire simulations.

4.1 Polynomial Chaos Expansion

Polynomial Chaos Expansion [36] provides a spectral decomposition of the function f, separating it into deterministic expansion coefficients f_i and stochastic components $\varphi_i(\boldsymbol{\xi})$, where φ_i are predefined random basis polynomials. This decomposition can be represented in a single-index notation, with $\boldsymbol{\xi}$ as the vector of d random variables $\{\xi_1, \ldots, \xi_d\}$, as shown in Eq. 4. For the selected case, we consider two random variables.

$$f(x) = \sum_{i=0}^{\infty} f_i(x)\varphi_i(\boldsymbol{\xi}) \tag{4}$$

The approach involves truncating the infinite summation at an appropriate high order p; in this test case, we use an order of 6.

$$f(x) \simeq \tilde{f}(x) = \sum_{i=0}^{p} f_i(x)\varphi_i(\boldsymbol{\xi}) \tag{5}$$

There are two main approaches to *Polynomial Chaos Expansion*: intrusive and non-intrusive. In this study, we use only the non-intrusive method; an example of the intrusive approach in fire modeling is provided in [35]. For non-intrusive PCE (NIPC), a surrogate model is developed and trained using detailed CFD simulations. In this case, training simulations are performed with FDS, and the polynomial coefficients are derived through a regression method [37]. Alternative methods include spectral projection [38]; however, in our previous study [35], using an Arrhenius model, spectral projection demonstrated slower convergence, requiring more training data when the number of input parameters was increased from two to four.

PCE offers a key advantage over other uncertainty quantification methods in that first-order $\mathbb{E}(y)$ and second-order $\mathbb{V}(y)$ statistical moments of the output can be analytically derived from the PCE coefficients [39]. This capability can improve model convergence by circumventing the costly process of numerically computing moments for the surrogate model.

4.2 Steckler Compartment Fire

The well-known *Steckler* room case follows the setup outlined in the FDS Validation Guide [24], so detailed descriptions of boundary conditions and material properties are omitted here. For the PCE approach, we assume two uncertain parameters with a

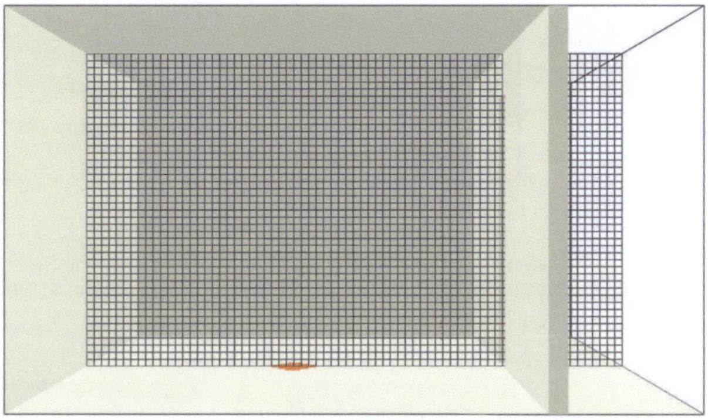

Fig. 12 Overview of simulation setup for the *Steckler* compartment

Table 1 Selected range of parameters for uncertainty analysis of compartment fire

Parameter	Max	Min	Mean
Heat release rate (kW/m^2)	1400	800	1000
Density (kg/m^3)	400	100	250

uniform distribution: the heat release rate of the burner and the wall density (Tab. 1). Figure 12 provides an overview of the simulation domain and key setup details.

In these simulations, the choice of uncertain parameters was not specifically optimized; heat release rate and wall density were selected as uncertain inputs. Often, a global sensitivity analysis accompanies uncertainty propagation to identify critical parameters that require close attention. However, this step is omitted in this study.

A total of 1000 simulations were performed with randomly selected input parameters (uniform distribution). The resulting flow fields—here, temperature distributions—were then used to train the surrogate model. The PCE model can be generated at varying orders; here, we compare orders from 2 to 6 for the temperature field along the symmetry plane aligned with the door. The *firePCE* tool, an in-house development, is applied to compute expectation and standard deviation values based on the PCE, with the ability to massively distribute calculations across multiple cores for efficiency.

Figure 13 visualizes the resulting uncertain temperature field at the symmetry plane, corresponding to the mesh slice location shown above in Fig. 12. The temperature range and contour lines are adjusted to emphasize regions of temperature uncertainty based on standard deviation.

In this compartment fire scenario, the selected input uncertainties have a minimal effect on the overall temperature distribution, particularly on the 50 dC contour line. In contrast, the 100 dC contour line exhibits a greater degree of propagated uncertainty, with its position varying by approximately 40 cm within the compartment.

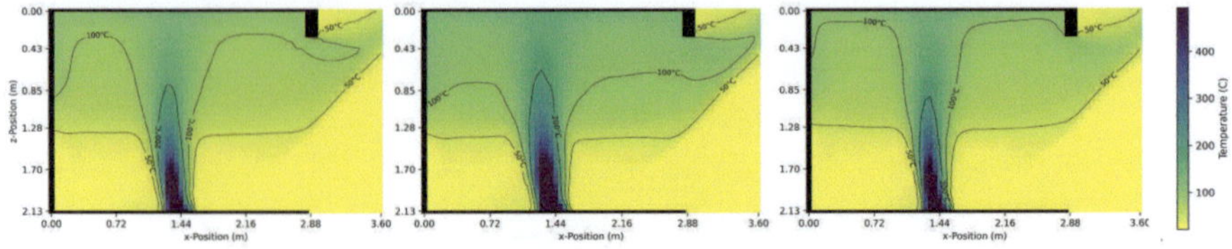

Fig. 13 Temperature distribution with uncertainties; left: expectation, middle: expectation minus standard deviation, right: expectation plus standard deviation; PCE order 4 with 1000 training samples

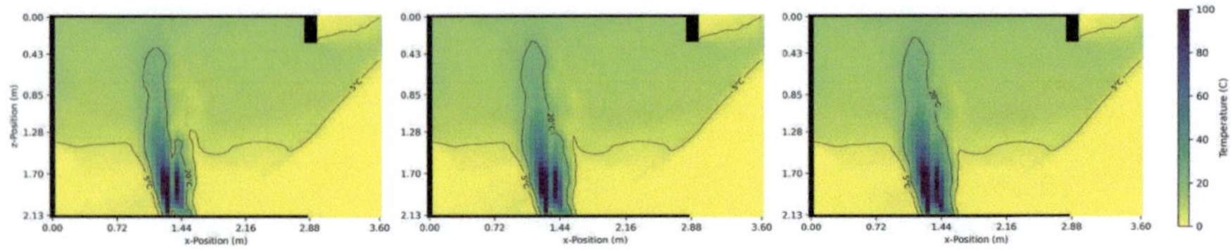

Fig. 14 Comparison of standard deviation depending on PCE order for a surrogate model trained by 1000 samples; left to right: 2nd, 4th and 6th order

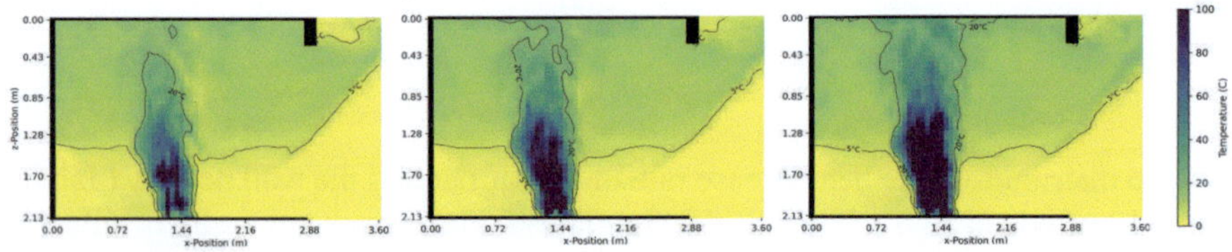

Fig. 15 Comparison of standard deviation depending on PCE order for a surrogate model trained by 100 samples; left to right: 2nd, 4th and 6th order

The effectiveness of PCE ultimately depends on the model's capability to accurately predict the flow field, which is influenced primarily by the amount of training data and the polynomial order. Figure 14 illustrates the impact of different polynomial orders—2, 4, and 6—on the standard deviation.

In this case, the impact of polynomial order is not clearly defined; this outcome can vary significantly for other setups (see, e.g., [35]). When applying any uncertainty (or sensitivity) analysis method, it is essential to assess its convergence, i.e., whether the output stabilizes as additional input samples are added. Here, the input data includes 1000 simulation samples, although a more comprehensive convergence study is typically necessary.

The training simulations were generated using a Latin hypercube sampling method. Figure 15 illustrates the effect of the number of training data on the predicted standard deviation, comparing results from 100 versus 1000 input simulations for 2nd and 6th order PCE. In this case, the impact of the number of training samples is more pronounced than the effect of polynomial order. Although the 5 dC standard-deviation iso-contour has a similar distribution within the compartment, there are larger differences near the outer boundaries and close to the burner flame.

It is important to note that applying an uncertainty propagation method does not improve the accuracy of simulation results when compared to experimental data; the results are only as accurate as the underlying simulation setup. This approach merely provides information on output uncertainty based on the specified uncertainties of selected inputs. Using PCE introduces additional errors, as it is a surrogate model trained on a set of training data. The required amount of training data depends on the specific simulation scenario, and model accuracy should always be validated using out-of-sample parameter values. PCE is one of several methods for incorporating uncertainty propagation into CFD modeling. Other approaches that have been successfully applied in detailed pyrolysis modeling include *Proper Orthogonal Decomposition* combined with interpolation of mode coefficients (commonly known as PODI), *Gaussian processes*, and *Autoencoders* based on neural networks. Like PCE, these methods generate surrogate models for detailed CFD simulations and have demonstrated faster convergence with fewer training samples in pyrolysis modeling. A comparison of these methods is beyond the scope of this work.

5 Conclusion

In this chapter, various challenges in fire spread modeling have been discussed and illustrated through selected simulation examples. While not an exhaustive review of current research, this chapter provides a brief introduction to the challenges of key cases in pyrolysis modeling, radiative heat transfer, and wildfire modeling. Additionally, an approach for managing uncertainties in CFD modeling was outlined, emphasizing the importance of understanding potential errors introduced by the use of surrogate models. This discussion highlights the complexities and critical areas for further investigation in advancing accurate fire spread simulations.

References

1. T. Rieke. “Arena SAP Garden im Olympiapark”. In: *Braunschweiger Brandschutz-Tage 2022. 36. Fachtagung – Forschung und Praxis*. Braunschweig: Jochen Zehfus, 2022. ISBN: 978-3-89288-228-2.
2. F. Brännström. “Application of Fire Simulation within the Railway Industry”. In: *14th International Conference and Exhibition on Fire Science and Engineering*. Interflam. London, July 4–6, 2016, p. 12.
3. C. Lautenberger, J. Torero, and C. Fernandez-Pello. “Understanding Material Flammability”. In: *Flammability Testing of Materials Used in Construction, Transport and Mining*. Elsevier, 2022, pp. 1–22. ISBN: 978-0-08-102801-8. https://doi.org/10.1016/B978-0-08-102801-8.00066-1.
4. S. Lin, Y. Liu, and X. Huang. “Climate-Induced Arctic-boreal Peatland Fire and Carbon Loss in the 21st Century”. In: *Science of The Total Environment* 796 (Nov. 20, 2021). https://doi.org/10.1016/j.scitotenv.2021.148924.
5. A.Witze. “Why Arctic Fires Are Bad News For Climate Change”. In: *Nature* 585 (2020), pp. 336–337. https://doi.org/10.1038/d41586-020-02568-y.
6. B. Byrne, J. Liu, K. W. Bowman, M. Pascolini-Campbell, A. Chatterjee, S. Pandey, K. Miyazaki, G. R. Van Der Werf, D. Wunch, P. O. Wennberg, C. M. Roehl, and S. Sinha. “Carbon Emissions

from the 2023 CanadianWildfires". In: *Nature* 633.8031 (Sept. 26, 2024), pp. 835–839. https://doi.org/10.1038/s41586-024-07878-z.
7. G. Accary and D. Morvan. "How Can CFD Contribute to the Understanding of Wildfire Behaviour?" In: *Computers & Fluids* 279 (July 2024), p. 106322. https://doi.org/10.1016/j.compfluid.2024.106322.
8. G. Maragkos, D. Zeinali, and B. Merci. "Influence of Convective Heat Transfer Modelling in CFD Simulations of Upward Flame Spread". In: *Fire Safety Journal* 122 (June 2021), p. 103347. https://doi.org/10.1016/j.firesaf.2021.103347.
9. T. Hehnen and L. Arnold. "PMMA Pyrolysis Simulation - from Micro- to Real-Scale". In: *Fire Safety Journal* 141 (Dec. 2023), p. 103926. https://doi.org/10.1016/j.firesaf.2023.103926.
10. P. J. Coelho. "The Role of Ray e Ects and False Scattering on the Accuracy of the Standard and Modiÿed Discrete Ordinates Methods". In: *Journal of Quantitative Spectroscopy & Radiative Transfer* 73 (2002), pp. 231–238. https://doi.org/10.1016/S0022-4073(01)00202-3.
11. B. Hunter and Z. Guo. "Numerical Smearing, Ray Effect, and Angular False Scattering in Radiation Transfer Computation". In: *International Journal of Heat and Mass Transfer* 81 (Feb. 2015), pp. 63–74. https://doi.org/10.1016/j.ijheatmasstransfer.2014.10.014.
12. C. Paul, S. Roy, J. Sailer, F. Brännström, M. M. Ahmed, A. Trouvé, H. Bordbar, S. Hostikka, and R. McDermott. "Detailed Radiation Modeling of Two Flames Relevant to Fire Simulation Using Photon Monte Carlo – Line by Line Radiation Model". In: *Journal of Quantitative Spectroscopy and Radiative Transfer* 329 (Dec. 2024), p. 109177. https://doi.org/10.1016/j.jqsrt.2024.109177.
13. H. Bordbar, S. Hostikka, P. Boulet, and G. Parent. "Numerically Resolved Line by Line Radiation Spectrum of Large Kerosene Pool Fires". In: *Journal of Quantitative Spectroscopy and Radiative Transfer* 254 (Oct. 2020), p. 107229. https://doi.org/10.1016/j.jqsrt.2020.107229.
14. G. J. Fiola, D. M. Chaudhari, and S. I. Stoliarov. "Comparison of Pyrolysis Properties of Extruded and Cast Poly(Methyl Methacrylate)". In: *Fire Safety Journal* 120 (Mar. 2021), p. 103083. https://doi.org/10.1016/j.firesaf.2020.103083.
15. P. Lauer, L. Arnold, and F. Brännström. "Inverse Modelling of Pyrolization Kinetics with Ensemble Learning Methods". In: *Fire Safety Journal* 136 (Apr. 2023), p. 103744. https://doi.org/10.1016/j.firesaf.2023.103744.
16. N. Bal. "Uncertainty and Complexity in Pyrolysis Modelling". PhD thesis. University of Edinburgh, 2012.
17. K. McGrattan, S. Hostikka, J. Floyd, R. McDermott, and M. Vanella. *Fire Dynamics Simulator Technical Reference Guide, version 6.7.4*. Nov. 2021. https://doi.org/10.6028/NIST.SP.1018.
18. C. Lautenberger and C. Fernandez-Pello. "Generalized Pyrolysis Model for Combustible Solids". In: *Fire Safety Journal* 44.6 (Aug. 2009), pp. 819–839. https://doi.org/10.1016/j.firesaf.2009.03.011.
19. Y. Wang, P. Chatterjee, and J. L. De Ris. "Large Eddy Simulation of Fire Plumes". In: *Proceedings of the Combustion Institute* 33.2 (2011), pp. 2473–2480. https://doi.org/10.1016/j.proci.2010.07.031.
20. J. Lachaud and N. N. Mansour. "Porous-Material Analysis Toolbox Based on OpenFOAM and Applications". In: *Journal of Thermophysics and Heat Transfer* 28.2 (Apr. 2014), pp. 191–202. https://doi.org/10.2514/1.T4262.
21. A. L. Sullivan. "Physical Modelling of Wildland Fires". In: *Encyclopedia of Wildfires and Wildland-Urban Interface (WUI) Fires*. Ed. by S. L. Manzello. Cham: Springer International Publishing, 2019, pp. 1–8. ISBN: 978-3-319-51727-8. https://doi.org/10.1007/978-3-319-51727-8_58-1.
22. M. M. Ahmed and A. Trouvé. "Large eddy simulation of the unstable flame structure and gas-to-liquid thermal feedback in a medium-scale methanol pool fire". In: *Combustion and Flame* 225 (2021), pp. 237–254. https://doi.org/10.1016/j.combustflame.2020.10.055.
23. S. Verma, M. M. Ahmed, and A. Trouvé. "The Structure of Line Fires at Flame Scale". In: *Wildland Fire Dynamics: Fire Effects and Behavior from a Fluid Dynamics Perspective*. Ed. by K. Speer and S. Goodrick. 1st ed. Cambridge University Press, June 30, 2022, pp. 36–62. ISBN: 978-1-108-68324-1. https://doi.org/10.1017/9781108683241.

24. K. McGrattan, S. Hostikka, J. Floyd, R. McDermott, and M. Vanella. Fire Dynamics Simulator Technical Reference Guide Volume 3: Validation, version 6.7.4. June 2022. https://doi.org/10.6028/NIST.SP.1018.
25. M. M. Ahmed, A. Trouvé, J. Forthofer, and M. Finney. "Simulations of Flaming Combustion and Flaming-to-Smoldering Transition in Wildland Fire Spread at Flame Scale". In: *Combustion and Flame* 262 (Apr. 2024),p. 113370. https://doi.org/10.1016/j.combustflame.2024.113370.
26. M. A. Finney, J. Forthofer, I. C. Grenfell, B. A. Adam, N. K. Akafuah, and K. Saito. "A Study of Flame Spread in Engineered Cardboard Fuelbeds: Part I: Correlations and Observations". In: (2013).
27. G. Accary, J. Darido, D. Morvan, L. Schneider, B. Betting, N. Frangieh, S. Meradji, and A. Simeoni. "Experimental and Numerical Investigation of a Single-Tree Fire". In: *Journal of Fire Sciences* 42.2 (Mar. 2024), pp. 142–170. https://doi.org/10.1177/07349041231218967.
28. A. Marcozzi, L.Wells, R. Parsons, E. Mueller, R. Linn, and J. K. Hiers. "Fast-Fuels: Advancing Wildland Fire Modeling with High-Resolution 3D Fuel Data and Data Assimilation". In: Environmental Modelling & Software 183 (Jan. 2025), p. 106214. https://doi.org/10.1016/j.envsoft.2024.106214.
29. I. T. Leventon, B. Batiot, M. C. Bruns, S. Hostikka, Y. Nakamura, P. Reszka, T. Rogaume, and S. I. Stoliarov. "The MaCFP Condensed Phase Working Group: A Structured, Global Effort towards Pyrolysis Model Development". In: *Obtaining Data for Fire Growth Models. ASTM International*, Mar. 2023. https://doi.org/10.1520/STP164220210111.
30. K. D. Steckler. "Flow Induced by Fire in a Compartment". In: *Symposium (International) on Combustion* 19.1 (1982), pp. 913–920. https://doi.org/10.1016/S0082-0784(82)80267-1.
31. I. Leventon. 2023. https://github.com/MaCFP/matl-db/tree/master/PMMA/Validation_Data/NIST_Gasification_Apparatus.
32. H. Scandelli. "Modélisation Multi-Échelle Des Milieux Poreux Réactifs à Haute Température Appliquée Aux Protections Thermique Spatiale et à La Pyrolyse de La Biomasse." PhD thesis. Bordeaux: Hesam Univeristy, Oct. 14, 2022.
33. S. Hostikka and F. Brännström, eds. *MaCFP Radiative Heat Transfer Phenomena*. 2023. https://iafss.org/macfp-radiativeheat-transfer-phenomena/ (visited on 01/29/2023).
34. IAFSS. *Measurement and Computation of Fire Phenomena (MaCFP Working Group)*. https://iafss.org/macfp/ (visited on 01/29/2023).
35. H. Jamil and F. Brannstrom. "Intrusive and non-intrusive uncertainty quantification methodologies for pyrolysis modeling." In: *Fire Safety Journal* 143 (2024), p. 104060. https://doi.org/10.1016/j.firesaf.2023.104060.
36. R. G. Ghanem and P. D. Spanos. *Stochastic Finite Elements: A Spectral Approach*. Berlin, Heidelberg: Springer-Verlag, 1991. ISBN: 0387974563.
37. M. Tatang, W. Pan, R. Prinn, and G. McRae. "An Efficient Method for Parametric Uncertainty Analysis of Numerical Geophysical Model". In: *Journal of Geophysical Research* 102 (Sept. 1997), pp. 21925–21932. https://doi.org/10.1029/97JD01654.
38. D. Xiu. *Numerical Methods for Stochastic Computations: A Spectral Method Approach*. July 2010. ISBN: 9780691142128. https://doi.org/10.2307/j.ctv7h0skv.
39. B. Sudret. "Polynomial Chaos Expansions and Stochastic Finite Element Methods". In: *Risk and Reliability in Geotechnical Engineering (Chap. 6)*. CRC Press, 2014, pp. 265–300. ISBN: 1-4822-2721-5.

Chemical Safety and Fire Defence as the Basis for Safe and Effective Fire Service Response in the Context of Safety Engineering

Julia Backhaus, Monique Glockmann, Philipp Haffner, Martina Henseleit, Katja Klicker, Katja Müller, Marie-Claire Ockfen, Daniel Schmitz, Bianca Wegener and Roland Goertz

Contents

Abstract

Li-ion batteries play a significant role in all areas of life. Their thermal decomposition caused by overcharging, along with the associated mechanisms and intervention options, is being studied intensively. The composition of the venting gases is of major importance. These gases contain not only toxic and corrosive components but also hydrogen and other flammable gases, meaning that thermal runaway may pose an explosion risk within the surrounding area. PVC

J. Backhaus · M. Glockmann · P. Haffner · M. Henseleit · K. Klicker · K. Müller · M.-C. Ockfen · D. Schmitz · B. Wegener · R. Goertz (✉)
Chair of Chemical Safety and Fire Defence, University of Wuppertal, Wuppertal, Germany
e-mail: goertz@uni-wuppertal.de

S. Bracke and R. Pieper (eds.), *Safety Engineering*,
https://doi.org/10.1007/978-3-658-47473-7_14

and other chlorinated polymers remain widely used. Their thermal decomposition is often associated with concerns about the potential formation of highly toxic substances, such as chlorinated dioxins and furans. Therefore, investigating the thermal decomposition process, characterizing the resulting products, and elucidating the reaction mechanisms are essential to understanding when special caution is required after a fire. Organophosphorus flame retardants are highly popular and widely used. New generations, such as DOPO derivatives, allow for straightforward production and a wide range of products. However, the thermal decomposition, along with its mechanisms and products, remains largely unexplored. Phosphorus-containing organic compounds can be extremely hazardous even in small quantities, particularly affecting the human body and nervous system. Numerous laboratory experiments and quantum chemical calculations are providing new insights into this area. Fluorinated surfactants (PFAS) have exceptional properties and were historically effective in aqueous film-forming foam (AFFF) for liquid fire suppression. However, PFAS are environmentally persistent, bioaccumulative, and toxic. Through an extensive statistical study and complex laboratory experiments, 523 chemicals were clustered based on their physicochemical properties and evaluated for extinguishing or sealing with foam. In no instance was AFFF found necessary. During real fire training, firefighters wearing breathing apparatus are exposed to fire by-products such as carcinogenic benzene. A study was carried out to investigate whether a significant concentration of benzene can be detected in the whole blood of trainees and instructors after real fire training. Initial results show no blood levels of benzol. To protect lives and property and minimize damage in fires and emergency situations, the fire department must arrive promptly with sufficient personnel and appropriate equipment. To meet this challenge, new system-theoretical approaches are being pursued and tested. Geographic Information Systems (GIS) and statistical methods, such as cluster analysis, play a key role, with the additional aim of harmonizing resource planning for fire services, emergency medical services, and disaster response.

Keywords

Lithium-ion batteries · Thermal decomposition · Thermal runaway · Overcharging · Venting gas · Polyvinyl chloride (PVC) · Combustion reaction · VCI combustion apparatus · GC-MS analysis · Fire smoke composition · Polymer matrix · Organophosphorus flame retardant · Aromatic compounds · Density functional theory (DFT) calculations · Decomposition mechanism · Chemical database · ChemInfo · Firefighting foam · AFFF (Aqueous Film-Forming Foam) · Clustering · Wood-fired training facility · Benzene concentration in whole blood · Fire service demand planning · Resource planning · Systems theory · Spatial analysis

1 Introduction

In the 50-year history of safety engineering at the University of Wuppertal, fire and explosion protection and the fire brigade have played an important role alongside other important topics. The Chair of Chemical Safety and Fire Defence was established in 2012. The overall objective of the Chair is to carry out research projects and offer courses that deal with the prevention, interruption or limitation of the effects of combustible and hazardous chemical reactions. With this in mind, the Chair has three main areas of research: Chemical Safety, Emergency response organization and management, and preventive fire protection.

Numerous practical problems in the sense of inter- and transdisciplinary research have led to the need to address aspects of fire investigation as a cross-cutting issue. The knowledge gained in this way has in turn been incorporated into the nature of the research projects as well as into academic teaching.

At the same time, the Fire Service Science Institute (FSI) was established to transfer scientific knowledge to the structure and work of the fire service. As an essential part of this work, the first dual study programme with integrated training for the higher fire service was launched in 2017 in cooperation with the Federal Armed Forces Fire Service and the Federal Ministry of Defence. Numerous graduates have since taken up leadership positions in the Federal Armed Forces Fire Service and in civilian fire services. In 2023, in cooperation with the Centre for Further Education at the University of Wuppertal, the transfer-oriented training programme will be expanded to include a specialist component as part of the promotion of middle-ranking civil servants to the higher ranks of the Federal Armed Forces Fire Service.

The following chapters give an insight into the ongoing research.

2 Thermal Decomposition of Lithium-Ion-Batteries

2.1 Thermal Runaway Due to Overcharging

Using lithium-ion batteries outside specified parameters for current, voltage, temperature, or under mechanical stress may trigger thermal runaway of the cells. This process involves a series of exothermic reactions at high temperatures, producing gaseous by-products that increase the internal pressure. If this pressure exceeds rupture threshold of the cell, the housing bursts, releasing gases and other components. These gases may ignite, forming a jet flame, and the explosive atmosphere can also ignite. [1, 2] Although lithium-ion batteries typically include a battery management system (BMS) to prevent overcharging by regulating current at the end-of-charge voltage, a defective or absent BMS can allow excess energy input, depleting lithium ions in the cathode. This raises temperature, causing instability and a decomposition cascade in the cell components. [3, 4]

To inform safe handling recommendations for burning lithium-ion batteries in risk assessments, understanding the decomposition process and fire behaviour is essential. Key parameters, like battery cell voltage, require monitoring. Figure 1 shows the voltage curve of a commercially available lithium polymer pouch cell (LiPo) during overcharging at different C-rates, where 1 C for a 2.5 Ah battery is 2.5 A.

Increasing the charging rate accelerates the onset of thermal runaway within the cell. This is indicated by a rapid elevation in cell voltage, followed by a subsequent decline to 0 V. After this event, the cell is rendered irreparably damaged. [5]

2.2 Ignition Characteristics and Temperature Curves

Thermal runaway in lithium-ion batteries can trigger ignition of released gases and the cell itself, with fire development influenced by factors like cell chemistry. Lithium iron phosphate cells (LFP) ignite less frequently, while nickel manganese cobalt oxide cells (NMC) are more hazardous. [6] Fig. 2 shows temperature curves for three identical LiPo pouch cells (800 mAh) during overcharging (3 C). Despite identical chemistry and charging, temperature behaviour differed: two cells ignited and one did not. This variability may be due to minor manufacturing differences or environmental factors. Pouch cells often behave inconsistently due to

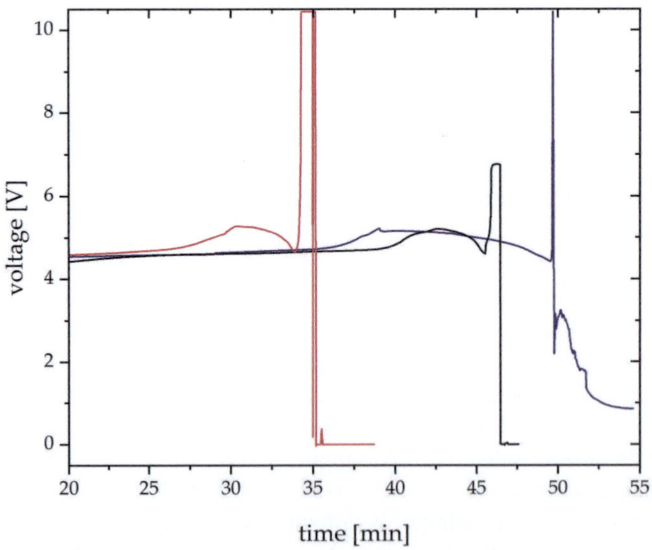

Fig. 1 Voltage curve of three identical lithium-ion batteries (LiPo, 800 mAh) during overcharging with different charging rates (in red 3 C, in black 2 C, in blue 1 C)

unpredictable casing rupture locations. A white mist, or venting, may be released initially, and with enough energy or heat, gases can ignite. Burning continues until fuel is depleted or intervention (e.g., cooling) occurs. Figure 3 shows venting and jet flame formation during thermal runaway of a pouch cell.

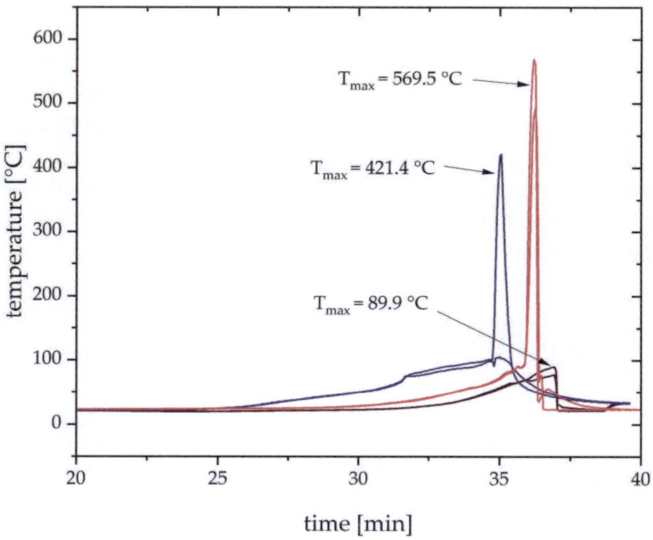

Fig. 2 Temperature curve of lithium-ion batteries (LiPo, 800 mAh) during overcharging and the resulting thermal runaway

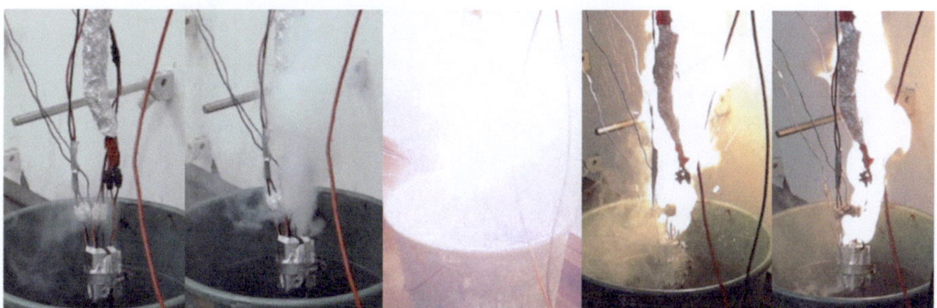

Fig. 3 Venting (left) and formation of a jet flame (right) during thermal runaway of a lithium-ion battery (LiPo, 800 mAh)

2.3 Venting Gas Analysis

To analyse the venting gases of lithium-ion batteries, they first have to undergo thermal runaway. Due to the risk of explosion and further damage, such as fire or the inhalation of fumes, these tests are conducted in a gas-tight, pressure vessel made of carbon steel that can withstand an overpressure of up to 40 bar. Figure 4 shows a schematic diagram of the test rig.

Lithium-ion cells from traction batteries can undergo thermal runaway in the pressure vessel through overcharging, overheating, or nail penetration. The emitted gas mixture is collected and quantified based on internal vessel pressure. Sampling is primarily accomplished by introducing gases into impingers filled with various solutions (e.g., carbonate buffer or organic solvents), which than can be analysed in detail afterwards. Organic analytes, such as electrolyte solvents and fire decomposition products, are identified using gas chromatography–mass spectrometry (GC-MS), while dissolved fluoride ion concentration is measured with ion chromatography (IC) to estimate hydrogen fluoride release. Additional gas samples are collected in a multi-foil bag and analysed by gas chromatography with a thermal conductivity detector (GC-TCD), with emphasis on hydrogen content due to its explosive potential. Figure 5 presents a chromatogram of the GC-TCD measurement of the venting gases released from a lithium manganese oxide (LMO) cell.

The analyser detects additional gases through a second channel, including carbon dioxide (CO_2), ethane (C_2H_6), ethene (C_2H_4), ethyne (C_2H_2), propane (C_3H_8), propene (C_3H_6), and propyne (C_3H_4). Table 1 presents the standardized composition of the venting gas (normalized to 100%) as determined by GC-TCD, while other components, though present, are not shown.

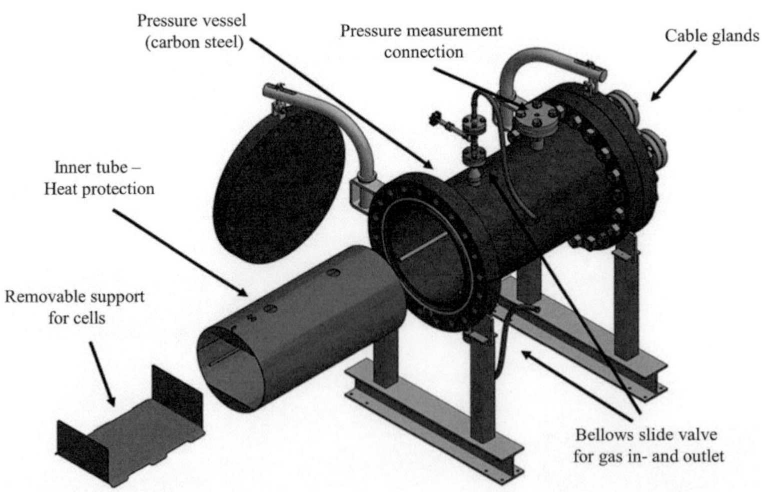

Fig. 4 Schematic three-dimensional representation of the pressure vessel (manufactured by AVN GmbH, Germany)

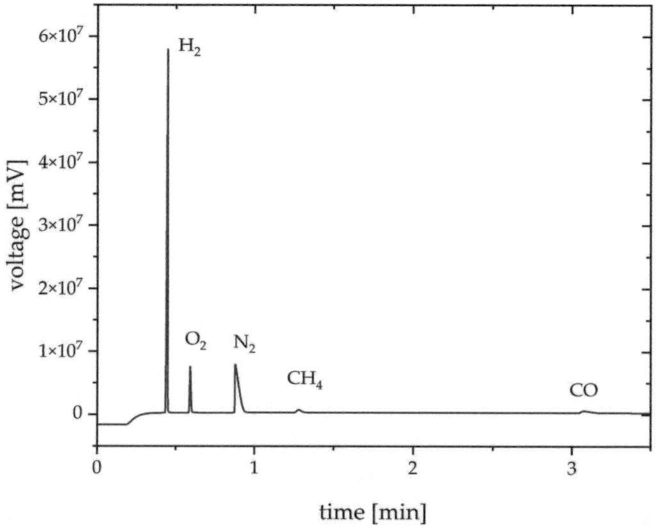

Fig. 5 Chromatogram of a venting gas sample from an experiment with a lithium manganese oxide cell

Table 1 Percentage Composition of Venting Gas in LMO Cell (Standardized to 100%)

Gas	Ratio of venting gas [Vol%]
Hydrogen (H_2)	28.6
Methane (CH_4)	4.6
Ethane (C_2H_6)	1.1
Carbon monoxide (CO)	19.4
Carbon dioxide (CO_2)	46.4

Hydrogen fluoride (HF) is another gas released during lithium-ion battery thermal runaway. [7] When venting gas is added to a carbonate buffer solution, hydrogen fluoride and hydrogen carbonate ions react to form carbonic acid and fluoride ions. The latter can be detected by ion chromatography. External calibration was used for quantification. Complete retention of HF in a single impinger may not occur, with some breakthrough potentially captured in a subsequent impinger. Using a geometric series, a theoretical total concentration of HF can be reasonably estimated. [8] The HF release varies with cell chemistry, with LMO cells showing the highest rates and NMC the lowest. Figure 6 displays the calculated HF volumes and total vent gas per cell type.

The total venting gas volume was calculated using the ideal gas law for a 170-liter container, accounting for temperature and pressure effects. Hydrogen fluoride

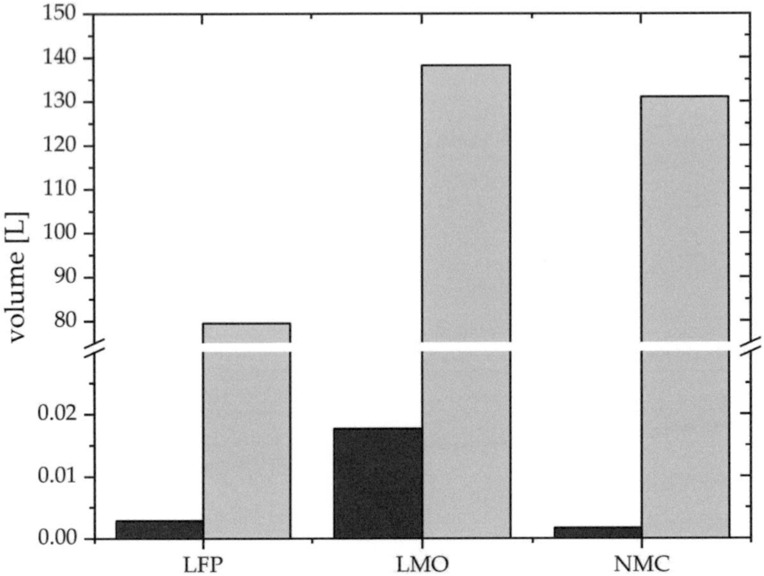

Fig. 6 Volumes of released HF (blue) and total Venting gas volumes (grey) for the cell chemistries LFP, LMO, NMC

levels in venting gas were 36.11 ppmv (LFP), 12.98 ppmv (NMC), and 127.75 ppmv (LMO). These values are lower than reported in literature, suggesting fluorine compounds persist in the solid residue, where fluoride salts have also been detected. Due to inconclusive results, alternative research approaches are being explored. It's also possible that hydrogen fluoride forms organic fluorine compounds upon release, warranting further investigation.

3 Investigating the Impact of Various Fire Conditions on Organic Decomposition Products of Polyvinylchloride (PVC) Flooring Materials

3.1 Introduction

In the event of fire, PVC undergoes thermal decomposition, which is influenced by fire conditions such as temperature, oxygen availability and burning time. These factors determine the type and quantity of organic decomposition products formed, including toxic compounds. [9]

The study of the organic decomposition products of PVC, as used in flooring, under different fire conditions is crucial to better understand the health and environmental risks and to develop appropriate fire safety measures.

3.2 Influence of Temperature on Combustion Reactions

In the scientific literature, a distinction is made between high-temperature and low-temperature combustion processes, given that the chemical reactions and products undergo significant changes depending on the temperature. The homolytic cleavage of carbon-carbon bonds occurs at approximately 431.85 °C, generating free radicals. In contrast, the cleavage of carbon-hydrogen bonds occurs at temperatures exceeding 626.85 °C. At elevated temperatures, thermal decomposition processes without oxygen participation become predominant, resulting in the formation of smaller hydrocarbons, including methyl and ethyl radicals. These processes are pivotal for the formation of flames and the decomposition of fuels. At lower temperatures, oxidative decomposition processes with direct oxygen participation are predominant. The temperature not only influences the reaction rate but also the chemical equilibrium. At higher temperatures, the equilibrium shifts in favour of certain products, changing the product distribution. In conclusion, temperature plays a significant role in determining the decomposition mechanisms and, consequently, the reaction products and combustion efficiency. [10, 11]

3.3 Influence of Ventilation on Combustion Reactions

In the field of fire dynamics, a distinction is made between two distinct types of fires: those that are fuel-controlled and those that are ventilation-controlled. In fuel-controlled fires, sufficient oxygen is available, allowing for nearly complete oxidation of the fuel. In this case, the quantity and composition of the fuel determine the fire behaviour, whereby oxygen is not a limiting factor. In contrast, ventilation-controlled fires occur when there is a lack of oxygen, which leads to inefficient combustion and incomplete oxidation. This produces products such as carbon monoxide (CO) and soot. Ventilation-controlled fires often occur in poorly ventilated rooms, increasing the risk of dangerous phenomena such as backdraft. [12–14]

3.4 Use of Combustion Apparatus

Fire research requires precise control of temperature and oxygen. Large-scale test facilities present a challenge in this regard, prompting the development of compact, controllable apparatus like the DIN 53436 fire apparatus. Originally conceived for toxicological investigations, this apparatus has been modified for chemical analyses without live animals. However, uneven temperature distribution and the apparatus' dimensions may impact precision. [15] A more advanced alternative is the VCI combustion apparatus (Fig. 7), which was developed in 1984. It offers superior temperature consistency, which enhances the reproducibility and quality of the results. It can operate with smaller sample quantities, reducing costs and material

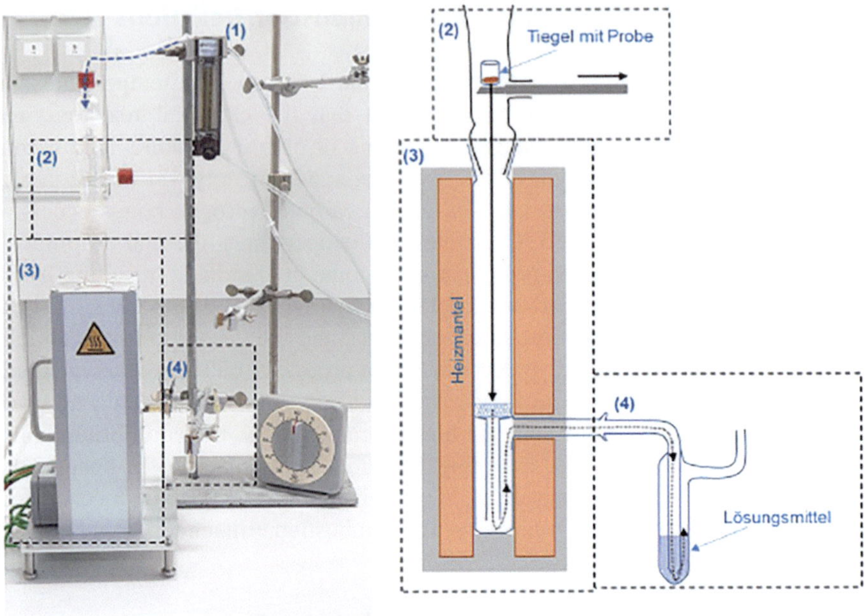

Fig. 7 VCI combustion apparatus and schematic view. [18]

consumption. It also allows precise control of combustion parameters such as temperature and air supply, facilitating the study of different combustion scenarios and making it a versatile tool for toxicological and safety analyses. [16, 17]

3.5 Research Results for the Combustion of CV Coating Extreme Mira 990D Using the VCI Combustion Apparatus and Subsequent GC-MS Analysis

A comprehensive analysis of fire smoke composition shows that temperature significantly affects the composition of organic compounds in smoke gases. The chemical composition of fire smoke is highly temperature dependent, with distinct profiles emerging across temperature variations. Higher temperatures promote the formation of high molecular weight polycyclic aromatic hydrocarbons (PAHs), which do not form at lower temperatures, underscoring the thermodynamic and kinetic conditions required for complex hydrocarbon structures. Additionally, oxygen supply notably impacts the smoke composition, particularly at lower temperatures. Aliphatic hydrocarbons form more readily in low oxygen conditions, as illustrated in Fig. 8 and 9 (highlighted in purple), indicating a preference for aliphatic compound formation with reduced oxygen. It is noteworthy that the elevated formation of aliphatics in the absence of oxygen suggests that the cyclisation of these aliphatic compounds to polycyclic aromatic hydrocarbons (PAHs) may already occur at relatively low temperatures in the presence of oxygen. Consequently, oxygen facilitates

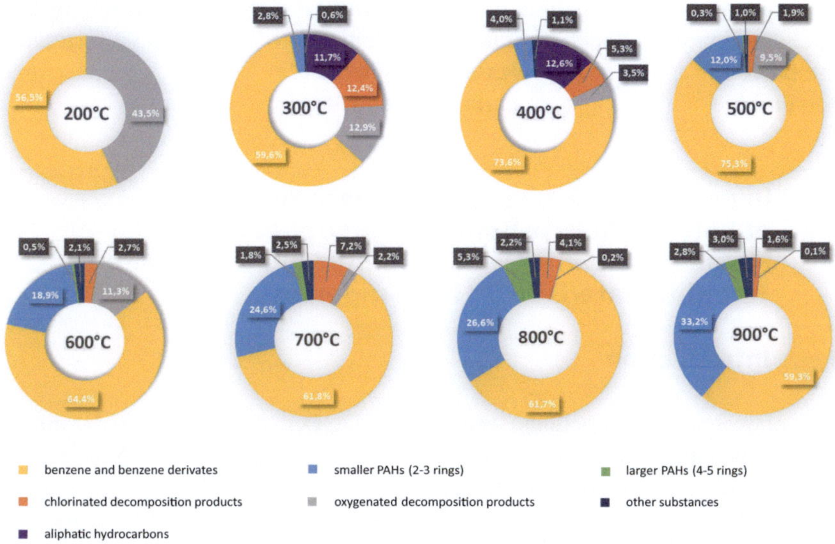

Fig. 8 Percentage fire smoke composition during the combustion of floor PVC under an air flow of 23.8 ml/min

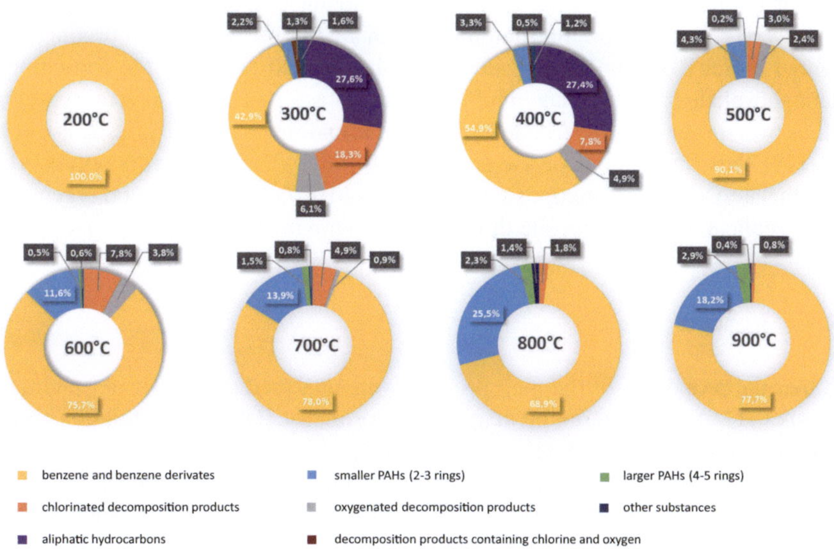

Fig. 9 Percentage fire smoke composition during the combustion of floor PVC under a nitrogen flow of 23.8 ml/min

the formation of PAHs, potentially by supporting oxidative mechanisms that enable the conversion of linear or branched aliphatics into cyclic aromatic structures. These processes appear to be less efficient in the absence of oxygen, which explains the dominance of aliphatic hydrocarbons in the absence of oxygen.

4 Investigation of the Thermal Decomposition of Polymer Matrix Bound Organophosphorus Flame Retardants Using a Sample of Polybutylene Terephthalate/DOPO-HQ Composite

4.1 Introduction

Organophosphorus compounds, especially DOPO derivatives (9,10-dihydro-9-oxa-10-phosphaphenanthrene-10-oxide), are emerging as effective, less-toxic flame retardants. These compounds decompose during fires, likely releasing phosphorus oxide radicals to inhibit hydrogen and hydroxide radicals from pyrolysis, thus reducing heat release and breaking the chain reaction. [19, 20] The by-product orthophosphoric acid may also dehydrate materials to form a protective carbon layer with flame-retardant properties. [21]

The synthesis of DOPO derivatives is relatively straightforward, allowing for a range of compounds noted for their flame-retardant properties over the past decade. However, research on fire inhibition mechanisms, decomposition products, and potential toxicity of DOPO derivatives remains limited. Notably, the formation of organophosphates, which may inhibit cholinesterase and cause severe symptoms such as overstimulation, gastrointestinal spasms, and respiratory paralysis, is of concern. [22] This study builds on He et al. (2019) [23], who developed a polybutylene terephthalate/DOPO-HQ composite (PBT/DOPO-HQ) with promising oxygen index results, replicating the composite through thermogravimetric analysis (TGA). [24]

4.2 Results

The thermal properties of the melt product were analysed using thermogravimetric analysis (TGA) and differential scanning calorimetry (DSC) to validate a simplified production process without an extruder (Fig. 10).

TGA curves indicate a continuous, single-stage mass loss for DOPO-HQ, PBT, and the composite. Mass loss for DOPO-HQ and PBT/DOPO-HQ begins around 300 °C and completes by 520 °C, with a gradual increase across an 80–100 °C range. For DOPO-HQ, mass is nearly fully lost within about 150 °C, unlike the PBT/DOPO-HQ mixture, which retains approximately 14% mass. Pure PBT decomposes later, starting around 370 °C and ending near 450 °C, leaving a 7% residue. The TGA curves suggest that adding DOPO-HQ slows PBT's thermal decomposition, with the composite curve resembling DOPO-HQ more than PBT and leaving more thermally stable residue, consistent with comparative data from He et al. (Table 2). [23]

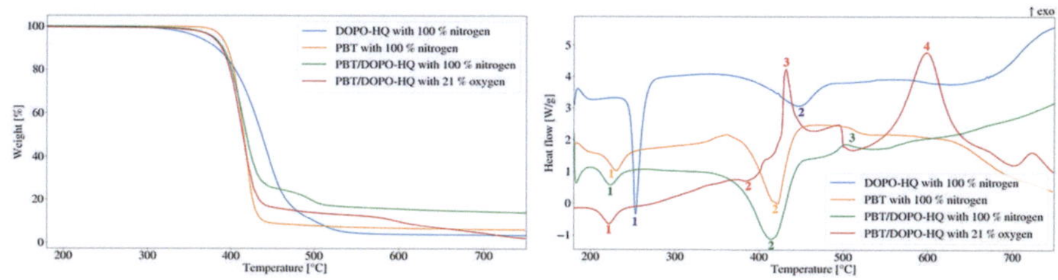

Fig. 10 TGA (l.) and DSC (r.) curves of DOPO-HQ, PBT and the produced PBT/DOPO-HQ composite

Table 2 Experimental values compared with the reference. [23]

Sample	$T_{5\%, ref}$	$T_{5\%, exp}$	$T_{max, ref}$	$T_{max, exp}$	Residue $_{ref}$	Residue $_{exp}$
PBT	408 °C	385 °C	442 °C	423 °C	3.97%	6.74%
DOPO-HQ	359 °C	361 °C	438 °C	442 °C	2.02%	3.45%
PBT/DOPO-HQ	383 °C	382 °C	439 °C	418 °C	6.87%	14.07%

a: $T_{5\%}$, the temperature at which 5% weight loss occurs.
b: T_{max}, the temperature at maximum weight loss rate.

DSC curves reveal endergonic processes between 200–250 °C and 370–420 °C for all three substances Fig. 10. The first range (200–250 °C) aligns with melting points of PBT (227 °C) and DOPO-HQ (245–255 °C) [9], while the second (370–420 °C) suggests evaporation or decomposition. The PBT/DOPO-HQ composite shows an exothermic reaction around 500 °C, likely from gas-phase radical recombination. In 21% oxygen, oxidative decomposition starts at 400 °C with a second exothermic peak near 600 °C. Due to the onset of vaporization and mass loss at 400 °C, this was set as the baseline for combustion furnace tests to simulate various fire ventilation conditions. For these tests, a 400–800 °C range under low-ventilation was used, following Theisen et al.'s guideline of a 20 mL/min air inflow and 750 °C for realistic fire conditions. [25] Thermal decomposition gases were acetone-washed and analysed by GC-MS (Fig. 11).

The GC-MS chromatogram of acetone wash solutions (via liquid injection) reveals all acetone-soluble fire gas compounds, except those with a retention time below 4 min due to the solvent cut, which excludes detection of carbon dioxide, benzene, and short-chain aliphatic phosphoric acid esters. Figure 12 highlights the most common compounds.

The decomposition products are primarily aromatic compounds such as toluene (2), styrene (8), benzoic acid (14), biphenyl (29), and naphthalene (25), common in most organic combustion processes. [26] Noteworthy are the less common compounds o-hydroxy biphenyl (15), dibenzofuran (16), ((butenyloxy)carbonyl) benzoic acid (17), and terephthalic acid (18), which are likely specific to the decomposed composite as they derive from the flame retardant DOPO-HQ and the polymer PBT. ((Butenyloxy)carbonyl)benzoic acid (17) and terephthalic acid

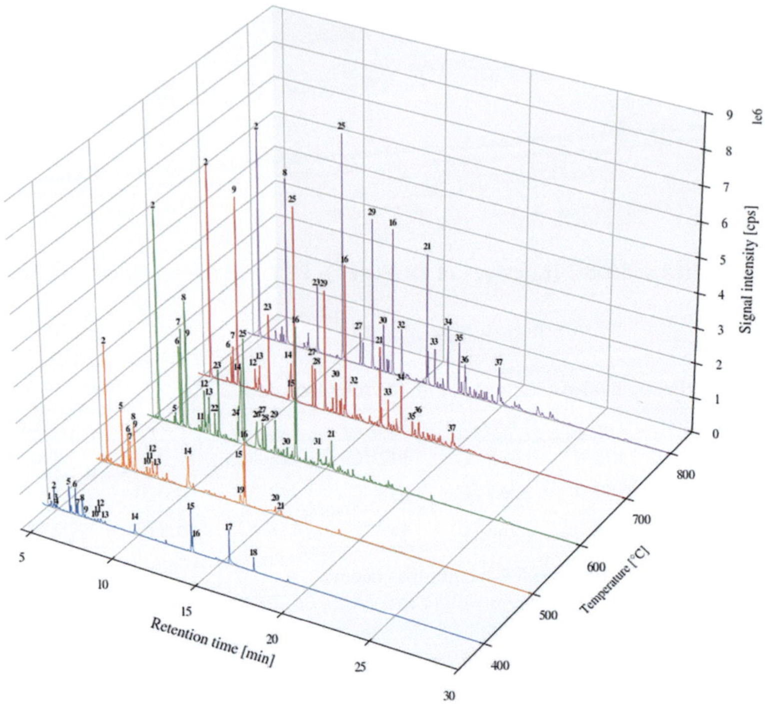

Fig. 11 Temperature-resolved display of the recorded chromatograms

(18) could be directly linked to the polymer matrix [27]. Based on the DOPO-HQ structure, Fig. 13 illustrates potential decomposition products.

To test this hypothesis, density functional (DFT) calculations were conducted using the Gaussian 16 program to determine the dissociation enthalpies for specific homolytic bond breaks within the DOPO-HQ molecule (Fig. 14, Table 3).

Calculations revealed that the phosphorus-oxygen bond is the weakest in the system, which is unusual given phosphorus's typical oxophilia. This weakness likely stems from the mesomeric stabilization of the resulting phenoxy radical. Phenoxyl radicals can persist when *ortho* and *para* positions are sterically or resonance-stabilized [28] Here, an ortho substituent provides both steric hindrance and resonance support, confirming the calculated bond energy. A decomposition mechanism was proposed, favouring homolytic cleavage of the phosphorus-oxygen bond (Scheme 1).

Thermal energy (73 kcal/mol) breaks the phosphorus-oxygen bond in compound 38, creating biradical 39, which likely remains undetected due to the furnace's energy input, capable of breaking 39's weaker bond (34 kcal/mol). This forms biradical 41, which reacts exothermically to produce o-hydroxy biphenyl (15) (−178 kcal/mol) or dibenzofuran (16) (−79 kcal/mol), matching DSC's exothermic process 3 (Fig. 10). Further study on intermediate 40 and possible phosphorus oxide radicals is planned, with additional Gaussian 16 calculations.

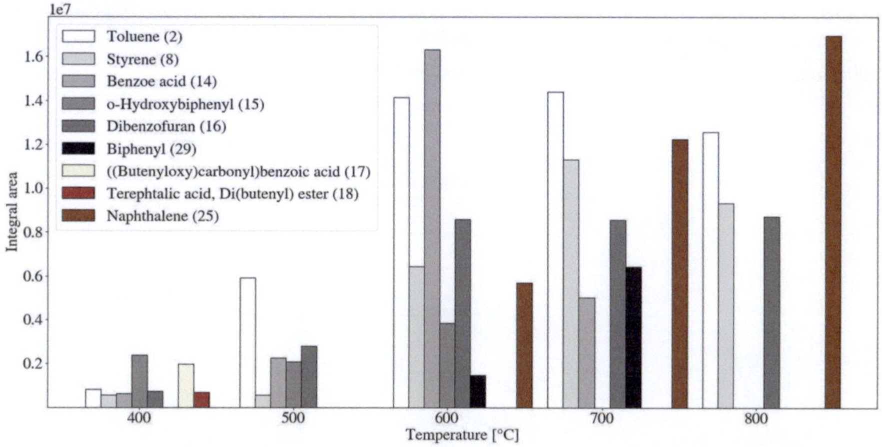

Fig. 12 Identified compounds of greatest interest

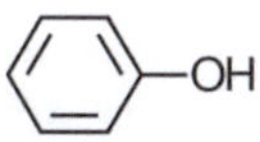

12

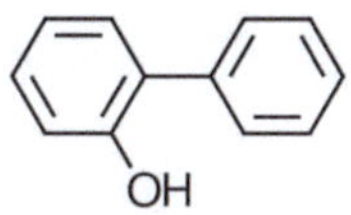

15

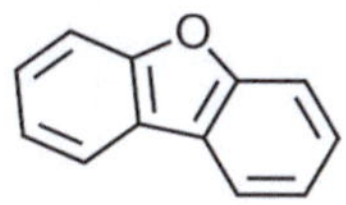

16

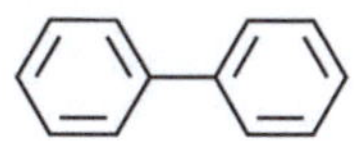

29

Fig. 13 Found compounds that presumably originate from the structure of DOPO-HQ

Fig. 14 Chemical structure of DOPO-HQ

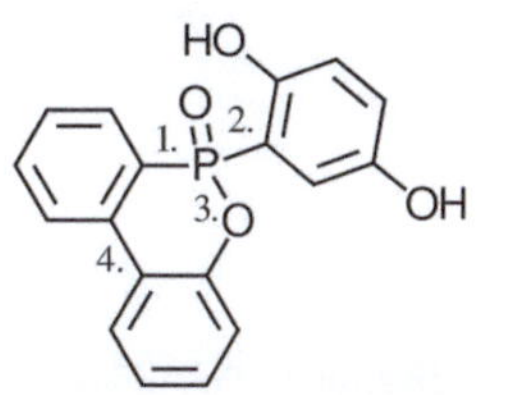

5 Re-Evaluation of Selected Chemicals with Regard to Recommended Firefighting Foams

5.1 Introduction

The ChemInfo database includes information on around 74,000 substances and mixtures, covering physico-chemical, ecotoxicological, and toxicological parameters as well as firefighting recommendations. [29, 30] ChemInfo previously

Table 3 Calculated dissociation enthalpies for possible homolytic bond breaks. Indicated energies were calculated at M06-2X/6-311G++(d,p) level

Split bond	Free dissociation enthalpies [kcal/mol]
1.	87
2.	84
3.	73 → weakest bond!
4.	108

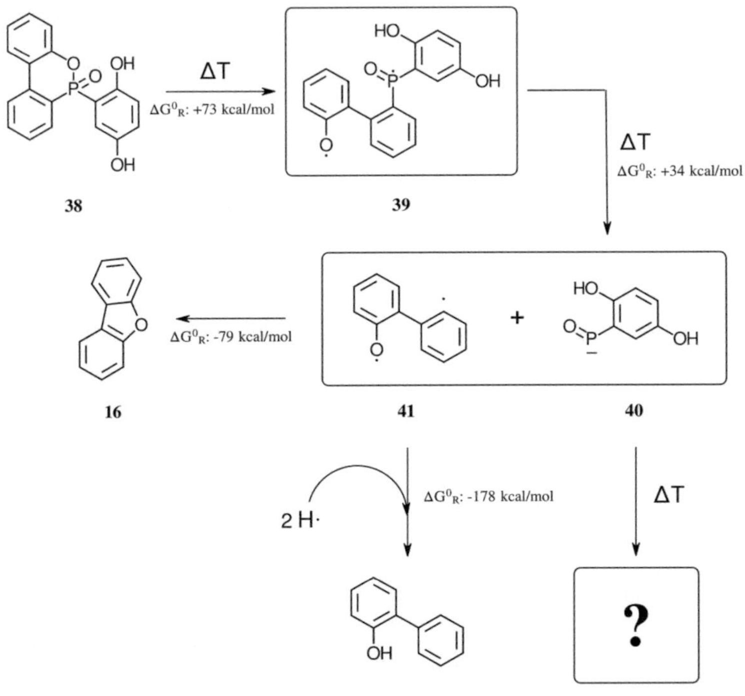

Scheme 1 Postulated decomposition mechanism of DOPO-HQ calculated at M06-2X/6-311G++(d,p) level

recommended aqueous film-forming foams (AFFF/A3F) for 550 chemicals. However, due to the environmental and health risks of per- and polyfluoroalkyl substances (PFAS) contained in these foams [31–33], the EU has initiated a ban on AFFF [34], necessitating an update of the ChemInfo entries. The re-evaluation of the firefighting foam recommendations was based on clustering the 550 substances by physico-chemical properties to predict similar fire and extinguishing behaviours. Experimental investigations and literature led to a flowchart for selecting alternative, fluorine-free foams for each cluster. The updated recommendations were added to the ChemInfo database.

5.2 Classification of the 550 Chemicals

The 550 chemicals were grouped by physical and chemical properties via cluster analysis. The relevant properties with sufficient data included: aggregate state, boiling point, vapor pressure, flash point, water solubility, density and substance class (determined using SMILES code). Data were complete for 68% (372 chemicals). The analysis was carried out using the scikit-learn library (v. 1.4.2). [35] The pre-processing steps included scaling of variables and converting them into dummy variable as well as a principal component analysis (PCA), which reduced 31 variables to 22 components. Cluster analysis included partitioning (KMeans), density-based (DBSCAN) and hierarchical (AgglomerativeClustering) methods with various fusion algorithms. An inertia analysis was used to identify a sufficient number of clusters. The comparison of the different clustering methods with a number of 20 clusters was carried out primarily according to chemical aspects. Hierarchical clustering with ward linkage and prior PCA provided the most meaningful grouping of the 372 chemicals. Figure 15 shows the resulting dendrogram. The remaining chemicals, with incomplete data, were categorized by chemical structure. Twenty-seven mixtures or component chemicals were not assigned to clusters due to unknown composition or already existing extinguishing agent recommendations for the pure substances. In total, 523 chemicals were grouped into the clusters shown in Table 4 based on physico-chemical properties.

5.3 Experimental Investigation

The next step was to assign appropriate extinguishing agents to each cluster, either through literature review or experimental testing. Special attention was paid to chemicals that react with water or foam concentrate components. Four test series were conducted with selected chemicals and three firefighting foams: synthetic fluorine-free foam (FFF), aqueous film-forming foam (AFFF), and alcohol-resistant foam (AR-FFF).

1. Time-resolved vapor permeability analysis of FFF and AFFF via gas chromatography
2. Determination of the spreading coefficient for FFF and AFFF
3. Small-scale fire and burn-back tests to assess extinguishing time and re-ignition resistance of FFF, AFFF, and AR-FFF
4. Foam stability testing of FFF and AR-FFF by measuring drainage rate

Compounds from each major substance class in Table 4, excluding solids, were tested. Based on cluster analysis similarities, results are expected to apply to other representatives within each class. The findings from these tests are integrated into a flowchart for selecting alternative extinguishing agents (Abschn. 5.4).

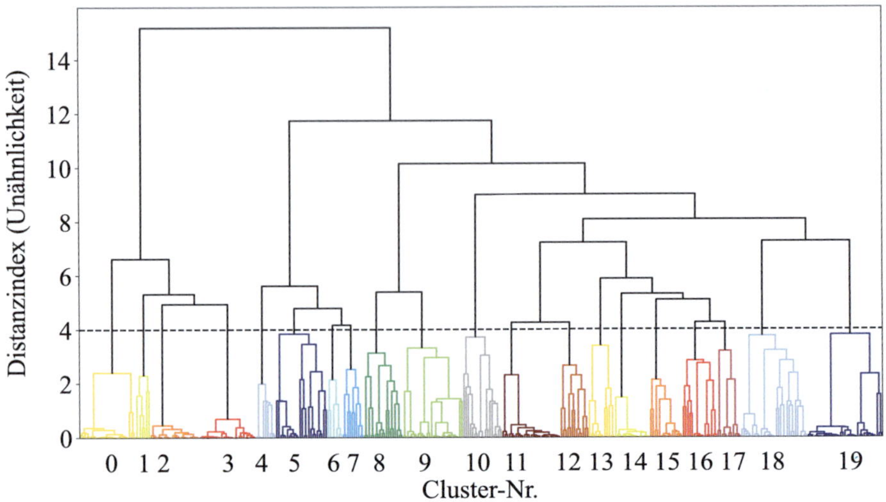

Fig. 15 Dendrogram of the hierarchical cluster analysis (ward linkage) with prior principal component analysis of the 523 chemicals containing data set

5.4 Re-Evaluation

The re-evaluation of firefighting foam recommendations followed the flow chart in Fig. 16, first focusing on water reactivity. For chemicals with highly exothermic reactions to water, affecting only 20 of the 523 chemicals (e.g., acid chlorides, phosphites), neither water nor foam are suitable. Alternatives like dry chemicals or inert gases are recommended. For the remaining 503, foam recommendations were tailored with water solubility as a key factor: In order to prevent immediate foam destruction caused by water-soluble substances (>10 g/L), a thickening agent such as xanthan gum is added to the foam concentrate to form a polymer film that acts as mechanical barrier. [36]

Within a substance class, the chemical behaviour and the influence on foam stability can be estimated very well based on the functional similarity of the compounds. Known foam-destroying agents include polar low-boilers, long-chain alcohols, alkylamines, alkyl phosphates and certain oils. [37, 38] Tests showed no foam degradation with *n*-octanol but high drainage rates for FFF with water-soluble *n*-butanol and 95% sulfuric acid (Abschn. 5.3). In contrast, polymer film formers (AR-FFF) maintained stability in sulfuric acid, suggesting alcohol-resistant foams with xanthan gum are ideal for organic acid fires. Tests with tripropylamine confirmed the foam-destroying effect of alkylamines. [37] While FFF was fully degraded by the tertiary amine, AFFF remained stable and extinguished the test fire nearly three times faster. AR-FFF also performed well with tripropylamine, offering high re-ignition resistance. AR-FFF should also be used to extinguish chemicals with high vapor pressure such as 1-propanethiol, as the vapor suppression by the polymer film is crucial. For the other tested chemicals, AFFF and FFF

Table 4 Results of the cluster analysis and trends in regard to the recommended firefighting foams (FFF: synthetic fluorine-free foam, AR: polymeric film former (alcohol-resistant foam))

No	Cluster	Recommended foam class
Hydrocarbons		
0	Aromatics	FFF
1	Gases	/
2	Cycloalkanes, aliphatics	FFF
3	Alkenes, alkynes	FFF
Nitrogen compounds		
4	Pyridines	AR
5	Primary amines Secondary/tertiary amines Anilines Nitriles Isocyanates Hydrazines Alcohol amines	AR (Reaction with water) AR AR AR AR (Reaction with water) AR AR
6	Nitro compounds, oximes etc	AR/FFF
7	Nitriles etc	AR
Halogenated compounds		
8	Halogenated oxygen compounds	AR/FFF (gentle foam application)
9	Halogenated hydrocarbons	FFF (gentle foam application)
Solids		
10	Solids	FFF
Alcohols		
11	Alcohols (sparingly soluble to water-soluble)	AR/FFF (FFF: high application rate)
12	Alcohols, cyclic ethers (freely soluble to very water-soluble)	AR
Sulphur compounds		
13	Thiols, sulphur-containing active substances (thiophosphates)	AR
Oxygen/phosphorus compounds		
14	Ketones	AR/FFF
15	Ether	AR/FFF
16	Acids, epoxides, phosphates, phosphonates etc	AR
17	Aldehydes	AR/FFF
18	Aromatic oxygen compounds	AR/FFF
19	Esters	AR/FFF
Other compounds		

(continued)

Table 4 (continued)

No	Cluster	Recommended foam class
W	Violent reaction with water	Dry chemicals/inert gases
F	Per- and polyfluoroalkyl substances	FFF
X	Pyrethrin I and II, tetraethyl lead	Individual recommendation

showed no significant differences in vapor suppression or extinguishing performance. Thus, no substance was identified requiring AFFF, allowing environmentally friendly firefighting foam recommendations.

6 Determination of Benzene Concentration in the Whole Blood of Firefighters After Live Fire Training

To improve the safety and effectiveness of breathing apparatus users during indoor firefighting, real fire training is essential. According to DIN 14097-1:2018-05, realistic operational tactics must be practiced in training facilities to ensure safety. The German fire department regulations No. 2 and 7 also emphasize realistic fire training as a vital part of training for breathing apparatus wearers. Such training facilities, which may use gas, liquid, or solid fuels, offer flexible design options depending on the type of facility. This flexibility allows for specific training objectives, with the fuel type enhancing the realism of fire scenarios. Wood-fired training facilities, for instance, create realistic fire conditions, allowing firefighters to train under authentic smoke and heat exposure. These settings provide opportunities to practice recognizing fire development, understanding flashover, and acclimatizing to high temperatures. Consequently, wood-fired training is crucial for meeting realistic training objectives.

Incomplete combustion of organic materials such as wood produces both inorganic gases, such as toxic carbon monoxide, and organic compounds, including carcinogenic benzene. [18, 39, 40] Benzene, an aromatic compound, has been detected in underventilated wood combustion [18] and recorded at high concentrations during live fire training. [41] Benzene enters the body primarily through inhalation, though dermal absorption also significantly contributes to internal exposure, especially since participants in live fire training wear self-contained breathing apparatus, minimizing inhalation exposure. [42–47]

Given the carcinogenic potential of compounds in fire smoke, firefighter exposure has been classified as carcinogenic. [48] Studies have assessed health risks related to benzene exposure using biomarkers in urine [41, 46] or exhaled breath.

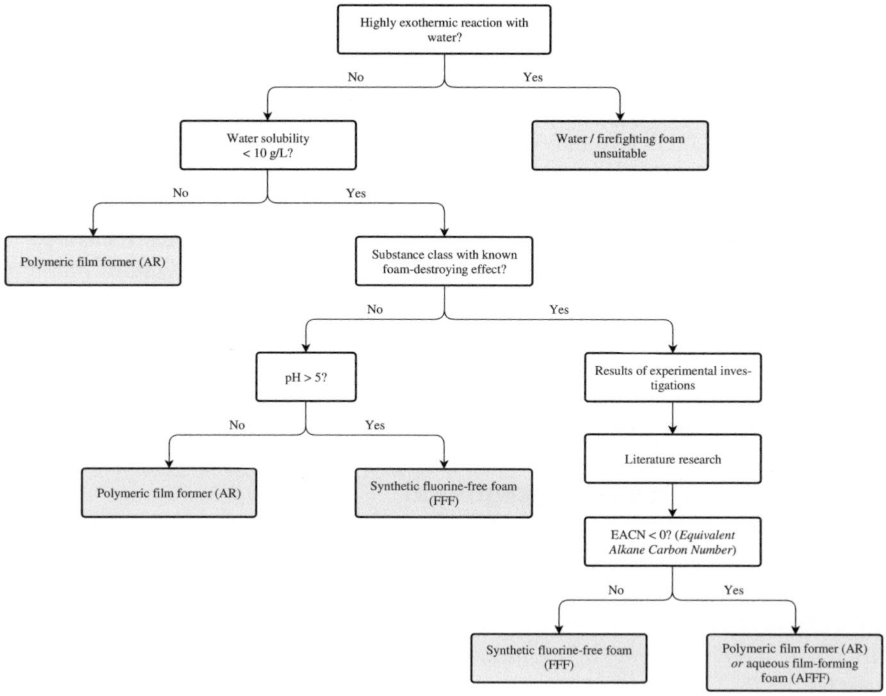

Fig. 16 Flow chart for the re-evaluation of the 523 chemicals regarding to firefighting foam recommendations

[45, 46] Although blood analysis is also used to assess exposure [42], it has mainly been applied to metal exposure in firefighting contexts. [49]

Therefore, a study was initiated to determine whether benzene is present at significant levels in the whole blood of trainees and instructors immediately after live fire training in a controlled environment.

The training took place in a fire training container complex with an observation room and a fire room. Two Euro pallets and fourteen chipboard panels were ignited in the fire room, while trainers and trainees, mostly seated in the observation room, observed fire and smoke behaviour. During the exercise, only one trainer was positioned outside the containers. The instructors adjusted ventilation to control oxygen supply, and participants observed fire development, smoke propagation, and the effects of extinguishing with a handheld nozzle.

Eleven firefighters from a German fire department participated in the study, with their blood benzene concentration measured before and after training. Participants were informed of the study details, provided written consent, and completed questionnaires covering gender, age, smoking status, prior fire smoke exposure, and their training role. The group included five trainers and six trainees, aged 25 to 47, all non-smokers. Ten participants wore full protective equipment, including breathing apparatus, while the trainer outside the container used a coverall and FFP2 mask.

Blood samples were drawn from each participant by qualified medical personnel via peripheral venipuncture. Initial samples were taken before training at the fire station to establish baseline values. After training, a second blood sample was taken immediately, with samples sent to an external lab for benzene analysis.

Preliminary results suggest that benzene levels in participants' blood are within safe limits. Further analysis will include air benzene levels measured in the training facility during the exercise and comparisons with literature findings. Additionally, the results will be evaluated concerning the attainment of steady-state levels and benzene's elimination half-life in the body.

7 Demand Planning for Public Fire Departments

7.1 Introduction and Background

Fire service demand planning in Germany is part of a continuous optimization process, akin to other societal systems. Since the establishment of municipal fire departments around 170 years ago, there has been ongoing effort to enhance their efficiency and effectiveness. Traditionally focused on past incidents, demand planning has shifted in recent decades to encompass a broader view of fire department infrastructure, personnel, and equipment to ensure operational readiness. [50] The introduction of New Public Management (NSM) in the 1990s added momentum to this development by requiring municipalities to quantify fire services with key performance indicators for benchmarking, aiming at more economical use of resources. [51]

Internationally, the U.S.-based RAND Corporation's "Fire Department Deployment Analysis" was a pioneering study, collaborating with New York City to use systems analysis and mathematical models for efficient resource planning. [52] In parallel, Germany's AGBF (1998) and the U.S. NFPA (2001) developed similar guidelines linking fire duration, survival rates, and property damage. Both standards are non-binding. [53, 54] However, methodological foundations for fire service demand planning remain inadequately detailed, with inconsistent terminology across states. A standardized understanding of terms like hazard, risk, and threat is essential, potentially modelled after occupational safety approaches. [55]

Currently, the goal of fire service demand planning, now legally defined in 12 federal states, is to establish local-specific performance capacities based on hazard analyses and derived planning scenarios.

7.2 Systems Theory Perspective

A shift in the perception of fire service performance could lead to a paradigm change, in which the mere availability of the fire service is defined as a type of "performance," analogous to technical systems, where every incident is seen as a disruption to system availability.

Scientific models, particularly systems theory approaches, play a central role in advancing methods for planning fire service structures. Ludwig von Bertalanffy's systems theory provides a universal framework for analysing technical, biological, and social science systems. In fire service demand planning, systems theory methods can also be used to model and mathematically represent complex relationships between various system elements—such as personnel availability, infrastructure, and incident occurrences [56]. Using stochastic processes, applications can be developed to quantify and simulate state changes within the "fire service" system, such as personnel or equipment failures [57].

An example of such models is the assessment of fire station availability as a time- and state-discrete Markov process. However, these models cannot fully address the unique needs of individual municipalities due to variations in local personnel and equipment availability. Figure 17 shows a system model of a fire department with two stations represented as a Markov chain with associated transition probabilities (Formula 1). It includes five states: full availability, partial availability, one station out of service, and total failure. [51]

$$\underline{\underline{M}} = \begin{bmatrix} 0 & 1 & 0 & 0 & 0 \\ 0{,}6 & 0{,}15 & 0{,}08 & 0{,}12 & 0 \\ 0 & 0{,}8 & 0 & 0 & 0{,}2 \\ 0 & 0{,}75 & 0 & 0 & 0{,}25 \\ 0 & 0{,}05 & 0{,}3 & 0{,}3 & 0{,}35 \end{bmatrix} \tag{1}$$

In addition to systems theory modelling, social changes must also be considered in fire service demand planning. The declining self-help capacity of the population, as well as rising expectations for public services, has led to increasing demand for government support. These societal changes must also be incorporated into demand planning to ensure a future-proof fire service structure.

7.3 Spatial Analysis

A key aspect of modern fire service demand planning is the use of Geographic Information Systems (GIS) and statistical methods for spatial analysis, enabling precise identification of hazard potentials and response hotspots, as well as simulation of incident rates based on land use. As part of the research project on Safety and Efficiency in Fire Service Demand Planning, new GIS-based spatial analysis methods were tested, along with multivariate statistical techniques like cluster analysis to classify spatial units by hazard potential and form homogeneous clusters. [58, 59]

The study of fire incident distribution revealed that fires correlate variably with human activity, making area size a better normalization factor than population size, as it accounts for factors like population density through land-use differentiation. [60] Based on hazard prevention data from North Rhine-Westphalia, a link was established between fire incidents and land use, enabling calculation of a land-use-specific fire rate (Fig. 18). This approach provides a standardized fire rate for comparing incidents across municipalities without distortions. [51] The calculation of the land-use-specific fire rate is carried out according to the following Formula 2:

$$B_{Ri} = \frac{B_i}{\mathrm{A}_i} \cdot 10 \tag{2}$$

where:

B_{Ri} = Land-use-specific fire rate for land area i, in the time interval [t, t+n]
B_i = Total number of fires on the specified area i, in the time interval [t, t+n]
A_i = Average total area of a specified land use i, in the time interval [t, t+n]
i = Index for the respective land area

To avoid distortions in inter-municipal fire rate comparisons, differences in land composition must also be considered. Therefore, it is necessary to calculate a standardized fire rate that projects fire intensities onto a hypothetical standard area. The choice of reference area for calculating the standardized fire rate can be flexible but must be consistent across all participating municipalities. A reference to

Fig. 17 Example model for describing a fire department with 2 stations as a Markov process

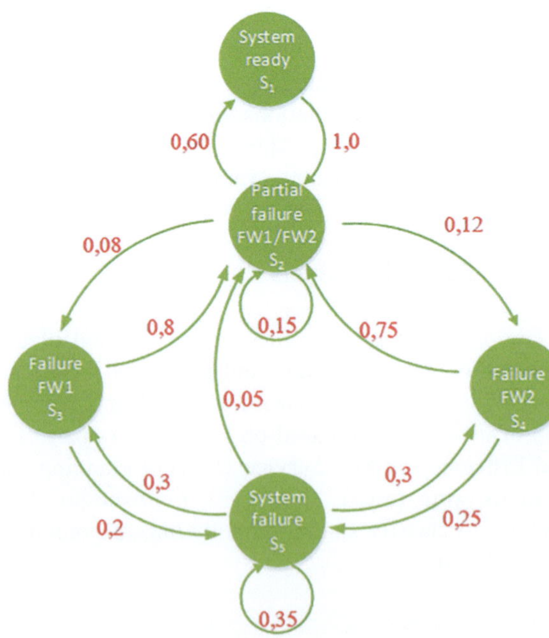

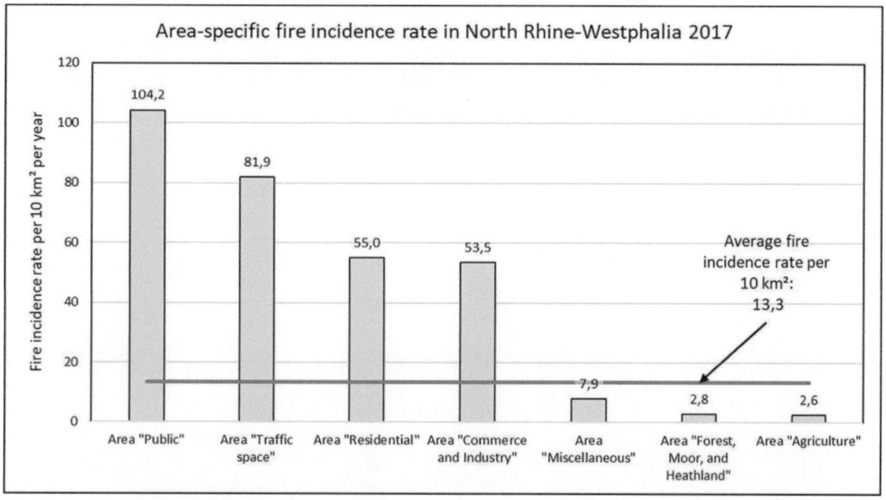

Fig. 18 Area-specific fire incidence rate in North Rhine-Westphalia

the actual land structure is advisable to prevent artificial results. For example, this standard area can be composed of the statewide shares of different land use types to create a comparable basis for municipalities. [57]

In the context of spatial analysis, which is part of the fire service demand planning process, the question of the appropriate spatial unit for the study inevitably arises. Traditionally, spatial analysis is conducted in a 1 km grid or based on smaller administrative divisions, such as municipal districts. However, a research project has tested approaches for studies with higher resolution. The analysis of hazard potential was carried out using a fixed grid with a 100 m edge length. It became apparent that, for example, in the evaluation of primary land use, traffic areas were not considered due to their low area share within the grid cells. Further considerations include the use of a hexagonal grid, which has proven to be a suitable alternative for simplifying the study area, especially when addressing questions of spatial and temporal accessibility [61]. The different approaches are compared in the following illustration Fig. 19.

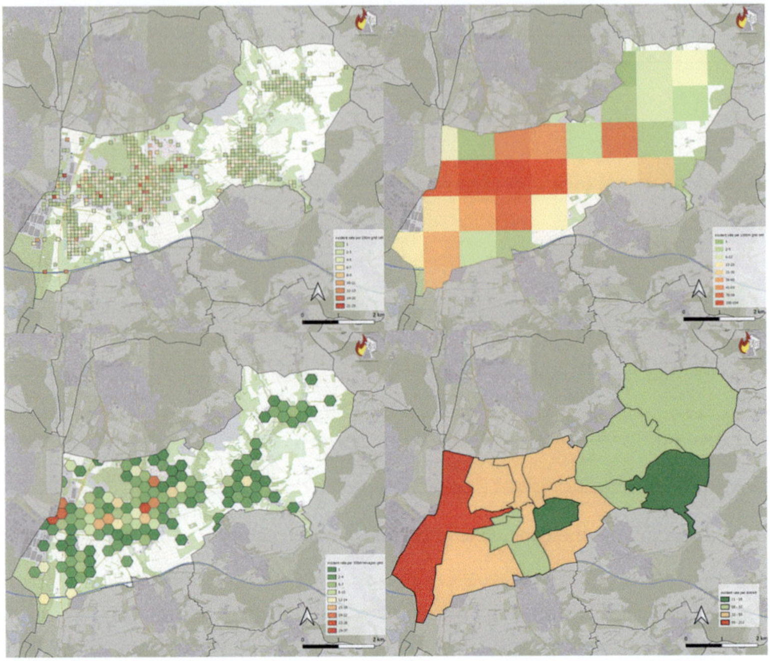

Fig. 19 Example comparison with four different analysis levels: 100/1000 m grid, 300 m hexagonal grid, and municipal districts

References

1. Q. Wang, B. Mao, S. I. Stoliarov, and J. Sun, "A review of lithium ion battery failure mechanisms and fire prevention strategies," *Progress in Energy and Combustion Science*, vol. 73, pp. 95–131, 2019, https://doi.org/10.1016/j.pecs.2019.03.002.
2. J. Sun *et al.*, "Toxicity, a serious concern of thermal runaway from commercial Li-ion battery," *Nano Energy*, vol. 27, pp. 313–319, 2016, https://doi.org/10.1016/j.nanoen.2016.06.031.
3. M. Lelie *et al.*, "Battery Management System Hardware Concepts: An Overview," *Applied Sciences*, vol. 8, no. 4, p. 534, 2018, https://doi.org/10.3390/app8040534.
4. R. Korthauer, *Handbuch Lithium-Ionen-Batterien*. Berlin, Heidelberg: Springer Berlin Heidelberg, 2013.
5. S. Ma *et al.*, "Temperature effect and thermal impact in lithium-ion batteries: A review," *Progress in Natural Science: Materials International*, vol. 28, no. 6, pp. 653–666, 2018, https://doi.org/10.1016/j.pnsc.2018.11.002.
6. G. Zubi, R. Dufo-López, M. Carvalho, and G. Pasaoglu, "The lithium-ion battery: State of the art and future perspectives," *Renewable and Sustainable Energy Reviews*, vol. 89, pp. 292–308, 2018, https://doi.org/10.1016/j.rser.2018.03.002.
7. D. Sturk, L. Rosell, P. Blomqvist, and A. Ahlberg Tidblad, "Analysis of Li-Ion Battery Gases Vented in an Inert Atmosphere Thermal Test Chamber," *Batteries*, vol. 5, no. 3, p. 61, 2019, https://doi.org/10.3390/batteries5030061.
8. L. H. Guenther *et al.*, "Second Life for Lithium-Ion Traction Batteries," *Sustainability*, vol. 16, no. 17, p. 7288, 2024, https://doi.org/10.3390/su16177288.

9. E. D. Owen, *Degradation and Stabilisation of PVC*. Dordrecht: Springer Netherlands, 1984.
10. J. Warnatz, "Chemistry of high temperature combustion of alkanes up to octane," *Symposium (International) on Combustion*, vol. 20, no. 1, pp. 845–856, 1985, https://doi.org/10.1016/S0082-0784(85)80574-9.
11. I. Glassman, *Combustion*. Waltham, MA: Academic Press, 2015.
12. V. Babrauskas, *Ignition handbook: Principles and applications to fire safety engineering, fire investigation, risk management and forensic science*. Issaquah, Wash.: Fire Science Publishers, 2003.
13. B. Karlsson, J. G. Quintiere, and N. Johansson, *Enclosure Fire Dynamics*. Boca Raton: CRC Press, 2022.
14. *DIN EN ISO 13943 Brandschutz*, 13943, Deutsches Institut für Normung e. V., 2023.
15. *Erzeugung thermischer Zersetzungsprodukte von Werkstoffen für ihre analytisch-toxikologische Prüfung*, 53436 2015–12, Deutsches Institut für Normung e. V., 2015.
16. Andreas Bräutigam, "Einflüsse der Brandgutzusammensetzung und der Verbrennungsbedingungen auf die Entstehung mittel- und schwerflüchtiger organischer Brandrauchinhaltstoffe," Dissertation, Ruhr-Universität Bochum, 2000.
17. R. Denig, "Brandsimulation durch Mikroverbrennung," *Z. Anal. Chem.*, vol. 330, no. 2, pp. 116–119, 1988, https://doi.org/10.1007/BF00470719.
18. Martin Meyer, "Untersuchung von Brandzersetzungsprodukten aus unterventilierten Verbrennungen von natürlichen und synthetischen sauerstoffhaltigen Polymeren," Dissertation, Bergischen Universität Wuppertal, 2021.
19. P. Wang, L. Chen, and H. Xiao, "Flame retardant effect and mechanism of a novel DOPO based tetrazole derivative on epoxy resin," *Journal of Analytical and Applied Pyrolysis*, vol. 139, pp. 104–113, 2019, https://doi.org/10.1016/j.jaap.2019.01.015.
20. B. A. Howell, "Thermal Degradation of Organophosphorus Flame Retardants," *Polymers*, vol. 14, no. 22, 2022, https://doi.org/10.3390/polym14224929.
21. Q. Wu, R. de Cassia Siqueira Curo Valle, J. Alexandre Borges Valle, F. B. Maestá, X. Meng, and M. J. L. Arias, "Recent Progress of DOPO-Containing Compounds as Flame Retardants for Versatile Polymeric Materials: Review," *World J. Text. Eng. Technol.*, no. 6, pp. 89–103, 2020.
22. H. Tang, Z. Zhu, R. Chen, J. Wang, and H. Zhou, "Synthesis of DOPO-based pyrazine derivative and its effect on flame retardancy and thermal stability of epoxy resin," *Polymers for Advanced Techs*, vol. 30, no. 9, pp. 2331–2339, 2019, https://doi.org/10.1002/pat.4674.
23. M. He, D. Zhang, W. Zhao, S. Qin, and J. Yu, "Flame retardant and thermal decomposition mechanism of poly(butylene terephthalate)/DOPO-HQ composites," *Polymer Composites*, vol. 40, no. 3, pp. 974–985, 2019, https://doi.org/10.1002/pc.24772.
24. B. Eskenazi, A. Bradman, and R. Castorina, "Exposures of children to organophosphate pesticides and their potential adverse health effects," *Environmental health perspectives*, vol. 107 Suppl 3, Suppl 3, pp. 409–419, 1999, https://doi.org/10.1289/ehp.99107s3409.
25. J. Theisen, W. Funcke, and S. Hamm, "Untersuchung der möglichen umweltgefährdung beim Brand von Kunststoffen: Forschungsbericht 104 09 222,"
26. E. Reizer, B. Viskolcz, and B. Fiser, "Formation and growth mechanisms of polycyclic aromatic hydrocarbons: A mini-review," *Chemosphere*, vol. 291, Pt 1, p. 132793, 2022, https://doi.org/10.1016/j.chemosphere.2021.132793.
27. J. Huang *et al.*, "Insights into the thermal degradation mechanisms of polyethylene terephthalate dimer using DFT method," *Chemosphere*, vol. 291, Pt 2, p. 133112, 2022, https://doi.org/10.1016/j.chemosphere.2021.133112.
28. P. Chaudhuri and K. Wieghardt, "Phenoxyl Radical Complexes," in *Progress in Inorganic Chemistry* (Progress in Inorganic Chemistry), K. D. Karlin, Ed., Wiley, 2001, pp. 151–216.
29. M. Wachsmuth and G. Knetsch, "Vom chemischen Stoffdatenpool „GSBL“ zum anwendungsorientierten Chemikalieninformationssystem „ChemInfo“," in *Umweltinformationssysteme – Wie verändert die Digitalisierung unsere Gesellschaft?*, U. Freitag, F. Fuchs-Kittowski, A. Abecker, and F. Hosenfeld, Eds., Wiesbaden: Springer Fachmedien Wiesbaden, 2021, pp. 229–239.

30. B. Weber, M. Herzberg, and M. Wachsmuth, "Gefahrstoffinformationen an Ort und Stelle," in *Umweltinformationssysteme - Wie trägt die Digitalisierung zur Nachhaltigkeit bei?*, F. Fuchs-Kittowski, A. Abecker, and F. Hosenfeld, Eds., Wiesbaden: Springer Fachmedien Wiesbaden, 2022, pp. 43–49.
31. S. Peshoria, D. Nandini, R. K. Tanwar, and R. Narang, "Short-chain and long-chain fluorosurfactants in firefighting foam: a review," *Environ Chem Lett*, vol. 18, no. 4, pp. 1277–1300, 2020, https://doi.org/10.1007/s10311-020-01015-8.
32. S. Y. Wee and A. Z. Aris, "Revisiting the "forever chemicals", PFOA and PFOS exposure in drinking water," *npj Clean Water*, vol. 6, no. 1, 2023, https://doi.org/10.1038/s41545-023-00274-6.
33. "Bayerisches Staatsministerium für Umwelt und Verbraucherschutz, Bayerisches Staatsministerium des Innern und für Integration, & Bayerisches Landesamt für Umwelt. (2018). Umweltschonender Einsatz von Feuerlöschschäumen." Accessed: Oct. 20, 2024. https://www.sfsg.de/fileadmin/downloads/SFSG/Leitfaden.pdf.
34. "European Chemicals Agency (ECHA). (2022). Annex XV restriction report – Per- and polyfluoroalkyl substances (PFASs) in firefighting foams." Accessed: Oct. 20, 2024. https://echa.europa.eu/documents/10162/4524f49c-ae14-b01b-71d2-ac3fa916c4e9.
35. Pedregosa, F., Varoquaux, G., Gramfort, A., Michel, V., Thirion, B., Grisel, O., Blondel, M., Prettenhofer, P., Weiss, R., Dubourg, V., Vanderplas, J., Passos, A., Cournapeau, D., Brucher, M., Perrot, M., & Duchesnay, E., "Scikit-learn: Machine Learning in Python," *Journal of Machine Learning Research*, 12(85), pp. 2825–2830, 2011.
36. X. Yu *et al.*, "Formation of stable aqueous foams on the ethanol layer: Synergistic stabilization of fluorosurfactant and polymers," *Colloids and Surfaces A: Physicochemical and Engineering Aspects*, vol. 591, p. 124545, 2020, https://doi.org/10.1016/j.colsurfa.2020.124545.
37. D. Exerowa and P. M. Kruglyakov, *Foam and Foam Films: Theory, Experiment, Application,* 1st ed. (Studies in interface science v.5). s.l.: Elsevier textbooks, 1997. http://gbv.eblib.com/patron/FullRecord.aspx?p=344651.
38. S. Ross and J. W. McBain, "Inhibition of Foaming in Solvents Containing Known Foamers," *Industrial and Engineering Chemistry*, 36(6), pp. 570–573, 1944.
39. Das Europäische Parlament und der Rat der Europäischen Union, *über die Einstufung, Kennzeichnung und Verpackung von Stoffen und Gemischen, zur Änderung und Aufhebung der Richtlinien 67/548/EWG und 1999/45/EG und zur Änderung der Verordnung (EG) Nr. 1907/2006: VERORDNUNG (EG) Nr. 1272/2008*, 2008.
40. *Vermeidung von Schäden durch Rauch und Brandfolgeprodukte - Gefahren, Risiken, Schutzmaßnahmen*, 3400, Gesamtverband der Deutschen Versicherungswirtschaft e. V., 2017.
41. J. Laitinen, M. Mäkelä, J. Mikkola, and I. Huttu, "Fire fighting trainers' exposure to carcinogenic agents in smoke diving simulators," *Toxicology letters*, early access. https://doi.org/10.1016/j.toxlet.2009.06.864.
42. Deutsche Forschungsgemeinschaft, *MAK- und BAT-Werte-Liste 2024*. German Medical Science GMS Publishing House, 2024.
43. B. Rossbach, D. Wollschläger, S. Letzel, W. Gottschalk, and A. Muttray, "Internal exposure of firefighting instructors to polycyclic aromatic hydrocarbons (PAH) during live fire training," *Toxicology letters*, early access. https://doi.org/10.1016/j.toxlet.2020.05.024.
44. K. M. Kirk and M. B. Logan, "Firefighting instructors' exposures to polycyclic aromatic hydrocarbons during live fire training scenarios," *Journal of occupational and environmental hygiene*, vol. 12, no. 4, pp. 227–234, 2015, https://doi.org/10.1080/15459624.2014.955184.
45. K. W. Fent *et al.*, "Firefighters' and instructors' absorption of PAHs and benzene during training exercises," *International journal of hygiene and environmental health*, early access. https://doi.org/10.1016/j.ijheh.2019.06.006.
46. K. W. Fent *et al.*, "Systemic exposure to PAHs and benzene in firefighters suppressing controlled structure fires," *The Annals of occupational hygiene*, early access. https://doi.org/10.1093/annhyg/meu036.

47. Institut für Arbeitsschutz der Deutschen Gesetzlichen Unfallversicherung (2024). "GESTIS-Stoffdatenbank: Benzol." Accessed: Oct. 21, 2024. https://gestis.dguv.de/data?name=010060.
48. P. A. Demers *et al.*, "Carcinogenicity of occupational exposure as a firefighter," *The Lancet. Oncology*, early access. https://doi.org/10.1016/S1470-2045(22)00390-4.
49. M. Engelsman, L.-M. L. Toms, A. P. W. Banks, X. Wang, and J. F. Mueller, "Biomonitoring in firefighters for volatile organic compounds, semivolatile organic compounds, persistent organic pollutants, and metals: A systematic review," *Environmental research*, early access. https://doi.org/10.1016/j.envres.2020.109562.
50. A. Ridder, "Risikologische Betrachtungen zur strategischen Planung von Feuerwehren: Empirische Befunde und Systematisierung von Zielsystemen," [Köln], 2015.
51. R. Goertz, H. Behrendt, and P. Haffner, "Sicherheit und Wirtschaftlichkeit in der Feuerwehrbedarfsplanung: Forschungsprojekt," BUW, Wuppertal, 2020.
52. W. Z. Hirsch, W. E. Walker, J. M. Chaiken, and E. J. Ignall, "Fire Department Deployment Analysis: A Public Policy Analysis Case Study- The RAND Fire Project," *Journal of the American Statistical Association*, vol. 76, no. 374, 1981.
53. National Fire Protection Association, *NFPA 1710: standard for the organization and deployment of fire suppression operations, emergency medical operations, and special operations to the public by career fire departments,* 2020th ed. Quincy, Massachusetts: NFPA, 2020.
54. AGBF bund, Ed., "Qualitätskriterien für die Bedarfsplanung von Feuerwehren in Städten: Empfehlungen der Arbeitsgemeinschaft der Leiter der Berufsfeuerwehren," Bonn, Nov. 2015.
55. A. Kahl and M. Bier, *Arbeitssicherheit: fachliche Grundlagen*. Berlin: Erich Schmidt Verlag, 2019.
56. D. J. Krieger, *Einführung in die allgemeine Systemtheorie* (Uni-Taschenbücher). München: Fink, 1996.
57. R. Goertz, H. Behrendt, and P. Haffner, "Sicherheit und Wirtschaftlichkeit in der Feuerwehrbedarfsplanung: Forschungsprojekt," Abschlussbericht, BUW, Wuppertal, 2020.
58. G. Bahrenberg, E. Giese, and J. Nipper, *Statistische Methoden in der Geographie, 2. Multivariate Statistik : mit 24 Tabellen,* 3rd ed. (Statistische Methoden in der Geographie). Stuttgart: Teubner, 2008.
59. K. Backhaus, B. Erichson, W. Plinke, and R. Weiber, *Multivariate Analysemethoden: eine anwendungsorientierte Einführung,* 15th ed. (Lehrbuch). Berlin and Heidelberg: Springer Gabler, 2018.
60. H. Behrendt, R. Goertz, and P. Haffner, "Standardisierte Normierung des Brandeinsatzaufkommens: Sachgerechter Leistungsvergleich für die Brandschutzbedarfsplanung," *Brandschutz: deutsche Feuerwehr-Zeitung*, no. 02, pp. 14–21, 2021.
61. P. Haggett, *Einführung in die kultur- und sozialgeographische Regionalanalyse: mit 64 Tabellen*. Zentralbibliothek der Wirtschaftswissenschaften in der Bundesrepublik Deutschland, *Berlin: De Gruyter.* https://www.wiso-net.de/document/ECON__021516235.

Industrial Science, Society and Safety

Learning from (Traffic) Psychology—The 4-E of Safety Management

Nicola Fricke and Nadine Schlueter

Contents

Abstract

Safety issues are a common challenge in organizations. While occupational health and safety developed a lot of methods and measures to improve safety standards in companies, psychology also developed approaches regarding safety management. One of these approaches—the 4E—is used in traffic psychology in order to increase the safety within the traffic environment. Translating this approach to a general safety perspective means that 1) enforcement, 2) education, 3) encouragement and 4) engineering measures are all used in a concerted way as a safety management strategy. Finally, the paper summarizes open research gaps which are the systematic investigation of the effects of the concerted introduction of the E's as well as possible further expansions of the concept based on the topic of safety culture.

N. Fricke (✉)
Department of Sociotechnical Systems, School of Mechanical Engineering and Safety Engineering, University of Wuppertal, Wuppertal, Germany
e-mail: nfricke@uni-wuppertal.de

N. Schlueter Department for Product Safety and Quality Engineering, School of Mechanical Engineering and Safety Engineering, University of Wuppertal, Wuppertal, Germany

S. Bracke and R. Pieper (eds.), *Safety Engineering*,
https://doi.org/10.1007/978-3-658-47473-7_15

Keywords

Safety management · Traffic safety · Psychology

1 Traffic Psychology and the 4-E Concept

Traffic psychology is a specific sub-discipline of psychology concerned with how humans behave and experience themselves within the context of the traffic environment as well as the involved psychological processes (see e.g. [1]). One topic of particular interest is increasing traffic safety. Similarly to other safety domains, multiple concepts and interventions for reaching the goal of increasing traffic safety can be found.

One such approach treats safety as a management task. The so-called safety management aims at identifying potential influencing factors which can afterwards be controlled whereby safety increases (see e.g. [2]). Within this traffic safety management approach, the concept of the 4-E has been introduced (see Fig. 1; based on [3]).

The 4-E are defined as the following:

Fig. 1 The 4-E of traffic safety management (based on [3]), www.colourbox.de

1) **Enforcement**: orders/prohibitions regulating behavior in traffic,
2) **Education**: informational and qualification measures for all road users,
3) **Encouragement**: incentives with the potential of changing the cost-benefit ratio, and
4) **Engineering**: changing the design of road infrastructure and transportation modes.

The concept thereby implies that successful safety management can be obtained through implementing and addressing measures of all four of the E in a concerted way. This is particularly important as there are many examples of interventions focusing on only one or two combinations of the four factors. While this is probably already a good start, the impact might be limited.

Examples for each of the 4-E in the traffic sector can easily be found: speed enforcement, public awareness campaigns such as "don't' drink and drive", insurance companies offering discounts for staying accident-free or the installation of street safety barriers such as speed bumps. Such measures are commonly used in other domains as well, even though a combined introduction and a systematic evaluation of the effects of single and their combined use remain an open research topic.

Relevant psychological concepts related to the 4-E will be explained in the following sections in order to support the combined introduction of measures for promoting "safe" behavior.

1.1 Psychological Processes Underlying Enforcement and Encouragement

There are basically three processes that describe learning as a means of changing behavior: classical conditioning, instrumental conditioning and observational learning (see e.g. [4]). While classical conditioning is mostly applied in settings in which negative associations are replaced by more beneficial ones—instrumental conditioning uses reinforcement strategies for shaping new/desired behavior [5] and is particularly interesting for strategies in the context of safety management. The concept divides positive and negative reinforcement which as a result will increase the likelihood of the occurrence of a specific behavior. Punishment—also divided into positive and negative—, on the other hand, will decrease the likelihood of a specific behavior ([6, 7]). Examples for positive/negative reinforcement and punishment in the traffic safety domain can be found in Table 1.

There are possible effects which support the combination of reinforcement- and punishment strategies as suggested within the 4-E concept for enhancing "safe" behavior. First of all, enforcement in the form of punishment will most likely lead to a decrease of unwanted behavior as the conditioning theory predicts. However, this does not imply that the behavior itself is unlearned. Instead, the unwanted behavior is suppressed [6] only as long as the enforcement/punishment persists. Secondly, accompanying negative emotional reactions towards the enforcement-authority might occur [6]. In order to avoid such negative effects and for deleting

Table 1 Positive/negative reinforcement and punishment (adapted from [7])

	Reinforcement	Punishment
Positive	Addition of a pleasant stimulus; e.g. receiving an insurance bonus for staying accident-free	Addition of an unpleasant stimulus; e.g. receiving a ticket for speeding
Negative	Removing an unpleasant stimulus; e.g. putting the seat-belt on in order to turn-off the warning signal	Removing a pleasant stimulus; e.g. handing-in the driving license for speeding

undesirable behavior, there is a need for offering other alternatives which are reinforced/encouraged [6]. Furthermore, the combined usage can change the cost-benefit ratio such that increasing the benefits of "safe" behavior and, at the same time, decreasing the costs [3] will result. This has direct implications and supports a combined usage of enforcement and encouragement intervention strategies.

The next section will further describe psychological models related to educational measures.

1.2 Psychological Models and Processes Related to Educational Interventions

The KAP-Model (Knowledge-Attitude-Practice) assumes that by using measures such as campaigns, leaflets and films, knowledge about e.g. hazards will be increased and followingly lead to prevention of accidents and critical situations (e.g. [8]). Unfortunately, the connections do not seem to be that easy. First of all, knowledge does not always influence attitudes or beliefs—as we will further discuss within the elaboration likelihood model. Secondly, a lack of correspondence between attitudes and behavior—based on study findings—questions this simple linear model (e.g. [9]).

Other models assume more sophisticated multifactorial influences on behavioral outcomes, such as the traditional theory of planned behavior [10]. The theory defines three influencing factors, which are attitudes, social norm and perceived behavioral control and which have a direct effect on behavioral intention, which again predicts behavior [10]. The model has been further revised into the reasoned action approach [11], adding behavioral, normative and control beliefs as well as background factors as relevant influencing factors. The model shows the complicated interplay of factors related to behavioral intentions.

Another relevant psychological mechanism, which should be taken into account when thinking about educational measures, is the elaboration likelihood model (ELM) of persuasion [12]. It can be used in the construction of persuasive communication, which is "a message advocating a particular side of an issue" ([9], p. 200). The ELM postulates two routes of information processing: one central route/systematic information processing and one peripheral route/heuristic information processing [12]. Which route is chosen for processing information depends on several factors, such as motivation or cognitive capacity. The hypotheses of the model specify that people with low motivation and/or low cognitive capacity use the peripheral route of information processing in contrast to people with high motivation and/or high cognitive capacity (e.g. [13]). The model has direct implications for the design of educational measures: if one assumes that the audience receiving the informational messages might have low motivation, presenting knowledge and facts might not result in the desired attitude change and behavioral outcomes.

Furthermore, educational safety trainings could be used as habit shaping means without the specific need of changing attitudes. Research supports the assumption that behavior turns into a habit if repeated often in relatively stable contexts [13]. Furthermore, a study investigated how long it takes for the repetition of a specific behavior to become a new habit and found a mean duration of 66 days [14]. This furthermore informs the need of designing trainings, which repeat the desired behavior daily for a long time, in order to lead to desired behavioral changes.

The next chapter discusses further psychological concepts relevant for engineering measures.

1.3 Psychological Concepts Relevant for Engineering Measures

One relevant concept for designing more usable systems and thus creating safer usage is the concept of affordances. The concept dates back to the ecological approach to visual perception [15], which states that there is a direct relationship between perception and action without cognitive steps in between. Within this approach there exists a theory of affordances [15]. The concept of affordances has been applied within the user-centered design community such as by Don Norman (2013) to the "design of everyday things" whereby he defines that affordances mean that an object directly implies how it can be used. A simple example is a stool that affords sitting [16]. The idea has further been applied and developed into the "self-explaining roads" concept aiming at creating a traffic environment which by design implies safe behaviors of users [17]. As an example, a narrow road will by design imply slower driving [2], which in consequence promotes safer behavior as a higher crash rate is related to increased speed (e.g. [18]). Thus, the concept of affordances when applied in safety-critical contexts, might help in eliciting desired behavior through carefully designed machines, systems and work places.

Furthermore, when addressing the engineering of new tools or work places or the change of existing ones, the participation of the workers in developing the systems and changes might be a key issue. On the one hand it follows the idea of human-centered systems design (see e.g. DIN EN ISO 9241-210) [19] in order to meet user needs and develop systems that are best fitted towards the characteristics and preferences of users. On the other hand, if systems are optimized for "safer" usage by e.g. usage of assistance systems, the introduction of such systems might be more accepted and used in the intended way in case of a preceding participatory process.

The next chapter describes possible applications of the 4-E concept to a general safety engineering perspective.

2 Translating the 4-E Concept to Safety Engineering: An Example

As discussed in the previous section, several psychological concepts are relevant for the application of the 4-E concept. Implications for transferring them to a general safety engineering perspective will be described in the following. First of all, the concept could be transferred and described as follows:

1) **Enforcement**: regulations such as punishment in case of safety violations,
2) **Education**: informational and qualification measures such as safety trainings,
3) **Encouragement**: incentives with the potential of changing the cost-benefit ratio, and
4) **Engineering**: changing the design of systems and tools.

What could an example of the combined 4-E safety management strategy look like? Let's imagine a manufacturing plant in which a concerted strategy shall be implemented in order to decrease the manipulation of safeguards. Based on the 4-E concept above this could be:

1) a disciplinary system for cases in which manipulation of safeguards is detected—this could be implemented on individual or team level,
2) safety training—focusing on repeating the usage of the systems with the safeguards active for a duration of 14 weeks, workers fill out a diary to report on the usage,
3) safety reward schema—workers/teams who actively report violations receive a reward such as extra free time or compensation for days which are manipulation-free,
4) participatory design of safeguards—installing safeguards which are developed together with the workers in order to be user-friendly and accepted.

3 Discussion

Dating back to Taylor and his idea of identifying the one best way of doing a specific task and then regulating and observing the workers to enforce this with rules and punishment in case of violations has proven insufficient in preventing accidents or making systems' outcomes "safer" (see e.g. [20]). While this finding might be true for the single aspect of enforcement, it might not persist when combined with encouragement measures.

Of course, one can argue that the proposed application of the 4-E in safety engineering is just another way of classifying interventions which already exist in

Fig. 2 One possible modification of the 4-E of safety management: adding Environment. www.colourbox.de

other forms. A recent systematic review specified seven main categories of intervention approaches: attitudinal, physiological, behavior based, safety climate, safety culture, structural and multifaceted [21]. Within this categorization, structural approaches include the engineering part and encouragement aspects. The problem with this classification though is that first of all, it includes topics on different levels, i.e. safety culture can generally be understood as including aspects of safety climate, therefore separating these dimensions does not seem useful. Secondly, multifaceted approaches include possibly all of the above-mentioned topics which is too unspecific. Furthermore, the approach does not account for the differing psychological processes lying behind the proposed impacts of the measures. Attitudinal approaches as an example are explained to be achieved via persuasive messages or teaching [21]. This is problematic as persuasive messages rely on elaboration processes such as specified in the elaboration likelihood model [12] which is explained in the previous section. Therefore, information presentation might not lead to any changes in attitudes if the messages are not properly designed and evaluated and furthermore, this possible attitude change might not lead to the desired behavioral outcome as can be explained by the aforementioned reasoned action approach [11].

The mentioned alternative classification [21] does have an advantage on the contrary: it specifies additional measures which are related to organizational aspects, e.g. safety culture. This aspect is currently only indirectly included in the 4-E concept such that the concerted introduction of measures along the four dimensions implies that safety is taken more seriously and can be seen as one aspect of safety culture. There are certain psychological processes, such as observational learning (see e.g. [7]) which assumingly are very important in the context of safety management. A role-model for example, showing how important safety is within the company and that it is preferred over productivity gains, seems to be a key issue in promoting safety. A modification and expansion of the 4-E concept such that organizational/environmental aspects relevant for safety culture are included (see Fig. 2), might be a useful modification. In conclusion, the systematic investigation of the effects of the combined usage of the 4-E as well as further additions to the concept based on psychological processes remain an open research topic.

References

1. Krems, J. & Baumann, M.: Verkehrspsychologie. In: A. Schütz, M. Brand, H. Selb & S. Lautenbacher (Eds.): Psychologie. Eine Einführung in ihre Grundlagen und Anwendungsfelder. Stuttgart: Kohlhammer, (2015).
2. Fahlbruch, B., Schöbel, M. & Marold, J.: Sicherheit. In P. Badke-Schaub, G. Flade, A.: Der rastlose Mensch. Konzepte und Erkenntnisse der Mobilitätspsychologie. Wiesbaden: Springer (2013).
3. Schlag, B. & Heger, R.: Ansätze einer psychologisch fundierten Straßengestaltung. In B. Schlag (Ed.), Verkehrspsychologie. Mobilität – Sicherheit – Fahrerassistenz. Lengerich: Pabst Science, (2004).
4. Geller, S. E.: The psychology of safety handbook. Boca Raton: CRC Press, (2001).
5. Anderson, J. R.: Learning and Memory. An integrated approach. New York: John Wiley & Sons, inc, (1994).
6. Becker-Carus, C. & Wendt, M.: Allgemeine Psychologie. 2., Berlin: Springer. (2017).
7. Myers, D. G.: Psychologie. Berlin: Springer, (2014).
8. Lund, J. & Aaro. L. E.: Accident prevention. Presentation of a model placing emphasis on human, structural and cultural factors. Safety Science, 42 (4), pp. 271–324, (2004).
9. Aronson, E., Wilson, T. D. & Akert, R. M.: Sozialpsychologie (8. Aufl.). München: Pearson Studium. (2014).
10. Ajzen, I.: The theory of planned Behavior. Organizational Behavior and Human Decision Processes, 50 (2), pp. 179–211. (1991).
11. Fishbein, M. & Ajzen, I.: Predicting and changing behavior. The reasoned action approach. New York: Psychology Press, (2011).
12. Petty, R. E. & Cacioppo, J. T.: Source factors and the elaboration likelihood model of persuasion. Advances in Consumer Research, 11, (1), pp. 668–672, (1984).
13. Jonas, K., Stroebe, W., & Hewstone, M.: Sozialpsychologie. Berlin: Springer, (2014).
14. Lally, P., van Jaarsveld, C.H.M, Potts, H.W.W. & Wardle, J.: How are habits formed: Modelling habit formation in the real world. European Journal of Social Psychology, 40, (6), 998–1009, (2010).
15. Gibson, J. J.: Wahrnehmung und Umwelt. Der ökologische Ansatz in der visuellen Wahrnehmung. München: U & S Psychologie, (1982).
16. Norman, D.: The design of everyday things. New York: Basic Books, (2013).
17. Theeuwes, J. & Godthelp, H.: Self-explaining roads. Safety Science, 19, pp. 217–225, (1995).
18. Arts, L. & Schagen, I.: Driving speed and the risk of road crashes: A review. Accident Analysis and Prevention, 38 (2),pp. 215–224, (2006).

19. DIN EN ISO 9241-210: Ergonomie der Mensch-System-Interaktion – Teil 210: Menschzentrierte Gestaltung interaktiver Systeme. Brüssel, CEN, (2019).
20. Dekker, S.: Foundations of Safety Science. A Century of Understanding Accidents and Disasters. Boca Raton: CRC Press, (2019).
21. Dyreborg J., Lipscomb H.J., Nielsen K., Törner M., Rasmussen K., Frydendall K.B., Bay H., Gensby U., Bengtsen E., Guldenmund F., Kines P.: Safety interventions for the prevention of accidents at work: A systematic review. Campbell Syst Rev., 18 (2), pp. 1–187, (2022).

Early Exit of the Baby Boomer Generation in Germany—Findings from the lidA-Study

Hans Martin Hasselhorn and Melanie Ebener

Contents

Abstract

This article examines the employment perspective of older workers in Germany (prospectively). It considers the reasons for leaving employment among those who take early retirement, or those who receive a disability pension (retrospectively). The data is taken from the 4th wave of the representative lidA-Study in 2022/2023 (www.lida-studie.de). The results indicate that most of the baby boomer generation (born in 1959 and 1965) would like to retire early, long before reaching their official retirement age and long before the point in time at which they believed they could continue working. At the same time, a small portion of the subsequent generation (those born in 1971), seems to be preparing themselves for a somewhat longer working life. The reasons for the preferred early retirement are mainly related to a desire for self-determination in life, followed by stressful work and health problems. In all three birth cohorts, there is a widespread willingness to remain in employment longer

H. M. Hasselhorn (✉) · M. Ebener
University of Wuppertal, Wuppertal, Germany
e-mail: hasselhorn@uni-wuppertal.de

S. Bracke and R. Pieper (eds.), *Safety Engineering*,
https://doi.org/10.1007/978-3-658-47473-7_16

than currently desired under certain circumstances—particularly if individuals were given more influence over how their work is organised. The retrospective survey data shows that the desire for self-determination also dominates the reasons for early retirement of those who have already left the workforce. In contrast, for those who have transitioned to a disability pension, demanding work is the primary factor in addition to health problems. A key conclusion is that if politics and economy want to keep older employees in the workforce for longer, they must find ways to make them *want* to work longer. This cannot be achieved without a fundamental consideration of working conditions.

1 Introduction

This paper deals with the transition of the baby boomer generation from working life to retirement. It begins with an overview of employment participation among those in their last years before reaching the statutory retirement age in Germany (Sect. 1). After explaining the study design and methodology of this article (Sect. 2), the empirical findings of the current 4th wave of the lidA-Study on the transition from working life to retirement are presented in Sect. 3. In the discussion section (Sect. 4), the results are contextualised and conclusions are drawn (Sect. 5).

1.1 Demographics and Skilled Labour Shortage

Demographic change has now entered a new phase: currently, about 9 million[1] working baby boomers in Germany are at a stage in life where early retirement is increasingly seen as a realistic option and as an alternative to continued employment. Most older workers in Germany choose to retire early, often well before they reach the statutory retirement age. The overall employment rate of the 55–64 age group in Germany has significantly increased over the past 25 years, reaching at 72% in 2021, and thereby almost the levels of leading Scandinavian countries (Eurostat 2023). The increase, however, has stagnated for those age cohorts nearing retirement age: the employment rate for 64-year-old men has remained relatively stable at about 40% since 2013, and for women, it has been around 30% since 2017 (BiB 2022).

[1] According to the Federal Statistical Office, the number of employed persons aged 55 to 64 in Germany in the second quarter of 2022 was 9.1 million (source: https://www.destatis.de/DE/Themen/Arbeit/Arbeitsmarkt/Erwerbstaetigkeit/Tabellen/ilo-quartal-geschlecht-alter.html, accessed on 23/02/2023).

It is assumed that the 2014 pension reform law introducing the “pension for particularly long-term insured persons” contributed to this stagnation, as it allows for deduction-free retirement from the age of 63[2] (Kaboth et al. 2020). Since 2015, about a quarter of all new pensions for socially insured workers have been attributed to this exit path, most recently (2021) with 269,000 cases (DRV 2022: 65). Additionally, about 200,000 employees retire early[3] with reductions each year, on average about 28 months before reaching the statutory retirement age (DRV 2022: 86). Furthermore, a large group retire significantly earlier[4] with a disability pension, accounting for about 160,000 people or 16% of all new pensions in 2021.

The low employment rate among older workers has raised concerns in politics. On the one hand, there is the issue of financial sustainability of the pay-as-you-go pension system (e.g., BMWI 2021), and on the other, the growing shortage of workers in Germany, which could limit Germany’s economic development (BMAS 2023). By the mid-2030s, a gap of up to seven million skilled workers is expected (BMAS 2023). This is closely related to the significant demographic challenge facing Germany, which will not diminish in the coming years. According to estimations by the Statistisches Bundesamt (Federal Statistical Office, 2022), the number of people of working age (20 to 66 years) will decrease by 1.6 to 4.6 million by the mid-2030s, while the number of people at retirement age (67 years and older) will increase by 4 million to at least 20 million.

1.2 “Early Exit Culture” Among Baby Boomers

The high number of people retiring before the statutory retirement age *also* reflect an “early exit culture” in Germany. De Vroom et al. (2004) describe the “early exit culture” as a widely accepted and institutionalised social norm that older workers have a right or even an obligation to leave the labour market before reaching statutory retirement age. Hult and Edlund (2008) diagnosed a “late exit culture” for Norway and Sweden based on European survey data, while Germany and Denmark were characterised by an “early exit culture”. Hess (2018) confirmed this early exit culture in Germany using data from 2008 and attributed it mainly to the formative early retirement programmes of the 1970s to 1990s. Population-based data from 2014/15 also shows that most workers in Germany would prefer to take early retirement, with younger age groups wishing to retire even earlier than older ones (Hasselhorn 2020). The representative lidA-Study, based on 2018 data, found that only 10% of the working baby boomer generation wanted to remain employed until the statutory retirement age (Hasselhorn et al. 2019: 18). Furthermore,

---

[2] Gradually increasing until the birth cohort of 1958 reaches 64.0 years and the birth cohort of 1964 reaches 65.0 years.

[3] Most recently (in 2021), 210,616 people, corresponding to 20.6% of all new insured pensions (DRV 2022: 86)

[4] Average age of 54.1 years (men, 2021) and 53.1 years (women, 2021) (DRV 2022: 132).

three-quarters of all participants reported living in an environment where the prevailing attitude was to retire earlier rather than later (Hasselhorn et al. 2019: 21).

1.3 Research Questions

What causes baby boomers to want to leave the workforce so early, despite overarching societal factors advising continued employment? The lidA-Study (living in work www.lida-studie.de) has been following the baby boomer generation in Germany since 2011, as they transition from working life to retirement. Current results from this representative survey, which uses randomly drawn samples, allow a deeper insight into the employment perspective of baby boomers today. The following questions will be answered in this article:

a. *How long do older workers in Germany want and are able to work?*
b. *What are the reasons given by those who wish to retire early?*
c. *What might motivate those who wish to retire early to stay employed longer?*
d. *What reasons do people who have already retired early give for their retirement?*

2 Methodology

2.1 LidA-Study

The lidA-Study is a cohort study in Germany since 2011, in which older employees are interviewed on topics such as work, health, and employment participation, by personal standardised interviews. The respondents are individuals born in 1959 and 1965, as well as for the first time in the 4th wave, individuals born in 1971. The lidA-Study so far has included four survey waves from the years 2011 ($N_{\text{Wave 1}}=6585$), 2014 ($N_{\text{Wave 2}}=4244$), 2018 ($N_{\text{Wave 3}}=3586$), and 2022/23 ($N_{\text{Wave 4}}=8884$). The interviews in waves 1 to 3 were conducted at home (CAPI - Computer Assisted Personal Interview), while in the 4th wave, they were conducted either at home or by phone ("CAPI by phone"). In this wave, the previous sample was supplemented to reach over 3000 individuals per birth cohort. In addition, almost as many individuals who were born in 1971 were interviewed for the first time. The survey period lasted from May 2022 to March 2023. The participants from wave 1 were socially insured employees on the cut-off date of December 31, 2009 and those from wave 4 were socially insured employees on December 31, 2021 (excluding civil servants, military personnel, self-employed, etc.). The sample was drawn randomly from 222 recruitment points across Germany, ensuring high representativeness for all socially insured employees in the 1959 and 1965[5] birth cohorts (Schröder et al. 2013, Steinwede et al. 2015). A detailed description of the

[5] For the current fourth survey wave 2022/23, no information on representativeness was available at the time of publication.

lidA-Study can be found in the lidA Cohort Profile (Hasselhorn et al. 2014). The analyses in this article are limited to data from wave 4.

2.2 Variables

Subjective Employment Perspective:
This refers to how employees assess their future employment participation. This article examines two aspects of the employment perspective with the following open questions:

- *Regardless of the legal retirement regulations, up until what age would you like to work? (desired retirement age)*
- *And what do you believe is the maximum age you can work up until?*

These two questions were developed as part of the lidA-Study, although they are similar to questions used in studies worldwide. In a slightly modified form, they were also used[6] in the nationwide GEDA 2014/2015-EHIS survey by the Robert Koch-Institute (Saß et al. 2017) across all age groups. The ages given in response to these two questions are referred to in this article as "target age".

Reasons for Wanting Early Retirement:
For employees who indicate they only want to work up to an age of 64 at most, we assume they are aware that this would be an early exit from working life. Their reasons for wanting *early retirement* were queried using a predefined list. Respondents were asked to indicate the role each reason played in specifying the early exit age (*no role/minor role/major role*). Multiple responses were possible. The list of 12 possible reasons is derived from the survey instrument of the second wave of the TOP study (Mergenthaler et al. 2017) by the Federal Institute for Population Research. The reasons are listed in Fig. 2.

Willingness to Continue Working:
The conditions under which a willingness to continue working, was queried as follows: "*There can be different conditions under which you might want to work longer than the<desired exit age>. Please indicate whether the following points could be a reason for you to want to continue working longer.*" Respondents were then presented with seven aspects (e.g., *good pay, flexible working hours*) to which they could respond with *Yes* or *No*. The reasons are shown in Fig. 3. The option *I definitely do not want to work longer* was also available.

Reasons for Early Exit:

[6] The age group analysis can be found at Hasselhorn (2020).

Individuals who indicated that they already receive an old-age or disability pension were asked 17 possible reasons for their early exit from work (these are fully listed in Fig. 4). For each reason, they were asked to indicate to what extent it contributed to their retirement (*contributed a lot/contributed somewhat/did not contribute*), allowing for multiple responses. The list of reasons is based on a corresponding question from the survey instrument of the second wave of the TOP study (Mergenthaler et al. 2017), although it was modified and expanded.

Socio-Demographic Variables:

Socio-demographic variables which were used included birth cohort (1959, 1965, or 1971), reported gender (male or female), and coded occupational information according to KldB 2010.

2.3 Statistical Methods

This publication is based exclusively on cross-sectional data from the current fourth survey wave and is descriptive in nature. However, where group differences in categorical or continuous variables are of interest, they are tested for significance using Chi^2 tests or t-tests. When comparing more than two groups, the Bonferroni correction was applied to adjust the significance level α to the number of tests. The statistical software used was IBM SPSS Statistics 28.

3 Results

3.1 Sample Description

In the 4th wave of the lidA-Study, 8884 people were surveyed. The survey period lasted from May 2022 to March 2023. Respondents born in 1959 were between 62.3 and 64.2 years old at the time of the survey. For simplicity, they are referred to in the article as "63-year-olds".[7] Similarly, this applies to the other two cohorts.

From the total group, three subgroups are considered in this article: employed persons, those who receive an old-age pension, and those who receive a disability pension. A total of 7159 people were classified as "employed", meaning they were employed for at least one hour per week, were not in the release phase of phased old age retirement ("Altersteilzeit") and did not receive a pension from their own income. We also analysed data from 370 participants who receive an old-age pension (76% of whom receive a pension with reductions) and 269 participants who receive a disability pension (75% of whom receive a full disability pension). The other groups from the lidA wave 4, such as the self-employed, those in the release phase of phased old age retirement and the unemployed, were not considered in this article.

Table 1 describes the characteristics of the employed participants in the lidA-Study wave 4, overall and by birth cohort. The three age groups do not differ

[7] This corresponds to the average age of the cohort and applies to 71% of all respondents in the cohort. At the time of the interview, 27% were 62 years old, and 2% were 64 years old.

Table 1 Sample characteristics of the employed participants in the lidA-Study wave 4 (significant differences between two groups are indicated by superscript letters referring to columns A to C)

Column	*A*	*B*	*C*	
Year Age approx Number	1959 63 years n = 1953	1965 57 years n = 2786	1971 51 years n = 2420	Total n = 7159
Gender				
Women	1043 (53.4%)	1480 (53.1%)	1260 (52.1%)	3783 (52.8%)
Men	910 (46.6%)	1306 (46.9%)	1160 (47.9%)	3376 (47.2%)
Weekly Working Hours (Total)				
Up to 34 h	623 (31.9%)^{BC}	777 (27.9%)	619 (25.6%)	2019 (28.2%)
35 to 44 h	1010 (51.7%)	1444 (51.8%)	1244 (51.4%)	3698 (51.7%)
45 + hours	320 (16.4%)	565 (20.3%)^{A}	557 (23.0%)^{AB}	1442 (20.1%)
Highest Educational Qualification				
Vocational Training (Business)	797 (42.9%)	1266 (46.9%)^{A}	995 (44.0%)	3058 (44.9%)
Vocational Training (School-Based)	190 (10.2%)	254 (9.4%)	211 (9.3%)	655 (9.6%)
Technical College	279 (15.0%)	444 (16.5%)	339 (15.0%)	1062 (15.6%)
University of Applied Sciences Degree	189 (10.2%)	276 (10.2%)	233 (10.3%)	698 (10.2%)
University Degree	322 (17.3%)^{B}	370 (13.7%)	381 (16.9%)^{B}	1073 (15.7%)
Other Qualification	10 (0.5%)	15 (0.6%)	19 (0.8%)	44 (0.6%)
No Educational Qualification	71 (3.8%)	72 (2.7%)	83 (3.7%)	226 (3.3%)
Employment Perspective				
Number (Percentage) of People who would like to work up to 64 Years of age at most	1110 (58.4%)	10,932 (70.7%)^{A}	1638 (69.8%)^{A}	4680 (67.0%)
"… Up to what age would you like to work?" Average (std. dev.) in years	64.5 ^{BC} (2.95)	63.0 ^{C} (3.16)	62.4 (4.36)	63.2 (3.66)
"… Up to what age are you able to work?" Average (std. dev.) in years	66.3 ^{BC} (3.24)	65.5 (3.57)	65.6 (4.45)	65.7 (3.83)

significantly by gender. Part-time employment is significantly more common in the older birth cohort, while longer weekly working hours of over 45 h are more common in the two younger cohorts. The distribution of occupational groups by educational qualification shows few significant differences between the cohorts.

The groups of those who receive an old-age pension or disability pension differ as expected in terms of birth year: 98% of all those who receive an old-age pension belong to the 1959 birth cohort, while only 60% of those who receive a disability pension belong to this cohort (see Table 2). There are no significant gender

Table 2 Sample characteristics of those who receive an old-age or disability pension in the lidA-Study wave 4 (significant differences between the groups are indicated by superscript letters referring to the group with higher percentages)

	A	*B*
	Old Age Pension	*Disability Pension*
Number	370	269
Old Age Pension		
with reductions	90 (24%)	––
without reductions	280 (76%)	––
Disability Pension		
partial	––	68 (25%)
full	––	201 (75%)
Birth Year		
1959	363 (98.1%)^{B}	161 (59.9%)
1965	3 (0.8%)	82 (30.5%)^{A}
1971	4 (1.1%)	26 (9.7%)^{A}
Gender		
Women	208 (56.2%)	161 (59.9%)
Men	162 (43.8%)	108 (40.1%)
Retirement Year		
Up to 2019	11 (3.1%)	167 (63.7%)^{A}
2020	32 (9.0%)	32 (12.2%)
2021	28 (7.8%)	39 (14.9%)^{A}
2022/23	286 (80.1%)^{B}	24 (9.2%)
Retirement decision made how many months before retirement		
Up to 12 months	205 (58.7%)	173 (84.4%)^{A}
more than 12 months	144 (41.3%)^{B}	32 (15.6%)
Retirement Timing		
Desired retirement timing	276 (76.9%)^{B}	92 (35.4%)
Would have preferred to retire earlier	42 (11.7%)	26 (10.0%)
Would have preferred to retire later	41 (11.4%)	142 (54.6%)^{A}

differences. Most people who receive an old-age pension entered retirement recently, in the 2022/23 survey period (80%). The decision on a disability pension was usually made at short notice (in 84% of cases within 12 months before retirement, for old-age pensions: 59%). In the group of people that receive an old-age pension, about three-quarters experienced the timing of retirement as desired. Among those with a disability pension, only about a third found the timing as desired, while more than half would have preferred to retire later.

3.2 How Long Do Baby Boomers Want to (and Can) Work?

Fig. 1a, b show the age up to which employed respondents would like to work and, believe they can work. The cumulative percentage of respondents who would or could work at a given age is shown. Some respondents provided a target age that was earlier than their actual age at the time of the survey; this should be understood as "I would have preferred to…".

For every target age at which someone would like to work (Fig. 1a), there is a higher percentage of individuals in the oldest cohort compared to the two younger cohorts, or more simply: the oldest would on average like to work longer. This difference is most pronounced at the target ages of 63 and 64. While 69% of the oldest cohort would still like to work up to the age of 64, less than half as many in the two younger cohorts expressed the same (1965 cohort: 33%, 1971 cohort: 32%). It is known from literature that older workers state a higher desired retirement age (Nilsson et al. 2011, Geuskens et al. 2012, Hasselhorn 2020). This higher desired

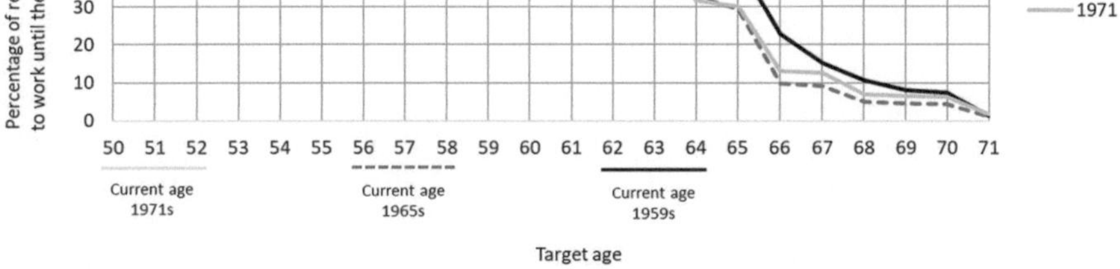

Fig. 1 Employment perspective of older workers by birth cohort. Number of respondents: $n_{1959}=1902$ (1a), 1832 (1b); $n_{1965}=2732$ (1a), 2589 (1b); $n_{1971}=2346$ (1a), 2262 (1b). The underlined target ages indicate the age range in which the respective birth cohort was at the time of the survey. **a** … up until what age would you like to work?; **b** … up until what age are you able you work?

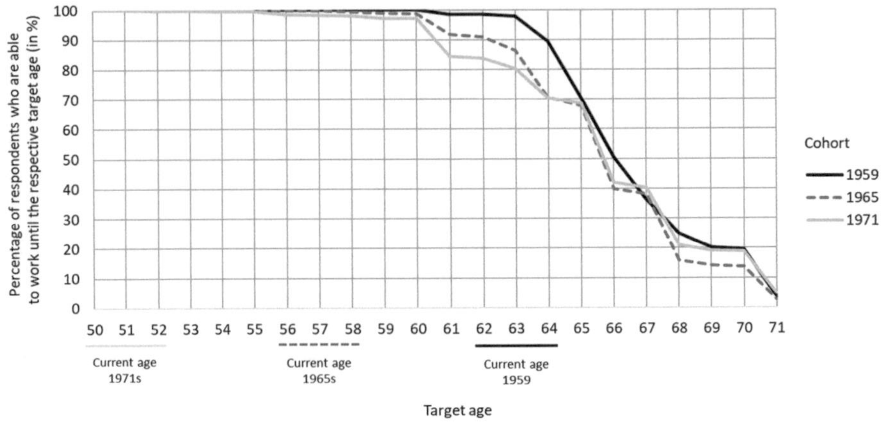

Fig. 1 (continued)

retirement age can likely be attributed to a "survivor effect" in the oldest cohort, as only those who were able to remain employed up to the age of 63, for example due to better health, are included in this cohort. Additionally, it is possible that younger people have a less clear idea of their work and life circumstances and their exit from the workforce is further in the future. Therefore they base their responses more on norms and "round" numbers, while older people derive their answers more from their specific situation (Griffin et al. 2013). This is supported by the findings that in the two younger cohorts, the percentage of individuals who would like to continue working drops more sharply after the target ages of 60, 63, and 65.

In all three cohorts, the percentage of people who believe they can still work (Fig. 1b), exceeds the percentage of those who would *still like to work* (Fig. 1a), regardless of the target age considered. The absolute differences are particularly pronounced at the target ages of 64 and 65. At 65 years old, for example, only 42% of the oldest cohort would still like to work, while 70% believe they could still do so.

Another difference between the two younger cohorts is evident: from the target age of 65 onwards, more people in the youngest cohort would like to continue working compared to the middle cohort. At the target age of 66, the difference is most pronounced, with 13% of the 1971 cohort versus 10% of the 1965 cohort expressing a desire to work (Fig. 1a). This raises the question of whether a (very small) share of the youngest cohort is already mentally adjusting to the increased statutory retirement age. A parallel effect can be observed from the target age of 65 onwards in the percentage of people who believe they can still work (Fig. 1b). This percentage in the youngest cohort reaches the level of the oldest cohort at the target age of 70, with about 19%.

3.3 Why Do so Many Employees Want to Retire Early?

The questions about the reasons for wanting to retire early were only asked to employees whose target age for desired retirement was under 65 years. As mentioned above, this includes 67% of all employed persons, with the percentage being predictably lower among those born in 1959 at 58%, compared to the other cohorts (1965: 71%, 1971: 70%; see Table 1).

The 12 reasons for wanting to retire early that were given to respondents, received varying levels of agreement (see Fig. 2). The reason, *more free time* received the highest level of support, with 83% of the overall group stating that "this reason played a major role", followed by *at some point, it has to end* at 64%. For half of the respondents, it was important that *sufficient financial security* had been achieved by this time, although only 36% considered the possibility of *a deduction-free old-age pension* to be a major factor. While *strenuous work*

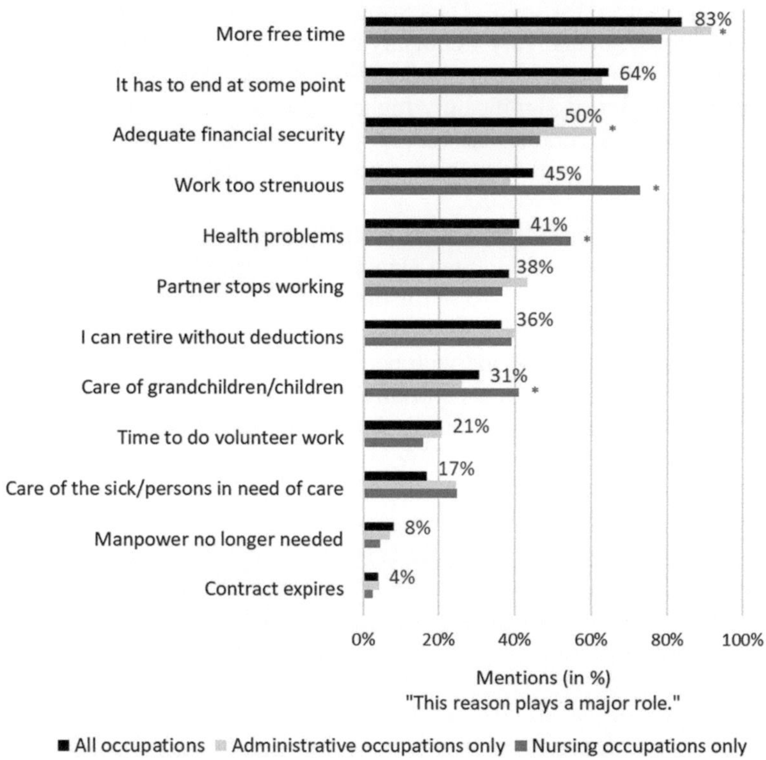

Fig. 2 Reasons for wanting early retirement among older workers aged 51, 57, and 63 years, who indicated they only want to work up to 64 years of age at most. Number of respondents: n = 4680 (all occupations), n = 169 (administrative occupations only), n = 293 (nursing occupations only). * = Subgroup (nursing or administration) has a significantly higher result than the other subgroup ($p < 0.05$)

and health problems were significant for about four out of ten respondents, family reasons (*caring for grandchildren/children, voluntary activities, or caregiving responsibilities*) appeared to be less relevant. Aspects related to the employment relationship (*being no longer valued, contract expiring*) were considered important by only a small proportion of respondents as a reason for wanting to retire early.

Comparisons by birth cohort and gender reveal significant differences (not shown). The reason *more free time* was cited significantly more often by the youngest birth cohort (85%) than by the oldest (81%), which seems plausible given the significantly longer weekly working hours of the youngest cohort (see Table 1). Similarly, it is understandable that the reason *at some point, it has to end* was cited more frequently by the oldest cohort (70%) than by the other two cohorts (1965: 63%, 1971: 61%), as it may often express a perceived entitlement acquired through many years of work. The fact that the *partner stops working* was a relatively common reason among the younger cohorts (1965: 41%, 1971: 43%) and significantly less so among the older cohort (27%) could be due to this issue already being resolved for some. The reason that *one can retire without deductions at the target age* was cited very frequently by those born in 1959 (52%), which is understandable as they are already in the relevant age phase and can better assess their situation. This was significantly higher than 29% of the middle cohort and 34% of the youngest cohort. The reason *more time for voluntary activities* was cited more frequently by the youngest cohort (24%) than by the middle (20%) or older (16%) cohort.

A differentiated analysis by gender partially reflects traditional gender roles: six out of twelve reasons were cited significantly more often by women than by men: *strenuous work* (48% vs. 41% for men) and *health problems* (43% vs. 39% for men), which is not surprising given the high number of older employed women in physically demanding jobs and with poor health in Germany (Burr et al. 2013). The reason that the *partner stops working* was also cited more often by women (43% vs. 33% for men), which is likely due to the usually older age of male partners. *Time for voluntary activities* (22% vs. 19% for men) and *caregiving responsibilities* were more often cited by women as reasons, specifically *caregiving for dependent persons* (21% vs. 12% for men) or caring for *grandchildren or children* (33% vs. 28% for men). On the other hand, men cited two reasons significantly more often than women: the possibility of *retiring without deductions* (38% vs. 35% for women) and the desire for *more free time* (85% vs. 82% for women).

There are also differences in the prominence of specific reasons across occupational groups. This is exemplified here by comparing nursing occupations on the one hand and administrative occupations on the other (see Fig. 2). The overall frequently cited reason of *more free time* was significantly more important for administrative occupations (91%) than for nursing occupations (78%). Likewise, achieving *sufficient financial security* was a relevant reason for significantly more people in administrative occupations (61%) than for those in nursing occupations (46%).

People in nursing occupations cited, *work too strenuous* (nursing: 73%, administration: 38%) and *health problems* (nursing: 54%, administration: 39%) significantly more often than those in administrative occupations; these reasons are not independent of each other. *Caring for grandchildren/children* was also a more important reason for significantly more people in nursing occupations (41%) than for those in administrative occupations (26%). This cannot be attributed to a gender difference between the two occupational groups, as both groups have nearly identical female proportions (nursing occupations 86%, administrative occupations 85%).

3.4 Reasons for Continuing to Work

As in the previous section, the following analysis only includes employed people who indicated that they only want to work up to the age of 64. Seventy-eight percent of these people would be willing to continue working longer under certain circumstances. In most cases, they cited several possible reasons for this “continued work”, with an average of about four reasons.

Fig. 3 shows the frequency of cited reasons. For 66% of these individuals, the reason to continue working would be *if they could freely decide how much they work*. The second most common reason was *if I could freely decide when I work*. Other aspects (*good pay, work not too strenuous, nice colleagues, interesting*

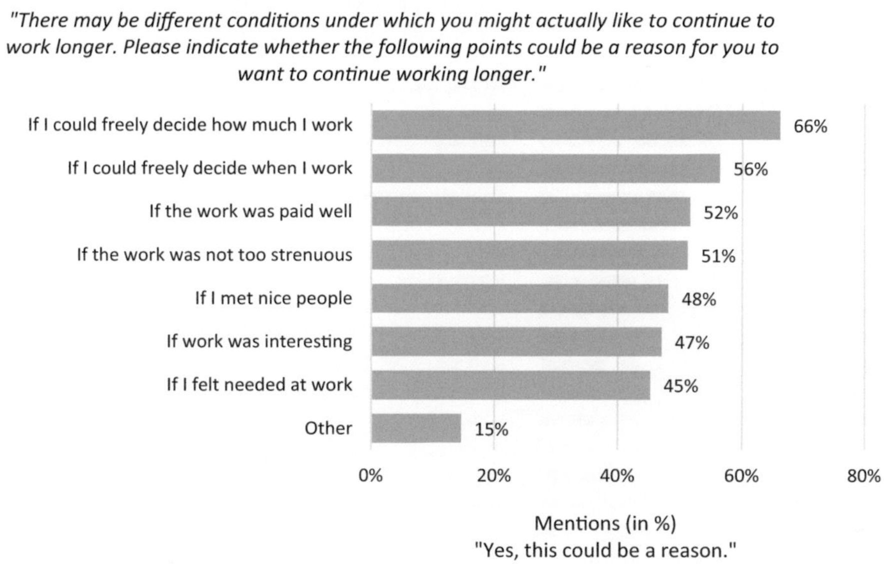

Fig. 3 Possible reasons for continuing to work among older workers aged 51, 57, and 63 years who indicated they only want to work up to 64 years of age at most. Number of respondents: n=4680

work, feeling needed at work), were seen as potential reasons to continue working by about half of the respondents. Only 15% indicated that other aspects might motivate them to continue working longer than initially desired.

The fact that nearly half (at least 45%) agreed with each individual statement indicates the high relevance of the selected aspects, all of which relate to work. The issue of self-determination appears to be particularly important. Additionally, the fact that only one in seven (15%) considers *other factors* to be decisive suggests that either the design of the work activity is indeed the primary factor in the decision to continue working, or that the nature of the question triggered a priming effect, where private aspects (e.g., health, family situation) were not considered.

There are also group differences (not shown): women were significantly more likely (80%) to be willing to continue working than men (75%), and more women agreed with each individual reason (except *if the work was interesting*). There is also an age gradient in the willingness to continue working (1959 cohort: 65%, 1965 cohort: 79%, 1971 cohort: 85%) and in each of the seven reasons: the younger the cohort, the more often the reason was given.

3.5 Why Do Baby Boomers Retire Early?

Although the lidA-Study cohorts were still of working age at the time of the survey (on average 51, 57, or 63 years old), many participants had already retired. Fig. 4 shows the distribution of reasons for early retirement in old-age

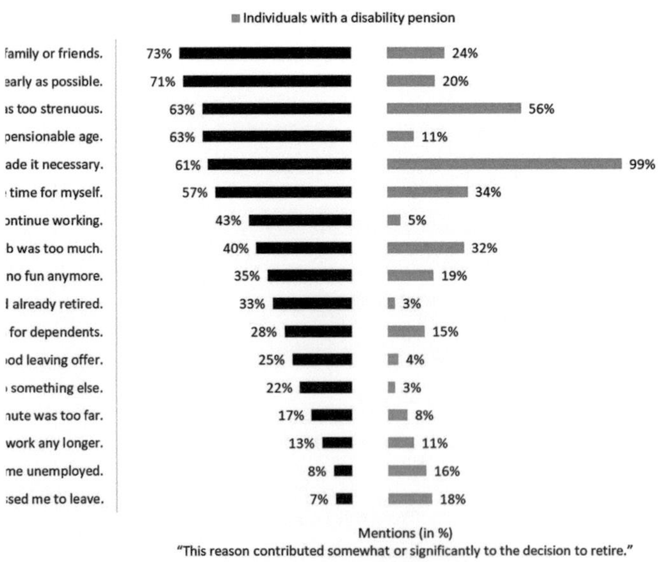

Fig. 4 Reasons for early retirement among those who receive old-age pensions (n = 370) or disability pensions (n = 269)

pensions (n = 370) and disability pensions (n = 269). Once again, most respondents cited multiple reasons simultaneously (old-age pension: ∅ 6.4 responses, disability pension: ∅ 3.7 responses).

In research on the transition to retirement, there is often a distinction made between factors that push workers out of their current work into retirement (push factors) and attractive factors from outside the working life that pull workers into retirement (pull factors) (Nicolaisen et al. 2012).

Among those who receive an old-age pension, the most frequently cited reasons fall into the category of *pull factors*. This suggests a desire for more self-determination in life (*spending more time with family and friends* 73%; *more time for myself* 57%), or reveals a fundamental attitude towards work: *I wanted to stop as early as possible* (71%). One in five wanted to do *something else*.

Push factors include aspects of work, such as *work being too strenuous* (63%), or *the work no longer being enjoyable* (35%). For some, the commute to and from work was too long (17%), 13% felt they were no longer needed at work, 8% cited unemployment as a reason for retirement, and 7% felt *pushed out of working life*. Another significant push factor was *health, which made retirement necessary* (61%). Private burdens cited as reasons include *caregiving responsibilities* (28%) and *the dual burden of household and work becoming too much* (40%). Finally, about one in three cited *the partner or spouse retiring* as a reason for their own retirement.

Some also took advantage of *opportunities for early retirement*. Sixty-three percent had *reached an age* that allowed retirement, and 22% had received a *favourable early retirement offer*. Additionally, 43% felt they no longer *financially needed to work*.

Push factors dominate among those who receive disability pensions; that is, the reasons are characterised by burdens: unsurprisingly, health reasons dominate (99%), followed by *work being too strenuous* (56%), and the *dual burden of household and work* (32%). The most frequently cited *pull factor* was *more time for myself* (34%).

The frequencies of all reasons differ significantly between the two pension groups shown in Fig. 4, except for *the dual burden of household and work* and *work being too strenuous*.

There were no significant differences in the reasons cited between those receiving a full and partial disability pension. In contrast, among those who receive old-age pensions, respondents with a pension with deductions cited four reasons significantly more often than those with a pension without deductions - particularly the *dual burden of household and work* (44% vs. 27%), followed by *work being too strenuous* (66% vs. 52%), *caregiving responsibilities* (31% vs. 18%), and lastly *more time for myself* (60% vs. 49%).

Seventy-three people felt pressured into retirement, especially into disability (18%) and, to a lesser extent, into old-age pensions (7%). With these cases, it was primarily due to the employer (44% of 73), direct supervisors (41%), or colleagues (29%). Less frequently, they felt pressured by family (partner or spouse 10%, children 9%), and slightly more often by relatives or friends (16%) (data not shown).

4 Discussion

This article examines the subjective employment perspective of older employees in Germany, as well as the reasons for early retirement or disability pensions. The results suggest that most baby boomers (born in 1959 and 1965) still wish to retire early, well before their statutory retirement age and before they are no longer able to work. At the same time, the results suggest that some of the generation following the baby boomers (born in 1971), are already adjusting to working longer. The reasons for wanting early retirement are mainly related to a desire for self-determination in life, followed by stressful work and health problems. In all three cohorts studied, there is a widespread willingness to continue working longer under certain conditions, particularly if they had more control over their work. It is shown that the wish for self-determination is also the dominant reason for those who have already retired early. For those with disability pensions, apart from health problems, strenuous work is the main reason. Selected results from Sect. 3 are discussed further below.

4.1 Interpreting the Figures on Employment Perspective

With regard to question a) on the employment perspective of older workers, there are two particularly interesting aspects: how the figures fit into existing data and how the three cohorts differ. To finish, the societal significance of the findings on employment perspectives is discussed.

4.1.1 No Difference from Existing Data

Several population-based studies on the subjective employment perspective of older workers in Germany have been conducted in the meantime. Their results were summarised in the article by Hasselhorn (2020). The results presented in this article are comparable to analyses based on data from the GEDA 2014/2015-EHIS survey by the Robert Koch-Institute. In this comprehensive survey, employees (not just those subject to social insurance) were also asked the two questions about employment perspective that are the focus here, albeit in a shorter form. The results of both studies are similar, both in terms of the percentage of those who would like to work up until age 65 or longer, and those who believe they are able to do so. Consequently, the current figures do not indicate any significant change over time in terms of employment perspectives in Germany.

4.1.2 Adjustment of the Younger Cohort to Societal Necessities?

Comparing the employment perspective of the three birth cohorts reveals the surprising finding that the younger 1971 cohort - included in lidA for the first time - overtakes the 1965 cohort from the target age of 65 onwards. That is, from this target age onwards, the percentage of those who would like to continue working or believe they can, is higher in the younger cohort than in the one six years older.

Such a possible generation effect has not been observed before. It deviates from the previous uniform observation in retirement research that the employment perspective of older workers is longer on average (i.e., has higher target ages) than that of younger workers. One explanation for the now deviating finding could be that slightly more members of the generation following the baby boomers have normatively taken on the new statutory retirement age of 67 years. At the same time, this could also be an expression of a perceived societal pressure to remain in employment longer. It should be considered that a parallel effect is also observed in the self-assessed ability to continue working. Both findings could therefore also reflect improved workability and health in selected groups. Further analyses are needed here, not least with the help of the planned lidA wave 5 survey in 2026.

4.1.3 What Does Employment Perspective Actually Express?

The subjective employment perspective could easily be seen as an *indicator of the actual later retirement age*. Indeed, this study shows that the reasons for wanting early retirement (prospective) correspond well with the reasons for actual early retirement in old-age pensions (retrospective). However, this does not mean that all respondents will retire at the date they currently wish to. Garthe and Hasselhorn (2023) found in longitudinal analyses with lidA data from 2014 and 2018 that about 40% of respondents had postponed their target age over four years. Nilsson (2011) also found a corresponding age effect in cross-sectional data among workers aged 55 and older. However, two studies suggest some predictive power of indicators of subjective employment perspective. Based on data from the German Ageing Survey (DEAS), Engstler (2019) found that half of the 55 to 61 year old respondents achieved their plans (retirement or continued work) within the next six years at the planned time (± six months), with the baseline survey conducted in 2008. Similarly, Örestig et al. (2013) found that over half of the 57- to 64-year-old workers surveyed in 2002/3 in Sweden, actually retire within a time span of ± one year around the *desired* retirement age.

Regardless of whether the widespread desire to retire early is a stronger or weaker indicator of the actual retirement age, it can be understood as a manifestation of a prevailing early exit culture. Hult and Edlund (2008) described that in "early exit cultures" such as Germany or Denmark, the appreciation of working is lower than in "late exit cultures" like Norway and Sweden. It is assumed that this culture of early exit not only negatively impacts older workers who would like to retire early but cannot, but also poses a challenge for businesses if these workers are difficult to motivate and less productive (Henkens and Leenders 2010).

4.2 Early Retirement—Why?

The results of questions b) to d), which deal with the reasons for wanting early retirement, as well as the reasons for actual early retirement and continued work, are discussed here across groups to provide a holistic view of the phenomenon of early retirement. Three thematic aspects are addressed before a methodological aspect is discussed regarding the role of subgroups.

4.2.1 Emphasis on Self-Determination

When older workers want to retire early, i.e., before the age of 65, the most prominent reason is *more free time,* which is clearly intrinsically motivated or—in the language of retirement research—a pull factor. Behind this, there may initially be *specific* ideas about how this free time should be spent on other areas of life. Some of these are reflected in the other response options and are supported (*time for voluntary activities, caring for grandchildren*), while others are not visible in this study (e.g., the desire to travel or enjoy life in other ways). Whether and to what extent the various reasons are related should be the subject of further analyses.

However, the high level of agreement (83%) among all respondents may also reflect a *general* notion of age-appropriate resting and the right to self-determination in post-working life, as van Dyk et al. found in a qualitative study of retirees in 2013. This interpretation is supported by the fact that the reason *at some point, it has to end*, which also expresses a perceived entitlement, is the second most common. These aspects can be seen as part of the previously mentioned early exit culture in the sense of de Vroom et al. (2004). According to them, after years of early retirement of older workers, early exit is seen as a social right, as a common good. Jansen (2013) empirically demonstrated in a comparison of European countries that the transition to retirement is influenced by these cultural aspects (in addition to material opportunities and institutional frameworks). The new figures presented here suggest that the issue of early exit culture remains relevant for Germany.

The possible reasons for wanting to continue working longer also prominently reflect the desire for (more) self-determination: the two most frequently cited reasons would be t*he ability to freely decide how much* and *when to work*. However, agreement with these reasons could also be due to older workers already anticipating that a work situation such as this would best allow them to successfully cope with their work, even with potentially declining strength.

4.2.2 Financial Reasons

Sufficient financial security, which half of the respondents cited as an important reason for wanting early retirement, and the possibility of a *deduction-free old-age pension* are not intrinsically motivating (as the *desire for free time* or the perceived entitlement to it) but are instead enabling factors. Among those who receive old-age pensions, the reason, *I no longer needed to work financially* was cited as a reason (among several) by nearly half of the respondents. While financial reasons are not as frequently cited as motivational reasons in this study, they may be crucial in achieving the wish for early retirement, especially when health is compromised (de Wind et al. 2013). Conversely, financial reasons can also keep older workers in the workforce *longer*: data from lidA wave 3 showed that the group of workers with a low income would like to remain employed longer than other income groups, regardless of their health status (Hasselhorn et al. 2020). In this case, the wish likely reflects a "need."

4.2.3 Work and Health

The next two reasons, *work too strenuous* and *health problems,* each cited as significant reasons for early retirement by at least 41% of respondents, are not identical but are related, as both factors influence each other (Hasselhorn and Müller 2021). While they do not top the list of reasons for the desired and also achieved early retirement, they are still important determinants of retirement timing. In retirement transition research, they are the most frequently studied factors influencing retirement timing (Scharn et al. 2018). Workers who can no longer manage their work because it is too strenuous and/or their health has deteriorated are likely to take advantage of an available realistic exit option. On the other hand, it is also true that if about 60% of those who want to retire early do not cite *health problems* or *too strenuous work* as reasons, it means that other reasons must be important, but these reasons have received less attention from the scientific community. Here, we first refer to the aspects mentioned under Sect. 4.2.1.

4.2.4 Consideration of Subgroups

The transition from working life to retirement is considered a complex process in which many factors, each playing an important role, influence each other. The complexity suggests that groups differ significantly in the retirement transition. In this article, we have primarily presented results for overall groups. We could only marginally show that subgroups differ in our thematic area. In social epidemiology, the importance of differentiated consideration by gender, migration status, and social status is often emphasised (Mielck and Wild 2021). We addressed differences between women and men in the results section for research questions b) and c). They were often significant and in line with expectations. However, our results can also relativize stereotypes when, for example, we report that 12% of older employed men consider caring for dependent persons as a reason to retire early (21% for women). However, when it comes to the transition from working life to retirement, there are also subgroups of particular political relevance beyond these three categories, namely occupational groups. The example of comparing occupational groups presented in Fig. 2 reveals a professional risk group. Those in nursing occupations cited the reasons for wanting to retire early, *work too strenuous* (73%) and *health problems* (54%), much more frequently than the comparison group in administration or the overall group. While the occupational health and employment risks of nursing staff are well known and widely discussed (BARMER 2020: 143ff, 162ff, 212ff), the size of the lidA wave 4 now allows for the identification of other occupational risk groups for the transition from working life to retirement.

4.3 Limitations of the Study

Limitations of this study include methodological aspects in particular:

This publication is based on data collected from May 2022 to March 2023, so it is influenced by the COVID-19 pandemic. The extent to which this affects the comparability of the results with previous surveys must be clarified in subsequent methodological studies. In this article, we only used data from the fourth survey wave, and the data and questions analysed here mostly dealt with specific issues, so it is unlikely that the results and interpretations have been influenced significantly.

Another question is the extent to which respondents' statements about reasons are reliable, or (also) stimulated by the survey situation. Especially when respondents who are considering continuing to work agree with more than half of the 7 explicit reasons on average, it must be questioned whether specific personal ideas actually underlie this, or whether plausible aspects are quickly agreed upon. Secondly, it remains open as to what extent the direct query of reasons/influences in the context of the transition to retirement leads to complete results, as only explicit motives can be accessed in this way (Ebener 2018). Finally, it should be noted that especially with a normative topic such as retirement transition, social acceptability may influence responses in personal interviews. Therefore, direct queries of reasons do not replace longitudinal collection and examination of assumed influencing factors, but are only intended to complement them.

5 Conclusions

Some central findings of this study suggest a need for action and potential. Firstly, it should be recognised socially and politically that the early exit culture continues in the baby boomer generation in Germany. This is at least partly an expression of the early retirement programmes of the 1970s to 1990s, as Hess (2018) describes. The wish for self-determination and the feeling of an acquired right to early retirement dominate the reasons for wanting early retirement. This was confirmed in the reasons for actual early retirement in old-age pensions. Years of dealing with this issue have shown that the question of retirement is very personal for each employee. Societal necessities due to demographic developments and the shortage of skilled workers are not at the forefront of individuals' considerations.

The results of this study are novel in three respects:

- For the first time, there are indications that a shift in thinking may be emerging in the generation following the baby boomers. The percentages of those who would like to work until 65 and beyond are slightly higher than in the older birth cohort. However, even these percentages are still low, and further analyses and future studies and surveys should verify this before it could have political consequences.
- The results also show that many older workers express a willingness to continue working longer than initially desired, particularly if working conditions were improved, especially through greater decision-making autonomy.

- Finally, it is clear that health is not the only reason why so many older workers want to and do retire early. The same applies to working conditions. Both factors are important and require political attention, but at least as important is the motivation of workers to remain employed. This aspect should receive more attention in research, economy and politics.

If policymakers and companies in Germany aim to keep older workers in the workforce longer, then they must succeed in making them *want* to work longer. This cannot happen without a fundamental consideration of working conditions.

Acknowledgements This article is the English translation of the German original publication: Hasselhorn H. M., Ebener M. (2023) Frühzeitiger Ausstieg der Babyboomer aus dem Erwerbsleben—Ergebnisse der lidA-Studie. Deutsche Rentenversicherung 02/2023, 152-174. Published with permission of the Journal Deutsche Rentenversicherung.

The lidA-Study wave 4 is funded by the German Pension Insurance Fund (DRV Bund funding code 0421/40-64-50-78), BARMER (4500129332), the Federal Ministry of Labour and Social Affairs (BMAS DA2-83000), and the German Social Accident Insurance (BG RCI P0102-000-2022).

References

BARMER Pflegereport 2020: Belastungen der Pflegekräfte und ihre Folgen. Schriftenreihe zur Gesundheitsanalyse – Band 26. Berlin. ISBN: 978–3–946199–57–1. https://www.socium.uni-bremen.de/uploads/News/2020/20201201_BARMER_Pflegereport_2020.pdf (Access on 28.02.2023).

Bundesinstitut für Bevölkerungsforschung [BiB] (2022): Renteneintritt der Babyboomer: Für viele ist schon mit 63 Schluss. Press Release 21/22 from 10.12.2022 https://www.bib.bund.de/DE/Presse/Mitteilungen/2022/pdf/2022-12-10-Erwerbsbeteiligung.pdf;jsessionid=BEE89A7E5A3E196CE7DFF18B13E8EA6C.intranet661?__blob=publicationFile&v=7 (Access on 23.02.2023).

Burr, Hermann/ Kersten, Norbert/Kroll, Lars/Hasselhorn, Hans Martin (2013): Selbstberichteter allgemeiner Gesundheitszustand nach Beruf und Alter in der Erwerbsbevölkerung. Bundesgesundheitsblatt, Gesundheitsforschung und Gesundheitsschutz, 2013, 56:349-358.

Bundesministerium für Arbeit und Soziales [BMAS] (2023): "Fachkräftesicherung ist Wohlstandssicherung". Rede des Bundesministers für Arbeit und Soziales, Hubertus Heil, im Plenum des Bundestages zur Fachkräftestrategie am 20. Januar 2023 https://www.bmas.de/DE/Service/Presse/Reden/Hubertus-Heil/2023/2023-01-20-rede-bundestag-fachkraeftestrategie.html (Zugriff am 23.02.2023).

Bundesministerium für Wirtschaft und Energie [BMWI] (2021): "Vorschläge für eine Reform der gesetzlichen Rentenversicherung" Stellungnahme des Wissenschaftlichen Beirats beim Bundesministerium für Wirtschaft und Energie, Mai 2021, https://www.bmwk.de/Redaktion/DE/Publikationen/Ministerium/Veroeffentlichung-Wissenschaftlicher-Beirat/wissenschaftlicher-beirat-vorschlaege-reform-gutachten.pdf?__blob=publicationFile&v=14 (Access on 23.02.2023).

de Vroom, Bert (2004): The shift from early to late exit: changing institutional conditions and individual preferences: the case of the Netherlands. In: Maltby, Tony/de Vroom, Bert/Mirabile, Maria Luisa/Øverbye, Einar (Hrsg.): Ageing and the Transition to Retirement: A Comparative Analysis of European Welfare States. Florence: Taylor and Francis; 2004: 120–153.

Deutsche Rentenversicherung [DRV] (2022): Rentenversicherung in Zeitreihen – Ausgabe 2022. Oktober 2022. DRV-Schriften Band 22.

de Wind, Astrid, Geuskens, Goedele, Reeuwijk, Kerstin G. et al. (2013): Pathways through which health influences early retirement: a qualitative study. BMC Public Health 13, 292. https://doi.org/10.1186/1471-2458-13-292 (open access).

Engstler, Heribert (2019): Wie erfolgreich sind ältere Arbeitskräfte in der zeitlichen Umsetzung ihrer Ausstiegspläne? Soziale Unterschiede der Übereinstimmung zwischen geplantem und realisiertem Alter der Erwerbsbeendigung. Zeitschrift für Gerontologie und Geriatrie 52 (Supplement 1), 14–24, https://doi.org/10.1007/s00391-018-1451-3. Open Access.

Ebener, Melanie (2018): Die Erfassung der Motivation, erwerbstätig zu sein, in arbeitswissenschaftlichen Studien. Dissertationsschrift. Online unter: https://elekpub.bib.uni-wuppertal.de/ubwhsmig/download/pdf/3556108?originalFilename=true (Zugriff am 23.02.2023).

Eurostat (2023): Employment and activity by sex and age - annual data. Online data code: LFSI_EMP_A (Zugriff am 23.02.2023).

Geuskens, Goedele/Oude Hengel, Karen M./Koppes, Lars L./Ybema, Jan Filip (2012): Predictors of the willingness and the ability to continue working until the age of 65 years. Journal of occupational and environmental medicine, 54 (5), 572–578. https://doi.org/10.1097/JOM.0b013e318248057a.

Griffin, Barbara/Loh, Vanessa, Hesketh, Beryl (2013): Age, gender, and the retirement process. In M. Wang (Ed.), The Oxford Handbook of Retirement (pp. 202–214). Oxford: Oxford University Press.

Hasselhorn, Hans Martin/Müller, Bernd Hans (2021): Arbeit und Gesundheit - Eine Bilanzierung aus 25 Jahren arbeitsepidemiologischer Forschung. In: Richter, Götz (Hrsg.). "Arbeit und Altern 2021. Herausforderungen für Wirtschaft und Gesellschaft", Baden-Baden: Nomos, 171–198, ISBN 978–3–8487–6838–7 https://www.baua.de/DE/Angebote/Publikationen/Fachbuecher/Arbeit-und-Altern.pdf (Access on 23.02.2023).

Hasselhorn, Hans Martin (2020): Wie lange wollen und können Erwerbstätige in Deutschland arbeiten? Deutsche Rentenversicherung 2020; 04: 485–506 https://www.deutsche-rentenversicherung.de/SharedDocs/Downloads/DE/Zeitschriften/DRV_Hefte_deutsch/2020/artikel_heft_4_hasselhorn.pdf?__blob=publicationFile&v=1 (Access on 23.02.2023).

Hasselhorn, Hans Martin/Ebener, Melanie/Vratzias, Athanasios (2020): Household income and retirement perspective among older workers in Germany—findings from the lidA Cohort Study. J Occup Health. 2020;62:e12130. https://doi.org/10.1002/1348-9585.12130 (open access).

Hasselhorn, Hans Martin/Borchart, Daniela/Brühn, Lesley et al. (Hrsg.) (2019): lidA – Idee, Studie, Ergebnisse – eine Kohortenstudie zu Arbeit, Alter, Gesundheit und Erwerbsteilhabe bei älteren Erwerbstätigen in Deutschland. Bergische Universität Wuppertal. https://arbeit.uni-wuppertal.de/fileadmin/arbeit/Brosch%C3%BCre_und_Flyer/lidA_Brosch%C3%BCre.pdf (Access on 23.02.2023).

Hasselhorn, Hans Martin/Peter, Richard/Rauch, Angela/Schröder Helmut/Swart, Enno/Bender Stefan/du Prel Jean-Baptist/ Ebener Melanie/March, Stefanie/Trappmann, Mark/Steinwede Jacob/Müller, Bernd Hans (2014): Cohort profile: The lidA Cohort Study—a German Cohort Study on Work, Age, Health and Work Participation. International Journal of Epidemiology 2014, 1736-1749, https://doi.org/10.1093/ije/dyu021 (open access).

Henkens Kène/Leenders, Monique (2010): Burnout and older workers' intentions to retire. Int J of Manpower 31(3):306–321. https://doi.org/10.1108/01437721011050594.

Hess, Moritz (2018): Erwartetes und gewünschtes Renteneintrittsalter in Deutschland. Z Gerontol Geriatr 51(1):98–104. https://doi.org/10.1007/s00391-016-1053-x (open access).

Garthe Nina/Hasselhorn Hans Martin (2023): Wollen und können ältere Beschäftigte länger erwerbstätig bleiben, wenn sich ihre Arbeit verbessert? Zbl Arbeitsmed 73:49-63. https://doi.org/10.1007/s40664-022-00490-w.

Hult Carl/Edlund, Jonas (2008): Age and labour market commitment in Germany, Denmark, Norway and Sweden. Work, Employment and Society 22(1):109–128. https://doi.org/10.1177/0950017007087419.

Jansen, Andreas (2013): Kulturelle Muster des Altersübergangs. Der Einfluss kultureller Normen und Werte auf die Erwerbsbeteiligung älterer Menschen in Europa. Kölner Zeitschrift für Soziologie und Sozialpsychologie 65:223–251.

Kaboth, Arthur/Drescher, Susanne/Brussig Martin (2020): Erwerbstätigkeit im Alter und der Übergang in Rente – ein Überblick aus der Arbeitsweltberichterstattung. In: Blank, Florian/Hofmann, Markus/Buntenbach,Annelie (Hrsg.). "Neustart in der Rentenpolitik: Analysen und Perspektiven", Baden-Baden: Nomos, edition sigma, 101–118, ISBN 978–3–8487–6572–0.

Mergenthaler, Andreas/Cihlar, Volker/Micheel, Frank/Sackreuther, Ines/Riedl, Sabine/Maruhn, Sabia (2017): TOP – Transitions and Old Age Potential: Methodenbericht zur zweiten Welle der Studie. BiB Daten- und Methodenberichte 2/2017. Wiesbaden: Bundesinstitut für Bevölkerungsforschung. https://d-nb.info/1157097561/34 (Access on 23.02.2023).

Mielck, Andreas/Wild, Verina (2021): Gesundheitliche Ungleichheit - Auf dem Weg von Daten zu Taten. Fragen und Empfehlungen aus Sozial-Epidemiologie und Public-Health-Ethik. Beltz Juventa. ISBN 978-3-7799-6241-0.

Nicolaisen, Magnhild/Thorsen, Kirsten/Eriksen, Sissel H (2012): Jump into the void? Factors related to a preferred retirement age: gender, social interests, and leisure activities. Int'l. J. Aging and Human Development; 75(3) 239–271, 2012.

Nilsson, Kerstin/Rignell-Hydbom, Anna/Rylander, Lars (2011): Factors influencing the decision to extend working life or retire. Scandinavian Journal of Work, Environment & Health, 37 (6), 473–480. https://doi.org/10.5271/sjweh.3181.

Örestig, Johan/Strand, Mattias/Stattin, Mikael (2013): A Wish Come True? A Longitudinal Analysis of the Relationship between Retirement Preferences and the Timing of Retirement. Population Ageing 6, 99–118, https://doi.org/10.1007/s12062-012-9075-7.

Saß, Anke-Christine/Lange, Cornelia/Finger, Jonas D et al. (2017): "Gesundheit in Deutschland aktuell" – Neue Daten für Deutschland und Europa. Hintergrund und Studienmethodik von GEDA 2014/2015-EHIS. Journal of Health Monitoring 2(1), 83–89, https://doi.org/10.17886/RKI-GBE-2017-012.

Scharn, Micky/Sewdas, Ranu/Boot, Cécile R. L./Huisman, Martijn/Lindeboom, Maarten/van der Beek, Allard J (2018). Domains and determinants of retirement timing: A systematic review of longitudinal studies. BMC Public Health, 18(1), 1083. https://doi.org/10.1186/s12889-018-5983-7 (open access).

Schröder, Helmut/Kersting, Anne/Gilberg, Reiner/Steinwede, Jacob (2013): Methodenbericht zur Haupterhebung lidA – leben in der Arbeit. FDZ Methodenreport 1/2013. Bundesagentur für Arbeit, Nürnberg. http://doku.iab.de/fdz/reporte/2013/MR_01-13.pdf. (Zugriff am 23.02.2023).

Statistisches Bundesamt (2022): Pressemitteilung zur Bevölkerungsentwicklung in D. Pressemitteilung Nr. 511 vom 02.12.2022. https://www.destatis.de/DE/Presse/Pressemitteilungen/2022/12/PD22_511_124.html (Zugriff am 23.02.2023).

Steinwede, Jacob/Kleudgen, Martin/Häring, Amando/Schröder, Helmut (2015): Methodenbericht zur Haupterhebung lidA – 2.Welle. FDZ-Methodenreport 7/2015. Bundesagentur für Arbeit, Nürnberg. https://doku.iab.de/fdz/reporte/2015/MR_07-15.pdf. (Zugriff am 23.02.2023).

Van Dyk, Silke/Lessenich, Stephan/Denninger, Tina/Richter, Anna (2013): Gibt es ein Leben nach der Arbeit? Zur diskursiven Konstruktion und sozialen Akzeptanz des "aktiven Alters". WSI-Mitteilungen 5, S. 321–328.

Working Environment Under Construction

Anke Kahl, Marina Bier, Florian Hafner, Nina Hanning, Magnus Nauth and Julia Waldorf

Contents

Abstract

Contemporary occupational health and safety that is accepted in science and practice acts directly as a technology companion. In addition to technical and technological innovations and changes, such as the introduction of new assistance systems, artificial intelligence in smart practical and communication tools, greater market saturation with sensor and laser technologies, and the introduction of new technologies in the course of decarbonization, social changes are also drivers or obstacles to occupational health and safety, which must be implemented at an early stage in the sense of a high level of protection. The targeted recruitment and support of older employees, the increased access of people with a migrant background to the labour market and the flexibilisation of work, e.g. through mobile forms of work, are selected examples of this. In addition, climate change and social climate protection measures pose special

A. Kahl (✉) · M. Bier · F. Hafner · N. Hanning · M. Nauth · J. Waldorf
Chair of Occupational Safety, University of Wuppertal, Wuppertal, Germany
e-mail: akahl@uni-wuppertal.de

S. Bracke and R. Pieper (eds.), *Safety Engineering*,
https://doi.org/10.1007/978-3-658-47473-7_17

challenges for occupational health and safety. The Chair of Occupational Safety and Health at the University of Wuppertal actively deals with selected fields of design of the above-mentioned changes and challenges in teaching, research and policy consulting. Interdisciplinary approaches to solutions, with close technical and methodological links, e.g. to civil protection and fire protection, but also to education and communication sciences or stakeholder in the circular economy, are required in order to develop practical design concepts.

Keywords

Occupational health and safety · Spontaneous volunteers · Climate adaptation · Policy consulting · Safety in indoor swimming pools · Fire fighting operational scenarios · Assistance systems · AI in occupational safety

1 Introduction (Anke Kahl)

The transformation in the world of work is challenging in terms of its complexity and speed as well as its impact in terms of scope and depth. A contemporary and modern occupational health and safety, which wants to do justice to its comprehensive approach to prevention, must not only keep pace with this change, but also actively act, design and provide timely supporting solutions that enable companies to respond appropriately to traditional, new or changed and possibly more complex risk and stress situations in order to ensure the safety and health of its employees in the changing world of work.

The dual occupational and health protection of the state and accident insurance institutions has the role of being an active companion for technological, social and legal changes (see examples in Table 1) and of shaping the "white" fields of action in a timely manner with suitable solutions.

Occupational safety-specific solutions in the changing world of work are not only a contribution to the employer's duty of care within the meaning of § 618 Civil Code (Bürgerliches Gesetzbuch—BGB) and to the objective of the Occupational Health and Safety Act (Arbeitsschutzgesetz -ArbSchG) in accordance with §§ 3–5 ArbSchG and the protection objectives anchored therein to assess and shape the working conditions. It is also a contribution to securing skilled workers and increasing the attractiveness of companies with regard to potential and healthy workers across all age groups.

Numerous studies [1, 2] also point out that for employees of Generation Z, in addition to the meaningfulness of the activity and the willingness of the employer to change, there is a desire for a long-term stable work-life balance, including flexibility. Modern occupational health and safety can also make an important contribution to the retention of skilled workers. A close integration with occupational health management through accepted offers also for young employees is required in order to make a substantial contribution to recruitment and against fluctuation.

Table 1 Examples of occupational safety-relevant challenges and options for action based on the DGUV Risk Observatory* (excerpt) (https://www.dguv.de)

Field of Change*	… on the concrete example	Effects on the occupational an health on e.g	Necessary needs & concretizations
New technologies	New composite materials, e.g. carbon fiber reinforced plastics	New or increased material hazards in recycling activities	New technical systems, new work equipment and organisational measures as well as PPE, research, guidance
	Use of limitless collaborative robots (cobots), e.g. in assembly processes	Increased accident and accident Risk of injury due to direct contact and possible influences, e.g. in the context of maintenance	Accident data is still missing, Guides, Research
Climate change, nature and Ressourcen-schutz, Decarbonization	Construction, operation and recycling of wind turbines	Increased hazards due to large-scale use of epoxy resins, work at great heights and under extreme weather conditions	Transfer knowledge into technical rules, Special fall protection, PPE, handling aids
	Construction, operation of dismantling new energy storage systems	New material and physicochemical hazards during use and recycling in the event of misuse and damage, Focus: Fire protection	New technical systems, new work equipment and organisational measures and PPE Strengthening of fire protection
	Vulnerability of society to climate-related negative events, e.g. increase in the number of cases	Targeted increase in resilience in the workplace, depending on the type of climate-related events (heat, extreme weather, invasive plants) and vulnerable groups	Transfer knowledge into technical rules, Comprehensive UV protection for vulnerable groups (e.g. working time regulations)

(continued)

Table 1 (continued)

Field of Change*	… on the concrete example	Effects on the occupational an health on e.g	Necessary needs & concretizations
Demographic trends and diversity	Ageing society, Increased demand for nursing staff of all qualification levels	High levels of sick leave and early retirement, among other things due to musculoskeletal disorders and psychological stress	Use of technical support for lifting and carrying activities, qualification of working time concepts
	Refugees and unskilled workers in the labour market	Higher risk due to language and qualification barriers, cultural peculiarities in the work environment, knowledge deficits	Simple guidelines for activities in the languages of the countries of origin
Social Affairs and Health	Attitudes and values of young employees (Gen Z)	Increased interest in maintaining one's own health and a safe working phase, strengthening the value system	Raising awareness of the topic of occupational health and safety through suitable formats already during training
	Accumulation of violent incidents on rescue workers / employees of aid organizations	Increased physical hazards and psychological stress (insults, insults)	Guidance on anti-violence and de-escalation training, personnel support for protection
New Work	Labour shortage Demand for work-life balance increases	Safe working environment in the private sector required, Guarantee of the legal. Minimum number of representative functions (e.g. first aid) for mobile work concepts	Concepts for safe mobile work, including special instruction concepts, Revising company work organisation
	Flexibilisation of work organisation, e.g. home office	Increased psychological stress, e.g. through isolation, accessibility, separation between work and private life, Less psychological stress, e.g. through better Compatibility with private life	Ergonomic design of the workplace in the private sphere, Guidance already available; Promoting the self-competence of employees is important, aspects of IT security / data protection

(continued)

Table 1 (continued)

Field of Change*	… on the concrete example	Effects on the occupational an health on e.g	Necessary needs & concretizations
Digitalisation and connectivity	IT- and software development, for example AR and VR	Training opportunities for real working conditions without hazards in AR or VR; Visualization of hazards in the environment through the use of AR; Remote maintenance through AR systems	Necessary competence development in the field of technology, consideration of legal requirements for data security and ergonomics in the use of AR/VR devices
	Internet of Things (IoT) Networking of systems in production processes	Error prevention/reduction through predictive maintenance, Real-time monitoring of environmental parameters, e.g. temperature, possibly increased psychological stress due to acceleration of production processes	Competence development in the field of system monitoring; Cybersecurity requirements
Legal New regulations and Changes	Cannabis Act, Canabis legalization	Changes in workplace law, Expanding addiction prevention	Specify rules for addiction prevention, guidelines for action, research (e.g. on the derivation of limit values)
	Circular Economy Act, e.g. Take-back obligations and waste hierarchy	Increasing number of recycling and cleaning activities, characterized by high aerosol loads, dusts, biological substances, etc. due to manual sorting proportions	Solutions Guidance Procedures, Research

In order to overcome these challenges in the long term, the existing scopes for action of the stakeholders involved must be actively and effectively involved. These range from policy consulting to science.

The Chair of Occupational Safety is dedicated under the direction of Prof. Dr.-Ing. habil. Anke Kahl has been both topics in the context of development and research: occupational safety-specific policy consulting as well as specific research questions that take up the change in the world of work.

In the following, selected topics of the chair, goals, concerns and research results as well as the topic of policy counselling are presented.

Field of adaptation to climate change
Field of Research: Extreme Weather Events/Spontaneous Helpers
Field of Research: Heat Stress.

Field of policy consulting
Field of workplaces in complex scenarios
Field of research: State of the art in swimming pools
Field of Research: Occupational Safety in Dynamic Scenarios of the Fire Brigade

Field of safety education
Field of Research: Use of AR in Vocational Training
Field of Research: AI Assistant in Occupational Safety and Health in SMEs

2 Field of Adaption to Climate Change

2.1 Research Topic: Extreme Weather Events & Spontaneous Volunteers (Marina Bier)

Due to climate change, weather extremes and the associated large-scale emergency and even disasters will occur more frequently in the future, such as heavy rainfall events, extreme heat or storms [3]. This development is also relevant for working environment and for the occupational health and safety of employees, both before, during and after an incident. In addition, it is necessary to build up resilience to such events in both companies and employees themselves. In order to illustrate the research activities of the Chair, examples of challenges that need to be taken into account are outlined below. Technical terms are used according to the conceptual model in occupational health and safety, see Fig. 1. This clearly presents the definitions and at the same time the relationship between key terms such as safety, assessment standard and danger.

Challenges ***Before an incident***, an employer or other responsible person (in accordance with Sect. 13 of the German Occupational Health and Safety Act (ArbSchG).) is faced with the challenge of deciding which preventive measures are required to maintain or quickly resume their buisness operations. Employees

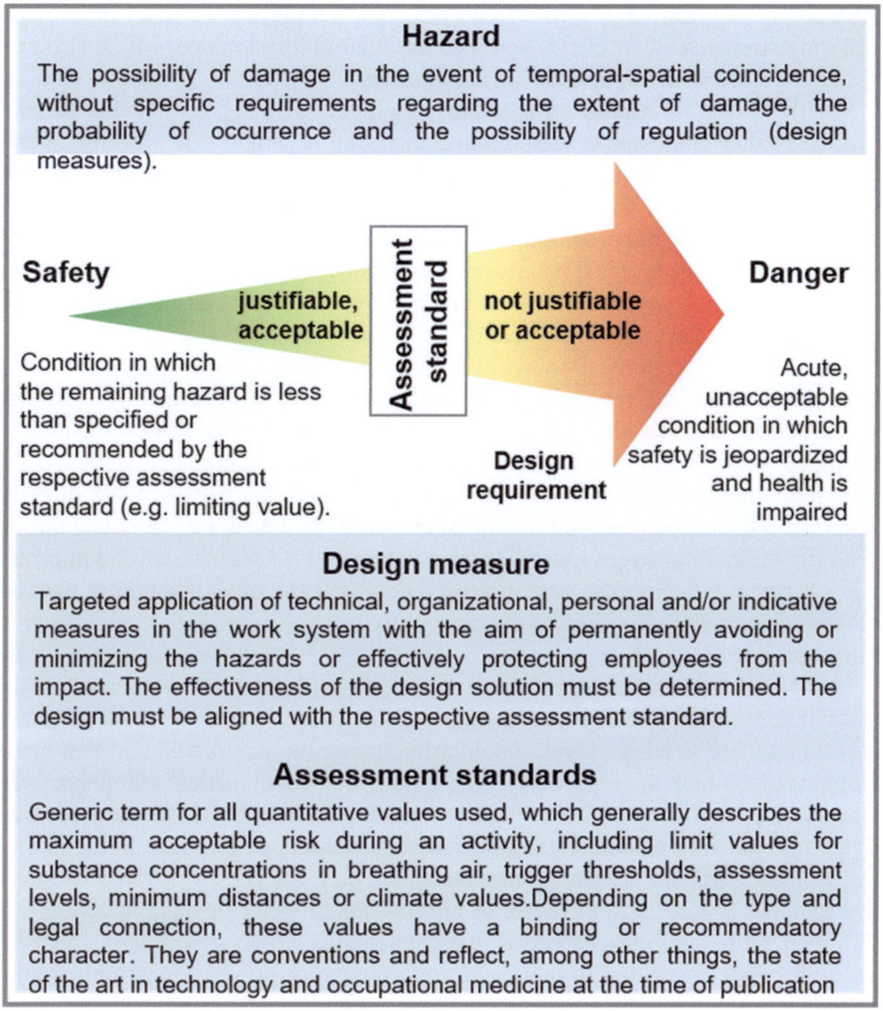

Fig. 1 Conceptual model in occupational health and safety [4]

may be required to carry out activities outside their normal work routine that could lead to additional *hazards* or unusual stresses, such as transporting important work equipment to higher floors to protect against flooding.

According to Sect. 5 ArbSchG, the employer has a legal obligation to "determine which occupational health and safety *[design] measures* are required by assessing the *hazards* associated with the employees' work". This assessment is known as risk assessment and is also required for infrequent activities. According to Sect. 12 ArbSchG, employees must also receive sufficient and appropriate instruction as one relevant, organizational *design measure* during their working hours, including on the results of the risk assessment. This type of instruction and other forms of knowledge transfer can also help employees to react more resiliently to extreme weather events and increase their ability to cope, which also creates an interface with the field of civil protection.

Shortly before or ***during an incident***, employers or other responsible persons face the further challenge of determining the point at which safe working is still possible or when work should be stopped or interrupted. This requires suitable *assessment standards*. In accordance with Sect. 9 ArbSchG, the employer must take measures to enable employees to get to safety in the event of imminent serious *danger*. Employees must be informed in advance as early as possible about the *danger* and the *design measures* to be taken, which they can carry out themselves, making effective hazard communication necessary.

After an incident, the response phase begins. This is characterized by disrupted infrastructure, such as blocked traffic routes due to fallen trees or mud, through to limited power supply, and unclear *hazards*, such as contaminated water. At the same time, the employer is under economic pressure to get back to work quickly. Nevertheless, occupational safety measures must be taken into account during clean-up work. Employees must also be provided with adequate work equipment in accordance with Sect. 5 Ordinance on Safety and Health Protection in the Use of Work Equipment (BetrSichV) that is safe to use, considering the intended conditions of use.

During the event and at the time of the response phase, past events such as the heavy rainfall event in Münster 2014 [5] and the flood 2021 in Germany [6], have shown that there is a high level of solidarity among the population and that many people want to help spontaneously. These so-called spontaneous volunteers, who want to help in advance without much advance planning and without being integrated into existing emergency organizations, are therefore exposed to high risks. Due to their altruistic actions in an emergency, they are largely covered by statutory accident insurance in the event of injuries and property damage in connection with their relief activities. However, the challenge is which *assessment standards* are useful and which *design measures* are effective to prevent or minimize injuries and damage to the health of spontaneous volunteers despite in such unplanned events.

Research activities As part of the research activities at the Chair of Occupational Safety, the challenges outlined above are considered and practical solutions are generated.

Significant *dangers* were identified on the basis of research into occupational accidents during natural disasters. Literature from the U.S. in particular was available for this purpose. In the event of flooding, drowning was the crucial *danger* during the event (approx. 76%, 47 fatalities during the period under review), of which almost half of the employees were in vehicles [7]. As *design measures*, safe behaviors to prevent commuting accidents are analyzed based on literature and support material for companies is developed for corresponding instructions.

An increase in occupational accidents was observed not only during but also after events, for example after Hurricane Maria in 2017 [8]. However, further research is needed here in order to conceptualize *assessment standards* and *design measures*.

In addition, influencing factors for improving the safety of spontaneous volunteers are being empirically investigated using the heavy rain-induced flood in 2021 in western Germany with the help of observations, analysis of accident figures, surveys and interviews with the aim of developing effective *design measures* for future events. Insights into the survey results can be found in the publication by Bier et al. [6].

Significant *hazards* for spontaneous volunteers arise from psychological stress, physical overload and environmental hazards, such as hazardous and biological substances, uneven surfaces that can lead to tripping or twisting, etc.. Serious *danger* were also drowning in particular, but also collapsing buildings, floating debris during the event as well as contaminated mud and pointed or sharp objects after the incident. Possible *design measures* to counter the *hazards* include the multiplication of knowledge by experienced or trained individuals and the provision of suitable personal protective equipment.

2.2 Research Topic: Heat Stress/Policy Workshop (Magnus Nauth)

The effects of climate change are becoming increasingly noticeable globally and jeopardise the safety and health of people worldwide. Various risk factors such as rising ambient temperatures, the increase in extreme weather events as mentioned in the article above, rising sea levels, droughts and water scarcity, as well as political and social conflicts have different regional impacts, but there is still practically no place on earth that is not affected by this change in one way or another [9].

Employees are often the first to be exposed to the effects of climate change, for longer periods of time and more intensively than most other population groups [10]. This results in changed requirements for occupational health and safety, which pose challenges for employees, employers and occupational safety specialists in order to effectively protect the health of employees.

Of the risk factors that are exacerbated or even newly occurring as a result of climate change, increased heat stress due to rising ambient temperatures is one of the factors that pose the greatest risk to employees, affecting around 2.41 billion workers [11]. Although Europe, and therefore also Germany, is not currently one of the regions in which excessive heat stress at work affects the majority of employees, the proportion of exposed employees in the total working population is increasing the most worldwide [12].

In order to meet these challenges for occupational health and safety in Germany, the Federal Ministry of Labour and Social Affairs (BMAS) launched a policy workshop entitled 'Work: Safe and Healthy—Climate Change' in spring 2023. The aim of the project is to review occupational health and safety in

Germany with regard to the changing risk situation caused by climate change in cooperation with a group of experts from representatives of occupational health and safety organisations, science institutes, employers' and employees' associations. The focus here is on the topics of 'Heat & UV radiation', 'Extreme weather', 'Sensitisation & compliance' and 'Vectors & hazardous substances', each of which will be discussed by the group of experts in separate workshop discussions. At the end of the policy workshop, regulatory adjustments, action aids for occupational safety stakeholders and their networking with each other are to be developed.

Challenges Many of the risk factors considered in the policy workshop are not new to occupational health and safety, but are exacerbated by climate change or now affect groups of employees who were not previously exposed to them. Previous solutions and measures are therefore proving to be inadequate or impractical in some cases.

Due to the fact that some groups of employees are still exposed to new hazards, educational work needs to be carried out here to sensitise employees and managers and convince them of the necessity of the measures. This is particularly important in the context of the policy workshop, as the BMAS has planned this as a participatory format in which a consensus is to be reached as a result of the policy workshop in dialogue with the social partners, the scientific community and the legislator. However, the information required for this regarding the affected parties and the corresponding solution approaches depends on various factors, meaning that interdisciplinary cooperation is required here.

As there are numerous occupational health and safety organisations in Germany, some of which have little coordination with one another, the people responsible for occupational health and safety in companies are sometimes overwhelmed by the wealth of information available. The policy workshop has therefore set itself the task of bringing together the various players and their endeavours for climate change-compliant occupational health and safety and preparing practical information for users.

Climate change and the associated hazards are also an active and constantly changing field of research with numerous intersections with other areas, so that new information must constantly be included in the considerations in order to meet the requirements for comprehensive protection of employees.

Research activities The 'ASUG-Klimawandel' policy workshop is organised and evaluated by a consortium of various project partners under the leadership of the BMAS, in which the Chair of Occupational Safety at the University of Wuppertal (BUW) is involved. The tasks performed by the Chair for the project include preparing the workshop discussions by researching international scientific literature on the respective key topics, analysing the results of these events and transferring them into suitable formats such as topic dossiers or regulatory adaptation proposals, conducting and evaluating interviews with experts beyond the circle of experts at the policy workshop, and designing and supporting the implementation of action concepts in selected pilot companies.

As the first workshop discussion on 'Heat & UV radiation' dealt with the topic with the greatest direct relevance for Germany, the preparation of the heat stress factor in particular was carried out extensively. To this end, a systematic literature search was carried out on international specialist literature, which supplemented and updated the findings of a review on this topic published by the Federal Institute for Occupational Safety and Health in 2022. A further component of this work was a comparison of the information provided by occupational health and safety organisations in various countries worldwide on the subject of heat stress at work, with a focus on risk factors, protective measures and guidelines. The result of this work is that the occupational health and safety organisations in Germany already cover a large part of the wealth of internationally available information, but there is still potential, for example, in the implementation of acclimatisation, implementation of buddy systems and self-paced work as effective protective measures.

3 Field of Policy Consulting (Anke Kahl)

The Federal Ministry of Labour and Social Affairs (BMAS) is advised on all questions of the safety and health of employees by state committees. In order to consider the associated, comprehensive topics from all relevant perspectives of the world of work, these committees include representatives of employers, trade unions, state authorities, statutory accident insurance and science as well as experts. Its members are specially appointed, expert representatives of the aforementioned banks.

The legal authorisation to establish state occupational safety committees is set out in § 18 para. 2 No. 5 of the Occupational Safety and Health Act (Arbeitsschutzgesetz—ArbSchG), which is completed in the respective occupational safety and health regulations. This authorisation is supplemented by § 24a of the ArbSchG.

Scientific expertise The integration of scientific expertise into this form of policy consulting is of central importance in the context of the relevant scientific disciplines, such as occupational science, occupational health and safety and occupational medicine, since new and future topics in particular are usually characterised by particularly complex challenges (e.g. on the topics of mobile work or climate adaptation) and thus current study situations and results can underpin the technical content level through quality-assured findings and conclusions. These ensure—on the basis of bidirectional and iterative communication strategies—a quality-assured entry level for the upcoming occupational safety policy decision-making processes. The scientific experts from different disciplines appointed by the BMAS bring their expertise to the five occupational safety and health committees, which have been established for years, as well as to the Committee for Safety and Health at Work (Ausschuss für Sicherheit und Gesundheit bei der Arbeit—ASGA), which has been newly established by the BMAS since 2021.

The Chair of Occupational Safety has been active in the field of policy consulting for over 15 years and is visible with its expertise, including through the expert work of the head of the chair in the Hazardous Substances Committee (Ausschuss für Gefahrstoffe—AGS) and since 2021 through the assumption of both the chairmanship of the Committee for Safety and Health at Work (ASGA) and the cross-committee-acting control group.

State Regulations In addition to the comprehensive advice of the BMAS on all questions relating to the respective occupational safety regulations, an essential task of the occupational safety and health regulations is the determination of the need for regulation and the development of cross-industry state rules. The function of the state regulations (also technical regulations) is to specify the requirements that are formulated by the detailed legal framework of the Occupational Safety and Health Act and the Occupational Safety and Health Ordinances as consciously open protection objectives.

Technical rules are not legal norms, but they are technically based on the level of protection of the "state of the art, occupational medicine and hygiene" as well as other secured occupational science findings. [13] For this purpose, the representatives of the above-mentioned Benches "the majority-specialist conviction of correctness and form rule content from it" [14].

With the technical clause "state of the art", the legislator has made a quite demanding level the benchmark for operational activities. [13]

Of great importance for the subject matter of the technical rules is that, on the basis of a hazard-related approach, compliance with a technical-scientific minimum standard is required.

This is intended, among other things, to ensure that occupational safety understands the progressive development of technology and science. [15]

A key position for the performance of tasks of the state committees in accordance with the requirements is the necessary degree of concretisation of the rules, in particular with regard to the presumption effect linked to the state rules. The presumption effect means that employers who have applied adequate occupational safety measures in accordance with the state regulations can assume that the requirements of the respective ordinance are met with legal certainty. Against this background, the rules of state committees must have an appropriate depth of content and high quality. [16]

Working methods of the occupational safety and health committees With regard to the subject-specific orientation of the occupational safety and health committees, it can first be stated that the five specialist committees—which have already been established for many years and are working very targeted—are geared towards the concretisation of protection objectives of individual or even several Occupational health and safety regulations with regard to their work mandate.

Among other things, the Committee for Hazardous Substances specifies the hazard-specific protection objectives and general requirements of the Hazardous Substances Ordinance by means of Technical Rules for Hazardous Substances and the Committee for Operational Safety specifies the protection objectives of several individual regulations by means of rules for operational safety, including the Industrial Safety Ordinance, the Noise and Vibration Protection Ordinance and the Ordinance on the protection of employees against threats from electromagnetic fields.

This vertical way of working was and is successful and enables the technical and content penetration and preparation of even demanding subject-specific design solutions and operational measures. As a result, however, this vertical way of working also meant that—with the all of a few individual cases—the methodological and technical-content range of the factor-specific committee work does not take up any interfaces with topics of other specialist committees, which only meets the operational reality to a limited extent. Targeted coordination processes between the specialist committees are neither planned on the basis of the rules of procedure of the committees nor have they been desired or admonished by the banks so far.

This way of working was and is not conducive to the creation of a coherent, i.e. content-coordinated set of rules. For the addressees of the set of rules—the employers—this lack of coherence with regard to the comprehensibility and operational practicability of the set of rules is a shortcoming. However, in order to fully fulfil the claim to coherence and freedom from contradiction for the users, it is not enough to look at the state rules alone.

The extensive and industry-oriented German Social Accident Insurance (Deutsche Gesetzliche Unfallversicherung—DGUV) and there regulations is the second pillar of the national regulatory strategy in dual occupational health and health and health law, which was already clearly formulated in the guideline paper on the reorganisation of the regulations and regulations in occupational safety and health from 2011. Among other things, the demand is that "the system of state rules and DGUV-rules is to be structured in terms of content and procedure in such a way that a coherent and coherent overall structure is created". [16]

Committee on Safety and Health at Work (ASGA) Not only the Covid-19 pandemic has more clearly shown the need for a possibility to establish cross-activity occupational health and safety rules and recommendations. [17]

The dynamic and continuous processes of change in the changing world of work also make the need for more coherence in the technical regulations more visible. Need for concretisation for the topics of mobile work, psychological stress and for changed working conditions due to climate change. Due to the diversity as well as the complexity of existing and partly new risk and burden profiles, the fields of action mentioned in one of the established specialist committees cannot be located thematically in one of the established specialist committees, so that in recent years the need for a new committee—of a cross-subject matter—has been suggested by the BMAS.

With the amendment of the Occupational Safety and Health Act (§ 24 a ArbSchG) in December 2020, the way was clear for the Committee for Safety and Health at Work (ASGA), which, in contrast to the specialist committees, does not issue individual regulations, but directly concretizes the Occupational Health and Safety Act by means of technical rules.

The ASGA is—in analogy with the established specialist committees—supported by the established five benches, each of which is staffed by 3 subject matter experts plus 3 deputies. He has given himself a work program in coordination with the counselling needs of the BMAS. With the establishment of the ASGA, the appointment periods of all six state occupational safety and health committees were harmonised in time, which is an essential prerequisite for being able to make structural changes in the entire committee system in the future, e.g. the establishment of a coordinated committee strategy with regard to joint prioritisation of tasks and the temporary mandates of experts.

The ASGA deploys project groups for the programmatically agreed topics, which—unlike the specialist committees—are not only staffed by experts in-house or appointed by the ASGA benches, but also by mandated persons of all five specialist committees. These bring different subject-specific perspectives and experiences into the work of the ASGA project group.

However, this so-called horizontal approach only takes into account the high standards of coherence and practicality if the complex and cross-committee topics to be processed result in a regulatory procedure that, on the one hand, considers the topics analogously to operational practice in a multiple perspective and, on the other hand, designs the necessary interfaces to the subject-specific rules in a factual and conflict-free way with the necessary depth of detail.

The technical and substantive range to be specified as well as the occupational health and safety concern of the proposed regulatory procedure will be decided in a project outline by consensus of the six committees. This describes the non-negotiable framework of the upcoming project work for all involved circles and members of the project groups.

Decisive for the thematic classification of the future ASGA rule, which is designed for consensus, is that already in this phase the strategic classification of the rule in the entire set of rules is thought out and anchored in the project outline. For this purpose, the ASGA has, among other things, established two types of rules, so-called roof rules and integrating rules.

The development of an umbrella rule is pushed forward when the majority of the specialist committees have already worked out rules on selected subject-specific aspects of the ASGA topic in focus. If this is the case, all content that does not have a subject-specific but a general or overarching character will be brought together in a structured ASGA rule in terms of subject matter and content.

The umbrella rule thus concretizes general design and/or process requirements on the basis of the ArbSchG, which result in particular from the Occupational health and safety regulations and the related rules. These technical and methodological concretisations are based on the Occupational health and safety regulations resp. Rules of the specialised committees without contradiction and completely converted into an umbrella rule (→“pull in front of the bracket”). These design and/or process requirements, which are thus combined, form the overarching technical basis for further specific requirements in the factor- or object-specific rules of the other committees. The coherence is achieved when all special rules are coordinated with each other and to the umbrella rule and the interfaces are clearly and unambiguously indicated for the users. [17]

The development of an integrating rule is supported if the cross-committee ASGA topic has so far been taken up by very few or none of the specialist committees and then only punctually.

The Integrating Rule then specifies a requirement defined in the ArbSchG, taking into account the essential aspects, including those that are (also) defined in the Occupational health regulations (ArbSch-VO) and, if necessary, also specified in the corresponding rules of other specialist committees.

Thus, this rule sets out in terms of technical content which requirements must be met at the operational level in order to assume that the corresponding protection objective of the ArbSchG is met. Insofar as requirements for partial aspects of the subject of the regulation are specified in the Occupational health and safety regulations, the technical and substantive specification is carried out in the Integrating Rule up to a depth required for the overall understanding. [17]

Further specific requirements in other rules on the factor- or object-related topic of the integrating rule are superfluous or can then be designed without contradiction to the integrating rule. The approval of all committees concerned is then required for the decision-making of an integrating rule by the ASGA. [18]

Currently, two ASGA project groups each develop a first rule per control type. Among other things, the topic of risk assessment is conceived as an umbrella rule, i.e. factor-specific aspects such as general procedure, operational organisation, expertise or documentation are specified, updated and merged in a new ASGA rule “risk assessment”. In the next step, the concern is to either remove this content in the existing rules of the specialist committees or to adopt it without contradiction. The subject-specific content, such as the use of assessment standards, e.g. trigger thresholds, workplace limit values, the use of measurement methods or even targeted operational measures, are then further specified in the rules of the specialist committees and adapted to the umbrella rule by means of suitable interfaces. The topic of “mental stress” is currently being developed as an integrating rule.

With a future coordination of the work programmes of the specialist committees and the ASGA on cross-committee topics, more flexible working methods and coordination processes, the establishment of new action aids and formats, e.g. for the efficient assessment of draft rules, embedded in an appreciative self-image, the ASGA will be able to make an effective contribution to the coherence of the state rules in close cooperation with the specialist committees. This aims to enable the state committee system to adapt more flexibly to the changes in the world of work in the next few years in the coming years.

4 Field of Workplaces in Complex Scenarios

4.1 Research Topic: State of Art and Occupational Health and Safety in Indoor Swimming Pools (Nina Hanning)

Swimming pool operations (here: indoor pools) are particularly complex workplaces in terms of occupational health and safety. The diverse activities range from water treatment and cleaning to looking after bathers and using technical installations. Various sources of hazards, such as chemical substances used in the course of water treatment and the handling of technical equipment, require careful analysis. The specific health and safety standards must also be taken into account. In particular, working in warm and humid environments and working in noisy environments present challenges in this field of work.

However, there are currently still gaps in the expertise and content with regard to the specific state of the art in the types of pools analysed, the relevant working environment conditions and the working and health conditions of the various employment and task groups in pools.

Challenges The state of the art evaluation standard forms a central engineering and business science basis for safety-relevant decision-making and selection processes in the context of product and occupational safety. Occupational health and safety law refers fundamentally to the state of the art in the Act on the Implementation of Measures of Occupational Safety and Health and its ordinances. It states that employers must base their occupational health and safety measures on the general principle that the measures to be planned must take into account the state of the art in technology, occupational medicine and hygiene as well as other established occupational science findings (§ 4 Act on the Implementation of Measures of Occupational Safety and Health). In this way, the legislators create a qualitative assessment standard that can be interpreted as a binding level of protection based on requirements.

Currently, only a few technical rules [19, 20] and technical publications [21, 22] address a specification of the state of the art in relation to swimming pool operations.

Although there are about 2800 indoor swimming pools [23] in operation in Germany, which are very diverse in terms of size, equipment and type of water treatment, government regulations do not explicitly refer to the safe operation of swimming pools. In principle, selected technical rules and DGUV regulations/rules refer to specific aspects of pool hygiene and technology [20, 24–26], but there is no overall concept for determining the state of the art in pool operations, taking into account the type and size of the pool and the water treatment process used in the specialist literature or in the dual occupational health and safety regulations.

A relevant requirement for the state of the art quality of swimming and pool water is the legal obligation under § 37 of the Infection Protection Act. In order to prevent harm to human health through its use, the swimming and pool water is treated and disinfected using chlorination and oxidising agents. [27, 28]

Organic substances and numerous pollutants continuously enter the pool water. Sweat, urea, skin flakes, cosmetics, etc., are introduced into the water by bathers and react with the chlorinating agents. [29] They react to form water-bound or airborne disinfection by-products. In addition to the known disinfection by-products trichloramine (NCl_3) and trihalomethanes (THM), it can be assumed that there are other, as yet unknown disinfection by-products that can affect the health of employees in swimming pools. [30]

The working environment conditions of employees in swimming pools are very diverse. However, very few studies or publications highlight the conditions that favour hazards, such as heat and cold, high humidity and noise [29, 31].

In addition, no data is available on the working and health conditions of employees in swimming pool operations. However, initial indications of the partially limited state of health of this group of employees have been supported for many years by various social partners and expert groups [32], as well as by the employees' own statements.

Research Activities The current state of research on the topic of indoor swimming pool operations focuses on the knowledge of the already well-studied disinfection by-products, trichloramine and trihalomethanes. Their formation, characterisation and effect as well as the effects and damage patterns following exposure are discussed. Data on other possible disinfection by-products is missing. There is also a lack of systematic studies on the working environment conditions and the actual state of the art in swimming pool operations.

The Chair of Occupational Safety at the University of Wuppertal is focussing on this research deficit. In addition to investigating whether the state of the art in swimming pool operations can be described precisely and comprehensibly, relevant factors are derived which influence the subjective perception of the working and health conditions of employees. These research findings, the results of measurement surveys of working environment conditions and the recording of work systems are used to evaluate whether a correlation can be described between compliance with the state of the art and the working and health conditions of employees in swimming pool establishments.

Specific survey instruments were developed for this purpose. Firstly, an extended checklist was designed to systematically record the state of the art in swimming pool operations. The focus here is on the water treatment process, the type of pool and the size of the establishment. On the other hand, a comprehensive online questionnaire was created to record the subjective perception of the working and health conditions of employees in swimming pool operations. In addition to questions on demographics and the field of work, this online questionnaire also includes questions on various aspects such as workload, job satisfaction, health complaints and perceived safety in the workplace. Moreover, measurement protocols were created for the technical measurement surveys of the working environment conditions, such as air humidity, air speed, heat radiation, lighting and noise, as well as for hazardous substance measurements of the substances trichloramine, trichloromethane, nitrosamines and chlorine.

These survey instruments were initially trialled in a pre-test in the swimming pools of the city of Wuppertal in a smaller sample. It turned out that the originally designed questionnaire was too long and complex for the target group of employees. In order to reduce the respondents' effort and achieve a higher response rate, the questionnaire was fundamentally revised. The new version was issued as a paper version and is considerably shorter and more focussed. This new version of the questionnaire only required participants to spend about 10 min completing it. The checklist for recording the state of the art and the protocols for the technical measurement surveys of the working environment conditions, on the other hand, proved to be practical and were retained in their original form. The hazardous substance measurements also revealed that the substance trichloramine can be regarded as a lead component. As a result, it was decided to focus the hazardous substance measurements of the main study specifically on this substance.

As part of the main study, 10 public indoor swimming pools in North Rhine-Westphalia, Lower Saxony and Rhineland-Palatinate were visited in the period from April to July 2024. A total of 63 returned questionnaires were analysed.

4.2 Reseach Topic: Occupational Safety in Dynamic Fire Service Scenarios (Florian Hafner)

Firefighting operations are characterized by a high degree of unpredictability and dynamism, as on-site conditions can often change abruptly and unforeseen hazardous situations may arise [33, 34]. This unpredictability necessitates a high level of flexibility and adaptability among firefighters to respond effectively to changing conditions. Sudden weather shifts, the unexpected spread of fire, or unforeseen structural instabilities can quickly exacerbate the situation at the scene. The increasing number of incidents, particularly in Germany, presents firefighters with increasingly more complex and challenging situations [35, 36].

Significant differences in the intensity and nature of operations are evident. For instance, technical rescue operations, such as extricating individuals from vehicles or responding to natural disasters, differ greatly from the demands of traditional fire suppression efforts [37]. This diversity of operational types significantly increases the complexity of incident management and places additional demands on the training and education of firefighters. The specific requirements vary not only in terms of the necessary technical equipment but also in relation to the physical and psychological pressures that firefighters face. These diverse operational environments and types not only complicate the comparability of incidents but also have a considerable impact on firefighter safety. In particular, the constant need to adapt to changing conditions on-scene requires a high level of situational knowledge and experience.

In this dynamic and often unpredictable environment, firefighters must perform numerous dangerous tasks, which significantly increases the risk of occupational accidents (OAs) [33, 38]. According to reports from the firefighter accident insurance (Feuerwehrunfallkasse), the accident rate among firefighters, with 45.0 reportable occupational accidents per one thousand full-time workers, is significantly higher than the national average of the German Social Accident Insurance (Deutsche Gesetzliche Unfallversicherung), which stands at 21.0 reportable occupational accidents per one thousand full-time workers [33]. This high accident rate highlights the particular dangers to which firefighters are exposed during their operations.

Firefighting operations occur in a variety of often very different work environments, which vary significantly depending on the type of operation [39]. While smaller incidents may involve hazards primarily associated with restricted working conditions, larger incidents can introduce additional risks from more complex sources of danger. These varying operational contexts create diverse conditions that considerably complicate the implementation of hazard assessment measures. The prospective hazard analysis is further complicated by the unpredictability of events, which can often only be concretely assessed once on-site [40]. Therefore, a precise and differentiated analysis of occupational accidents is essential to ensure firefighter safety. Identifying hazards across different phases of an operation—from the approach to the scene, through the primary incident response, to the conclusion of the operation—can help pinpoint specific phases where risks are elevated, and targeted measures can be developed to mitigate accidents.

One of the central research questions focuses on identifying those phases of an operation in which occupational accidents occur most frequently. This also includes examining how these phases differ in terms of the severity and frequency of accidents. This investigation is crucial to developing preventive measures aimed at reducing occupational accidents and integrating these measures into operational workflows.

Challenges One of the biggest challenges in analyzing occupational accidents in the fire service is the lack of standardized data collection. While professional fire departments independently document occupational accidents, firefighter accident insurance funds responsible for volunteer fire departments record this data using different criteria, making comparative analysis difficult. Particularly problematic is the lack of precise recording of the time and specific situation of the accident, complicating the assignment of accidents to particular phases of the operation. Inadequate data collection hinders both the comprehensive analysis of risk factors and the development of targeted preventive measures.

Furthermore, the high variability of incidents poses an additional challenge. Both the type of operation and external factors such as weather conditions, time of day, or environmental influences can significantly alter the risk situation [38, 39]. This variability makes it difficult to formulate generalized statements about accident risks, as the specific operational circumstances vary greatly. Additionally, the physical and psychological burden on firefighters plays a critical role. High stress, unforeseen events, and the potential danger to one's own life contribute to a significantly elevated risk of accidents during operations [41]. This calls for specific measures to strengthen firefighters' resilience and stress management capabilities.

Research activities To address the central research questions, an extensive retrospective analysis of accident and incident reports is being conducted. These data are systematically evaluated to identify patterns and clusters of accidents within specific operational phases. The developed evaluation tool facilitates structured data collection and analysis, allowing for both quantitative and qualitative insights. In addition, interviews with experienced fire department leaders are conducted to define operational phases and identify key parameters. These interviews are carried out using a literature-based, semi-structured guide.

The qualitative content analysis of these interviews aims to identify the characteristics that define an operational phase and to derive insights into the relevant influencing factors. Finally, the insights gained from analyzing the accident and incident reports are compared with the defined phases and their characteristics to identify accident hotspots across different operational phases.

The overarching goal of the research is to identify those operational phases in which occupational accidents occur most frequently and to analyze potential causes of these accidents. Based on these findings, measures aimed at improving firefighter safety will be developed, including targeted training and education programs as well as the implementation of specific protective measures aimed at significantly reducing the risk of accidents during critical phases.

5 Field of Safety Education

5.1 Research Topic: Use of AR in Vocational Education (Julia Waldorf)

The work environment is in a transition [42] and is constantly challenged by the use of new technologies, the shortage of skilled employees and further developments. In order to ensure the ability to work in such a dynamic environment, the development of skills in vocational education is crucial [43]. This also applies to the development of fundamental occupational health and safety knowledge [44] as accident statistics show that apprentices and young employees are exposed to a higher risk of accidents than older employees [45–47]. Apprentices can be characterised by inexperience and a lack of occupational safety knowledge [48] or incautiousness [47]. Literature suggests that further reasons for the higher accident rates among young employees are the more risky and physically demanding tasks they carry out [49]. According to studies, the main accident hotspots for young employees are mainly in the assembly and cleaning of machines [47] or mechanical hazards. Accidents can also be caused by psychological strain, such as a lack of control over the pace of work. [50]

The main accident hotspots can be the starting point for a holistic approach to conveying knowledge, and thus a well-founded understanding of occupational safety is to be built up among the apprentices.

Traditional teaching approaches can reach their limits, especially when it is about teaching complex and dangerous work processes without increasing the risk for learners. Not only the pandemic has led to an expansion in digital education programmes [51]. A new approach can be provided by the technology of augmented reality (AR), the expansion of reality through virtual elements [52]. Whereas virtual reality (VR), which is already well established in the gaming sector, creates a completely virtual environment, AR adds objects such as images, videos or 3D objects to the existing reality [53]. These can be displayed for example via smartphones, tablets or special AR glasses. AR offers numerous areas of application with a wide range of opportunities and challenges that require further research.

Use of AR AR systems have already been implemented in many areas with different success. Selected examples of the diversity of AR are outlined below.

Education and training are among the most common areas of application for AR [54]. AR is also an integral part of the production environment, car navigation and the tourism sector [53, 55]. In the field of advertising, products are virtually simulated using AR and enhanced with additional content [56]. In architecture, buildings are visualised in the planned environment; in gaming, AR provides objects which are projected from the game into reality [57].

This section of the paper focuses on the application of AR for knowledge transfer in the vocational training. In this field, AR is able to create very complex and adaptable learning environments. A key advantage is that experiences in AR are close to reality [56]. AR thus offers the potential for teaching occupational health and safety topics and for practising how to deal with hazardous situations without risk [57–59].

Several literature reviews have identified the benefits of AR in education [54, 58, 60–62].

According to literature, AR can positively influence *learner outcomes* through improved performance, increased motivation or better understanding [58–61]. Image-based AR can develop learners' spatial skills, practical skills and conceptual understanding [63]. The possibilities for interaction, communication and design of the AR systems can be mentioned as *pedagogical impacts* [60, 61]. AR systems can increase learner engagement and enjoyment [58–60]. *Further advantages* highlighted in the studies include the fact that the invisible and abstract can be visualised and that AR systems are easy to use [58, 60, 61].

The aim is to use media such as AR in teaching and learning situations where they can add value to the learning process from a didactic perspective [64]. Especially through the possibility of visualising invisible or abstract objects [59, 60] the use of AR for health and safety issues is apparent. AR creates the opportunity to maintain the actual environment while interacting with virtual objects in real time [65]. This characteristic of AR simplifies the transfer of knowledge into reality. In addition, AR environments can be adapted to learners in order to convey individualised and customised content [62].

There are already applications of AR in the field of training in the commercial sector—for example for welding training, in aviation training or for fire drills. In the education sector, technically and didactically well-designed applications are still rarer, as a number of challenges need to be faced. [56]

Challenges AR offers a promising addition to vocational training and, in particular, the teaching of occupational safety-related learning content. However, to enable a widespread use, some of the challenges outlined below need to be faced.

One of the main challenges for the use of AR is the necessary technical infrastructure [59]. This includes a sufficient internet connection, suitable end devices and the necessary software. Many vocational schools and companies are not adequately equipped to introduce AR technologies, for example the Wi-Fi at German vocational schools is often inadequate [66, 67]. The development as well as the implementation of AR can also generate high costs, especially for the creation of 3D content [59].

On the other hand, it is possible that virtual models such as in AR can be a support in educational institutions with limited resources, for example, if the use of expensive materials can be saved. Wang et al. also underline the possibility of saving material costs through the use of AR [59].

The didactic and methodological possibilities of digital media such as AR should be considered for educational purposes [65] and aspects of usability as an important requirement for successful use [68] must be applied in order to reduce the cognitive load on learners.

Even in the case of a didactically well-designed learning environment, the teachers who use the AR systems must also have sufficient technical skills and be qualified to use them. They not only need to be familiar with the technology, but also be able to integrate it into lessons or training in a useful way [66].

It is also worth considering the characteristics of the learners, as success in using a digital learning programme depends on previous experience and individual personality characteristics [69]. Technological affinity as a positive attitude towards technology [70] and technology acceptance as the actual intention to use a technology [71] are important requirements for succesful learning with AR [72]. The prior knowledge of the learners must also be considered in order to enable a suitable and possibly adaptive learning programme and to use the benefits of the technology.

Regarding occupational health and safety, it must be taken into account that digital media can lead to additional stress factors, such as increased screen time [73]. Depending on the application device, AR glasses can also cause motion sickness (also known as simulator sickness), an effect that has been increasingly researched in the field of VR, but which can also occur with AR [74]. Symptoms such as fatigue, dizziness or headaches occur as a result of a discrepancy in information between the visual and vestibular sensory systems [75].

The restriction of the field of vision by AR devices and the distraction of users can cause additional hazards. It is therefore important to differentiate use as training from actual use in the workplace.

Furthermore, the novelty effect hast o be taken into account [62]. This describes the initial enthusiasm and attention for new technologies, which fades over time [76]. It remains to be examined whether AR can also have a positive learning effect in the long term.

Research activities The number of publications on the topic of AR in education is constantly increasing [47, 54]. However, the state of research has not yet progressed very far in the area of occupational safety-related learning content for education. The use of AR in education currently takes place mainly in a natural science context [61]. However, the use of digital media such as AR in metal technology training should be increasingly promoted in order to develop vocational competences [77].

The Chairs of Occupational Safety and Didactics of Technology at the University of Wuppertal are examining this need for research in an interdisciplinary project. In this project, an AR-supported learning arrangement is being developed and evaluated for learning occupational safety-relevant content in vocational training for the metal sector. It involves the development of a workflow-based learning environment that can be used on a smartphone, tablet or Microsoft HoloLens. The relevant hazards, the characteristics of the target group and the didactic structuring of the learning content are taken into account. Local companies are involved in the development and evaluation. As a commonly used machine in the first learning fields of the metalworking apprenticeship, the column drill forms the prototype, followed by other machines. An initial target group analysis has shown that the male-dominated sample of apprentices can be characterised as rather technical-afine [80], suitable for the generation [78]. Despite there is no general prior experience with AR, it can be assumed that the apprentices will be able to handle an AR learning arrangement intuitively after a short briefing [79].

This project aims to answer the research question of whether AR is able to increase learning effectiveness. In this context, it is also important to examine whether AR itself or the didactic implementation is the key to learning success [65].

Overall, it can be concluded that AR technology is constantly evolving and offers a potential in the field of education despite the current technical and resource-related challenges. The requirement for this is to consider didactic principles and the expertise of teachers as well as the relevant characteristics of learners. In addition to the question of whether AR is able to increase learning effectiveness, the long-term effects of this technology in the field of education must also be investigated. Furthermore, hazards associated with the increased use of AR must be investigated. Finally, the question remains how cost-effective and widespread AR systems can be developed to enable widespread use in vocational schools and training companies, even if requirements and costs have already decreased [54, 60].

5.2 Reseach Topic: AI Assistant in Occupational Safety for SME (Florian Hafner)

Small and medium-sized enterprises (SMEs) as well as microenterprises often face significant challenges in the field of occupational safety and health (OSH). Frequently, they lack sufficient resources and expertise to fully implement legal regulations and internal safety standards. These deficits result in a higher frequency of accidents and hinder both the implementation of safety measures and the comprehensive conducting of risk assessments [80]. The underlying causes are primarily insufficient resources and a lack of specialized knowledge [81]. Studies by the Federal Institute for Occupational Safety and Health (Bundesanstalt für Arbeitsschutz und Arbeitsmedizin) indicate that many guidelines on risk assessment are too complex and impractical, and that many respondents possess little to no basic knowledge regarding OSH legal requirements [81]. Only about 50 percent of small enterprises conduct risk assessments [82].

Challenges SMEs face several specific challenges in the area of occupational safety. On the one hand, many companies lack the resources to conduct risk assessments regularly and to consistently comply with safety standards [81]. On the other hand, available guidelines often prove to be too complex, which complicates practical implementation [81]. The large number of legal requirements and sector-specific regulations further contribute to uncertainty, especially for microenterprises that are unable to employ in-house OSH experts.

Research Activities To address these challenges, the research project SafetyAssist is developing an AI-driven assistance solution. This solution is based on advanced large language models (LLMs) and aims to help SMEs implement OSH regulations more easily and effectively. The objective of the research is to create a digital platform that answers company-specific questions about OSH and provides up-to-date information on legal regulations.

As part of the development, a comprehensive of analysis requirements is first conducted to identify typical risk and exposure factors in SMEs. These are categorized into physical, chemical, ergonomic, and psychosocial risks and translated into a format understandable for the AI system. The next step involves integrating the collected knowledge into an existing LLM, which will be adapted to the specific needs of OSH and the SME target group.

A particular focus is placed on the automatic updating of the database to ensure that legal changes and new regulations are promptly reflected in the system's responses. The front-end of the platform will be designed to be user-friendly and accessible, facilitating easy use across different devices. Upon completion of the development phase, the system will undergo a comprehensive evaluation phase to ensure that it meets the needs of SMEs.

References

1. Kochhan Chr, Cichecki L (2024) Recruiting a New Generation of University Graduates – How Companies Need to Communicate with Gen Z, Springer Gabler
2. Faas T (2024) A question of age? Employment orientation and Gen Z Friedrich Ebert Foundation, Berlin
3. IPCC: Climate Change 2022: *Impacts, Adaptation and Vulnerability*. Contribution of Working Group II to the Sixth Assessment Report of the Intergovernmental Panel on Climate Change [H.-O. Pörtner, D.C. Roberts, M. Tignor, E.S. Poloczanska, K. Mintenbeck, A. Alegría, M. Craig, S. Langsdorf, S. Löschke, V. Möller, A. Okem, B. Rama (eds.)]. Cambridge University Press. Cambridge University Press, Cambridge, UK and New York, NY, USA, 3056 pp., (2022)
4. Kahl, A. (Hrsg.): Arbeitssicherheit – Fachliche Grundlagen. Erich Schmidt Verlag, Berlin. (2019)
5. Fathi, R. and Rummeny, D. and Fiedrich, F.: Organisation von Spontanhelfern am Beispiel des Starkregenereignisses vom 28.07.2014 in Münster. *Notfallvorsorge*. 2, 27–34. (2017)
6. Bier, M., Fathi, R., Stephan, C., Kahl, A., Fiedrich, F., & Fekete, A.: Spontaneous volunteers and the flood disaster 2021 in germany: development of social innovations in flood risk management. *Journal of Flood Risk Management*. (2023)
7. Fayard, G. M.: Fatal work injuries involving natural disasters, 1992-2006. *Disaster Medicine and Public Health Preparedness,* 3(4), 201–209. (2009)
8. Ramírez-Martínez, L., Chamah-Nicolás, M., Nieves-Plaza, M., Ruiz-Rodríguez, J., Ruiz-Medina, P., Ramos-Melendez, E. O.: Epidemiology of traumatic falls after Hurricane Maria in Puerto Rico. *Injury Epidemiology*, 7(1), 1–8. (2020)
9. Lee, Hoesung; Calvin, Katherine; Dasgupta, Dipak; Krinner, Gerhard; Mukherji, Aditi; Thorne, Peter W. et al. (2023): IPCC, 2023: Climate Change 2023: Synthesis Report. Contribution of Working Groups I, II and III to the Sixth Assessment Report of the Intergovernmental Panel on Climate Change [Core Writing Team, H. Lee and J. Romero (eds.)]. IPCC, Geneva, Switzerland.
10. Kiefer, Max; Rodríguez-Guzmán, Julietta; Watson, Joanna; van Wendel de Joode, Berna; Mergler, Donna; Da Silva, Agnes Soares (2016): Worker health and safety and climate change in the Americas: issues and research needs. In: Revista panamericana de salud publica = Pan American journal of public health 40 (3), S. 192–197.
11. Flouris, A., Azzi, M., Graczyk, H., Nafradi, B., and Scott, N., eds. 2024. Heat at Work: Implications for Safety and Health. A Global Review of the Science, Policy and Practice. ILO.
12. Chemicals and climate change in the world of work. Impacts for occupational safety and health: research report (2023). [1 ed.]. Geneva: ILO.
13. Faber U (2010) Kurzgutachten – Rechtliche Bedeutung von technischen Regeln für den Arbeitsschutz im Betrieb. Köln
14. Kohte W, Pangert R, Pernack E.-F, Tannhauer J (2024) Zur rechtshierarchischen Einordnung von Regeln und Empfehlungen für Arbeitsstätten. In: sicher ist sicher, Heft 1 2024, Erich-Schmidt Verlag GmbH & Co. KG, Berlin
15. Gesetzesbegründung des ArbSchG, BT-Drcks. 13/3540, S. 13, sowie bereits Bücker, Feldhoff, Kohte. Vom Arbeitsschutz zur Arbeitsumwelt (1990) Rn. 257; Gusy, „Antizipierte Sachverständigengutachten" im Verwaltungs- und Verwaltungsgerichtsverfahren, N+R 1987, S. 156 ff.
16. Leitlinienpapier zur Neuordnung des Vorschriften- und Regelwerks im Arbeitsschutz (2011)
17. Villwock O, Prauss B. (2023) Aufgaben des neuen Ausschusses für Sicherheit und Gesundheit bei der Arbeit und sein Verhältnis zu den speziellen Arbeitsschutzausschüssen, Bundesministerium für Arbeit und Soziales, in: beck-Online. Die Datenbank, ARP 2023, 51
18. Ausschuss für Sicherheit und Gesundheit bei der Arbeit (ASGA). (2023) Dachregeln und Integrierende Regeln – Definitionspapier. unpublished

19. (2018) EmpfBS 1114 “Anpassung an den Stand der Technik bei der Verwendung von Arbeitsmitteln”
20. (2018) TRGS 460 “Vorgehensweise zur Ermittlung des Standes der Technik”
21. Born M, Kahl A, Lützenkirchen F et al. (2018) Stand der Technik bei der Absackung – ein weiteres Praxisbeispiel zur Anwendung der TRGS 460. Gefahrstoffe – Reinhaltung der Luft:203–206
22. Wolf M, Born M, Lechtenberg-Auffarth E et al. (2014) Stand der Technik – Anwendung im Gefahrstoffrecht. Gefahrstoffe – Reinhaltung der Luft:347–353
23. Deutsche Gesellschaft für das Badewesen e. V. (2024) https://baederatlas.com. https://baederatlas.com/. Accessed 18 Oct 2024
24. Deutsche Gesetzliche Unfallversicherung DGUV Regel 107–001 „Betrieb von Bädern“, August 2018
25. Deutsche Gesetzliche Unfallversicherung (2018) DGUV Information 207–018 “Beurteilung von Gefährdungen und Belastungen am Arbeitsplatz in Bäderbetrieben”
26. Deutsche Gesetzliche Unfallversicherung (2020) DGUV Information 213–040 „Gefahrstoffe bei der Aufbereitung von Schwimm- und Badebeckenwasser“
27. (2000) Gesetz zur Verhütung und Bekämpfung von Infektionskrankheiten beim Menschen (Infektionsschutzgesetz – IfSG)
28. Jacobs JH, Spaan S, van Rooy GBGJ et al. (2007) Exposure to trichloramine and respiratory symptoms in indoor swimming pool workers. Eur Respir J 29:690–698. https://doi.org/10.1183/09031936.00024706
29. WORLD HEALTH ORGANIZATION (2006) Guidelines for safe recreational water environments. VOLUME 2 SWIMMING POOLS AND SIMILAR ENVIRONMENTS
30. Richardson SD, Ternes TA (2005) Water analysis: emerging contaminants and current issues. Anal Chem 77:3807–3838. https://doi.org/10.1021/ac058022x
31. Hesami Arani M, Jaafarzadeh N, Khaleghi Dehabadi P et al. (2020) Health and safety hazards identification and risk assessment in the swimming pools using combined HAZID and ALARP. Environ Health Eng Manag 7:151–160. https://doi.org/10.34172/EHEM.2020.18
32. (2024) Bäder. https://oeffentliche-private-dienste.verdi.de/mein-arbeitsplatz/baeder. Accessed 18 Oct 2024
33. Deutsche Gesetzliche Unfallversicherung e. V. (2020) Feuerwehrleute leben gefährlich
34. Gerhold P (2012) Gefährdungsbeurteilung im Feuerwehr- und Rettungsdienst. Ein umfassender Handlungsleitfaden für den Sicherheitsbeauftragten. Campus. Stumpf + Kossendey, Edewecht
35. (2024) Gefahrenabwehrbericht 2022: So viele Freiwillige bei der Feuerwehr wie nie | Land. NRW. https://www.land.nrw/pressemitteilung/gefahrenabwehrbericht-2022-so-viele-freiwillige-bei-der-feuerwehr-wie-nie. Accessed 21 Oct 2024
36. (2024) Statistik zur Gefahrenabwehr in Nordrhein-Westfalen. Jahresbericht 2023
37. BRANDSchutz (ed) (2019) Das Feuerwehr-Lehrbuch. Grundlagen – Technik -– Einsatz, 6., überarbeitete Auflage. Kohlhammer, Stuttgart
38. Arbeitsgemeinschaft der Leiter der Berufsfeuerwehren in NRW (2017) (Muster-) Gefährdungsbeurteilung für die Feuerwehr: Bereiche Einsatz, Ausbildung und Übung. https://www.vdf-nrw.de/uploads/tx_bitloftvdfnrwdownload/Mustergefaehrdungsbeurteilung.pdf. Accessed 30 Jun 2020
39. (1999) Feuerwehr-Dienstvorschrift 100 – Führung und Leitung im Einsatz. FwDV 100
40. Cimolino U (ed) (2020) Einsatzleiter-Handbuch Feuerwehr. Einsatzlagen und Führungshilfsmittel von A bis Z. Ecomed-Verl.-Ges, Landsberg
41. Heller U (2018) Unterweisung zu psychischen Belastungen im Feuerwehrdienst: Grundlagen und Arbeitshilfen. In: Arbeitsgemeinschaft der Feuerwehr-Unfallkassen (ed) Für die Zukunft gewappnet? Sicherheit und Gesundheit in der Feuerwehr. Dokumentation des FUK-Forum “Sicherheit” 2017, 1. Auflage. Hanseatische Feuerwehr-Unfallkasse Nord, Kiel, Schwerin, Hamburg, pp 86–91

42. Brenscheidt, S., Siefer, A., Hinnekamp, H., Hünefeld, L., Meyer, S.-C. & Backhaus, N. (2021). Arbeitswelt imWandel: Zahlen - Daten -Fakten. https://www.baua.de/DE/Angebote/Publikationen/Praxis/A105
43. Dehnbostel P (2012) Berufliche Kompetenzentwicklung im Kontext informellen und reflexiven Lernens–Stärkung der Persönlichkeits-und Bildungsentwicklung. In: Barre K, Hahn C, Dehnbostel P et al. (eds) Kompetenz. Fragen an eine (berufs-)pädagogische Kategorie. Helmut-Schmidt-Univ., Fak. für Geistes- und Sozialwiss, Hamburg
44. Kahl A (ed) (2019) Arbeitssicherheit. Fachliche Grundlagen. Erich Schmidt Verlag, Berlin
45. Ajslev J, Dastjerdi EL, Dyreborg J et al. (2017) Safety climate and accidents at work: Cross-sectional study among 15,000 workers of the general working population. Safety Science 91:320–325. https://doi.org/10.1016/j.ssci.2016.08.029
46. DGUV (2023) Arbeitsunfallgeschehen 2022
47. Salminen S (1996) Work-Related Accidents Among Young Workers in Finland. Int J Occup Saf Ergon 2:305–314. https://doi.org/10.1080/10803548.1996.11076359
48. Koch P, Nienhaus A (2022) Arbeitsschutzwissen und Gesundheitskompetenz. Zbl Arbeitsmed 72:165–174. https://doi.org/10.1007/s40664-022-00465-x
49. Nielsen ML (2012) Adapting 'The Normal' – Examining Relations between Youth, Risk and Accidents at Work. NJWLS 2:71. https://doi.org/10.19154/njwls.v2i2.2358
50. Hanvold TN, Kines P, Nykänen M et al. (2019) Occupational Safety and Health Among Young Workers in the Nordic Countries: A Systematic Literature Review. Saf Health Work 10:3–20. https://doi.org/10.1016/j.shaw.2018.12.003
51. DGUV (2022) Trendsuche der DGUV – Trendbericht für das Jahr 2021
52. Milgram P, Kishino F (1994) A Taxonomy of Mixed Reality Visual Displays. IEICE TRANSACTIONS on Information and Systems E77–D:1321–1329
53. Azuma, Ronald T. (1997) A Survey of Augmented Reality. Presence: Teleoperators and Virtual Environments 6:355–385
54. Müller S, Kruse S (2022) Systematisches Review: Augmented Reality in der technischen Bildung. 42-61 Seiten / Journal of Technical Education (JOTED), Bd. 10 Nr. 2 (2022): Journal of Technical Education (JOTED). JOTED 10:42–61. https://doi.org/10.48513/joted.v10i2.249
55. Adelmann R (2020) Augmented Reality in der industriellen Praxis. In: Orsolits H, Lackner M (eds) Virtual Reality und Augmented Reality in der Digitalen Produktion. Springer Gabler. in Springer Fachmedien Wiesbaden GmbH, Wiesbaden, pp 7–32
56. Prange M (2021) Virtual Reality und Augmented Reality in der Bildung – Ein Überblick zum Thema. Digitale Bildung für Lehramtsstudierende:235–246. https://doi.org/10.1007/978-3-658-32344-8_32
57. Janaćković, Goran Lj, Dejan Vasović et al. (2015) Augmented reality in safety education and training. International Journal of Engineering Education 31:713–725
58. Sırakaya M, Alsancak Sırakaya D (2022) Augmented reality in STEM education: a systematic review. Interactive Learning Environments 30:1556–1569. https://doi.org/10.1080/10494820.2020.1722713
59. Wang M, Callaghan V, Bernhardt J et al. (2018) Augmented reality in education and training: pedagogical approaches and illustrative case studies. J Ambient Intell Human Comput 9:1391–1402. https://doi.org/10.1007/s12652-017-0547-8
60. Akçayır M, Akçayır G (2017) Advantages and challenges associated with augmented reality for education: A systematic review of the literature. Educational Research Review 20:1–11. https://doi.org/10.1016/j.edurev.2016.11.002
61. Garzón J, Pavón J, Baldiris S (2019) Systematic review and meta-analysis of augmented reality in educational settings. Virtual Reality 23:447–459. https://doi.org/10.1007/s10055-019-00379-9
62. Söbke H, Zander S, Londong J (2018) Augmented Reality als Lernmedium: Potenziale und Implikationen. awIC 3. https://doi.org/10.14464/awic.v3i0.282

63. Cheng K-H, Tsai C-C (2013) Affordances of Augmented Reality in Science Learning: Suggestions for Future Research. J Sci Educ Technol 22:449–462. https://doi.org/10.1007/s10956-012-9405-9
64. Lauer L, Peschel M, Marquardt M et al. (2019) Augmented Reality (AR) in der Primarstufe–Entwicklung einer AR-gestützten Lehr-Lerneinheit zum Thema Elektrik. Naturwissenschaftliche Kompetenzen in der Gesellschaft von morgen. Jahresband der GDCP.
65. Lauer L, Peschel M (2022) Praxisideen für Augmented Reality (AR) im naturwissenschaftlich-orientierten Sachunterricht. In: Brandt B, Bröll L, Dausend H (eds) Digitales Lernen in der Grundschule III. Fachdidaktiken in der Diskussion, 1st ed. Waxmann, Münster
66. Lauer L, Peschel M (2022) Praxisideen für Augmented Reality (AR) im naturwissenschaftlich-orientiertenSachunterricht. In: Brandt B, Bröll L, Dausend H (eds) Digitales Lernen in der Grundschule III. Fachdidaktiken inder Diskussion, 1st ed. Waxmann, Münster
67. Heimes L, König D, Komorowski T et al. (2024) Digitale Kompetenz und Partizipation – Memes, Schönheitsfilter und Fake News in der Berufsschule. berufsbildung – Zeitschrift für Theorie-Praxis-Dialog:45–47
68. Sheikh S, Bin Heyat MB, AlShorman O et al. (2021) A Review of Usability Evaluation Techniques for Augmented Reality Systems in Education. In: 2021 Innovation and New Trends in Engineering, Science and Technology Education Conference (IETSEC). IEEE
69. Attig C, Wessel D, Franke T (2017) Assessing Personality Differences in Human-Technology Interaction: An Overview of Key Self-report Scales to Predict Successful Interaction. In: Stephanidis C (ed) HCI International 2017 – Posters' Extended Abstracts. 19th International Conference, HCI International 2017, Vancouver, BC, Canada, July 9–14, 2017, Proceedings, Part I, vol 713. Springer-Verlag, s.l., pp 19–29
70. Franke T, Attig C, Wessel D (2019) A Personal Resource for Technology Interaction: Development and Validation of the Affinity for Technology Interaction (ATI) Scale. International Journal of Human–Computer Interaction 35:456–467. https://doi.org/10.1080/10447318.2018.1456150
71. Davis FD, Bagozzi RP, Warshaw PR (1989) User Acceptance of Computer Technology: A Comparison of Two Theoretical Models. Management Science 35:982–1003. https://doi.org/10.1287/mnsc.35.8.982
72. Tangocci E, Hartmann C, Bannert M (2023) Immersives Lernen in der Berufsschule. MedienPädagogik 51:268–288. https://doi.org/10.21240/mpaed/51/2023.01.21.X
73. Deutsche Gesetzliche Unfallversicherung Trendsuche der DGUV – Trendbericht für das Jahr 2021
74. Kaufeld M, Mundt M, Forst S et al. (2022) Optical see-through augmented reality can induce severe motion sickness. Displays 74:102283. https://doi.org/10.1016/j.displa.2022.102283
75. Platho C, Tristram S, Kupschick S (2023) Einsatzmöglichkeiten von VR-Brillen in der experimentellen Verkehrssicherheits-und Mobilitätsforschung. Berichte der Bundesanstalt für Straßenwesen : M, Mensch und Sicherheit, Heft 337. Fachverlag NW in Carl Ed. Schünemann KG, Bremen
76. Miguel-Alonso I, Checa D, Guillen-Sanz H et al. (2024) Evaluation of the novelty effect in immersive Virtual Reality learning experiences. Virtual Reality 28. https://doi.org/10.1007/s10055-023-00926-5
77. Eder A (2015) Akzeptanz von Bildungstechnologien in der gewerblich-technischen Berufsbildung vor dem Hintergrund von Industrie 4.0. Journal of technical education 3. https://doi.org/10.25656/01:11573
78. Kleinjohann M, Reinecke V (2020) Zielgruppe Generation Z: Charakteristika. In: Marketingkommunikation mit der Generation Z. Springer Gabler, Wiesbaden, pp 13–23
79. Waldorf, Julia; Kahl, Anke (2023): Zielgruppengerechter Arbeitsschutz für Auszubildende? Eine AugmentedReality (AR)-Lernumgebung für die metalltechnische Berufsausbildung. Tagungsdokumentation: SmarteTechnologie und AR in der Arbeitswelt. Hg. v. Bundeanstalt für Arbeitsschutz und Arbeitsdmedizin (BAuA).Dortmund. Online verfügbar unter https://d-nb.info/1340489392/34#page=33.

80. Kohte W, Kaufmann S (2021) Lotsen und Netzwerke in der betrieblichen Gesundheitspolitik. Düsseldorf: Hans-Böckler-Stiftung
81. Sczeseny C, Keindorf S, Droß P, Jasper G (2014) Kenntnisstand von Unternehmen und Beschäftigten auf dem Gebiet des Arbeits- und Gesundheitsschutzes in KMU. Bundesanstalt für Arbeitsschutz und Arbeitsmedizin
82. Lück P et al. (2018) Grundauswertung der BIBB/BAuA-Erwerbstätigenbefragung 2018

Meeting Future Challenges in the World of Work: Ergonomic Strategies, Methods and Tools to Attract Skilled Talent and Create Sustainable, Productive and Industry-Leading Environments

Christoph Mühlemeyer, Ann Kathrin Wissemann, Ralf Pieper, Patrick Serafin and André Klußmann

Contents

Abstract

In an era of rapid change, organizations must adopt advanced ergonomic strategies to attract skilled talent and foster sustainable, industry-leading environments. The growing importance of ergonomics highlights the necessity of

C. Mühlemeyer (✉) · A. K. Wissemann · P. Serafin · A. Klußmann
Institut Für Arbeitsmedizin, Sicherheitstechnik Und Ergonomie E. V. (ASER), Wuppertal, Germany
e-mail: muehlemeyer@institut-aser.de

R. Pieper
Bergische Universität Wuppertal, Wuppertal, Germany

S. Bracke and R. Pieper (eds.), *Safety Engineering*,
https://doi.org/10.1007/978-3-658-47473-7_18

adhering to established meta-methodological design approaches, while continuing to prioritize fundamental research in areas such as anthropometry and human performance. Networks like KomNet are redefining service and knowledge management by efficiently addressing initial inquiries, guiding users to develop tailored solutions for the specific challenges of their organizations. In sectors like the event industry, targeted research ensures the availability of skilled talent, addressing unique occupational demands. BDS-Instrument exemplifies a holistic approach, seamlessly blending ergonomics with operational goals. Meanwhile, modern ICT tools raise critical questions around ethics and privacy, emphasizing the need for responsible technology deployment. Interactive learning and training, grounded in validated ergonomic methods, enhance research and education, providing added value for students and supporting their professional development. These concrete examples demonstrate how to meet the future challenges of work, ensuring productivity, innovation, and human well-being in a dynamic global landscape. The article is dedicated to Karl-Heinz Lang.

Keywords

Holistic Approach · Shopfloor Transition · Digital Ergonomics · BDS-Instrument · KIM · KPI · ICT · F4D · KomNet · Cobots · Physical Strength

1 In Memory of Karl-Heinz Lang

It is with great sadness that we share the news of Karl-Heinz Lang's passing after a long illness. His loss leaves a profound void in our lives and in our team at ASER Institute. Karl dedicated his life to building ASER, shaping it into the invaluable resource it is today. Through his tireless commitment and leadership, he made it possible for us to help countless individuals with their work-related concerns and challenges.

His remarkable expertise, built over decades, was an asset to all of us, and his legacy will continue to inspire our work. As a colleague, director, and board member, Karl was a friend to all of us, and we feel deeply privileged and grateful for the time we shared with him. His passion and spirit will be sorely missed, but his memory will remain with us as we carry forward his vision. The entire team at ASER Institute mourns his passing, and our heartfelt condolences go out to his family.

2 The Ever-Growing Importance of Ergonomics

2.1 Future Challenges

The shortage of skilled talent is driving a value creation loss of €118 billion annually (approx. 3.2% of Germany's gross domestic product) and poses a greater challenge than rising energy costs, according to German chamber of commerce and industry forecasts. Additionally, the burden of 201,723 recognized cases of occupational disease, 844,284 reportable workplace accidents, and 888.9 million days of work disability is pushing labor costs ever higher. The following Table 1 provides an overview of reported occupational accidents, occupational diseases, and their associated economic losses in Germany during 2022 [1].

Future competitiveness hinges on the ability to accurately predict, prevent, and proactively interrupt the progression of chronic conditions.

2.2 Significant Opportunities

Independent studies on Return on Investment (ROI) reveal impressive cost–benefit ratios for health-related interventions, ranging from 1:2 to as high as 1:5.91 for healthcare costs and 1:2.5 to 1:4.85 for absenteeism [2–7]. Each prevented sick day saves €2830 per employee in production downtime and preserves €4967 per employee in gross value added [8]. These preventive efforts underscore the powerful financial impact of health initiatives, reinforcing both productivity and profitability across the organization.

3 Meta-Methodological Design Concept

Established knowledge in occupational and safety sciences forms the foundation for employee safety and health protection. Building on this, anthropometry and biomechanics enable improved human–machine interaction through optimized movement, while work physiology focuses on reducing fatigue and enhancing

Table 1 Occupational Safety and Health (OSH) Statistics for Germany, 2022

Occupational Safety and Health (OSH) Statistics	Value
Reportable occupational accidents	844,284
Fatal occupational accidents	533
Recognized occupational diseases	201,723
Fatal occupational diseases	2164
Days of work lost	888.9 million
Economic production loss	€118 billion

performance. The interdisciplinary integration of these scientific insights allows for comprehensive optimization of working conditions, supporting sustainable organizational growth [9].

3.1 Beyond Legal Anchoring and Compliance

Employers are required to implement measures that ensure not only the safety of their employees but also their overall well-being and productivity. These core principles are anchored in EU and national regulations, aiming for continuous improvement of working conditions [10–12]. This involves a comprehensive assessment of work environments based on established ergonomic principles [13]. By prioritizing the humane design of work, employers not only fulfill their legal obligations but also create a workplace that promotes health, well-being, and sustainable performance. As management expert Peter Drucker is once said to have remarked, "The best way to predict the future is to create it". In the context of workplace design, this means proactively shaping environments that support and enhance human capabilities. Implementing these measures can lead to significant improvements like enhanced employee satisfaction and engagement, reduced absenteeism and turnover, increased productivity and work quality and lower incidence of work-related health issues (see Sect. 2.2).

3.2 Overarching Design Fields

The overarching primary design fields [13–16] within the following assessment criteria encompass establishing an optimized structure for work organization and occupational health and safety, fostering ongoing employee qualification, and developing age-sensitive, inclusive workplace designs. Additional focal points include proactive measures to eliminate and prevent both direct and indirect discrimination and robust protections for informational self-determination, ensuring stringent data privacy standards (see Table 2).

Harmlessness and tolerability. Strategies to eliminate or minimize workplace hazards. This includes proactive hazard control at the source and designing a workplace that prioritizes safety at every level.

Feasibility. Creating workplaces that prioritize occupational safety and health; developing ergonomically optimized workspaces that enhance usability and adapt to human dimensions; and establishing standards for workplace setup and operation, including provisions for visual display unit workplaces and remote work arrangements.

Reasonableness and freedom from impairment. Core design fields include enriching and expanding tasks to foster engagement, crafting safety- and

Table 2 Early milestones for assessment criteria of human-centered work design

Level	Kirchner, 1972; Rohmert & Rutenfranz, 1975	Hacker, 1978	Luczack & Volpert, 1987
1	Feasibility	Feasibility	Harmlessness and tolerability
2	Tolerability	Harmlessness	Feasibility
3	Reasonableness	Freedom from impairment	Reasonableness, freedom from impairment, autonomy,
4	Satisfaction	Personal development	Satisfaction and personal development
5			Social compatibility and participation in design

health-oriented operational structures—particularly in work organization and process flows—and implementing protective regulations for individuals with disabilities, ensuring compliance for example with German workplace ordinance (ArbStättV) [17] and Social Code (SGB) Book Nine (IX) [18] and to support accessibility and inclusion.

Satisfaction and personal development. Key action areas include implementing a comprehensive corporate prevention policy that emphasizes fostering positive social relationships, cultivating a strong safety culture among employees, and enacting preventive strategies to safeguard against both direct and indirect discrimination or disadvantage.

Social compatibility. Key design fields include promoting employee participation and empowering the works council with meaningful co-determination in decision-making processes. This approach ensures that employee voices are integrated into organizational development, fostering a culture of collaboration and shared responsibility.

4 Research on Anthropometry and Human Performance

Research on anthropometry and human performance is crucial for effective workplace design. It helps ensure ergonomic layouts, safe limit compliance in diverse work environments and enables standardization in ergonomic product and workplace design. The following study gathered data on isometric hand force and hand preference in a representative sample of the German population (N = 1207; 428 females, 779 males). Variables included age, gender, hand preference, and occupation. Maximum isometric forces were measured during one-handed pulling (supported by the other hand), gripping, and turning a screwdriver (supination) for both hands [19].

4.1 Fundamental Insights into Human Strength

Previous studies show that males reach peak strength at 20–25 years, decreasing to about 60% by age 70, with females at approximately 67% of male strength. Hand preference generally results in a stronger preferred hand, with side differences up to 27%, impacting performance and safety in tasks that may require the use of the non-dominant hand [20–24].

4.2 Methods

A standardized method inventory was developed to analyze hand force, including a background questionnaire, force measuring stand, and 3D hand scanner. Maximum isometric forces were measured for pulling, gripping, and screwdriver turning (supination) in both hands. Descriptive statistics, with results displayed as boxplots, report median values with the 25th and 75th percentiles for dispersion. For standardized measurement of maximal isometric force, participants exerted maximum voluntary effort (MVE) in a height-adjustable test stand across three force exertion types: pulling, gripping, and torque (supination). Body postures were standardized based on prior studies to ensure consistency, with specific setups for each exertion type, including a modified jamar-dynamometer for grip strength. Force was quickly increased to maximum within 1 s and held for 3 s, with data recorded as the largest 2-s moving average, discarding measurements with deviations over 20% [19].

4.3 Results

The results for pulling strength, separated by gender with females ($n=395$) on the left and males ($n=671$) on the right, are presented in the following boxplot (Fig. 1).

Females exhibited roughly 67% of male maximum force across pulling, grip, and torque measurements. Maximum force remained stable across the 20–59 age groups, with individual variability reflected by a coefficient of variation between 22 and 29% across age, gender, and exertion type. For analyzing hand preference ($N=749$; 257 females, 492 males), the preferred hand was generally stronger, with side differences ranging from $<1\%$ to 11% depending on task, hand preference, and gender. However, 33–50% of participants were stronger with their non-dominant hand, indicating high variability within handedness types [19].

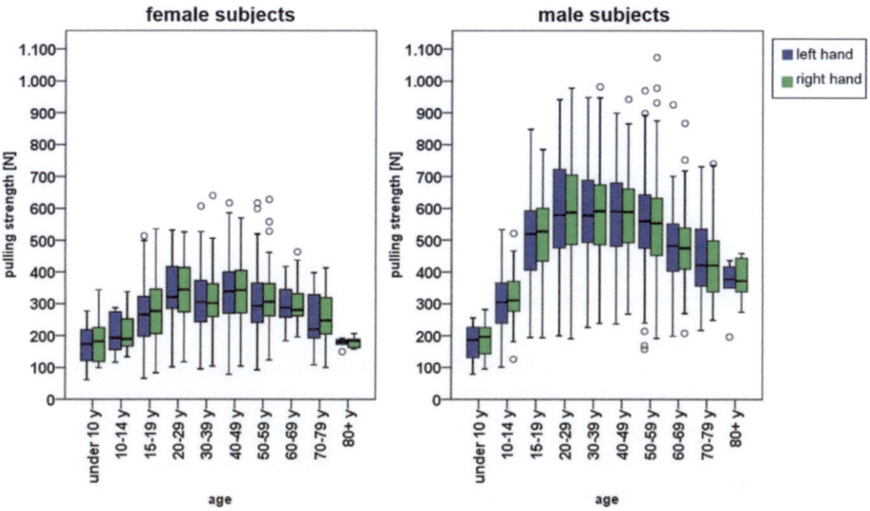

Fig. 1 Pulling strength of participants; females (n=395, left chart) and males (n=671, right chart)

4.4 Discussion, Conclusion, Limitation

Data on hand forces and factors like age, gender, and hand preference are critical for ergonomic design. The study included 1207 participants (428 females, 779 males) aged 5 to 91, with age nearly normally distributed except for underrepresentation in the 30–39 age group. Thanks to the relatively large sample size, many variables were well quantified. Furthermore, previous study results were also confirmed. For example, the study confirms that women have approximately 67% of men's maximum isometric force, with little age-related strength decline observed among 20- to 59-year-olds, contrary to previous literature. Gender strongly influences maximum isometric force, highlighting its importance for ergonomic design, though age and hand preference effects appear minimal within the studied age range. However, the study's cross-sectional design, lack of repeated measures, and limited assessment of hand preference represent methodological limitations, affecting the generalizability of results to broader populations [19].

5 Service and Knowledge Management Redefined with Kompetenznetz Nordrhein-Westfalen (KomNet)

In today's dynamic landscape, effective service and knowledge management is no longer just a strategic advantage; it's a necessity. Both private enterprises and public administrations face increasing pressure to deliver tailored solutions promptly. As the volume of available knowledge grows exponentially, the challenge lies not in its abundance, but in its efficient accessibility and application. Customers, whether individuals or businesses, expect seamless, personalized service experiences. This demand is evident in initiatives like the EU's service directive, which led to the introduction of a unified "citizen's telephone number 115" in Germany. To meet these expectations, organizations must harness the power of knowledge management. By optimizing knowledge sharing, retrieval, and application, they can enhance decision-making, improve problem-solving, and ultimately deliver superior service [25].

5.1 Development History

Germany's North Rhine-Westphalia (NRW) state, in collaboration with ASER, established KomNet in the 1990s. The pilot phase in 1997–1998 revealed that complex consulting services, while valuable, weren't widely utilized through traditional call center models. A dedicated "expert hotline" was deemed too costly and less efficient than a helpdesk approach. To address these challenges, KomNet was structured as an extended user helpdesk This innovative platform serves as a blueprint for citizen-centric knowledge management, seamlessly integrating problem-solving expertise with a user-friendly service approach. Since 1997, KomNet has been a trusted source of core competencies from diverse experts in the fields of work science and economics. Since 2001, KomNet has consistently tracked customer satisfaction [25]

5.2 Workflow

KomNet functions as a virtual "knowledge and service community", connecting the specialized knowledge and practical experience of partner organizations' experts across various topics. Even complex real-world questions find high-quality solutions through KomNet's platform. Taking it a step further, the system anonymizes these dialogues, making them publicly available as a free, barrier-free knowledge base for the broader public [26]. Users can access this valuable resource quickly, conveniently, and free of charge (visit www.komnet.nrw.de; see Fig. 2).

Since 1999, KomNet users have been able to rate the usefulness of individual dialogues. If a statement is given, that a dialogue is "wrong or inapplicable " at the user´s option, the dialogue is checked by the dialogue editorial staff and the user is informed about the result of this verification. However, content quality alone isn't enough; usability and accessibility are paramount for user adoption.

By making a subset of the helpdesk's knowledge base publicly accessible, KomNet significantly enhances its service value. This empowers internet-savvy users to find solutions independently, 24/7, and boosts the overall efficiency of the system [27, 28]. KomNet was structured as an extended user helpdesk, incorporating the key elements listed in Table 3.

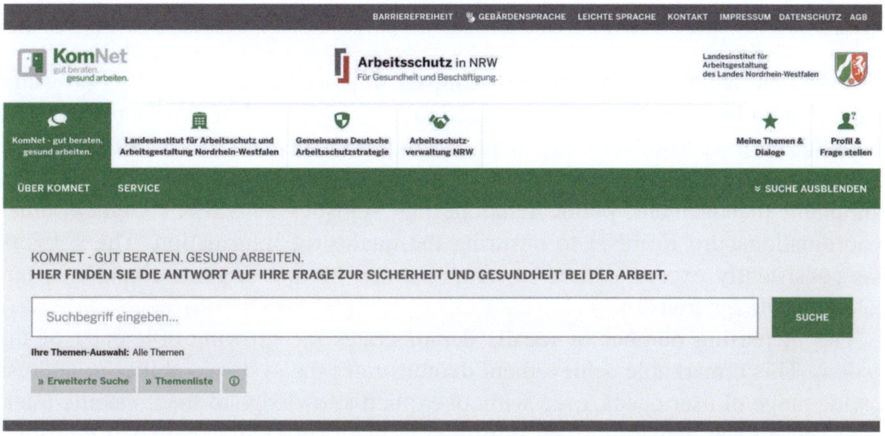

Fig. 2 Service and knowledge management redefined with KomNet (www.komnet.nrw.de)

Table 3 An insight into KomNet's key elements

Role/Team/Network	Responsibility
Expert Pool	A network of experts providing high-quality knowledge
Competence Center	A dedicated team that curates and delivers tailored knowledge solutions
Dialogue Editorial Staff	A team that transforms individual discussions into public knowledge

5.3 Results

In September 2024, KomNet's knowledge base was accessed 322,563 times monthly, excluding automated queries, and contained 5285 dialogues. Additionally, the system fields 162 new questions, means that, for every new direct query, the knowledge base was accessed 1991.1 times, highlighting the significant value and relevance of its content.

Since 2001, KomNet has consistently tracked customer satisfaction. To date, 142.975 ratings regarding the knowledge base have been collected, with 82.2% of users finding the retrieved information "very helpful" or "helpful." Following each query, users are also invited to assess the service based on four key criteria regarding to their own individual questions. With 21,301 ratings collected, also a clear picture of user experience regarding the internal processes has emerged. An impressive 96.9% of users rate the clarity and relevance of the system's responses as "good" or "excellent."

5.4 Conclusion

The valuable feedback loop enables KomNet to continually improve its offerings and maintain high levels of user satisfaction. The high level of satisfaction reflects the dedication of KomNet's expert pool, competence center at Landesinstitut für Arbeitsschutz und Arbeitsgestaltung NRW (LIA), dialogue editorial staff and complaint management, public relations and statistics at ASER's Gemeinsames Koordinationsbüro KomNet to ensuring the quality of information. The network has consistently exceeded user expectations and became a game-changer in targeted knowledge transfer.

The increasing number of recalls demonstrates the growing efficiency of the system. This remarkable achievement demonstrates the system's ability to address a wide range of user needs, even without explicit knowledge of their specific interests. The proven demand-driven information platform facilitates expert consultations beyond its initial scope. Its success extends beyond NRW, demonstrating its potential to revolutionize knowledge sharing across various disciplines.

6 Research for Securing Skilled Talent in Event Industry

The event industry, Germany's sixth-largest employer with over 1.1 million employees, faces an acute skilled talent shortage following nearly three years of pandemic-induced restrictions and short-time work. Over three-quarters of companies are now scrambling to fill positions, particularly in technical, production, project management, and sales [29].

The pandemic has led to a brain drain and decreased interest in the industry due to economic uncertainty and the added strain of physical and mental health risks associated with demanding work schedules and unpredictable workloads. The R.I.F.E.L. 2022 study [29] revealed a staggering number of event companies in Germany desperately seeking to fill vacancies. The data underscores the industry's urgent need for skilled talent across various roles (see Fig. 3).

6.1 Methods

The objectives of the research for securing skilled talent in event industry within the Fachkräftesicherung 4D (F4D) project are to secure skilled talent by (re) attracting returning professionals and promoting young talent.

It aims to design inclusive, human-centered jobs that are accessible to individuals of all ages and genders. Additionally, the project seeks to enhance the competitiveness and resilience of the industry, enabling it to adapt and thrive in a dynamic market.

Insights will be consolidated through the development of a digital platform with interactive tools, while scientific research (ASER Institute) and industrial research (GEWITEB) will focus on industry-specific methods and the integration of emerging technologies (see Fig. 4).

6.2 Kick-Off

At the hybrid kick-off meeting on August 29, 2024, in Wuppertal-Vohwinkel, the project team, comprised of work scientists and industry representatives, engaged in a dynamic discussion on challenges and solution strategies for securing skilled labor in the event industry (Fig. 5).

6.3 First Results

Milestones achieved include the analysis of studies on the specific requirements of the event industry and the development of a new, semi-standardized interview guide. We successfully initiated a call for practical participants for interviews through industry associations and professional contacts. Additionally, at BOE International we were presented at an industry trade fair where we held a presentation slot on fwd: stage to further engage with key stakeholders, facilitating valuable discussions.

The new F4D-platform prototype has also been launched, with a dedicated internal section being developed to provide exclusive access for project participants. Using the semi-standardized interview guide, standardized interviews have been conducted and will continue to be conducted via videoconferences, incorporating artificial intelligence (AI) tools to enhance the process. An in-depth data analysis is also upcoming, promising valuable insights for the project.

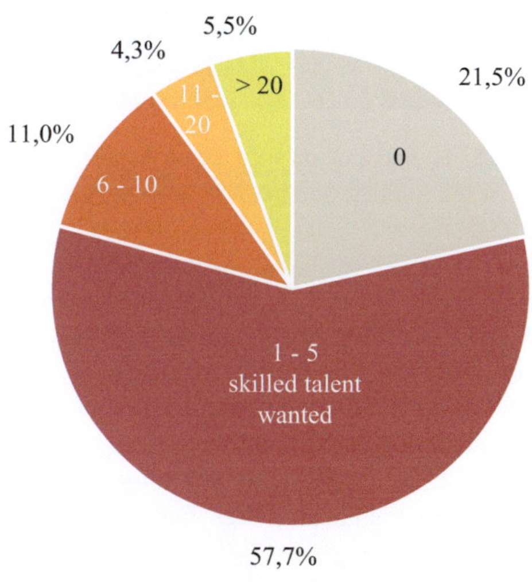

Fig. 3 Demand for skilled talent in Germany's event industry

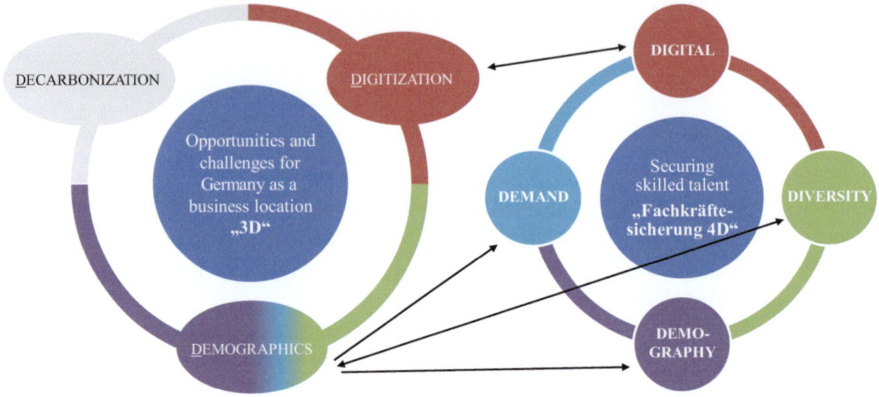

Fig. 4 The "three major D's" (3D; strategy of federal government) and securing skilled talent 4D

Fig. 5 Hybrid kick-off meeting on August 29, 2024, in Wuppertal-Vohwinkel

Our future workforce depends heavily on our ability to attract and retain skilled talent. Therefore, a preliminary forward-looking platform concept for workforce development has been developed, as shown in Fig. 6. The programming requirements outlined in the specifications have been assessed, with a particular focus on complex logical relationships in processes planned for future automation. Specifically, intelligent learning algorithms, predictive models, decision support, and visual analytics have been examined for integration into the designed F4D-platform.

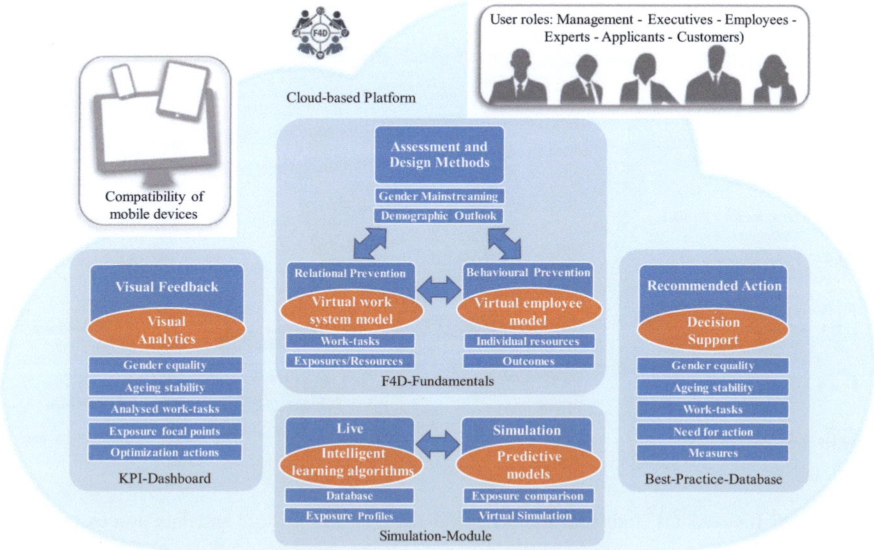

Fig. 6 Concept of an industry-specific platform for securing skilled talent

7 Optimization with a Holistic Approach

The digital, modular instrument of the Belastungs-Dokumentations-System (BDS-Instrument), honored with multiple awards, including the EU-OSHA Award [30], serves as a robust solution for ensuring compliance across diverse corporate domains. Integrating an extensive array of methods, procedures, analyses, and metrics drawn from over 350 international publications, it enables organizations not only to meet legal and technical requirements but to exceed them where necessary [9].

By employing the BDS-Instrument, companies can efficiently navigate complex regulatory frameworks while maintaining rigorous technical standards, thereby enhancing legal security. Key regulations addressed include Directive 89/391/EEC, various international laws, as well as national statutes such as the Occupational Health and Safety Act (ArbSchG), Workplace Ordinance (ArbStättV), Ordinance on Industrial Safety and Health (BetrSichV), Technical Rules for Workplaces (ASR), Ordinance on Occupational Health Care (ArbMedVV), and the UVV 1 Principles of Prevention [9] (Fig. 7).

Specialized analyses, including e. g. those required by the ArbMedVV [31] and detailed assessment for cases such as occupational disease BK 2116 (hip osteoarthritis), support targeted, expert assessments that elevate organizational health management to new levels.

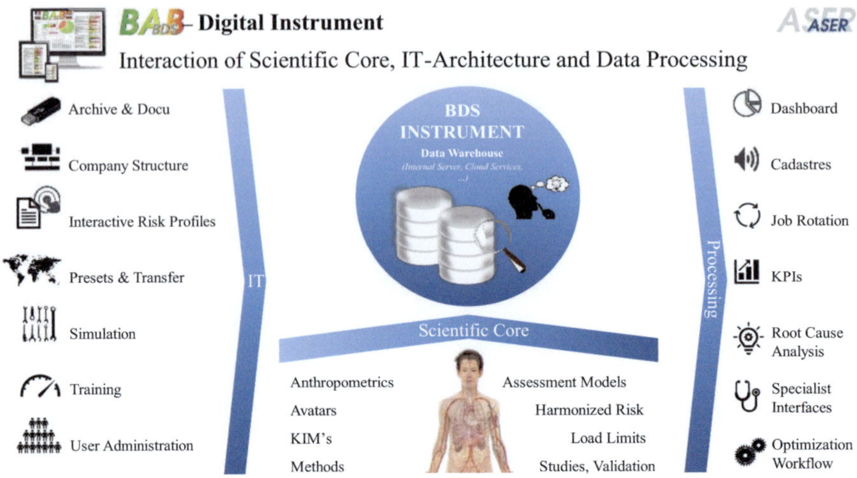

Fig. 7 BDS-Instruments' interaction of scientific core, IT-architecture and data processing

7.1 Evaluation of BDS-Instruments' Scientific Quality Criteria

The BDS-Instrument has undergone continuous evaluation across multiple studies over the past decades. However, ongoing evaluations remain highly valuable, as each review provides new insights. With increasingly large datasets and advanced analytical methods, the overall understanding of the instrument can be further refined. This iterative process is particularly beneficial when high-quality studies contribute to a more comprehensive and nuanced assessment.

Methods. The evaluation structure for evaluating validity, reliability and objectivity was derived from scientific testing literature. Extensive validity studies included assessment algorithms (n = 2053), differentiation of exposure-pairs (n = 3897) and prediction accuracy of work-related diseases (BDS as leading indicator) based on a cross-sectional industrial epidemiological study using regression analysis (n = 368). The inter-rater reliability was tested based on 600 individual decisions statistical evaluated by quantified agreement rate with unweighted Fleiss' kappa (κ) and intra-class correlation coefficients (ICC 3.1). The intra-rater reliability was tested by repeated use of the instrument after 1.5 years [9]. The criterion validity assessment is based on a successfully completed cross-sectional occupational epidemiological study with an internal control group [32], following a pre-developed study protocol and data protection concept [33, 34, 35]. 183 workplaces were analyzed, evaluated, and documented, and 807 employees underwent physical examinations and interviews. Additionally, their health status (complaints, diseases) was documented based on clinical examinations and standardized physician-conducted interviews as per the cross-sectional study protocol. For 368 pseudonymized participants, results of the workplaces' exposure situations were available through workplace condition assessments using BDS-Instrument in the form of ergonomic full-shift analyses for 32 different exposure types.

Results. Statistically significantly increased relative health risks (OR with 95%-CI) of employees working on dangerous workplaces identified by instrument's algorithms according to the harmonized risk concept were found. The parallel test showed comparable aggregation results for the avatars on the impact of the spinal column and neck structure as well as on the upper and lower limbs. The test of intra-rater reliability showed the importance of an exact analysis of subtime exposure aspects [9]. The following Table 4 shows the odds ratio (OR) with a 95% confidence interval (CI) and significance level for low back pain (LBP), also referred to as lumbalgia or lumbago.

Discussion. In the cause-effect relationship between load handling and the collective diagnosis of lumbalgia/lumbago, stronger effects are assumed in the working population, as the employees who were present and thus tended to be healthy were examined. The company epidemiological cross-sectional study is limited by the type of study. Therefore, follow-up research should also focus on the secondary quality criteria and on outcomes at medium doses workplaces (BDS-Instruments' level 4).

7.2 Assessment and Simulation of Collaborative Workplaces

It is crucial to have scientifically and practical methods in place for assessing safety and health risks while improving workflows. These methods should enable cross-industry analysis of both traditional and emerging workplaces within a robust, flexible framework [36, 37]. Therefore, the BDS was applied to a cobot workplace in the packaging area of cosmetics industry, as briefly demonstrated below.

Introduction. A cobot's inherently safe design can meet essential machinery directive 2006/42/EC standards, enabling safe, ergonomic integration within the European Economic Area. In Germany, cobot-workplaces must comply with the occupational safety and health regulations, covering both classic hazards (e.g., mechanical, electrical) and health risks from physical or mental strain, ensuring comprehensive employee safety and wellbeing in collaborative settings [38, 39].

Table 4 Odds ratio (w 95%-CI) and significance level for low back pain (lumbalgia/lumbago)

Lumbalgia or lumbago	OR	95%-CI	p-value
Manual handling of Loads			
BDS-Instruments' Safety zone	–	–	–
BDS-Instruments' Threshold zone	1.81	0.91–3.54	0.086
BDS-Instruments' Hazard zone	2.15**	1.33–3.50	0.002

Methods. In the analyzed cobot-workplace, a female employee handles a sub-task in the packaging process for personal care items by manually transferring tubes from input boxes to target packaging without any tools. This takes place indoors, within an eight-hour shift in an industrial hall, where the employee and cobot operate asynchronously in the same area; the cobot lifts lids and places them on a conveyor belt without directly interacting with the employee's task [40]. When applied, the BDS-Instrument enables simulation and comparison between the cobot-enhanced and purely manual packaging scenarios, allowing for a detailed ergonomic evaluation of physical and mental health impacts [41, 42, 43].

Results. Both setups exhibit multiple intolerable risks, highlighting the need for redesign to meet musculoskeletal disorders (MSD) risk standards (e.g., AMR 13.2) and maternity protection requirements [36, 37] (Table 5).

The eight-hour shift exposure profile identified significant ergonomic risks in both scenarios. Without the cobot, physical overload included 7 h of static standing and frequent trunk torsion (>10°), while mental overload stemmed from a 7s cycle time and monotonous tasks. Introducing the cobot increased physical strain, with key indicator method for manual handling operations 2012 (KIM-MHO) score rising from 50.3 to 53.3 due to higher hand movement frequency. Mental strain worsened due to a reduced cycle time (5s, −29%) and fewer interaction opportunities.

Conclusion. The findings reveal that collaborative environments do not inherently guarantee healthy working conditions, even when designed with this intent. To navigate the shopfloor transition effectively, companies must adopt an interdisciplinary strategy for analysis, assessment, and design. Given shorter product lifecycles and rising workplace complexity, integrating diverse expertise, utilizing KPIs, addressing compliance risks, and driving continuous improvement are essential for a safe, competitive, and audit-ready shopfloor [36, 37].

7.3 Professional Optimization in Automotive Industry

The BDS-Instrument enables the precise identification of low-exposure roles, making it ideal for light-duty, rehabilitation, and job rotation placements. It

Table 5 BDS-Instruments' Results, Key Performance Indicators (KPIs) and Reports (extract)

BDS-Instruments' Results, Key Performance Indicators (KPIs) and Reports	w/o cobot	Cobot-interaction
KIM-MHO (2012 risk concept) result (risk score)	50.3	53.6
Physical Overload factor/employee (BDS report)	6	6
Cycle time (duration)	7s	5s
Repetition of work tasks (BAB level)	7	7
Connection to the technical process (BAB level)	4(5)	6(7)
Age stability (BDS KPI)	No	no
Mental Overload factor/employee (BDS report)	6	7
MSD risk (AMR 13.2 criteria)	Yes	Yes
Ordinance on Occupational Health Care—Sect. 5	Optional	Optional
Irresponsible Risk (§9 MuSchG criteria)	Yes	Yes
Resulting Overall Production Ergonomics Rank	1 of 2	2 of 2
Generally recommended for shopfloor	No	No

includes comprehensive registers for noise, lighting, and climate conditions, while also addressing psychological stressors through interactive analytics to pinpoint high-exposure areas. For Industrial Engineering (IE), BDS offers detailed potential analyses, facilitating the identification of improvement areas in industrial production processes, including ergonomic optimizations [9].

Seamlessly integrated into internal quality and efficiency initiatives, the BDS-Instrument enhances the management of occupational health and work-related illnesses. It provides robust ergonomic documentation, ensuring compliance and transparency in response to customer inquiries. Built on over 50 years of applied research and continuous development, BDS-Instrument delivers cross-industry expertise, empowering companies to modernize workflows, address workforce challenges, and drive sustainable growth [9].

At Continental AG, the BDS-Instrument plays a pivotal role in advancing digital tools and innovative work practices, streamlining operations, and enhancing efficiency. It supports the creation of inclusive work environments, promoting equal opportunities across genders and age groups, while optimizing workflows to boost productivity. By aiding in talent retention and reducing turnover, sick days, and healthcare costs, BDS significantly strengthens Continental AG's competitiveness, delivering sustainable value and positioning the company as a leader in workplace innovation and employee well-being. Notably, the group Human Relations and ergonomics experts at the headquarters in Hanover are playing a key role in maintaining the network, promoting best practices, and providing comprehensive reporting to management. In collaboration with the significant commitment of the key professionals in the business areas of Automotive, Tires, and ContiTech, the success in driving this highly important topic forward is made possible [55–57].

8 Ethics and Privacy in Advanced Information and Communications Technologies (ICT Tools)

8.1 Introduction

In an age defined by rapidly advancing technology, the benefits of ICT innovations for societal development come with significant ethical and privacy challenges. This section examines these issues within the scope of research on ICT tools. As digitalization progresses, vast amounts of personal data are being collected, stored, and easily accessed, heightening both expectations for positive data-driven outcomes and risks of potential misuse of sensitive information. New ICT tools, for instance, aim to enhance the health and productivity of aging workforces by collecting personal and sometimes sensitive health data. While intended for beneficial purposes, such data collection raises concerns around tracking, profiling, and the unforeseen use of personal information. Thus, ethical considerations and robust data protection measures are essential in both ICT research and organizational applications. This section outlines key challenges and core requirements for addressing ethics, privacy, and data protection, including compliance with privacy standards and the role of Ethics Advisory Boards [44].

8.2 Example Horizon 2020

Horizon 2020 projects are subject to stringent ethical guidelines, ensuring research integrity and compliance with international, EU, and national laws. The Annotated Model Grant Agreement (AGA) outlines fundamental ethical principles, including gender equality, conflict of interest management, and confidentiality. Additionally, the Charter of Fundamental Rights of the European Union (CFR) and the European Convention on Human Rights (ECHR) provide a robust legal framework. Emanuel et al. identified seven core principles essential for the ethical conduct and evaluation of research, initially applied to clinical trials but adaptable to various research contexts, including the development of ICT solutions. These principles align closely with the General Data Protection Regulation (GDPR) principles for personal data [45].

In the realm of modern ICT research, these principles present unique challenges. For example, ensuring informed consent for data collection and usage in digital environments requires careful consideration. Data privacy and security are paramount, necessitating robust measures to protect sensitive information. Ethical algorithms and AI development are crucial to prevent bias and discrimination. Transparency and accountability in research processes are essential to maintain public trust. Finally, the responsible dissemination of research findings is vital to avoid potential misuse and ensure societal benefit. By adhering to these ethical principles, researchers can contribute to the development of innovative ICT solutions that respect human rights, privacy, and dignity [46–48].

8.3 Managing Ethical Challenges

Ethics by Design. To ensure the ethical conduct of ICT research, a proactive Ethics by Design approach is crucial. This involves integrating ethical considerations into every stage of the research process, from planning to implementation and beyond.

A key component of this approach is the establishment of an Ethics Advisory Board (EAB). This board, composed of ethical experts and project members, plays a vital role in identifying potential ethical and privacy issues, advising on compliance with relevant regulations, and reviewing research protocols. Before any research activity involving human participants or personal data, the study design must be submitted to the EAB for review and recommendations [49, 50].

By prioritizing ethical considerations and adhering to relevant regulations, researchers can ensure the responsible development and deployment of ICT solutions that respect human rights, privacy, and dignity. The following Table 6 shows an ethics checklist to identify the need for ethics approval. If one of the answers to question 1 to 5 is "Yes" and neither 6 nor 7 is "Yes", a formal verification of the requirement of an ethics approval is advised under consideration of aspects 8 and 9.

Additionally, ethics approvals from independent ethics committees may be necessary, particularly for research involving sensitive personal data or potentially risky procedures. To determine the need for such approvals, researchers can utilize ethics checklists to identify potential ethical concerns.

It is essential to consider the regulatory landscape, especially regarding medical device regulations such as the EU Medical Device Regulation (MDR). If an ICT solution falls within the scope of the MDR, additional ethical and regulatory requirements, including clinical evaluations and ethics committee approvals, may apply.

Risk-based approach. The GDPR mandates a risk-based approach to data protection, requiring organizations to assess the severity and likelihood of risks to individuals' rights and freedoms. This risk assessment is particularly crucial for high-risk data processing activities, such as those involving sensitive personal data or innovative technologies. To conduct a comprehensive risk assessment, organizations should consider factors like the type of data processed, the purpose of processing, the potential impact on individuals, and the technical and organizational measures in place to protect the data. A Data Protection Impact Assessment (DPIA) is a valuable tool for systematically evaluating these risks [51] (Table 7).

The DPIA process involves several stages, including identifying the data processing activities, assessing the potential risks, implementing appropriate safeguards, monitoring the effectiveness of these measures, and documenting the entire process. It's important to note that the DPIA is not a one-time exercise. Organizations should regularly review and update their DPIA to reflect changes in technology, regulations, and their own data processing activities. By proactively

Table 6 Ethics Checklist to identify the need for ethics approval

No	Question	Y	N
1	Does your research involve human participants?		
2	If yes, are they potentially vulnerable individuals or groups (e.g., children, students, employees)?		
3	Does your research involve the collection of confidential or sensitive information?		
4	Does your research involve controversial subject matter (e.g., sensitive, embarrassing or intrusive topics)?		
5	May your research cause any harm to participants, researchers or others (e.g., invasive techniques, physical interventions)?		
*If all answers above are "No", formal ethics approval is **not** required*			
6	Has the research activity already approval from an ethics committee?		
7	Is the research activity part of a larger study that has already received approval from an ethics committee?		
*If answer to question 6 **or** 7 is "Yes", further ethics approval is **not** required*			
8	Have health and safety risks due to the research been considered?		
9	Are data protection compliance measures taken for any personal data collected by the research?		

Table 7 Risk assessment checklist to identify the need for DPIA

No	Question	Y	N
1	Does the activity involve systematic monitoring of publicly accessible spaces?		
2	Are the persons unaware of collecting and processing of their data?		
3	Does the activity use automated data collection from open sources?		
4	Does the activity involve usage of new technologies?		
5	Is there a combination of a person's data from more than one source?		
6	Is the personal data used for a purpose other than its original intention?		
7	Would the processing of personal data be perceived as disruptive in this way?		
8	Might data processing concern specific categories of personal data, e.g., health?		
9	Does the data processing involve decisions or assumptions about individuals?		
10	Would the data subject consider the processed information to be private?		

addressing data protection risks, organizations can ensure compliance with the GDPR and safeguard the privacy of individuals.

Processing sensitive data. When conducting research involving human participants, it is imperative to prioritize ethical principles such as informed consent, privacy, and data security. Researchers must obtain voluntary informed consent from all participants, ensuring they fully understand the study's objectives, procedures, potential risks, and benefits [52, 53].

For vulnerable populations, such as employees in hierarchical organizations, special care must be taken to avoid coercion and ensure true voluntariness. Clear measures should be implemented to protect participant confidentiality and minimize the potential for adverse consequences, including job-related repercussions (see Fig. 8).

To safeguard privacy, researchers should employ appropriate data protection techniques, such as anonymization or pseudonymization. Anonymization removes any personal identifiers from the data, making it impossible to link the data to specific individuals. Pseudonymization involves replacing personal identifiers with codes or pseudonyms, while keeping a separate secure record linking the pseudonyms to the original identifiers. By adhering to these ethical principles and employing rigorous data protection measures, researchers can conduct responsible and trustworthy human subject research [44, 52–54].

Fig. 8 General iterative process of data protection impact assessment (DPIA)

9 Interactive Learning and Training of Validated Methods

At the Faculty of Mechanical Engineering and Safety Engineering, together with the associated institutes, we continuously advance teaching and research, for example through the development and fostering of interactive learning of methodological competence. The academic discipline of "safety and quality regulation" has submitted, and will continue to submit, research proposals within the framework of "Freiraum" by the "Stiftung Innovation in der Hochschullehre," where ideas for teaching can be developed and tested. Through a series of lectures and laboratory sessions in both our Bachelor's and Master's programs, we instruct students on how to systematically analyze real-world operational challenges using scientific methods, with the goal of alleviating pressures on social security systems through effective prevention.

With our "Interactive Methods" concept, we aim to develop and test an innovative teaching model that makes workplace exposures and the simulation of health impacts engagingly interactive. By integrating these methodologies with automation analysis, sensor technology, motion capture, and virtual reality, we will significantly enhance students' ability to absorb, process, store, and apply the course content.

Looking to the future, the demand for professionals equipped with advanced analytical and technological skills is on the rise. Students in Safety Engineering at the Bergische University of Wuppertal will particularly benefit from this approach, gaining hands-on experience with cutting-edge tools and techniques that are essential in today's workplace. This immersive learning experience will not only prepare them for the complexities of modern safety challenges but also empower them to drive innovation in occupational health and safety practices in their future careers.

Acknowledgements The research project "Fachkräftesicherung 4D: Digital Innovation for Gender- and Age-Sensitive Work Design in the Event Industry" is funded within the framework of the NEXT.IN.NRW innovation competition by the Ministry for Economic Affairs, Industry, Climate Action, and Energy of the State of North Rhine-Westphalia and by the European Union through the European Regional Development Fund (EFRE/JTF Program NRW 2021–2027). This acknowledgement pertains to the section of this overview article that introduces the "Fachkräftesicherung 4D (F4D)" research project.

The presented force measurement study was part of the 'Design4All' project, funded by the Ministry of Economy, Energy, Industry, Small and Medium Businesses and Trades of North Rhine-Westphalia (Germany) in conjunction with the 'NRW Ziel 2 Program' of the European Regional Development Fund (ERDF). This acknowledgement pertains to the section of this overview article that introduces the results of strength measurements in the German population.

Ageing@Work is a transnational European Union funded HORIZON 2020 research project that has focused on developing a series of highly adaptive, personalized ICT tools for the working and living environments of ageing workforces, to allow them to remain healthy, productive and competitive for longer.

References

1. BMAS/BAuA (2023). Sicherheit und Gesundheit bei der Arbeit – Berichtsjahr 2022. Unfallverhütungsbericht Arbeit. 1. Auflage. Dortmund: Bundesanstalt für Arbeitsschutz und Arbeitsmedizin, 2023. ISBN: 978-3-88261-761-0, Seiten: 214, Papier, PDF-Datei, https://doi.org/10.21934/baua:bericht20230817. Download von www.baua.de/suga.
2. Aldana, S. G. (2001). Financial impact of health promotion programs: A comprehensive review of the literature. American Journal of Health Promotion, 15(5), 296–320.
3. Chapman, L. S. (2000). Meta-evaluation of worksite health promotion economic return studies: 2000 update. American Journal of Health Promotion, 15(1), 1–10.
4. Chapman, L. S. (2005). Meta-evaluation of worksite health promotion economic return studies: 2005 update. American Journal of Health Promotion, 19(6), 1–11.
5. Golaszewski, T. (2001). Shining lights: Studies that have most influenced the understanding of health promotion's financial impact. American Journal of Health Promotion, 15(5), 332–341.
6. Pelletier, K. R. (2001). A review and analysis of the clinical and cost-effectiveness studies of comprehensive health promotion and disease management programs at the worksite: 1998–2000 update. American Journal of Health Promotion, 16(2), 107–116.
7. Pelletier, K. R. (2005). A review and analysis of the clinical and cost-effectiveness studies of comprehensive health promotion and disease management programs at the worksite: Update VI 2000–2004. Journal of Occupational and Environmental Medicine, 47(10), 1051–1058.
8. Bundesanstalt für Arbeitsschutz und Arbeitsmedizin (BAuA). (2022). Volkswirtschaftliche Kosten durch Arbeitsunfähigkeit 2021. In Sicherheit und Gesundheit bei der Arbeit – Berichtsjahr 2021 (S. 1–2). Dortmund: BAuA. https://www.baua.de/DE/Themen/Monitoring-Evaluation/Zahlen-Daten-Fakten/pdf/Kosten-2022.pdf?__blob=publicationFile&v=4.
9. Mühlemeyer, C. (2024). Weiterentwicklung und Evaluierung eines arbeitswissenschaftlichen Instruments zur ganzheitlichen Analyse, Beurteilung und Gestaltung arbeitsbedingter Belastungen und Gefährdungen von Arbeitssystemen. Forschungsbericht – Nr. 43, Institut ASER e. V., Wuppertal, 2024, ISBN 978-3-936841-39-8.
10. Pieper, R. (2022). Arbeitsschutzrecht – Kommentar für die Praxis (7. Aufl.). Bund Verlag.
11. Pieper, R. (voraussichtlich 2024/25). Physische Arbeitsbelastungen. In J. vom Stein, T. Rothe & B. Schlegel (Hrsg.), Gesundheitsmanagement und Krankheit im Arbeitsverhältnis – Handbuch. C.H. Beck.
12. Pieper, R. (2024). Arbeitsstättenverordnung – Basiskommentar (5. Aufl.). Bund Verlag.
13. Luczak, H. & Volpert, W. (1987). Arbeitswissenschaft: Kerndefinition – Gegenstandskatalog – Forschungsgebiete. Eschborn: RKW-Verlag
14. Hacker, W. (1978). Allgemeine Arbeits- und Ingenieurpsychologie. Psychische Struktur und Regulation von Arbeitstätigkeiten (2., überarb. Aufl.). Bern: Huber. (Schriften zur Arbeitspsychologie, 20).
15. Kirchner, J. H. (1972). Arbeitswissenschaftlicher Beitrag zur Automatisierung. Analyse und Synthese von Arbeitssystemen. Berlin West, Köln, Frankfurt/M.: Beuth-Vertrieb. (Schriftenreihe "Arbeitswissenschaft und Praxis", Bd. 23).
16. Rohmert, W., & Rutenfranz, J. (1975). Arbeitswissenschaftliche Beurteilung der Belastung und Beanspruchung an unterschiedlichen industriellen Arbeitsplätzen. Bonn: Bundesminister für Arbeit und Sozialordnung, Referat Öffentlichkeitsarbeit.
17. German Workplace Ordinance (Arbeitsstättenverordnung – ArbStättV) of 12 August 2004 (Federal Law Gazette I p. 2179), last amended by Article 5 Paragraph 1 of the Ordinance of 18 October 2017 (Federal Law Gazette I p. 3584).
18. German Social Code (Sozialgesetzbuch – SGB) Book Nine (IX) – Rehabilitation and Participation of Persons with Disabilities – as published on 23 December 2016 (Federal Law Gazette I p. 3234), last amended by Article 13 of the Act of 24 June 2022 (Federal Law Gazette I p. 959).

19. Serafin, P., Mühlemeyer, C., Levchuk, I., Lang, K.-H., Gebhardt, H. J., & Klussmann, A. (2015). Physical strength of a German population sample: Differences in age, gender and hand preference. Occupational Ergonomics, 12(1–2), 49–59. https://doi.org/10.3233/OER-150223.
20. Schmauder M, Solf JJ (1992). Einfluss der Händigkeit bei der Handhabung von Arbeitsmitteln [Effect of handedness when handling work equipment]. Schriftenreihe der Bundesanstalt für Arbeitsschutz-Forschung- Fb661, Bundesanstalt für Arbeitsschutz (Ed.). Bremerhaven: Wirtschaftsverlag NW.
21. Schmauder M, Eckert R, Schindhelm R (1992). Forces in the hand-arm system: Investigations of the problem of lefthandedness. Int J Ind Ergonom 12: pp. 231–237. https://doi.org/10.1016/0169-8141(93)90029-D.
22. Hettinger T (1983). Isometrisches Muskeltraining [IsometricMuscle Training]. 5th Edition. Stuttgart, New York: Thieme.
23. Strasser H, Wang B (2007). Effects of hand preference and direction of rotation on screwdriver torque strength and physiological costs of muscles involved in arm pronation and supination, in: Strasser H. (Ed.): Assessment of the Ergonomic Quality of Hand-held Tools and Computer Input Devices. Amsterdam: IOS Press, pp. 67–74.
24. Bohannon RW (2003). Grip strength: A summary of studies comparing dominant and nondominant limb measurements. Percept Motor Skill 96: pp. 728–730. https://doi.org/10.2466/PMS.96.3.728-730.
25. Lang, K.-H., Sassmannshausen, A., Mühlemeyer, C., Klussmann, A., Gebhardt, H. J., Nover, H., & Deilmann, M. (2009). KomNet and REACH-Net – Examples of successful consulting services concerning occupational safety and health. In International Ergonomics Association (Ed.), Proceedings of the 17th World Congress on Ergonomics, IEA 2009 [CD-ROM]. Zurich: International Ergonomics Association.
26. Deilmann, M., Lang, K.-H., Müller, B. H., Saßmannshausen, A., & Tielsch, R. (2002). KomNet – Instrument zur Verbesserung der Arbeitsqualität [KomNet – Instrument for the improvement of working quality]. Sicherheitsingenieur, 33(8), 38–42.
27. Deilmann, M. (2008). Integrierte Services zum Themenbereich Beschäftigungsfähigkeit – Neues Arbeiten in NRW: Erfahrungen und Ausblick [Integrated services in the area of Employment ability – new working in North Rhine-Westphalia: experiences and forecast]. In R. Pieper & K.-H. Lang (Eds.), Sicherheitswissenschaftliches Kolloquium 2006–2007 (Forschungsbericht Nr. 18). Wuppertal: Institut ASER e. V.
28. Lang, K.-H., Schäfer, A., Gebhardt, H. J., & Müller, B. H. (2004). Instrumente für den Arbeitsschutz für die menschengerechte Gestaltung von Erwerbs- und Eigenarbeit [Occupational safety and health tools for the ergonomic design of employed and voluntary or private work]. Zeitschrift für Arbeitswissenschaft, 58(1), 61–65.
29. Research Institute for Exhibition and Live-Communication (R.I.F.E.L.). (2022). HR- & Gehaltsreport 2022. R.I.F.E.L. Institut. https://www.rifel-institut.de/forschung/.
30. Continental AG. (2017, 12. Juli). Continental erhält Best Practice Award der Europäischen Agentur für Sicherheit und Gesundheit am Arbeitsplatz (EU-OSHA). Pressemitteilung. Abgerufen am 7. November 2024 von https://www.continental.com/de/presse/pressemitteilungen/continental-erhaelt-award/.
31. Waldminghaus, A. K. (2016). Entwicklung eines Konzepts zur betrieblichen Umsetzung der Verordnung zur arbeitsmedizinischen Vorsorge (ArbMedVV) [Bachelorthesis]. Bergische Universität Wuppertal, Fakultät für Maschinenbau und Sicherheitstechnik, Abteilung Sicherheitstechnik, Fachgebiet Sicherheits- und Qualitätsrecht.
32. Bundesanstalt für Arbeitsschutz und Arbeitsmedizin (BAuA). (2019). MEGAPHYS – Mehrstufige Gefährdungsanalyse physischer Belastungen am Arbeitsplatz. Band 1 (1. Auflage, Projektnummer: F 2333). https://www.baua.de/DE/Angebote/Publikationen/Berichte/F2333.pdf?__blob=publicationFile&v=11.
33. Klussmann, A., Steinberg, U., Brandstädt, F., Schust, M., Hartmann, B., Gebhardt, H., Serafin, P., Ditchen, D., & Liebers, F. (2016). Study protocol of the validation of new key

indicator methods. In Preventing work-related musculoskeletal disorders in a global economy. PREMUS 2016. Ninth International Scientific Conference on the Prevention of Work-Related Musculoskeletal Disorders. Book of abstracts (p. 139). Institute for Work & Health Canada.
34. Klussmann, A., Liebers, F., Gebhardt, H., Rieger, M. A., Latza, U., & Steinberg, U. (2017). Risk assessment of manual handling operations at work with the key indicator method (KIM-MHO) – determination of criterion validity regarding the prevalence of musculoskeletal symptoms and clinical conditions within a cross-sectional study. BMC Musculoskeletal Disorders, 18(1), 184. https://doi.org/10.1186/s12891-017-1542-0.
35. Klußmann, A., Liebers, F., Schust, M., Brandstädt, F., Hartmann, B., Serafin, P., Schäfer, A., & Gebhardt, H. (2020). Hintergründe und Konzept zur Entwicklung und Validierung der neuen Leitmerkmalmethoden. In Gesellschaft für Arbeitswissenschaft (Ed.), Bericht zum 66. Kongress der Gesellschaft für Arbeitswissenschaft „Digitaler Wandel, Digitale Arbeit, Digitaler Mensch?" (Beitrag C.5.1). GfA-Press. ISBN 978-3-936804-27-0
36. Mühlemeyer, C. (2020). Assessment and Design of Employees-Cobot-Interaction. In T. Ahram (Ed.), Advances in Intelligent Systems and Computing: Vol. 1018. Human Interaction and Emerging Technologies (pp. 773–778). Springer International Publishing. https://doi.org/10.1007/978-3-030-25629-6_120.
37. Mühlemeyer, C. (2022). Beurteilung und Gestaltung von Cobot-Arbeitssystemen. In R. Pieper & K.-H. Lang (Eds.), Sicherheitswissenschaftliches Kolloquium 2018–2021+ und weitere Foren im zweiten COVID-19-Pandemiejahr (Band 15) (pp. 162–170). Institut für Arbeitsmedizin, Sicherheitstechnik und Ergonomie e. V., Institut ASER e. V. ISBN 978-3-936841-38-1
38. German Institute for Standardization. (2011). Ergonomics principles in the design of work systems (ISO 6385:2016). German version EN ISO 6385:2016. DIN.
39. Joint Ministerial Gazette. (2015). Hazards at the Interface of Humans – Work Equipment – Ergonomic and Human Factors, Work Systems. TRBS 1151. GMBl, No. 17/18, p. 340.
40. Bauer, W., Bender, M., Braun, M., Rally, P., & Scholtz, O. (2016). Leichtbauroboter in der manuellen Montage – einfach einfach anfangen. Fraunhofer IAO.
41. Müller, B.H., & Hettinger, Th. (1981). Interpretations- und Bewertungsverfahren arbeitswissenschaftlich-ergonomischer Felddaten. Zeitschrift für Arbeitswissenschaft, 35, 82–86.
42. Peters, H. (1986). Verfahren zur Beurteilung arbeitsbedingter Belastungen BAB. In Erfassung und Bewertung arbeitswissenschaftlicher Daten. Schriften des IfaA (Hrsg.), Nr. 62, S. 64–78. Köln.
43. Klußmann, A., Mühlemeyer, C., Lang, K.-H., Dolfen, P., Wendt, K.-D., Gebhardt, H., Neumann, B., & Schäfer, A. (2013). Praxisbewährte Methoden zur Bewertung und Gestaltung physischer Arbeitsbelastungen. In Leistung und Lohn – Zeitschrift für Arbeitswirtschaft, BDA – Bundesvereinigung der Deutschen Arbeitgeberverbände (Hrsg.), Nr. 541–545. Heider-Verlag.
44. Serafin, P., Wissemann, A. K., Gebhardt, H. J., Mühlemeyer, C., Schäfer, A., & Lang, K.-H. (2022). Considering Ethics and Privacy Issues in Modern ICT-Tools. In V. G. Duffy et al. (Eds.), Digital Human Modeling and Applications in Health, Safety, Ergonomics and Risk Management. Healthcare Applications (LNCS 13521, pp. 534–545). Springer Nature Switzerland AG. https://doi.org/10.1007/978-3-031-17902-0_38.
45. Emanuel, E. J., Wendler, D., & Grady, C. (2000). What Makes Clinical Research Ethical? JAMA, 283(20), 2701–2711. https://doi.org/10.1001/jama.283.20.2701.
46. European Commission. (2018). H2020 Programme AGA – Annotated Model Grant Agreement (Version 5.1). https://ec.europa.eu/research/participants/data/ref/h2020/grants_manual/amga/h2020-amga_en.pdf.
47. Council of Europe. (1950). European Convention for the Protection of Human Rights and Fundamental Freedoms, as amended by Protocols Nos. 11 and 14.
48. European Parliament, & Council of the European Union. (2016). Regulation (EU) 2016/679 of the European Parliament and of the Council of 27 April 2016 on the protection of natural

persons with regard to the processing of personal data and on the free movement of such data, and repealing Directive 95/46/EC (General Data Protection Regulation). Official Journal of the European Union, L 119/1.
49. Council for International Organizations of Medical Sciences (CIOMS). (2016). International Ethical Guidelines for Health-related Research Involving Humans. CIOMS. ISBN 978-929036088-9.
50. European Parliament, & Council of the European Union. (2017). Regulation (EU) 2017/745 of the European Parliament and of the Council of 5 April 2017 on medical devices, amending Directive 2001/83/EC, Regulation (EC) No 178/2002 and Regulation (EC) No 1223/2009 and repealing Council Directives 90/385/EEC and 93/42/EEC. Official Journal of the European Union, L 117/1.
51. Regulation (EU) 2016/679 of the European Parliament and of the Council of 27 April 2016 on the protection of natural persons with regard to the processing of personal data and on the free movement of such data, and repealing Directive 95/46/EC (General Data Protection Regulation), OJ 2016 L 119/1.
52. Article 29 Data Protection Working Party. (2017). Guidelines on Data Protection Impact Assessment (DPIA) and determining whether processing is "likely to result in a high risk" for the purposes of Regulation 2016/679. 17/EN WP 248 rev.01. European Commission. https://ec.europa.eu/newsroom/document.cfm?doc_id=44137.
53. European Commission. (2018). Ethics in Social Science and Humanities. http://ec.europa.eu/research/participants/data/ref/h2020/other/hi/h2020_ethics-soc-science-humanities_en.pdf.
54. Council for International Organizations of Medical Sciences (CIOMS). (2016). International Ethical Guidelines for Health-related Research Involving Humans. CIOMS. ISBN 978–929036088–9. https://cioms.ch/wp-content/uploads/2017/01/WEB-CIOMS-EthicalGuidelines.pdf.
55. Personalwirtschaft. (2022). Sonderheft BGM 2022. Personalwirtschaft Digital. https://digital.personalwirtschaft.de/de/profiles/a84dbeddfb04-personalwirtschaft-digital/editions/sonderheft-bgm-2022/preview_pages (Preview).
56. Sifa-Sibe. (2017). Mit Ergonomie zu mehr Qualität. Sicherheitsingenieur. https://www.sifa-sibe.de/gesundheitsschutz/ergonomie/mit-ergonomie-zu-mehr-qualitaet/.
57. Hannen, P. (2017). Ergonomie-Netzwerkkonferenz 2017: Continental will 100 Prozent Ergonomie. Sicherheitsingenieur DIGITAL, 08/2017.

Perspectives of Product Quality and Safety Engineering in a Changing Society

Manuel Löwer and Nadine Schlüter

Contents

Abstract

In times of social change, from Industry 4.0 to an Industry 5.0 and Society 5.0, the topics of product quality and safety are also strongly influenced. While the implementation of networked production and the Internet of Things through to the Internet of Production continues to progress, scientists from all disciplines are researching issues and solutions that a globally networked society poses. The following article is intended to provide an overview of where we currently stand from Industry 4.0 to Industry 5.0 and Society 5.0 and how this affects the various aspects of product quality and safety. Three different perspectives are taken into consideration. From an abstract meta-level point of view on

M. Löwer · N. Schlüter (✉)
Department Product Safety and Quality Engineering, University Wuppertal, Wuppertal, Deutschland
e-mail: schlueter@uni-wuppertal.de

S. Bracke and R. Pieper (eds.), *Safety Engineering*,
https://doi.org/10.1007/978-3-658-47473-7_19

technologies and industries, followed by a more concrete example of changes in companies. Eventually, the question of how a person's individually perceived safety with regard to a product can be captured.

Keywords

Product safety · Quality · Safety engineering · Society 5.0

1 From Industry 4.0 to Industry 5.0 and Society 5.0—A Meta Perspective

As part of Industry 4.0, innovative technologies such as the Internet of Things (IoT), Artificial Intelligence (AI) and automation have been introduced to increase the productivity and efficiency of production systems. Industry 4.0 is therefore a significant contribution from research to fundamentally develop industry into a new evolutionary stage. In addition to the technological discoveries, ways of thinking and working are also changing. Companies are making their processes more dynamic in order to adapt more quickly to changing requirements and continue to be able to offer competitive products with the corresponding safety and quality on the global market [17]. Above all, the possibilities of digitalization are the driver of change here: data integration, analysis of large amounts of data, data-based decision-making and transparency along the entire value chain are another aspect of Industry 4.0 [1]. In addition to these developments and innovations, standards must be established or adapted in parallel to enable and efficiently control the interaction of the individual system elements of Industry 4.0 (e.g. systems, machines, etc.). This in turn opens up new business opportunities [2].

Companies that successfully implement corresponding Industry 4.0 technologies and concepts can increase their efficiency, reduce costs and better serve customer requirements thanks to digitalization and networking [1].

Today's technical systems are therefore characterized by ever closer interaction between mechanics, electronics and software. This change affects not only products but also production systems in particular (as shown in Fig. 1). Advances in information and communication technologies (ICT), data analysis and Artificial Intelligence (AI), for example, are enabling the development of so-called autonomously adaptable Industry 4.0 systems that expand the capabilities of today's production systems. A key capability of Industry 4.0 systems is therefore adaptability. It contributes to the resilience of manufacturing processes and systems by compensating for machine, plant and workforce failures [15].

At present, autonomously adaptable Industry 4.0 systems are still a conceptual vision that is considered a forward-looking development and represents the transition to Society 5.0. The aim is to create production systems that are increasingly complex and rich in variants, while at the same time being characterized by short innovation cycles. The developments and innovations required for this can be divided into four areas, as shown in Fig. 2.

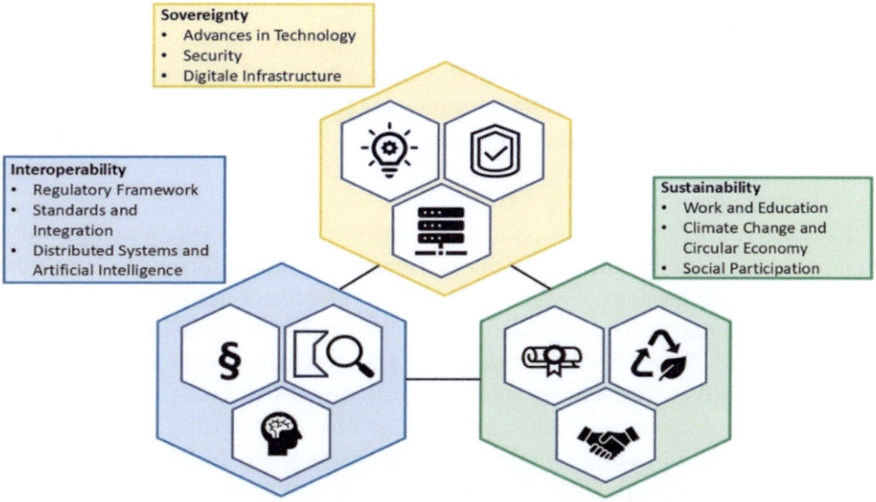

Fig. 1 Mission statement 2030 for Industry 4.0 according to [3]

Fig. 2 Actual Industry 4.0 developments and innovations according to [4]

Based on the current Industry 4.0 reference architecture—RAMI 4.0—cyber-physical systems and their implementation are currently being realized, which are characterized by process efficiency, digital connectivity and the use of AI. In the course of this, innovative business models are being and have already been implemented that can handle the effects of dynamic markets and thus minimize costs while maximizing shareholder profit. This also means that the costs for the environment, resources and society are currently not being taken into account to a sufficient extent. It is precisely the inadequate consideration of sustainability and the resulting consequences that is leading to a further development of technologies and standards towards a Society 5.0 [5].

In Society 5.0, competitiveness and sustainability will be made possible within a common industrial framework. This will make alternative forms of sustainability and resilience feasible. Workers will make even greater use of digital devices, leading to a human-centric approach to technology. The accountability of companies in terms of their overall value creation will continue to increase to the point that each industry sector will be able to see the extent to which progress towards the well-being of people, resilience and sustainability is solid [5].

This will result in changes for our current industry, which are set out in Fig. 3.

These social and technological changes will also open up completely new possibilities in terms of product quality and security. For example, edge computing technology enables data processing at the edge of the network, reduces latency and improves data protection. In terms of product quality and safety, data protection guidelines can thus be implemented more efficiently and continuous measurement of quality and safety features can be autonomously recorded directly at the point of failure. Digital twins not only enable the simulation of systems in real time, but also the prediction and corresponding proactive reduction of defective products in the value chain. Artificial Intelligence already supports the analysis of complex

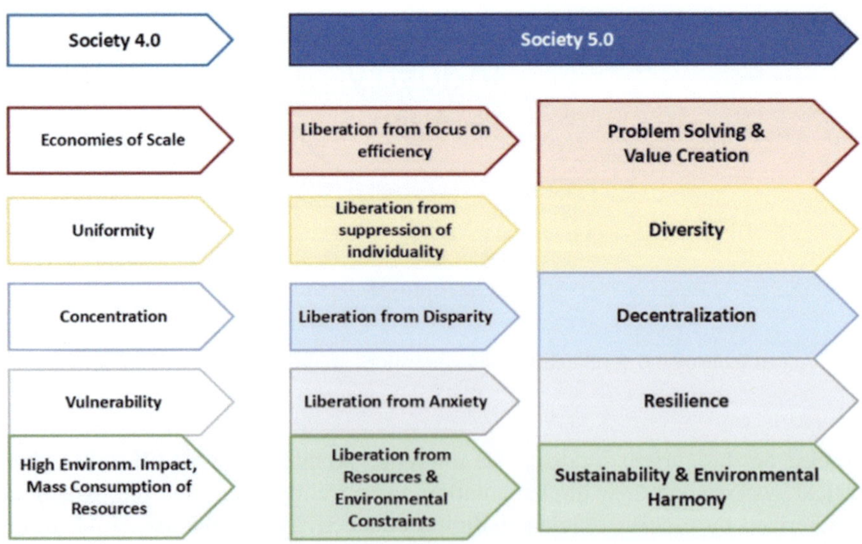

Fig. 3 From Society 4.0 to Society 5.0 according to [5]

systems and identifies causal relationships that would not be possible with previous methods of product quality and safety. By feeding these findings back into the development of not only products but also production processes, potential is generated that strengthens the resilience of future Society 5.0 systems in particular. However, this requires high bandwidth and low latency through advanced data transfer technologies that outperform 5G, as well as secure communication between the networked devices [5]. In the context of product safety, this means guaranteeing data and information flows in products as an essential, fundamental task.

Product safety and quality will have to take into account an even wider range of requirements than before. Customer orientation is not only changing towards the consideration of unspoken, latent, innovative customer needs, but also towards a human-centered world of needs in which the improvement of quality of life and the simultaneous realization of individual needs are sought [20]. This also demands solutions for social problems and economic growth from product quality and security, as the integration of cyberspaces with physical space entails a high degree of networking not only of products, but also of problems that the products are created to solve.

In other words: in future, product safety and quality will no longer end at the product system boundary. The entire environment of the product, redefined by Society 5.0 (the networks in physical space in terms of climate, environment and sustainability, i.e. also with cyberspaces in terms of redundancy, resilience, data protection, etc.) must be taken into account. The focus of problem-solving extends beyond value creation to the creation of value and the solution of social challenges [5]. Thus, the value creation of the company, including its relationship to the environment and networking with society in terms of product safety and quality, must be considered as a system. This means an expansion of the system boundaries, which will increase complexity exponentially. What this means in specific cases will be explained in more concrete use cases later on. Section 2, for example, looks at product safety and quality change in shop floor management. This is followed in Sect. 3 by a transfer to the level of processes. In Sect. 4 the subjective perception of an individual regarding product safety is questioned. All in all, a short sum up is given in Sect. 5.

2 Product Safety and Quality Transformations—from the Perspective of Shop Floor Management

How the further development from Industry 4.0 to Industry 5.0 or Society 5.0 is shaping up and to what extent this is leading to innovations in terms of product safety and quality in production will be explained below using the example of shop floor management.

Eight key technologies are taken into account as influences or drivers of further developments, as illustrated in Fig. 4.

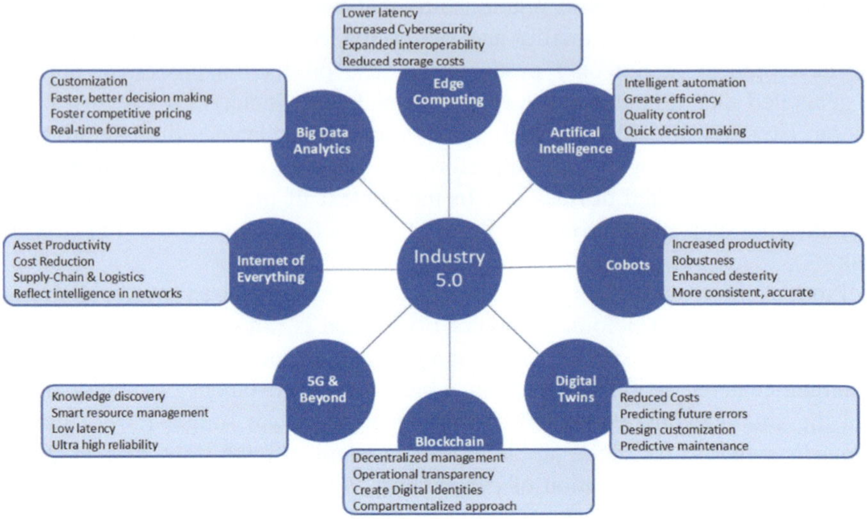

Fig. 4 Key Enabling Technologies in the transition to Industry 5.0 according to [5]

Six of the eight drivers—Big Data Analytics, Edge Computing, Artifical Intelligence, Blockchain, Digital Twins and 5G & Beyond—are key technologies that have already been explained in Sect. 1. There is also the key technology of cobots. In other words, robots that work together with workers as colleagues. This will lead to more consistent and accurate workflows in production, making processes more robust. As only robust, high-quality processes will enable the creation of high-quality products, it is to be expected that the rates of defective parts will fall in the area of quality assurance. The eighth key technology is the "Internet of Everything". An expansion of the Internet of Things (IoT) to include the networking of all elements in a supply chain, including the implementation of AI in this network.

Current shop floor management is characterized by the networking of individual production elements, be it technical systems or people. The focus in terms of safety and quality is customer-centric along the entire product life cycle. Sensors supply real-time data to ERP systems and there is immersive collaboration between man and machine. Today's shop floor management thus differs from earlier production systems in which robot-supported, automated processes in production were aimed solely at satisfying the customer, without taking into account the disposal or recycling phase or any disadvantages for the environment and climate [5], as visualized in Fig. 5.

Based on this shop floor situation, the eight key technologies mentioned above are changing the factories of the future towards a fully networked supply chain, the aim of which is to provide customers, society and the environment with

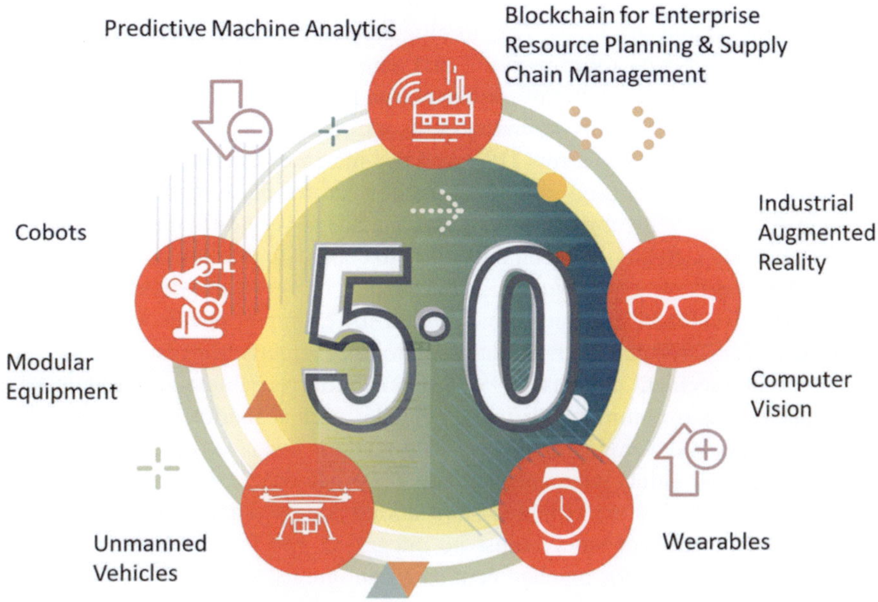

Fig. 5 Towards Shop Floor 5.0 according to [5]

a product that best meets their needs. To this end, ERP systems will have even more data at their disposal through the use of wearables by employees, autonomous transport vehicles, cobots and computer vision. Predictive machine analytics will make health management for machines even more efficient and, with the support of industrial augmented reality, will enable employees in the maintenance process to take responsibility for their own actions. Cobots and the design of modular equipment will also increase flexibility in the production process. The flow of information required for this will be secured by blockchain technology [5].

With regard to safety for people, the environment and society, the procedures in occupational safety, product safety and quality must also be changed in such shop floor systems. Although safety awareness in the industry for human health has already improved considerably, the focus here must be significantly expanded to include the entire production quality, including the environment, society and climate. Whereas occupational safety was previously the sole responsibility of safety engineers, an interdisciplinary approach with the knowledge of experts from various fields is now required in order to raise safety performance and the risks perceived by employees and society to a level that is free of concern. Digitalization not only supports this, but is also a new source of error. Safety strategies need to be adapted accordingly with regard to complex data analyses and the interaction between humans and machines. Due to climate change, production systems must be designed to be even more resilient and safety concepts must be made more dynamic in order to cope with unforeseen events [6].

The increasing complexity of the systems will also further increase the risk of unexpected interactions and errors [18]. And this increases the risk of accidents. The traditional linear models used to date are inadequate for these highly complex and dynamic systems. Predicting all possible error paths is very difficult, which is why there is currently a greater focus on the ability to cope with unforeseen events [7].

In order to proactively avoid the occurrence of errors again, concepts are currently being developed in basic research that, for example, enable the identification of complex error chains in product-environment interactions for the first time by networking accident databases [19], implementing variable and dynamic error chain analyses and continuously improving the user scenarios to be taken into account in product development [7].

However, such safety concepts are not only necessary for product systems, but also for production systems and shop floor systems. The measurement, analysis and optimization of safety must be integrated into the production management system or shop floor management as a continuous process. One example of this is safety performance measurement systems (SPMS). These are structured approaches for monitoring and improving safety performance in production [8]. The measurement systems record previously designed and selected safety indicators. These can be, for example:

1. Percentage of tasks completed,
2. frequency of safety inspections,
3. employee prioritization of safety,
4. number of near miss reports and
5. stop-work authority of employees [8].

The measured values for these safety indicators are transferred to data collections, processed and analyzed in accordance with the strategy for feedback and regular updating of the production system. SPMS thus not only capture negative safety events, but also positive practices that are promoted accordingly and optimized in terms of the interaction between safety and other business dimensions. Real-time feedback using innovative data analytics and big data technologies can increase the efficiency of monitoring, with feedback of data analysis results to employees in the shortest possible time to initiate immediate corrective actions. Five basic requirements apply to safety performance measurement systems (SPMS) [8]:

1. Monitoring of day-to-day variability,
2. real-time feedback,
3. learning from experience,
4. management of trade-offs and
5. adaptability.

However, as already explained in Sect. 1, Society 5.0 no longer focuses solely on production in terms of safety and quality. The interactions with the environment and society must also be taken into account. This means that many more processes, including those outside the company itself, as well as their interactions with the specific manufacturing processes must be taken into account. This networking of processes, both internally with data from within the company and externally with indirect data sources, leads to new challenges for management systems that are intended to control and guide the quality and safety of products as well as company processes. A consideration of these challenges follows in Sect. 3.

3 Ensuring Product Quality—from a Process Perspective

As already indicated at the end of Sect. 2, the factors influencing the safety and quality of a company no longer end at the factory gate. The system boundaries are expanding exponentially in terms of complexity. The entire product life cycle as well as the entire value chain, supply chain and administrative processes must be taken into account in terms of quality, safety, environment and sustainability [18].

Today, integrated management systems (IMS) are used to understand, analyze and better design this enormous number of processes, which also interact with each other, i.e. are networked. These differ from the IMS propagated in the 1980s in that the topics of quality, environment and occupational health and safety with regard to the manufacturing processes and the product were mapped in the management system at that time. Today, IMS can fulfill various topics and management system standards, depending on the needs of the company. In addition to the classic topics of quality, environmental protection and occupational health and safety, the topics of energy, sustainability, IT security and compliance are often added [9].

The future IMS uses the stored models, data and links to the workflows and databases of the company's other IT systems to integrate global trends with local realities through an analytical framework. It takes into account the interactions between technology, tasks, culture, structure, power, strategy, regulation, society and markets to create a comprehensive understanding of the complex dynamics of our business system. This is only possible through advancing digitalization, which enables the use of artificial intelligence. Only AI tools are capable of processing the dynamic, highly complex interdependencies for us humans in an economically efficient manner. While the use of image analysis algorithms in Quality Assurance (QA) is increasing rapidly, large language models are being used to analyze and design process instructions. As dependence on the IT infrastructure increases the risk of cybersecurity threats, this also implies new and—in some cases—more stringent requirements for security management and therefore the IMS. It is also important to use AI to counteract the standardization and bureaucratization of decision-making, while enabling employees to contribute their human expertise and social interactions in order to prevent alienation [10]. All in all, Quality

Management has to find solutions to improve the objective but also perceived safety and quality of employees regarding the use of AI and working with cobots. Especially cobots are challenge due to the increase of mutual work of humans and robots. A topic that is focused in the following chapter.

4 Perceived Product Safety– from an Individual Perspective

The term "perceived" implies a subjective assessment that is carried out by an individual person. Applied in the field of safety, it rather seems unconventional or even illicit to use subjective data to measure the safety level of a product. As engineers, safety is given in case the existing risk is lower than the accepted risk [11]. And these risks are calculated by as accurate objective data as possible.

But as an individual human being, a product might be as safe as possible from an engineering point of view, but we are not going to use it, if we feel unsafe while using it. Human beings develop feelings of safety based on several personal needs. Based on Maslow's theory there is an inseparable relationship between the fulfilment of a person's needs and its perceived safety. From a psychological point of view there have to be considered the needs of physiological, safety, love, self-esteem and self-actualization [12]. These facts become even more important now that we are facing Industry 5.0 technology like cobots. Humans are working and achieving output together with roborts, working in a mutual process. Those robots, called cobots, not only have to be safe in terms of product safety. They also have to generate the feeling of being safe in the eyes of the employee who is working with the cobot.

In order to ensure not only objective product safety but also perceived product safety by a worker or customer, influences on perceived safety have to be taken into account in the designing phase of a product. Such a model for evaluating perceived safety is created by [13]. They are focusing on Level 3 autonomous driving and the influences that affect the perceived safety of occupants. Their goal is to positively influence the perceived sense of safety of vehicle occupants by using driver assistance systems.

Their model includes 17 influencing factors and also their interactions, based on the following three research questions:

1. What influences on perceived safety are vehicle drivers exposed to?
2. Can the influencing factors of perceived safety be described in a model?
3. How can perceived safety be assessed in autonomous vehicles?

After a comprehensive literature study 17 different safety influence factors have been determined for the current state of the art. In order to handle those factors and their relations, they have been clustered in five different categories [13]:

1. Control
2. Feeling of Protection
3. Measurement
4. External Influences and
5. Driver's Posture

Those categories and their factors have been modeled in a first concept, that is shown in Fig. 6.

The "Control" category includes influencing factors regarding human machine interface (HMI) and advanced driver assistance system (ADAS). As HMI is a display inside the vehicle that enables communication between the car and the driver, it is crucial to include acoustic, haptic and visual factors regarding HMI. ADAS covers all driver assistance systems and all its requirements have to be fulfilled in order to assure the safety of the system. In order to measure the grade of safety of both, HMI and ADAS, in an objective way, all requirements have to be fulfilled which can be checked e.g. by sensors.

The category "Driver's Posture" is not only including objective factors but also subjective one. The design of a driver seat is influenced by the dimensions of a person (Anthropometry), the required freedom of movement, accessibility of controls and visibility of the environment. All these factors can be measured in an objective way but are also influencing the subjective feeling of comfort [14]. So, adjustable seats and the process of adjusting have to be included, because they have a positive influence on perceived safety [15].

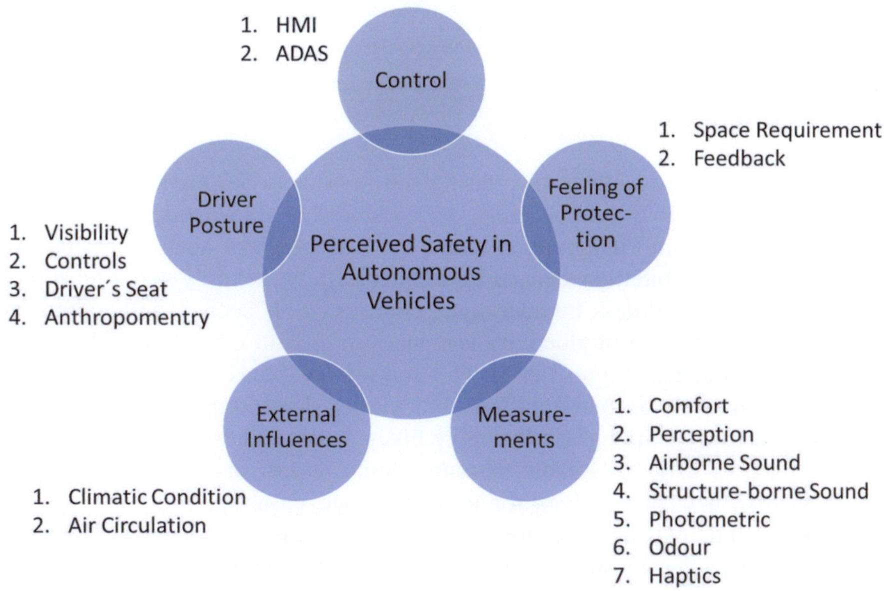

Fig. 6 Perceived Safety in autonomous vehicles according to [13, p. 186]

Haptics, Odour, Photometric, Structure-borne sound, Airborne sound, Perception and Comfort are all part of the "Measurement" category. Haptics and Odour are proven to influence perceived quality and comfort [16]. Photometric measures focus on the lighting within the car, considering light from outside as well as inside of the cockpit. Sounds or noise—structure-borne or airborne—can be measured objectively by sensors but are also perceived in different ways by individuals.

External Influences dan be caused by climate conditions and air circulation. Temperature and humidity inside the vehicle are difficult to determine when it comes to optimum range for a person and perceived comfort.

Last but not least, the feeling of protection includes the space that is required by the driver while in the cockpit. Also, feedback on vehicle's controls and displays are a very important factor in order to creating a feeling of safety.

All mentioned factors are influencing a driver's perception of safety on their own. But furthermore, interrelations between the factors can intensify the influence of safety perception but might also neutralize each other's effects. As a first model, the influence factors can be measured in an objective way via requirements fulfillment or even real-time via sensors, while subjective impressions are captured via interviews. Based on the first results, the conceptual model of 17 influencing factors for perceived safety in autonomous driving have to be analyzed regarding relevance as well as correlations [13].

The suggested model can be used as a source of information on which influencing factors can affect perceived safety. The model summarizes all objective influencing factors on the driver's perceived safety in autonomous cars plus subjective factors. Future research should focus on objective assessments using a catalogue based on the model and weighting factors.

5 Conclusion

Summing up, the transition from Industry 4.0 to Society 5.0 marks a significant reorientation towards a human-centric and sustainable framework for industrial advancement. This shift integrates cutting-edge technologies—such as the Internet of Things, Artificial Intelligence, and big data analytics—not only to optimize production but also to address broader societal needs. With Society 5.0, companies are expected to balance productivity and innovation with ethical considerations, environmental impact, and resilience. This transition extends the concept of product quality and safety beyond traditional boundaries, demanding that organizations address these aspects throughout the entire lifecycle and supply chain.

As industries adopt adaptable systems, digital twins, and predictive analytics, they open new possibilities for measuring and ensuring safety and quality in real-time, thus enhancing resilience against unforeseen challenges. However, this interconnected environment requires that safety measures also adapt to the complexities introduced by digitalization. The focus thus shifts from merely safeguarding individual processes to creating a robust, networked infrastructure that supports both social well-being and environmental sustainability.

In embracing Society 5.0 principles, companies can establish a foundation for future-ready operations that prioritize the enhancement of human life, leveraging technology not only for efficiency but also for safety and security. Transformations towards an industry that is accountable, adaptable, and aligned with the evolving expectations of our society and the environment are pressed ahead already. This includes considering the perceived safety as well as the objectively measurable safety. Thus, parallel to this, product quality and safety engineering has to keep up with these enhancements.

References

1. Pistorius, J.: Industrie 4.0 – Schlüsseltechnologien für die Produktion. Grundlagen, Potenziale, Anwendungen. Springer Vieweg, Berlin, Heidelberg, (2020).
2. Bundesministerium für Wirtschaft und Energie (BMWi): Leitbild 2030 für Industrie 4.0. Digitale Ökosysteme global gestalten. URL https://www.bmwk.de/Redaktion/DE/Schlaglichter-der-Wirtschaftspolitik/2019/10/kapitel-1-5-leitbild-2030-fuer-industrie-40.html, last accessed 2024/11/05.
3. Plattform Industrie 4.0: POSITIONSPAPIER. Leitbild 2030 für Industrie 4.0 – Digitale Ökosysteme global gestalten, 2019. URL: www.plattform-i40.de/IP/Redaktion/DE/Downloads/Publikation/Positionspapier%20Leitbild.pdf?__blob=publicationFile&v=8, last accessed 2024/11/05
4. Forschungsbeirat der Plattform Industrie 4.0/acatech – Deutsche Akademie der Technikwissenschaften (Hrsg.): Themenfelder Industrie 4.0 – Forschungs- und Entwicklungsbedarfe zur erfolgreichen Umsetzung von Industrie 4.0, 2. überarbeitete Fassung, https://doi.org/10.48669/fb40_2022-04, (2022).
5. Mourtzi, D., Angelopoulos, J. and Panopoulos, N.: A Literature Review of the Challenges and Opportunities of the Transition from Industry 4.0 to Society 5.0. In: energies, Vol. 15, MDPI, https://doi.org/10.3390/en15176276, (2022).
6. Swuste, P., Groeneweg, G., van Gulijk, C., Zwaard, W., Lemkowitz, S., Oostendorp, Y.: The future of safety science. In: Safety Science, Vol. 125, Elsevier, (2020).
7. Julitz, T.; Schlüter, N.; Löwer, M.: Scenario-based Failure Analysis of Product Systems and their Environment. In: Proceedings of the 33rd European Safety and Reliability Conference (ESREL 2023) Published by Research Publishing, Singapore, September pp.3282 – 3289 https://doi.org/10.3850/978-981-18-8071-1_P685-cd, (2023).
8. Peñaloza, G. A., Saurin, T. A., Formoso, C. T., Herrerac, I. A.: A resilience engineering perspective of safety performance measurement systems: A systematic literature review. In: Safety Science, Vol. 130, Elsevier, (2020).
9. Rebelo, M. F., Santos, G., Silva, R.: Integration of management systems: towards a sustained success and development of organizations, Journal of Cleaner Production, Volume 127, pp. 96–111, ISSN 0959–6526, https://doi.org/10.1016/j.jclepro.2016.04.011, (2012).
10. Le Coze, J.-C., Dupr'e, M.: Safety as a network, digital and global reality. In: Safety Science, Vol. 156, Elsevier, (2022).
11. DIN EN ISO 12100:2011–03, Berlin: Beuth Verlag GmbH, (2011).
12. Raue, M., Streicher, B. and Lermer, E.: Perceived Safety, Cham, Springer International Publishing, (2019).
13. vom Stein, N., Löwer, M.: Unveiling Trust: A Model-Based Approach to Understanding and Enhancing Perceived Safety in Autonomous Vehicles, in: Advances in Human Factors of Transportation, Vol. 148, https://doi.org/10.54941/ahfe1005208, pp. 181–190, (2024),

14. Bubb, H., Bengler, K., Grünen, R. E. and Vollrath, M.: Automobilergonomie, Springer Fachmedien, Wiesbaden, (2015).
15. Akalin, N., Kristoffersson, A. and Loutfi, A.: Do you feel safe with your robot? Factors influencing perceived safety in human-robot interaction based on subjective and objective measures, International Journal of Human-Computer Studies, vol. 158, p. 102744, (2022).
16. Zhang, L., Helander, M. G. and Drury, C. G.: Identifying Factors of Comfort and Discomfort. In: Sitting: Human Factors - The Journal of the Human Factors and Ergonomics Society, vol. 38, no. 3, pp. 377–389, (1996).
17. Forschungsbeirat der Plattform Industrie 4.0/acatech – Deutsche Akademie der Technikwissenschaften (Hrsg.): Engineering autonom wandelbarer Industrie 4.0-Systeme, https://doi.org/10.48669/fb40_2024-2, (2024).
18. Schlüter, N.: Generic Systems Engineering. A Methodical Approach to Complexity Management, 1. Edition 2024, Springer Berlin, Heidelberg, ISBN 978-3-662-67993-7, https://doi.org/10.1007/978-3-662-67994-4, (2024).
19. Julitz, T.; Tordeux, A.; Löwer, M.: COMPUTER-AIDED DESIGN OF FAULT-TOLERANT HARDWARE ARCHITECTURES FOR AUTONOMOUS DRIVING SYSTEMS. In: Proceedings of the Design Society, Volume 3: ICED23, https://doi.org/10.1017/pds.2023.105, (2023), pp. 1047–1056.
20. Löwer, M.; Katzwinkel, T.; Limbach, D.: Integration agiler Produktentwicklung und Ökodesign bei KMU – Lösungsstrategien für umweltverträgliche Produkte und Produktionsprozesse im Kontext von Kleinunternehmen. In: Industrie 4.0 Management, 10.30844/IM, (2023), pp. 46–50.

Security and Resilience: Protection of Infrastructures

Kai-Dietrich Wolf and Daniel Lichte

Contents

Abstract

In today's dynamic and uncertain security landscape, protecting critical infrastructures has become increasingly important. These backbone systems provide essential services and goods to modern society, yet are also vulnerable to intentional disruption, leading to significant interruptions in service delivery. We argue, that an integrated approach that combines security and resilience is necessary for holistic protection of critical infrastructures, in particular because security and resilience have complementary goals: security aims at minimizing the probability of

K.-D. Wolf (✉)
University of Wuppertal, Wuppertal, Germany
e-mail: wolf@iss.uni-wuppertal.de

K.-D. Wolf · D. Lichte
Institute for Security Systems, University of Wuppertal, Velbert, Germany

D. Lichte
German Aerospace Center (DLR), Sankt Augustin, Germany

S. Bracke and R. Pieper (eds.), *Safety Engineering*,
https://doi.org/10.1007/978-3-658-47473-7_20

successful attacks, while resilience concepts focus on enabling systems to cope effectively with resulting interruptions. Aiming at the assessment of measure effectiveness, we propose a risk-based framework that brings together a model for both, vulnerability to physical attacks and potential cascading consequences. Both models use distributed parameters to reflect uncertainties, e.g. regarding the effectiveness of countermeasures. Applying this approach to a simple case study, we show that equivalent measure combinations can be found within both domains, although the contributions to risk mitigation are varying in different scenarios. In the analysis, we highlight the paramount importance of considering uncertainties when selecting security and resilience measures and emphasize the need for efficiency analysis (cost–benefit assessments) to find optimal or appropriate solutions, especially in the context of limited budgets. We conclude that the proposed framework is flexible and can serve as a starting point for further research and application.

Keywords

Critical infrastructure protection · Security and resilience integration · Risk-based vulnerability assessment

1 Introduction

In today's dynamic and uncertain security landscape, protecting critical infrastructures has become increasingly important. These backbone systems provide essential services and goods to modern society, yet are also vulnerable to intentional interruptions, leading to significant disruptions in service delivery (Nick et al. 2023; McDaniels et al. 2007). The escalating threat posed by terrorism, extremist sabotage, and even state-sponsored actors underscores the need for effective protection strategies. This paper presents a concept that integrates security and resilience approaches to provide a comprehensive framework for safeguarding critical infrastructures against various threats.

Critical infrastructures are often inherently vulnerable due to their geographically distributed and coupled nature and their disruption can lead to substantial consequences (McGill et al. 2007; Trump et al. 2018; Nan et al. 2014). Recent events have demonstrated the reality and diversity of these threats. Threats cannot always be averted by security measures if, for example, a terrorist background is present or state actors are involved. In particular, so-called "hybrid threats" pose a challenge to conventional security measures. Therefore, approaches to mitigate the effects of attacks on critical infrastructure must be considered. In this context, the concept of resilience plays an important role (Tamminga 2015). Governments worldwide have acknowledged the importance of protecting critical infrastructures, reflected in evolving regulations such as the EU's CER Directive and national legislation like Germany's KRITIS-DG (Bundesministerium des Innern und für Heimat 12/21/2023). Both of these pieces of legislation call for action in the areas of security as well as resilience. However, a pressing issue arises from the imprecise distinction between security and resilience.

Security primarily focuses on preventing successful attacks by anticipating potential scenarios and implementing measures to thwart attacks. A comprehensive understanding of this concept requires the consideration of multiple factors, including the likelihood of threats, vulnerability, and damage potential. In this context, several models have been developed to quantify physical security risk, each with its strengths and limitations.

One approach to quantifying physical security risk is through semiquantitative risk analysis (Harnser Group 2010). This method involves assigning scores to various factors influencing the likelihood of threats, vulnerability, and potential damage. By aggregating these scores, a comprehensive risk ranking can be obtained, enabling informed decision-making. An analytically more sophisticated approach is probabilistic risk analysis (McGill et al. 2007; Sandia National Laboratories 1989). This method involves the use of probability distributions to quantify the likelihood of threats, the security system's vulnerability and resulting damage potential. By considering multiple scenarios and factors, a comprehensive quantitative physical security risk model can be achieved.

Several frameworks have been established to enable the assessment of physical security risk. The Estimate of Adversary Sequence Interruption (EASI) approach is a widely used model that focuses on vulnerability assessment introduced by Bennett (1977) and described by e.g. Garcia (2005). Systematic Analysis of Vulnerability to Intrusion (SAVI) including the Adversary Sequence Diagram (ASD) (Sandia National Laboratories 1989) is another framework that combines elements of probabilistic risk analysis and semiquantitative methods. In recent years, research has focused on developing new approaches to quantify physical security risk. In this context, the approach compiled by Garcia (2008) has gained attention, as it provides a structured methodology for identifying potential intrusion paths and analyzing the effectiveness of security measures. This method involves experts estimating probabilities and time parameters in respective scenarios. It is further developed e.g. by Lichte and Wolf (2017).

One of the challenges associated with security assessment is uncertainty regarding the effectivity of vulnerability reducing measures (Lichte et al. 2021). In order to assess vulnerability, experts must estimate the time required to overcome barriers and the probability that attackers will be detected. However, it is rarely possible to rely on extensive prior knowledge. Additionally, the likelihood of a particular attack scenario depends on the attacker's decision and is therefore epistemic in nature. The central question that arises in the context of uncertainty is to what extent an analysis of the security risk can provide good support for decisions in the presence of large uncertainties. Quantitative methods offer the advantage that uncertainties can be considered analytically and can thus provide a basis for evaluating the quality of decisions.

In contrast, resilience considers systems' ability to withstand various disruptions, including unforeseen ones, through adaptability and rapid recovery (Bi et al. 2023; Wied et al. 2020). The concept of resilience describes an infrastructure system's ability to cope with disruptions, including extreme situations that may be unforeseen or unexpected (Mentges et al. 2023). This capacity encompasses various sub-capabilities, such as:

- Maintaining critical services during hazardous conditions (NIAC 2010)
- Recovering functionality after overcoming hazardous conditions (Mottahedi et al. 2021)
- Adapting to experiences from past hazardous events (Francis and Bekera 2014)

The growing emphasis on resilience is accompanied by the recognition that not every disruption can be prevented. Instead, it is essential to learn how to cope with the consequences of such events (Halekotte et al. 2025). In this context, critical infrastructure resilience becomes a key factor in ensuring the continuity and reliability of essential services, despite the uncertainty and unpredictability of threats. By developing strategies for coping with disruptions, we can minimize the impact of hazardous conditions and ensure that critical infrastructure remains functional, even in extreme situations.

There are various approaches for assessing the resilience of an infrastructure. A general distinction can be drawn between outcome-based (performance-based) and property-based (attribute-based) approaches (Mugume et al. 2015; Pagano et al. 2019; Förster et al. 2019; Wang et al. 2023). Outcome-based approaches measure resilience based on the outcome of one or multiple disruptive events that a system (real or virtual) had to deal with. In the context of infrastructure protection, many outcome-based resilience assessments rely on the use of performance curves (sometimes also called resilience curves). A performance curve shows the evolution of a system's performance over the course of a particular disruption—and there are a plethora of summary metrics, that can be extracted from these curves to describe how well the system has handled the corresponding event (Poulin and Kane 2021; Yodo and Wang 2016; Trucco and Petrenj 2023). While many approaches base their assessment of resilience on a single performance curve, Halekotte et al. (2025) argue that in order to assess a system's resilience, multiple curves need to be considered. Property-based approaches assess resilience based on the status of certain system properties which are assumed to build the system's capacity to cope with disruptive events (Asadzadeh et al. 2017; Cutter 2016). Most common within the class of property-based approaches are composite resilience indicator frameworks, which aggregate information on a variety of indicators following a specific hierarchy to obtain a small set of indices displaying the resilience status of the system of interest.

In summary, this means that security aims to minimize a system's vulnerability to successful attacks, while resilience concepts focus on enabling systems to effectively cope with resulting disruptions, i.e., to lose as little performance as possible. Taken together, it becomes clear that security and resilience must complement each other in order to achieve comprehensive protection of infrastructure against attacks. This is particularly true, because security measures may not be very effective against some threat scenarios while resilience strategies, aiming at coping and mitigating consequences, may fail in the context of largely uncertain cascading effects.

This paper presents a framework that combines these complementary approaches. By minimizing the likelihood of successful attacks (security) and enhancing systems' ability to absorb and adapt to disruptions (resilience), we demonstrate how this integrated approach can lead to improved critical infrastructure protection. Our concept is illustrated by a case study, highlighting the practical implications and discussing further avenues for research inspired by this synthesis.

2 Methods

2.1 Performance-Based Resilience Curve

The concept of resilience is often illustrated using performance curves, a framework initially proposed by Bruneau et al. (2003) and later expanded by Cimellaro et al. (2009). These time-dependent curves depict the decline in performance, which eventually recovers to its original level over time following a sudden disruption. The degree of resilience in a system is characterized by a less pronounced dip and a quicker recovery (Pimm 1984). Additionally, some researchers suggest that systems can learn from disruptions, leading to an improved performance that exceeds their original capabilities, sometimes surpassing 100% (Francis and Bekera 2014).

Resilience loss *RL* during a disruption is often approximated by the formulation first proposed by Bruneau et al. (2003):

$$RL = \int_{t_b}^{t_e} [100 - Q(t)]dt \tag{1}$$

In Fig. 1, the initial performance of a system at time t_b at the start of a disruption is scaled to 100%. The actual performance of the system over the elapsed time from the start of a disruption at t_b to its end at t_e, is described by the performance level function $Q(t)$. The difference between the initial state of the performance and the actual performance then results in the total resilience loss.

Demmer et al. (2022) propose logistic functions to describe the absorption phase and thereafter the recovery phase. Using this approach, it is possible to model the performance of a system during a disruption in one resilience cycle with the following analytic equation:

$$Q(t) = P + \frac{L_1}{\left(1 + e^{-k_1 * (t - t_1)}\right)} + \frac{L_2}{\left(1 + e^{-k_2 * (t - t_2)}\right)} \tag{2}$$

Herein, P is the initial performance scaled to 100%. L_1 describes the depth of the performance drop. Measures aiming to minimize the performance drop, i.e. L_1 are often referred to as increasing a system's robustness (Shinozuka et al. 2004). k_1 refers to the speed of loss of performance, sometimes called rapidity (Pimm 1984). Measures to reach a more graceful degradation of the system, i.e. lowering the

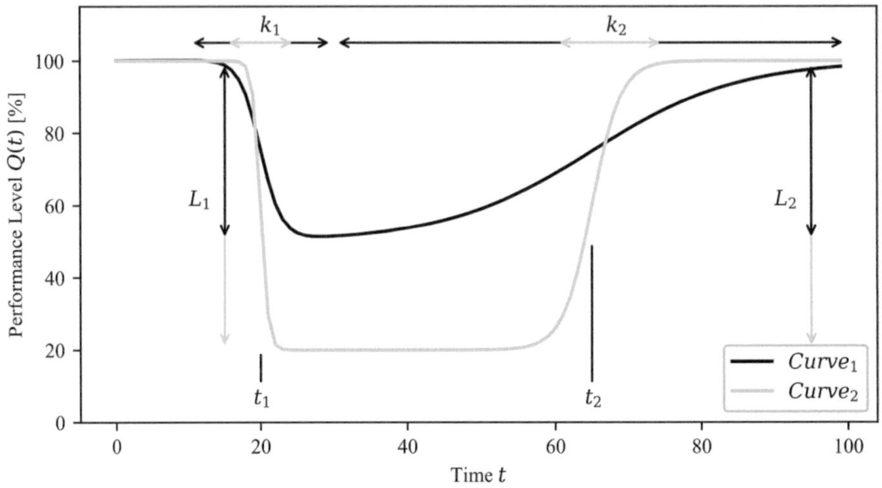

Fig. 1 Curves for $Q(t)$ describing the performance of a system during a disruption over time based on double logistic function. Curve 1 in grey line and Curve 2 in black line for two different disruptions

rapidity are thus considered by lowering k_1. The parameter t_1 depicts the time of the inflexion point. Because absorption and recovery phases are considered independently of each other but within the same mathematical expression, the same description is valid for the recovery phase. L_2 is the amount of recovered performance, k_2 characterizes the speed and t_2 again is the time of the recovery phase inflexion point. Measures aiming to lower the recovery time are reflected in the model by changing t_2.

2.2 Vulnerability Modeling and Assessment

The vulnerability against physical attacks is commonly described by the probability that an attack sequence cannot be interrupted before its completion (Garcia 2005). A physical protection system serves to reduce this vulnerability by implementing measures to detect and delay an intruder and interrupt the attack in time. The analysis of such a system includes on the one hand the identification of possible intrusion paths, and on the other hand the calculation of the vulnerability for the identified paths. A path describes the spatial and temporal sequence of an attack and can be modeled based on the barriers an intruder must overcome. Generally, multiple attack paths are possible. These can be derived from an Adversary Sequence Diagram (ASD), which visualizes a graph of zones and barriers of a site. Figure 2 shows the schematic of an ASD.

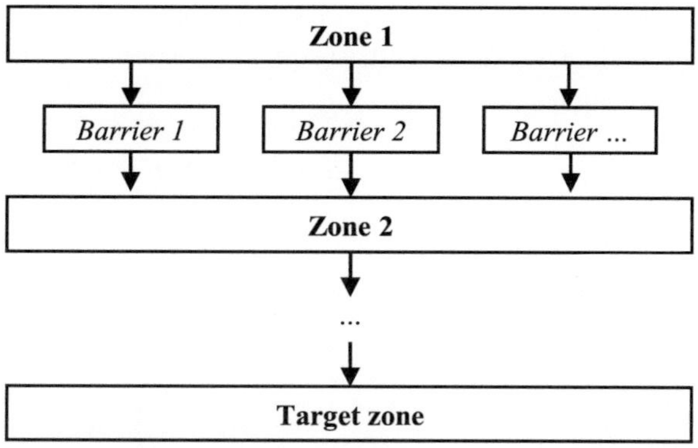

Fig. 2 Schematic of an Adversary Sequence Diagram

The effectiveness of security measures can be described by capabilities of protection, observation and intervention associated to each barrier as visualized by Fig. 3 for one path.

In its simplest form, the capability of observation is parametrized by a probability of detection P_D. Protection and intervention capabilities are described by a time game of delaying the intruder long enough to intervene in time. Here, protection capability is parametrized by the time t_P an intruder is delayed, and intervention capability is parametrized by the time t_I needed to respond and interrupt the attack. Then, the probability for timely intervention P_I is

$$P_I = P(t_I < t_{RP}) \tag{3}$$

where t_{RP} is the residual protection time, given by the sum of protection times of all remaining barriers along the path. The residual protection time at the i-th of n barriers is:

$$t_{RP,i} = \sum_{j=i}^{n} t_{P,j} \tag{4}$$

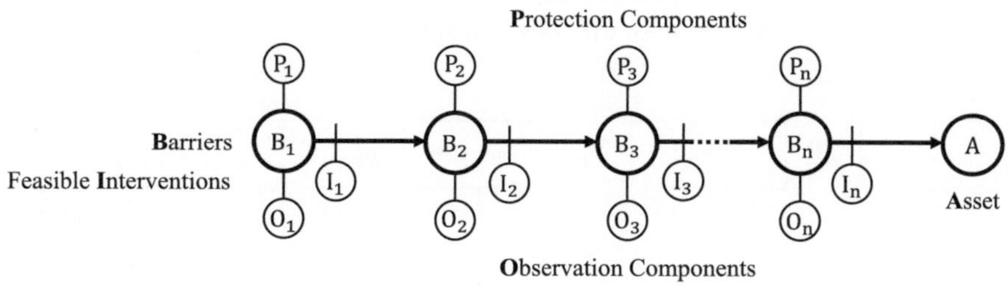

Fig. 3 Capabilities of security measures based on Garcia (2008). Source: Lichte and Wolf (2017)

Uncertainties regarding t_P and t_I are considered by describing these parameters as random variables with specified probability distributions, for instance normal distributions.

A barrier is defined vulnerable if detection or timely intervention do not occur. The vulnerability of a barrier is the probability that the barrier is vulnerable:

$$V_B = 1 - P_D \cdot P_I \tag{5}$$

A path is vulnerable if all barriers along the path are vulnerable. Accordingly, the path vulnerability is

$$V_P = \prod_{i=1}^{n} V_{B,i} \tag{6}$$

3 Approach

The basis of our approach is the classic three-part formulation of the contributions to the risk function in physical security risk assessment:

$$R = T \times V \times C \tag{7}$$

In this formula, the risk is composed of the threat likelihood T, the vulnerability V and the possible resulting consequences C.

For the threat likelihood T, we take a simplified view. For each scenario under consideration, we assume a given probability of occurrence of the scenario. This then applies for a fixed observation period of 20 years.

Following the description of the relationships between security and resilience introduced in Sect. 1, the effectiveness of measures can be integrated into this formula as shown in Fig. 4. It can be assumed that in some cases, security measures not only reduce vulnerability, but also contribute to reducing the probability of attack, in the sense of preventing attacks. However, these relationships are explicitly not the subject of our discussion. This means that security measures are able to reduce the vulnerability of the system to attacks, while resilience aims to minimize the consequences of a successful attack.

Consequently, we link the assessment of security and resilience by incorporating the effectivity of measures of both contributions into the risk function. The

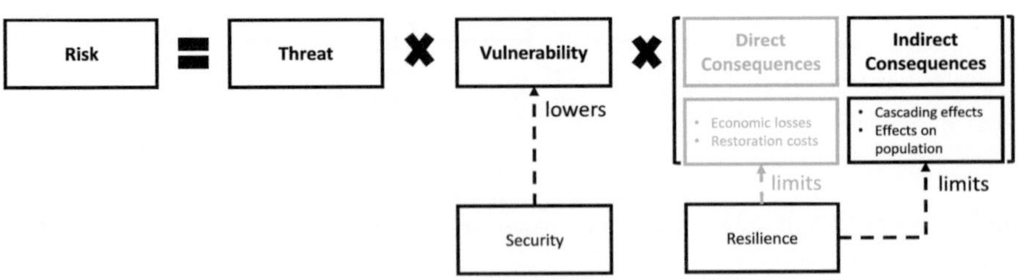

Fig. 4 Used risk model with incorporated effectivities for security and resilience measures

effectivity is assessed in two different models. In the context of vulnerability V, we use the approach introduced in Sect. 2.2, which is based on the EASI model. Using this approach, we analyze the effectivity of security measures to thwart an attack. To illustrate the effect of measures, we will vary their effectiveness in the following example.

The consequences C of attacks that are successful despite security measures can be limited by additional measures to enhance resilience. Thus, we integrate the description of the resilience of the critical entity under consideration into the consequences as part of the risk function. For this purpose, we use the abstract model introduced in Sect. 2.1. It is based on a double-logistic function that depicts the performance of the critical entity in the event of disruptions (Demmer et al. 2022). In our approach, we assume that the particularly severe indirect consequences, e.g. loss of essential services for the population and cascading failure of other critical supply infrastructures, are directly correlated with the loss of resilience of the affected infrastructure. Although this correlation cannot be assumed to be linear, we do not examine it in depth here, as a simplified view is sufficient to show the advantages of taking uncertainties into account in our approach. For the same reason, the direct effects, which are generally more reliably determined, are not considered here.

By varying the model parameters in the individual models, our approach allows us to analyze possible configurations of security measures and resilience-enhancing measures. In principle, the choice of parameters to be varied can be restricted as desired. By freely combining the parameter variations, the different configurations can then be compared with regard to the resulting risk and the contributions of vulnerability V and resilience loss RL.

In order to consider uncertainties regarding the effectiveness of the measures in both areas, we describe the parameters by random variables. We use normal distributions for the vulnerability model and beta distributions for the resilience model. The shape of the beta distribution is determined by its two shape parameters, α and β, which influence its skewness and kurtosis.

We combine the results of the different risk contributions for a risk assessment using a Monte-Carlo simulation. Therefore, we first combine the risks of the individual scenarios into an aggregated risk R_{agg}:

$$R_{\text{agg}} = \sum_i T_i \cdot V_i \cdot RL_i \tag{8}$$

Here we define a desired risk level r for which a suitable configuration of measures is sought. We then select the equivalent effective configuration for the defined r from all possible configurations for further analysis. Here, we then decompose the aggregated risk R_{agg} of these solutions into their respective contributions to vulnerability V and consequences, i.e. resilience loss RL.

4 Case Study

In the following, we demonstrate our approach by analyzing a small fictitious energy network that inherits a critical transformer substation with the core processes of transforming and transmitting electricity. The network is simplified by a double logistic model of performance as introduced in Sect. 3. The layout of the substation under consideration is shown schematically in Fig. 5.

For our case study, we define a total of four different attack scenarios *A*, *B*, *C*, *D*. These are each characterized by different attack modes, which are described in Table 1. As shown there, these scenarios also differ in terms of their direct damage to the substation.

To simplify matters, we assume the likelihood of occurrence for the different scenarios shown in Table 1: An attack damaging only the transformer is assumed to occur twice as likely as attacks damaging both core processes. We assume these likelihoods for a time interval of observation of 20 years.

Figure 6 shows an ASD reflecting the substation layout and including the existing security barriers. The substation's security system consists of three barriers: the perimeter (*P*), the outer skin of the control technology building (*B*) and the enclosure of the transformer (*E*). The barriers form two protection layers for both assets. Three variants are available for each of the barriers P_i, B_i and E_i in our case study, the properties of which are defined in Table 2. These can be combined independently at the individual barriers to form an overall system, resulting in a total of $n = 27$ possible configurations. We further assume a fixed probability

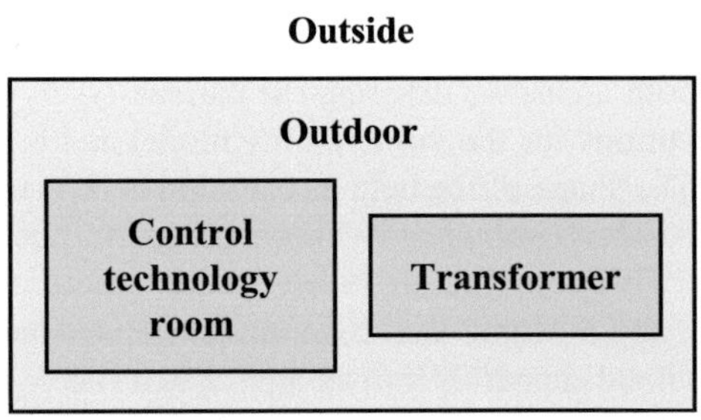

Fig. 5 Substation layout

Table 1 Description of considered attack scenarios A, B, C, D

Scenario	Attack mode	Damage mode	Likelihood
A	Transformer, weak tools	D_1	0.01
B	Transformer, medium tools	D_1	0.01
C	Control technology room, medium tools	D_2	0.005
D	Control technology room, powerful tools	D_2	0.005

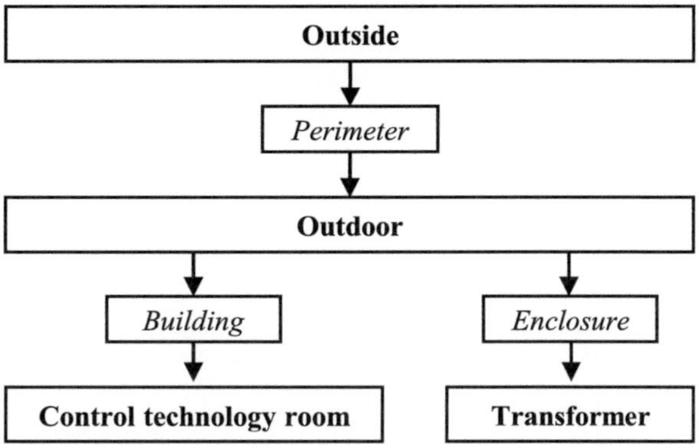

Fig. 6 Adversary Sequence Diagram (ASD) for the considered substation

Table 2 Mean μ and standard deviation σ of normal distributions for t_P for variants of protection at the barriers perimeter (P_i), building (B_i), enclosure of transformer (E_i) for scenarios A, B, C, D

Variant	Scenario A		Scenario B		Scenario C		Scenario D	
	μ (s)	σ (s)	μ (s)	σ (s)	μ (s)	σ (s)	μ (s)	σ (s)
P_1, E_1	120	24	90	18	90	18	30	6
P_2, E_2	600	120	300	60	300	60	210	42
P_3, E_3	900	180	600	120	600	120	420	84
B_1	300	60	150	30	150	30	60	12
B_2	540	108	420	84	420	84	180	36
B_3	1200	240	900	180	900	180	720	144

of detection $P_D = 0.9$ and time needed for intervention $t_I \sim \text{Norm}(900\text{s}, 60^2\text{s}^2)$. Following the principle of the weakest path, the vulnerability of the security barriers is calculated using the approach described in Sect. 3.

The two models are linked to each other via the scenarios: the direct damage caused in the substation by a successful attack causes different disruptions in the power grid. These are represented by the fact that in its performance model, the initial loss of performance L_1 for damage mode D_2, in which both core processes transformation and transmission fail, is higher and the recovery phase described by t_2 takes longer. In damage mode D_1, only the transforming process fails.

The response of the system without resilience-enhancing measures is shown in the performance curves in Fig. 7, leading to the resilience loss $RL_{D1} = 7599$ and $RL_{D2} = 10666$ for damage mode D_1 and D_2, respectively. Accordingly, the system and its response to D_1 and D_2 as characterized by the respective curves Q(t) (Eq. (2)) are parametrized as described in Table 3.

To calculate the effectiveness of the resilience enhancing measures, we define three options for the parameters $L_{1,i}$ (robustness), $k_{1,i}$ (rapidity) and $t_{2,i}$ (time-to-recovery), where each i-th option has different values as given in Table 4. They are

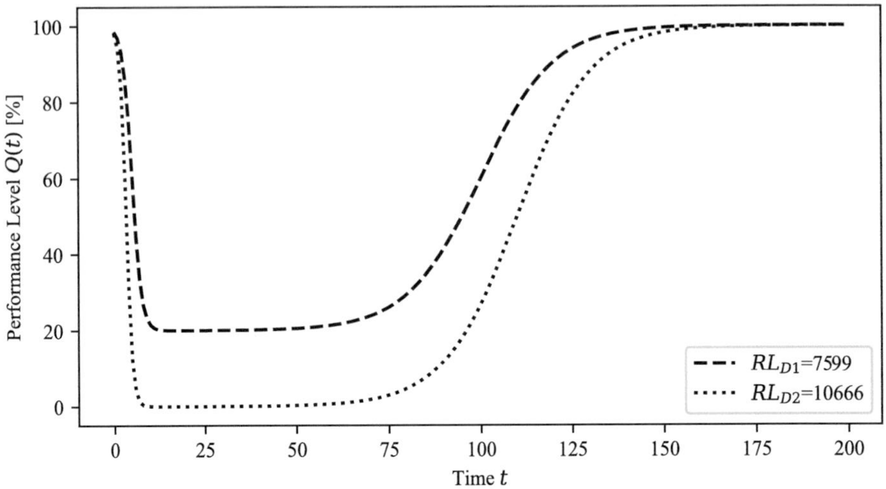

Fig. 7 Initial Response of the transmission system without resilience-enhancing measures

Table 3 Initial parametrization of the transmission system model for both damage modes

Damage mode	Parameter					
	L_1	k_1	t_1	L_2	k_2	t_2
D_1	80	0.8	20	80	0.1	100
D_2	100	1.2	20	100	0.1	110

based on the initial configuration of the model behavior for D_1 and D_2. We use a beta distribution with four parameters α, β, a, b, where a and b represent the minimal and maximal value respectively. Each row represents two distinct distributions.

The variants available for the individual parameters in the case study can be freely combined, resulting in $n = 27$ possible configurations, analogous to the consideration of security. We use beta distributions to account for uncertainties about the true effectiveness in a complex system.

5 Results

In this section, we combine measures to reduce vulnerability and increase resilience in order to determine their efficiency in relation to the risk. To this end, we first perform an analytical calculation of the vulnerability of the security system of the substation for possible combinations of security measures. This is followed by an independent calculation of the effectiveness of the resilience-enhancing measures. Measures in both domains, security and resilience are then combined in the

Table 4 Considered parameter variations for resilience-enhancing measures in the transmission system model using beta distributions

Variant	Damage mode D_1				Damage mode D_2			
	α	β	a	b	α	β	a	b
$L_{1,1}$	2	5	72	78	2	5	90	98
$L_{1,2}$	3	5	50	60	3	5	85	90
$L_{1,3}$	4	5	20	30	4	5	70	85
$k_{1,1}$	2	4	0.4	0.8	2	4	0.4	0.8
$k_{1,2}$	1.5	6	0.2	0.4	1.5	6	0.2	0.4
$k_{1,3}$	1.5	10	0.1	0.2	1.5	10	0.1	0.2
$t_{2,1}$	6	2	100	110	6	2	90	100
$t_{2,2}$	5	1.7	85	100	5	1.7	70	80
$t_{2,3}$	5	1.4	70	85	5	1.4	60	70

final step using the introduced risk-based approach. We show the result of the risk assessment of the configurations in different plots.

5.1 Vulnerability

The vulnerability of the substation's security system is analytically calculated using the model described in Sect. 3 and the values introduced in Sect. 4. For the $n = 27$ possible combinations, results are obtained for scenarios A, B, C, D respectively.

Figure 8 visualizes the obtained results for vulnerability V in the scenarios A, B, C, D. It is clearly recognizable that the vulnerability decreases with the successive improvement of a measure at a barrier, e.g. for P_i in configurations $i \in \{1,2,3\}$. At the same time, vulnerability also decreases if the overall intensity of the measures is increased. At the maximum strength of the assumed measures, it can be seen that the residual vulnerability in scenarios A, B is very low, while scenarios C, D can be prevented much less effectively. The reason for this is the stronger access means assumed in these scenarios (see Sect. 4).

5.2 Resilience

The resilience loss of the whole transmission network when confronted with the damage modes D_1 and D_2 is obtained from numerical integration of the performance curve introduced in Sect. 3, taking into account the considered uncertainties. For this purpose, we use Monte-Carlo simulation based on the distributed parameter variations shown in Sect. 4.

The resulting resilience loss RL for all $n = 27$ possible configurations is visualized by violin plots in Fig. 9. The results show a different behavior for the two

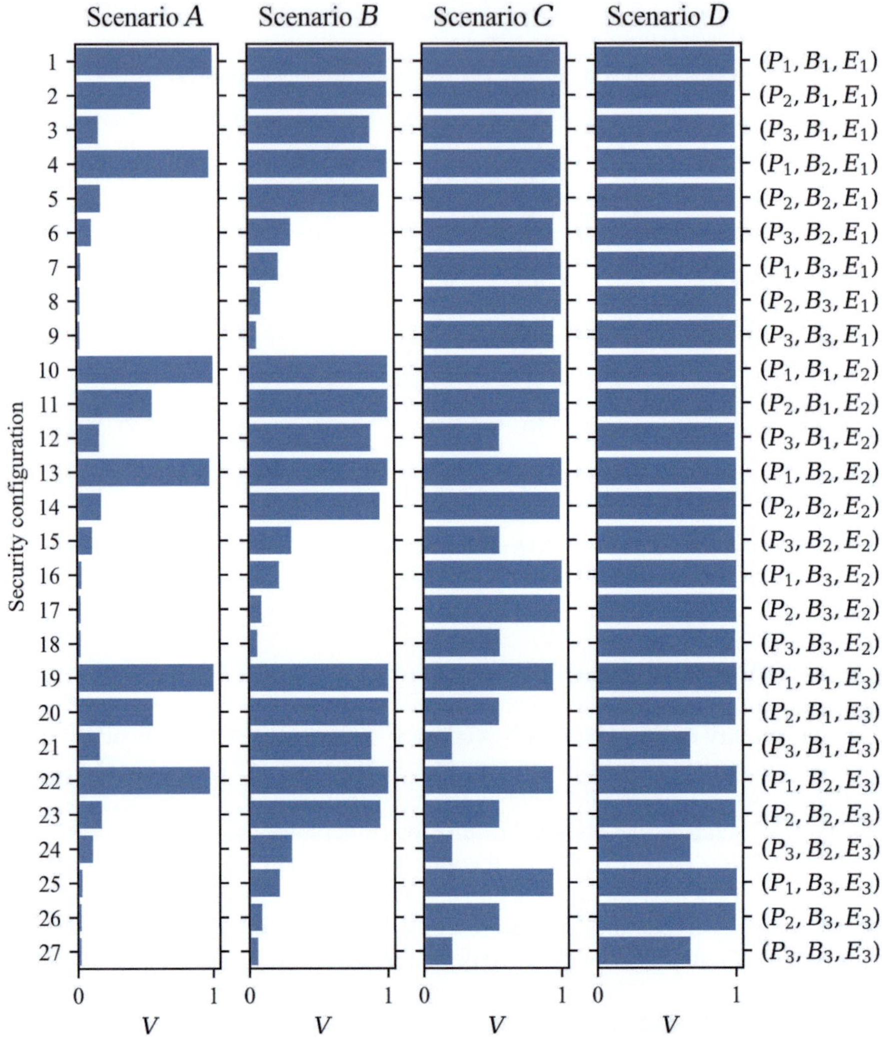

Fig. 8 Vulnerability of security system configurations in the scenarios A, B, C, D

damage modes. The result for D_1 shows a strong influence of the variation of the parameter L_1, i.e. of robustness increasing measures that reduce the initial performance loss. This influence is less pronounced for D_2. Instead, there is a greater overlap in the results when each parameter is varied. Overall, as predetermined by the simple form of the performance model, the resilience loss is significantly higher for damage mode D_2, as intended by the deliberate selection of parameters.

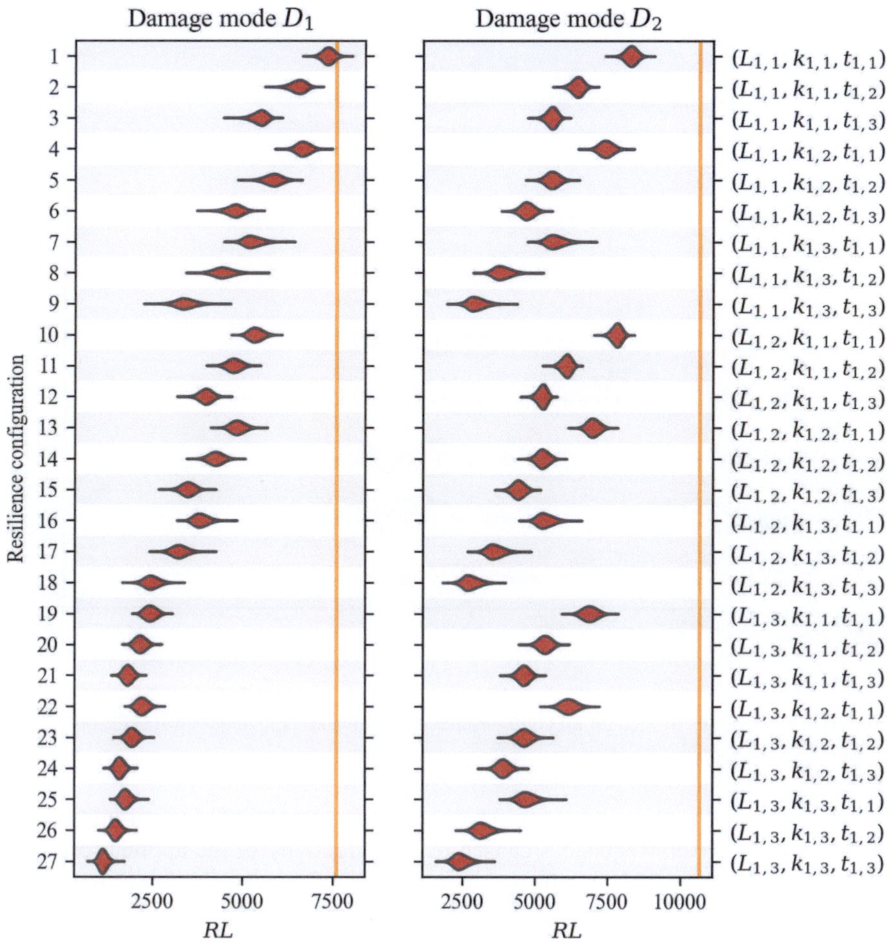

Fig. 9 Violin plots of *RL* for possible measure combinations for D_1 and D_2, Orange lines indicate *RL* for the initial system

5.3 Risk Analysis

Finally, we combine the individual results into a risk assessment. To achieve this, we aggregate the risk across all scenarios in accordance with the approach presented in Sect. 3. In order to enable an analysis of the results, we define a desired mean risk level $r \approx 80$ that allows a comparison of equally or similarly effective configurations of measures across all scenarios. The comparable configurations from varying security and resilience measures are visualized via violin plots in Fig. 10.

It can be seen that there are different configurations that have very similar effectiveness in terms of the risk result, although the variance, visible in the

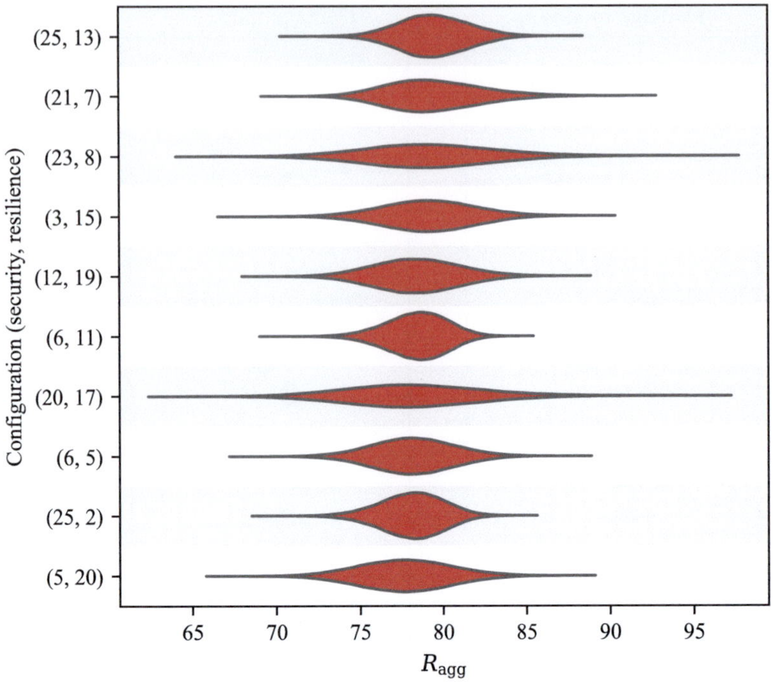

Fig. 10 Distributions of R_{agg} for all configurations with mean risk level $r \approx 80$

width of the violin plots, can vary. This variance represents the uncertainty about the severity of the consequences caused by the effectivity of the applied resilience enhancing measures. This is reflected in the spread of the distribution of the aggregated risk R_{agg} (see violin plots in Fig. 10). The configurations (23, 8) and (20, 17) in particular show a greater variance here. It should be noted that the amount of uncertainty associated with the effectiveness of vulnerability reduction measures is not visible here. This is because these uncertainties are already accounted for in the vulnerability model itself. There, the actual binary decision as to whether the system is vulnerable or not is transformed into a probabilistic result, i.e., the vulnerability V.

Figure 11 resolves the results by scenario into individual risk matrices. The vulnerability, against a successful attack is plotted on the abscissas of the matrix, while the resilience loss RL for the resulting damage mode is plotted on the ordinate. The figure illustrates that, due to the different configurations, the respective risk contributions of vulnerability (security) and consequence (resilience) are very different. This means, for example, that for configuration (21, 7) the vulnerability V is low in scenario A and C, but the corresponding RL is relatively high. The configuration (25, 2) shows similar risk contributions, where the effectivity of measures reducing the vulnerability is even higher and the risk contribution of RL is consistently high in all scenarios. Conversely, for configuration (5, 20), the vulnerability

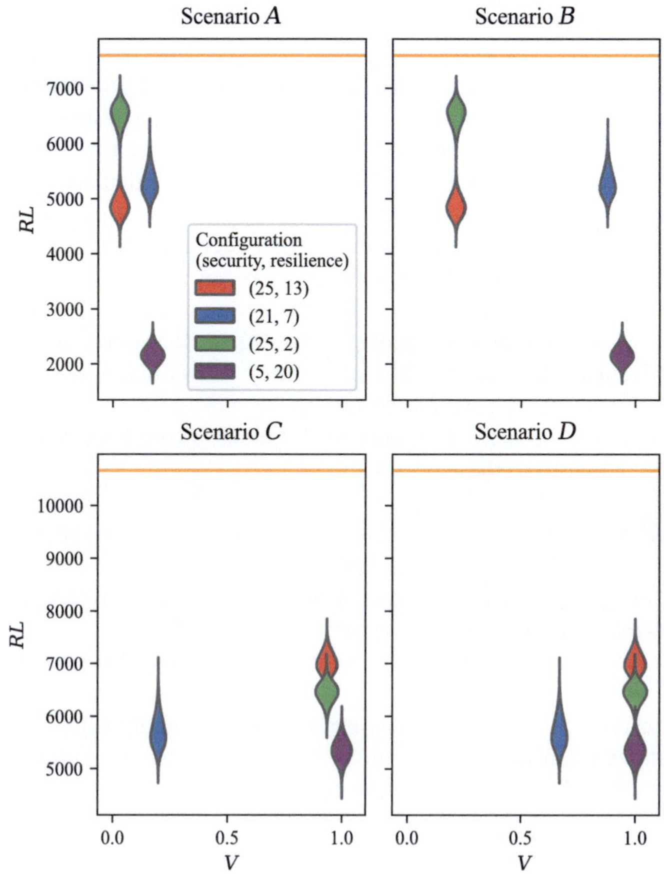

Fig. 11 Risk matrices resolved for vulnerability and *RL* for configurations (25, 13), (21, 7), (25, 2), (5, 20). Orange lines indicate *RL* for initial system. From top left to bottom right: Scenario A, B, C, D

is comparably high in all scenarios, but *RL* is limited by strong resilience measures effective in all scenarios. Additionally, considering configuration (25, 13) it appears that the effectivity of measure combinations varies over the different scenarios, as we assume differing attack paths as well as damage modes D_1, D_2. Moreover, a greater variance can be observed here for configuration (21, 7). In scenario *A*, for example, there is the possibility that (21, 7) is more effective than configuration (25, 13), i.e. that (25, 13) is not certain to be better than (21, 7), as indicated by overlapping boxes. Scenario *C* and *D* show a similar result for the configurations (25, 13) and (25, 2).

6 Discussion

The results show that there are equivalent combinations of vulnerability-reducing and resilience-enhancing measures with respect to the aggregated risk of all scenarios (see Fig. 10). At the same time, the contribution of the measures to risk reduction varies significantly in some cases, which is particularly evident in the decomposition for the individual scenarios in Fig. 11. Here it can be seen that the effectiveness of measures also varies greatly between the scenarios considered. It is important to note that the shown analysis does not include any efficiency considerations, i.e. cost–benefit analysis. The results also show the influence of the uncertainties considered in our approach with respect to the loss of resilience due to the effectiveness of measures. The case study shows that results with a better mean value in some cases are not certain to be more effective in relation to the aggregated risk R_{agg} or at least in individual scenarios (see Sect. 5.3). Three main points can be drawn from these observations:

1. There are probably threat scenarios that can be covered by both security and resilience. Here, the rational combination of both measures in a comprehensive concept can presumably save costs. Combinations of both concepts therefore appear to make sense in order to achieve the most effective protection possible. However, configurations that only use one concept are also possible. As these are generally maximum configurations with regard to either security or resilience measures, this is likely to involve a high-effort approach.
2. Uncertainties regarding the measure effectiveness and consequences can lead to undesirable effects when selecting configurations based on the overall risk as well as risks in individual scenarios. These uncertainties must therefore be considered when evaluating possible measure configurations. This applies in particular because extremely high consequences can occur in rare cases. In these cases, the focus may be on certain mitigation or limitation of these impacts. A low variance can be an evaluation criterion in the assessment.
3. Although the results of the case study show the possible equivalence of effectiveness for the aggregated risk of different combinations of measures, they do not provide any information about optimal or appropriate solutions. To this end, it is essential to include the efficiency, i.e. to weigh up which measures are appropriate in terms of costs and prevented consequences. This requires an additional cost-benefit analysis. Typically, there will be budgets for the costs insofar as the costs are constrained, which results in an optimization problem.

A detailed cost–benefit analysis of the coexistence of security and resilience is therefore necessary for effective and efficient comprehensive protection of infrastructures. However, this is associated with a number of challenges that require further research:

- Threat likelihood is subject to epistemic uncertainty: the probability of future attacks is difficult to determine as it results from human decisions and depends on many influences.

- The effectiveness of the measures is subject to aleatoric (stochastic) uncertainty: the employed models simplify reality. Resilient effects in complex socio-technical systems such as critical infrastructures are difficult to describe and the effectiveness of security measures in real attacks is unclear due to a lack of evidence.
- Consequences arise in several dimensions, some of which are difficult to assess and fraught with uncertainty: In addition to direct and indirect economic costs due to infrastructure disruptions, there are cascading effects on other critical infrastructures. These also entail resulting risks to health and life, for example. The assessment of the different dimensions might not only differ in terms of metric but also regarding their weighting in a conscious configuration choice.
- Varying system response to applied measures can lead to significant risk contributions from scenarios with low likelihood but high consequences: Uncertainties regarding measure effectivity must be considered in a comprehensive risk assessment to enable an informed risk-based choice of measure configurations.
- Appropriateness of protective measures is not a hard optimization criterion: A decision for certain configurations that is appropriate to the boundary conditions and value-based ideas does not have to be optimal in terms of effectiveness or efficiency, i.e. a pure cost-benefit analysis may not be expedient here.

7 Conclusion

In the face of hybrid threats, structurally induced vulnerability and potential catastrophic consequences, the comprehensive protection of critical infrastructures is of great importance. A closer look reveals that this goal can only be achieved by taking a complementary approach to security and resilience. An effective and appropriate selection of possible measures from both domains can generate synergetic advantages in terms of protection. At the same time, there are sometimes major uncertainties regarding attack scenarios, possible consequences and the effectiveness of measures. These must consequently be considered when choosing potential measures.

In this paper, we have proposed an approach that uses the risk definition framework from physical security risk analysis to bring together two models that consider vulnerability to physical attacks and potential consequences. The vulnerability is considered using a probabilistic model of security systems that evaluates their effectiveness in reducing vulnerability. To describe the consequences, we propose a model that uses logistic functions with distributed parameters to quantify the response to a damaging scenario, i.e. a successful attack. The variation of certain parameters incorporates the effectiveness of resilience-enhancing measures. Using these models, we calculate an aggregated risk and decompose it into the individual components of vulnerability and resilience loss for analysis.

We applied this approach to a simple case study, where we defined freely combinable variants for the security and resilience measures. We modeled these using distributed parameters so that we could take uncertainties into account. Our results show that there are different combinations of measures whose aggregated risk is equivalent, although the contributions to risk mitigation may vary. In addition, it becomes apparent that the effectiveness of measures can differ significantly in different scenarios. The uncertainties considered lead to considerable variance in risk reduction, so that a clear preference for similarly effective configurations of measures may be difficult. This can be particularly important against the background of scenarios that occur with a very low probability but have extreme effects. Resilience strategies, aiming at coping and mitigating consequences, may fail in the context of largely uncertain cascading effects and thus may not be a favorable choice here. In this case, a low variance in measure effectiveness can be an assessment criterion. On the other hand, security measures may not be very effective against some threat scenarios.

In addition to the evaluation of the effectiveness of possible measures from both areas proposed here, a comprehensive consideration of these points also makes it clear that an evaluation of efficiency is absolutely essential in order to find optimal or appropriate solutions, especially as this selection takes place against the background of limited budgets.

The approach chosen here shows initial interesting results with regard to the equivalence of measures in security and resilience and the necessary consideration of uncertainties. We are thus taking the first steps in the direction of a comprehensive and complementary approach to the protection of critical infrastructures. At the same time, we are aware that further detailing of the models for considering vulnerability and impacts is necessary. In particular, more detailed models of the cascading consequences of disruptions to critical infrastructures are needed. In addition, an analysis of efficiency, i.e. cost–benefit assessments, is essential for an optimal selection of possible measures. This challenging path should be taken together with the operators of critical infrastructures, as we have already begun to do. This article can be seen as a first step in this direction, which highlights the challenges described. The proposed flexible approach brings together security and resilience in a complementary analysis for the first time, while accounting for uncertainty, and can serve as a starting point for further steps in research and application.

References

Asadzadeh, Asad; Kötter, Theo; Salehi, Pourya; Birkmann, Joern (2017): Operationalizing a concept: The systematic review of composite indicator building for measuring community disaster resilience. In *International Journal of Disaster Risk Reduction* 25, pp. 147–162. https://doi.org/10.1016/j.ijdrr.2017.09.015.

Bennett, Harold A. (1977): EASI (Estimate of Adversary Sequence Interruption) - An Evaluation Method for Physical Security Systems. In *NUCLEAR MATERIALS MANAGEMENT* 6 (3), pp. 371–379.

Bi, Wei; MacAskill, Kristen; Schooling, Jennifer (2023): Old wine in new bottles? Understanding infrastructure resilience: Foundations, assessment, and limitations. In

Transportation Research Part D: Transport and Environment 120, p. 103793. https://doi.org/10.1016/j.trd.2023.103793.

Bruneau, Michel; Chang, Stephanie E.; Eguchi, Ronald T.; Lee, George C.; O'Rourke, Thomas D.; Reinhorn, Andrei M. et al. (2003): A Framework to Quantitatively Assess and Enhance the Seismic Resilience of Communities. In *Earthquake Spectra* 19 (4), pp. 733–752. https://doi.org/10.1193/1.1623497.

Bundesministerium des Innern und für Heimat (12/21/2023): Entwurf eines Gesetzes zur Umsetzung der Richtlinie (EU) 2022/2557 und zur Stärkung der Resilienz von Betreibern kritischer Anlagen. Available online at https://www.bmi.bund.de/SharedDocs/gesetzgebun1539–6924gsverfahren/DE/Downloads/referentenentwuerfe/KM4/KRITIS-DachG-2.pdf?__blob=publicationFile&v=6.

Cimellaro, Gian Paolo; Fumo, Cristina; Reinhorn, Andrei M.; Bruneau, Michel (2009): Quantification of Disaster Resilience of Health Care Facilities. Available online at http://www.buffalo.edu/mceer/catalog.host.html/content/shared/www/mceer/publications/MCEER-09-0009.detail.html.

Cutter, Susan L. (2016): The landscape of disaster resilience indicators in the USA. In *Natural Hazards* 80 (2), pp. 741–758. https://doi.org/10.1007/s11069-015-1993-2.

Demmer, Tobias; Kahlen, Jens; Lichte, Daniel; Wolf, Kai-Dietrich (2022): Towards the Prediction of Resilience; An Equation-based Resilience Representation. In Maria Chiara Leva, Edoardo Patelli, Luca Podofillini, Simon Wilson (Eds.): Proceedings of the 32nd European Safety and Reliability Conference. 32nd European Safety and Reliability Conference. Singapore: Research Publishing Services, pp. 1417–1424. https://doi.org/10.3850/978-981-18-5183-4_R23-04-456-cd

Förster, Peter; Schachtebeck, Per M.; Feuerle, Thomas; Hecker, Peter; Branlat, Matthieu; Herera, Ivonne; Woltjer, Roger (2019): An Approach for Attribute- and Performance-Based Evaluation of Interdependent Critical Infrastructures. In Electronic Navigation Research Institute (Ed.): Air Traffic Management and Systems III. Singapore: Springer (Lecture Notes in Electrical Engineering), pp. 35–50.

Francis, Royce; Bekera, Behailu (2014): A metric and frameworks for resilience analysis of engineered and infrastructure systems. In *Reliability Engineering & System Safety* 121, pp. 90–103. https://doi.org/10.1016/j.ress.2013.07.004.

Garcia, Mary Lynn (2005): Vulnerability Assessment of Physical Protection Systems: Elsevier.

Garcia, Mary Lynn (2008): The Design and Evaluation of Physical Protection Systems. 2nd ed. Burlington, MA, USA: Elsevier.

Halekotte, Lukas; Mentges, A.; Lichte, D. (2025): Do we practice what we preach? The dissonance between resilience understanding and measurement. In *International Journal of Disaster Risk Reduction* 118, p. 105265. https://doi.org/10.1016/j.ijdrr.2025.105265.

Harnser Group (2010): A Reference Security Management Plan for Energy Infrastructure: European Commission. Available online at http://ec.europa.eu/energy/infrastructure/studies/doc/2010_rsmp.pdf.

Lichte, Daniel; Witte, Dustin; Termin, Thomas; Wolf, Kai-Dietrich (2021): Representing Uncertainty in Physical Security Risk Assessment. In *European Journal for Security Research*. https://doi.org/10.1007/s41125-021-00075-3.

Lichte, Daniel; Wolf, Kai-Dietrich (2017): Quantitative Multiple-Scenario Vulnerability Assessment Applied to a Civil Airport Infrastructure. In: Proceedings ESREL 2017, vol. 27. Portoroz, Slovenia: CRC Press.

McDaniels, Timothy; Chang, Stephanie; Peterson, Krista; Mikawoz, Joey; Reed, Dorothy (2007): Empirical Framework for Characterizing Infrastructure Failure Interdependencies. In *Journal of Infrastructure Systems* 13 (3), pp. 175–184. https://doi.org/10.1061/(ASCE)1076-0342(2007)13:3(175).

McGill, William L.; Ayyub, Bilal M.; Kaminskiy, Mark (2007): Risk Analysis for Critical Asset Protection. In *Risk Analysis* 27 (5), pp. 1265–1281. https://doi.org/10.1111/j.1539-6924.2007.00955.x.

Mentges, Andrea; Halekotte, Lukas; Schneider, Moritz; Demmer, Tobias; Lichte, Daniel (2023): A resilience glossary shaped by context: Reviewing resilience-related terms for critical infrastructures. In *International Journal of Disaster Risk Reduction* 96, p. 103893. https://doi.org/10.1016/j.ijdrr.2023.103893.

Mottahedi, Adel; Sereshki, Farhang; Ataei, Mohammad; Nouri Qarahasanlou, Ali; Barabadi, Abbas (2021): The Resilience of Critical Infrastructure Systems: A Systematic Literature Review. In *Energies* 14 (6), p. 1571. https://doi.org/10.3390/en14061571.

Mugume, Seith N.; Gomez, Diego E.; Fu, Guangtao; Farmani, Raziyeh; Butler, David (2015): A global analysis approach for investigating structural resilience in urban drainage systems 81, pp. 15–26. https://doi.org/10.1016/j.watres.2015.05.030.

Nan, Cen; Sansavini, Giovanni; Kröger, Wolfgang; Heinimann, Hans Rudolf (2014): A Quantitative Method for Assessing the Resilience of Infrastructure Systems. In *PSAM 2014 - Probabilistic Safety Assessment and Management*, p. 12.

NIAC (2010): A Framework for Establishing Critical Infrastructure Resilience Goals Report: Final Report and Recommendations by the National Infrastructure Advisory Council.

Nick, Florence Catherine; Sänger, Nathalie; van der Heijden, Sophie; Sandholz, Simone (2023): Collaboration is key: Exploring the 2021 flood response for critical infrastructures in Germany. In *International Journal of Disaster Risk Reduction* 91, p. 103710. https://doi.org/10.1016/j.ijdrr.2023.103710.

Pagano, Alessandro; Sweetapple, Chris; Farmani, Raziyeh; Giordano, Raffaele; Butler, David (2019): Water Distribution Networks Resilience Analysis: a Comparison between Graph Theory-Based Approaches and Global Resilience Analysis. In *Water Resources Management* 33 (8), pp. 2925–2940. https://doi.org/10.1007/s11269-019-02276-x.

Pimm, Stuart L. (1984): The complexity and stability of ecosystems. In *Nature* 307 (5949), pp. 321–326. https://doi.org/10.1038/307321a0.

Poulin, Craig; Kane, Michael B. (2021): Infrastructure resilience curves: Performance measures and summary metrics. In *Reliability Engineering & System Safety* 216, p. 107926. https://doi.org/10.1016/j.ress.2021.107926.

Sandia National Laboratories (1989): SAVI: Systematic Analysis of Vulnerability to Intrusion. Sandia National Laboratories. Albuquerque, NM, USA.

Shinozuka, Masanobu; Chang, Stephanie E.; Cheng, Tsen-Chung; Feng, Maria; O'Rourke, Thomas D.; Saadeghvaziri, M. Ala et al. (2004): Resilience of Integrated Power and Water Systems.

Tamminga, Oliver (2015): Zum Umgang mit hybriden Bedrohungen. Auf dem Weg zu einer nationalen Resilienzstrategie. Berlin: Stiftung Wissenschaft und Politik (SWP-aktuell, 92/2015).

Trucco, Paolo; Petrenj, Boris (2023): Characterisation of resilience metrics in full-scale applications to interdependent infrastructure systems. In *Reliability Engineering & System Safety* 235, p. 109200. https://doi.org/10.1016/j.ress.2023.109200.

Trump, Benjamin D.; Florin, Marie-Valentine; Linkov, Igor (Eds.) (2018): IRGC Resource Guide on Resilience (Volume 2): Domains of resilience for complex interconnected systems. EPFL International Risk. Lausanne.

Wang, Yuntao; Zhang, Chi; Chen, Albert S.; Wang, Guoqiang; Fu, Guangtao (2023): Exploring the relationship between urban flood risk and resilience at a high-resolution grid cell scale. In *The Science of the total environment* 893, p. 164852. https://doi.org/10.1016/j.scitotenv.2023.164852.

Wied, Morten; Oehmen, Josef; Welo, Torgeir (2020): Conceptualizing resilience in engineering systems: An analysis of the literature. In *Systems Engineering* 23 (1), pp. 3–13. https://doi.org/10.1002/sys.21491.

Yodo, Nita; Wang, Pingfeng (2016): Engineering Resilience Quantification and System Design Implications: A Literature Survey. In *Journal of Mechanical Design* 138 (11). https://doi.org/10.1115/1.4034223.

MIX
Papier aus verantwortungsvollen Quellen
Paper from responsible sources
FSC® C105338

If you have any concerns about our products,
you can contact us on
ProductSafety@springernature.com

In case Publisher is established outside the EU,
the EU authorized representative is:
Springer Nature Customer Service Center GmbH
Europaplatz 3, 69115 Heidelberg, Germany

Printed by Libri Plureos GmbH
in Hamburg, Germany